Applied Functional Analysis
An Introductory Treatment

Editorial Board

A Jeffrey, Engineering Mathematics,
  University of Newcastle upon Tyne (*Main Editor*)
R Aris, Chemical Engineering and Materials Science,
  University of Minnesota
W Bürger, Institut für Theoretische Mechanik,
  Universität Karlsruhe
J Douglas Jr, Mathematics,
  University of Chicago
K P Hadeler, Institut für Biologie,
  Universität Tübingen
W F Lucas, Operations Research and Industrial Engineering,
  Cornell University
J H Seinfeld, Chemical Engineering,
  California Institute of Technology

Applied Functional Analysis
An Introductory Treatment

R D Milne
University of Bristol

Pitman Advanced Publishing Program
BOSTON · LONDON · MELBOURNE

PITMAN PUBLISHING LIMITED
39 Parker Street, London WC2B 5PB

PITMAN PUBLISHING INC.
1020 Plain Street, Marshfield,
Massachusetts

Associated Companies
Pitman Publishing Pty Ltd, Melbourne
Pitman Publishing New Zealand Ltd, Wellington
Copp Clark Pitman, Toronto

First published 1980

AMS Subject Classifications: (main) 46BXX, 46CXX, 46EXX, 46FXX
(subsidiary) 49GXX, 35A15, 35A35, 47A60, 49D15

Library of Congress Cataloging in Publication Data
Milne, Ronald Douglas, 1930–
  Applied functional analysis.

  (Applicable mathematics)
  Bibliography: p.
  Include index.
  1. Functional analysis. I. Title. II. Series.
QA320.M54   515'.7   79-16736
ISBN 0 273 08404 6

© R D Milne 1980
All rights reserved. No part of this publication may be reproduced, stored in a retrieval system, or transmitted in any form or by any means, electronic, mechanical, photocopying, recording and/or otherwise without the prior written permission of the publishers.

To Dorothy, Alisoun and Douglas

Contents

Preface ix
List of Symbols xiii

1 Preliminaries 1
1.1 Introduction 1
1.2 Preliminaries 3

2 Algebraic theory of vector spaces and linear transformations 16
2.1 Introduction 16
2.2 The definition of a vector space 17
2.3 Subspaces 21
2.4 Linear independence, basis and dimension 28
2.5 Linear transformations 35
2.6 Linear functionals and the dual space 49
2.7 The algebraic dual of a linear transformation 59
2.8 Linear transformations on finite-dimensional spaces 64
2.9 Application examples 100

3 Metric, normed and inner product spaces 110
3.1 Introduction 110
3.2 Metric spaces 111
3.3 Normed vector spaces 134
3.4 Linear transformations on normed vector spaces 143
3.5 The dual of a normed space 155
3.6 Inner product spaces 180
3.7 Linear transformations on inner product spaces 211
3.8 Linear transformations on finite-dimensional inner product spaces 231
3.9 An outline of spectral theory 247
3.10 Application examples 256
Appendix 3.A The Hölder and Minkowski inequalities 271
Appendix 3.B Topological and metric spaces 274
Appendix 3.C The Lebesgue integral 276

4 Calculus of operators and operator equations 281
  4.1 Introduction 281
  4.2 Calculus of operators 283
  4.3 The variational calculus 301
  4.4 Direct solution of positive operator equations 326
  4.5 Projection solution of operator equations 364
  4.6 The finite element method 381
  4.7 Application examples 403
  Appendix 4.A. Order of magnitude notation 420

5 Distributions 422
  5.1 Introduction 422
  5.2 Distributions 424
  5.3 Integral transforms 448
  5.4 Sobolev spaces 459
  5.5 Application examples 470

References 481

Answers to selected exercises 489

Index 493

Preface

The language and concepts of functional analysis are increasingly finding their way into the current literature in applied mathematics, science and engineering. There are good reasons for this trend and it will undoubtedly gain momentum. However, it does mean that the applied mathematician, scientist or engineer who has no familiarity at all with functional analysis is, to some degree, handicapped in his work.

Functional analysis does not provide the practitioner with a magical technique allowing him to solve problems which have not yielded to a more traditional approach. Rather its attraction and utility lie in its ability to unify and clarify a whole range of apparently disparate mathematical concepts, leaving the central ideas exposed. Functional analysis does this by combining abstraction with a concise and systematic terminology.

Those who wish to use functional analysis must, of course, be prepared to learn enough of the subject to become familiar with the jargon and absorb a certain degree of abstraction. To those who are primarily concerned with applying mathematics, the central question is, 'How much need one learn before functional analysis becomes a useful tool?' It is probably true to say that no great depth of understanding is necessary in order to be able to read and appreciate many of the research papers which use functional analysis to illuminate or solve applied problems. The majority of the many texts on functional analysis are written by pure mathematicians for pure mathematicians, contain far too much intricate technical detail for the person interested in applications and, naturally enough, develop the material with full rigour.

The treatment of functional analysis in this book is designed to allow the applied mathematician, scientist or engineer, whose background in classical analysis may not be very solid, to learn, without a vast investment in time and effort, enough of the subject to be able to appreciate its character and potential usefulness. Accordingly, the scope of the basic material, which is dealt with in Chapters 2 and 3, is deliberately limited; for example, general topological spaces and Lebesgue integration are not included, although short, very informal accounts of these topics appear in appendices. In a similar spirit, theorems are not always proved and several of what might be described as the major theorems of functional analysis are barely mentioned. Naturally, adequate references are given to enable the reader to pursue a topic in which he finds particular interest. In order to ease the reader's way, an informal discussion precedes each major step in the theory, thereby providing some motivation for the introduction of the forthcoming ideas.

Thus Chapters 2 and 3, together with Chapter 1, which contains preliminary material, provide a straightforward, mildly rigorous first course in functional analysis supplemented with worked examples many of which have their origin in applications. Chapters 4 and 5 deal, respectively, with the approximate solution of operator equations and the theory of distributions (or generalized functions). Both these topics are most naturally described and evaluated in functional analytic terms and illustrate admirably the impact of functional analysis on modern applied mathematics. These would be sufficient reasons for the selection of these topics; their inclusion in preference to other equally important applications is simply due to the fact that it is in these areas that I gained my interest and experience in functional analysis through applications to problems in control and fluid and solid mechanics.

It is encouraging to note that, in recent years, a number of books have appeared which treat functional analysis from an applied point of view (*see*, for example, items [5, 6, 18, 33, 36, 145] in the list of references). There are also signs that functional analysis may become a recognized part of undergraduate courses in applied mathematics and perhaps also of the more theoretical engineering courses (*see*, for example, [4]).

This book should be suitable for final-year and graduate students in applied mathematics and for graduate students in the general sciences and engineering. In a sense the book is both terminal and introductory; by itself it should enable the reader to cope confidently with the language and approach of modern applied mathematics, but hopefully it will also encourage some to pursue the subject much further.

A decimal numbering system is used for theorems, definitions and examples, thus Definition 3.5.2 is the second definition in Section 5 of Chapter 3. Sets of exercises appear at the end of most sections and these are numbered as, for example, Exercise 2.8,3, the comma being used in place of the point to distinguish these from example numbers:

for the most part each exercise contains a number of questions on one topic.

Many of the exercises begin with the words 'Show that'. The reader should interpret this command to lie somewhere between 'convince yourself that' and 'prove that', depending on inclination. Rigorous proofs can usually be found in the cited references. Hints for the solution of exercises are often given and for those requiring numerical answers which cannot be directly verified a list of answers has been provided.

I am grateful to several of my colleagues at Bristol University who read various parts of the draft typescript, particularly to Jerry Wright and David Griffel. Their efforts do not, however, absolve me from responsibility for any errors, obfuscations or misconceptions that remain in the text.

I also wish to thank Eleanor Gibbins and Sheila Hook who cheerfully and very skilfully typed the various drafts of the manuscript.

May 1979   R D Milne
                                                                                             University of Bristol

List of symbols

Chapter 1 contains a summary of the symbolism pertaining to sets, relations and functions.

A^*	adjoint of the operator A, 212
$\text{Adj}\,[t_{ij}]$	adjugate of the matrix $[t_{ij}]$, 96
$a'_x(x, y)$	partial strong derivative of the functional $a(x, y)$, 349
$B(x_0, r)$	ball of radius r, centre x_0, 117
$\mathscr{B}[0, 1]$	vector space of functions of bounded variation on $[0, 1]$, 164
C or $\mathscr{C}$	the complex number field
$\mathscr{C}_n$	complex n-space, 19
$\mathscr{C}[0, 1]$ or $\mathscr{C}^{(0)}[0, 1]$	space of continuous functions on $[0, 1]$, 137
$\mathscr{C}^{(m)}[0, 1], \mathscr{C}^{(m)}(\Gamma)$	space of m-times continuously differentiable functions on $[0, 1]$, on the region Γ, respectively, 184
$\text{Co}(S)$	convex hull of the set S, 26
Γ	a bounded region in $\mathscr{R}_n$, 184
$\mathscr{D}$	space of testing functions (Chapter 5 only), 425
$\mathscr{D}(F)$	domain of the function/operator F, 11
$D^k x$	kth-order mixed partial derivative of x, 184
$D_j x$	first partial derivative of x with respect to the jth variable, 283
$\det\,[t_{ij}]$	determinant of the square matrix $[t_{ij}]$, 79
$\dim \mathscr{V}$	dimension of the vector space $\mathscr{V}$, 30
$d(x, y)$	metric or distance function, 112
$d(x, \mathscr{W})$	distance from x to subspace $\mathscr{W}$, 166
$d_p, d_\infty, \tilde{d}$	particular metrics, 113
$dT(x; s)$	weak differential of T at x in the direction of s, 287
δ	the delta functional, 425
δ^j_i	the Kronecker delta, 14
$\partial \Gamma$	boundary of the region Γ, 184
e^j	jth standard basis vector, 32
$\mathscr{E}_n$	Euclidean or unitary n-space, 183
f^{*-1}	convolution inverse of the distribution f, 440
$f \times g$	direct product $\Big\}$ of the distributions f, g, 436
$f * g$	convolution
θ	the zero vector, 18
θ^*	the zero functional, 50
$[g_{ij}], [g^{ij}]$	fundamental matrices, 72

LIST OF SYMBOLS

$\mathcal{H}, \mathcal{H}^T$	Hilbert spaces, 184, 330
$\mathcal{H}^{(m)}, \mathcal{H}_0^{(m)}, \mathcal{H}_0^{(-m)}$	Sobolev spaces, 463
I	the identity operator, 12
inf	infimum or greatest lower bound, 9
$\mathcal{L}(\mathcal{V}, \mathcal{U})$	space of bounded linear transformations $\mathcal{V} \to \mathcal{U}$, 147
$\mathcal{L}_p[0,1], \mathcal{L}_\infty[0,1]$	normed spaces of integrable functions on $[0, 1]$, 137
l_p, l_∞	normed sequence spaces, 136
l_E	extension of the linear functional l, 161
λ_ρ	eigenvalue of a linear transformation, 44
$\mathcal{N}(T)$	null space of the linear transformation T, 39
ν	outward normal to the region Γ, 190
ν	nullity of a linear transformation (in Chapter 2), 39
ξ_i	covariant
ξ^i	contravariant $\Big\}$ coordinates of the vector x, 52
$\{\xi_i\}$	column
$\lfloor \xi_i \rfloor$	row $\Big\}$ n-tuple, 64
O, o	order of magnitude symbols, 420
$\mathcal{O}$	zero vector space, 19
$\emptyset$	empty set, 4
Pf	pseudofunction, 428
R or $\mathcal{R}$	the real number field
$\mathcal{R}_n$	real n-space, 19
$\mathcal{R}(F)$	range of the function/operator F, 11
ρ	rank of a linear transformation (in Chapter 2), 39
S^0	$(\subset \mathcal{V}^*)$ annihilator of $S \subseteq \mathcal{V}$, 54
$^0S^*$	$(\subset \mathcal{V})$ annihilator of $S^* \subseteq \mathcal{V}^*$, 174
(S, d)	metric space, 112
$\mathcal{S}$	space of testing functions (in Chapter 5 only), 449
sup	supremum or least upper bound, 9
T'	dual (or conjugate) of the transformation T, 60
T^n	n-fold product of the transformation T, 41
$T_{\mathcal{W}}$	restriction of T to the subspace $\mathcal{W}$, 48
$T'(x)$	strong derivative of T at x, 289
$T'^w(x)$	weak derivative of T at x, 289
$T^{(n)}(x)$	nth strong derivative of T at x, 298
$[t_{ij}]$	rectangular or square matrix, 37
$[t_{ij}]^T = [t_{ji}]$	transpose of the matrix $[t_{ij}]$, 69
$\text{tr}[t_{ij}]$	trace of the matrix $[t_{ij}]$, 59
U	unitary operator, 226
$V_\alpha^\beta[x]$	total variation of x on $[\alpha, \beta]$, 163
$\mathcal{V}$	vector space, 18
$\mathcal{V}_n$	n-dimensional vector space, 32
$\mathcal{V}^*$	dual (or conjugate) of the vector space $\mathcal{V}$, 50
$\mathcal{V}^{(\rho)}$	eigenspace of a linear transformation, 44

LIST OF SYMBOLS

$\mathscr{V}/\mathscr{W}$	quotient space of $\mathscr{V}$ with respect to $\mathscr{W}$, 25
$(\;,\|\cdot\|)$	normed vector space, 135
x^*	element of $\mathscr{V}^*$, 50
$x_g^{(k)}$	generalized eigenvector, 85
$x + \mathscr{W}$	$\mathscr{W}$-coset of x, 25
$1_+(t)$	the unit step function, 427
$\|\cdot\|$ or $\|\cdot\|_{\mathscr{V}}$	norm function (on vector space $\mathscr{V}$), 135
$\|\cdot\|_p$	p-norms, 136
$\|\cdot\|_\infty$	uniform norm, 136
$\|\cdot\|$	pseudonorm, 141
$\|\cdot\|_T$	energy norm, 330
$\|T\|_s$	spectral norm of the transformation T, 241
$\oplus$	direct sum of subspaces, e.g., $\mathscr{U} \oplus \mathscr{W}$ is the direct sum of $\mathscr{U}$ and $\mathscr{W}$, 24
$\mathscr{W}^c$	direct complement of subspace $\mathscr{W}$, 24
$\bar{S}$	closure of the set S, 275
$[S]$	subspace generated by the set S, 23
$[y,x]$ or $y(x)$	value of the linear functional, y at x, 52
$\langle\cdot,\cdot\rangle, \langle\cdot,\cdot\rangle_{\mathscr{H}}$	inner product, e.g., $\langle x, y\rangle$ is the inner product of x and y, 181
$\langle\cdot|\cdot\rangle_T$	energy inner product, 330
$\perp$	orthogonal, e.g., $x \perp y$ means that x is orthogonal to y, 185
$\mathscr{W}^\perp$	orthogonal complement of the subspace $\mathscr{W}$, 185
$x^{(k)} \to x$	strong convergence or convergence with respect to the metric of $x^{(k)}$ to x, 136
$x^{(k)} \rightharpoonup x$	weak convergence of $x^{(k)}$ to x, 169
$x^{(k)} \overset{*}{\rightharpoonup} x$	weak* convergence of $x^{(k)}$ to x, 171

1 Preliminaries

1.1 Introduction

Seen against the background of the long history of mathematics, functional analysis is a recent development. As a recognizable branch of mathematics it dates from the early years of this century, but its origins lie much further back.

Functional analysis is an abstraction and extension of several areas of mathematics, notably those of classical analysis, geometry and algebra, all of which underwent extensive and rapid development during the eighteenth and nineteenth centuries. Much of this development was sparked off by Fourier's somewhat heuristic treatment of the differential equation governing heat conduction in solids wherein he effectively introduced Fourier series. This work, coupled with the somewhat earlier and apparently irreconcilable approaches to the vibrating string problem by Euler and D'Alembert, showed the need for a generalization and extension of the idea of function which, up to that time, had embodied little more than what we should now mean by a smooth curve [3].

Allied with this development was the need to make rigorous the idea of limit which of course was (and is) the foundation stone of calculus. This, in turn, meant that the concept of real number should be properly formulated and understood.

Branches of mathematics other than calculus were also undergoing rapid change or development in the nineteenth century. Geometry was emerging from the straitjacket of Euclid and becoming an analytical subject divorced from the drawing of figures and the world of immediate physical experience. Geometry gradually became more abstract and axiomatic in approach until it developed into that branch of mathematics which we now call topology. The work of Cayley on matrices and the introduction of vectors formed the early beginnings of what has now become the extensive branch of mathematics called algebra. Roughly speaking algebra has generalized the familiar operations of addition (and/or multiplication) of real numbers and applied them, again using an axiomatic approach, to other entities: in this way various structures are built up which owe their existence solely to the chosen axioms (subject of course to consistency). It is nowadays conventionally accepted that an algebra does not deal with limiting processes and so has no need of the concept of limit; that is to say that, whatever operations are defined for the algebra, those operations are combined a finite number of times only.

In classical analysis the notion of function is closely associated with real numbers; for example, we speak of 'functions of a real variable'. The modern view of a function is that it defines a correspondence between pairs of real numbers. But why should the objects to be placed in correspondence be restricted to real numbers? In functional analysis this restriction is removed by generalizing the concept of function to that of operator, where now a correspondence can be defined between pairs of arbitrary objects or entities. Since we are now dealing with collections of arbitrary entities the language of sets is the natural one to use and an operator is defined as a correspondence between members or elements of prescribed sets. This means that such elements may themselves be functions and from this stance we would, for example, view a differential equation as embodying an operator which transforms the function appearing on the left-hand side into the function appearing on the right. In the classical view these functions would be seen as the aggregate of their values, but in functional analysis the function (correspondence) itself is treated as an entity in distinction to its value. Thus we may speak of a set of continuous functions or a set of differentiable functions without specifying any particular member, in the same way as we talk of the set of real numbers.

This generalization of function is not, however, sufficient on its own to give us functional analysis because we have not specified a topology (geometry) or structure (algebra) for the sets under consideration. The real number system has a topology which allows the concept of limit to be rigorously defined and it also carries the structure of ordinary arithmetic. For the idea of operator to be useful it is necessary to introduce at least some of these properties into our sets of objects. Functional analysis adopts the topology of 'distance' between objects, a concept closely analogous to the distance of ordinary space but, of course, defined axiomatically. Thence a limit concept can be introduced and a calculus or analysis developed both for objects and operators as we do for numbers and functions in classical analysis. In particular the question of convergence of a sequence of objects or operators can be meaningfully discussed.

So much for topology. Functional analysis borrows its structure from ordinary vector algebra, taking over the operations of the sum of vectors and the multiplication of a vector by a number: it also takes over the scalar or dot product (but has no use for the vector product). Conventionally, when we impose this structure on our sets of objects we call them spaces (a term first introduced by Fréchet in 1906), so that if the objects are functions we speak of function spaces.

This particular amalgamation of function, topology and algebra is, roughly, what we mean by functional analysis. It is a generalization of classical analysis set up as an abstract axiomatic system, from which theorems may be deduced; the theorems of classical analysis are naturally special cases of these. Functional analysis has an important unifying role, often bringing under one umbrella whole groups of results which may appear to be quite disparate. Much of the benefit of unification derives from the economic and concise notation and nomenclature which is an integral part of functional analysis.

1.2 Preliminaries

In this section we wish to review, rather briefly, the basic mathematical concepts of sets, relations and functions. The treatment is necessarily simple and heuristic since these concepts can fairly be said to be at the heart of mathematics and a full discussion would be much too extensive and involved. (An informal and attractively presented view of these matters can be found in [1].) For the reader who is already familiar with this material the following will serve to introduce the notation and terminology used throughout the book. Somewhat amplified versions of the material presented here can be found in, for example, [2, 7, 9, 10, 25].

Sets

By a *set* we shall mean a collection of objects viewed as a single entity: the objects in the set are referred to as *elements*, *members* or *points* of the set. We shall denote a set by an italic capital letter $A, B, S, X, \ldots$ and the elements of the set by lower case italic letters $a, b, s, x, \ldots$. If s is an element of the set S then we write $s \in S$; if s is not an element of S we write $s \notin S$.

A set may be designated by displaying the elements in braces; for example, $\{2, 4, 6, 8\}$ denotes the set of positive even integers less than 10. Alternatively, and more commonly, we specify the property P which characterizes the elements of the set by writing

$$S = \{x \mid P(x)\},$$

which is to be read, 'S is the set of objects x such that x satisfies the property P'. Needless to say the property P must be meaningful for each x: in this sense we may specify that x is an element of some (larger) 'universal' set U and write

$$S = \{x \in U \mid P(x)\}.$$

Thus if U were the set of real numbers R and S were the set of rationals lying between 2 and 10 then we would write

$$S = \{x \in R \mid x \text{ rational}, 2 < x < 10\},$$

the comma being read as 'and'. When the context makes it clear that a particular universal set is being considered, explicit reference to it will be omitted.

When the universal set is the set of real numbers R (*the real line*) a commonly accepted notation is used to denote *intervals of the real line*; if $a, b \in R$ are such that $a < b$ then we define

$$[a, b] = \{x \mid a \leqslant x \leqslant b\},$$
$$(a, b) = \{x \mid a < x < b\},$$

referred to respectively as the *closed* and *open* intervals from a to b. The symbol $[a, b)$ denotes the closed–open interval $a \leqslant x < b$. In this book we shall as a matter of convenience use the unit closed or open intervals $[0, 1]$ or $(0, 1)$ as the archetypal intervals of the real line. It is always possible to translate results arrived at for the unit interval to any interval of finite length by employing a suitable scaling. We write $[a, \infty)$ to denote the semi-infinite interval $a \leqslant x < \infty$ and very often use $(-\infty, \infty)$ to denote the whole of R.

Two sets A and B are said to be equal if they consist of exactly the same elements, and this is denoted by writing $A = B$.

From a given set we may form new sets, called *subsets* of the given set; for example, the set of even integers is a subset of the set of integers. A set A is a subset of a set X—written $A \subseteq X$ (or $X \supseteq A$)—if every element of A is also an element of X. This allows for the possibility that A and X might be equal: if $A \neq X$ then we write $A \subset X$ and speak of A as a *proper subset* of X. The set with no elements is called the empty (or void) set and is written $\emptyset$.

Let us now turn to a description of the elementary operations that can be performed with sets.

(i) The *union* of the sets A and B is the set of all elements which belong to A or to B or to both; this set is denoted by $A \cup B$ or $B \cup A$; thus

$$A \cup B = \{x \mid x \in A \text{ or } x \in B\}.$$

(ii) The *intersection* of the sets A and B is the set of all elements which belong to both A and B: this set is denoted by $A \cap B$ or $B \cap A$; thus

$$A \cap B = \{x \mid x \in A, x \in B\}.$$

If $A \cap B = \emptyset$ the sets A, B are said to be *disjoint* (or nonintersecting).

(iii) The *difference* of the sets A and B (or the *complement* of B relative to A) is the set of all elements of A that do not belong to B: this set is denoted by $A - B$; thus

$$A - B = \{x \mid x \in A, x \notin B\}.$$

When A is some universal set, say U, then it is common to call $U - B$ simply *the complement of* B and it is then written as B'.

The operations of forming unions and intersections are associative; thus, for example,

$$A \cup (B \cup C) = (A \cup B) \cup C.$$

In addition we have the (symmetrical) distributive laws:

$$A \cap (B \cup C) = (A \cap B) \cup (A \cap C),$$
$$A \cup (B \cap C) = (A \cup B) \cap (A \cup C).$$

Another way of combining sets is to form the Cartesian product. Given two sets A and B, the set of all ordered pairs (a, b) with $a \in A$ and $b \in B$ is called the *Cartesian product* of A and B, and is denoted by $A \times B$. Note the difference between the *ordered pair* (a, b) and the set $\{a, b\}$; in the latter we could equally well write $\{b, a\}$ whereas in the former we obtain a different pair if we write (b, a). It follows that $A \times B$ is not the same set as $B \times A$. Perhaps the most obvious example of a Cartesian product is the coordinate plane $R \times R$ in which each point is in correspondence with a number pair, namely

6 PRELIMINARIES

its coordinates. The definition of product readily extends to the case of n sets for any positive integer n.

Example 1.2.1

(a) If $N = \{1, 2, 3, \ldots\}$ denotes the set of natural numbers then the sets $\{3, 5\}$ and $\{x \in N \mid x^2 - 8x + 15 = 0\}$ are equal.

(b) The set of *integers* is
$$Z = \{0, 1, -1, 2, -2, 3, -3, \ldots\}.$$
The set of *rational numbers* is
$$Q = \left\{\frac{m}{n} \mid m, n \in Z, n \neq 0\right\}.$$

(c) The set $\{1, 2\}$ has four subsets, namely $\emptyset$, $\{1\}$, $\{2\}$ and $\{1, 2\}$; the first three are proper subsets.

(d) Two sets A and B are equal if and only if both $A \subseteq B$ and $B \subseteq A$.

(e) (i) Set operations can be pictorially represented by means of a *Venn diagram*. Here the universal set is represented by a rectangular region and its elements by the points of the region—subsets are then represented by areas within the rectangle. Figures 1.1(a) and (b) illustrate the union and intersection of two subsets A and B while Figure 1.1(c) illustrates the distributive law
$$A \cap (B \cup C) = (A \cap B) \cup (A \cap C).$$

(ii) Here is a formal proof of the distributive law illustrated in Figure 1.1(c). Let $x \in A \cap (B \cup C)$, then $x \in A$ and $x \in B \cup C$; thus either $x \in A$ and $x \in B$ or $x \in A$ and $x \in C$, that is to say $x \in (A \cap B) \cup (A \cap C)$. We have thus shown that
$$A \cap (B \cup C) \subseteq (A \cap B) \cup (A \cap C).$$
Now suppose $y \in (A \cap B) \cup (A \cap C)$, then $y \in A \cap B$ or $y \in A \cap C$: it follows that $y \in A$ and $y \in B \cup C$, that is to say $y \in A \cap (B \cup C)$. Hence
$$(A \cap B) \cup (A \cap C) \subseteq A \cap (B \cup C).$$
The result (d) above shows the sets to be equal.

(f) (De Morgan's laws) We have
 (i) $(A \cap B)' = A' \cup B'$;
 (ii) $(A \cup B)' = A' \cap B'$.

PRELIMINARIES 7

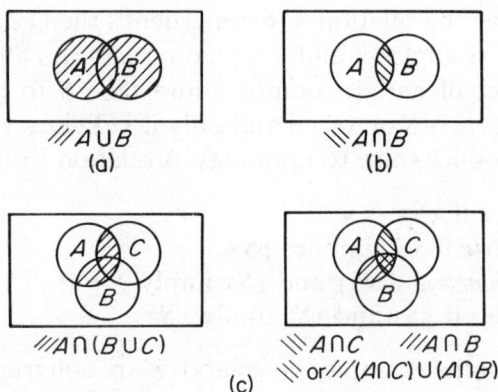

Fig. 1.1

These can be seen by drawing a Venn diagram or proved directly. The details are left to the reader.

(g) If $\{S_\rho\}$ is an arbitrary collection of sets where the index ρ ranges over some index set (*see* p. 14) the union of this collection is denoted $\cup_\rho S_\rho$ and is the set of all elements x such that x belongs to at least one of the S_ρ. Similarly $\cap_\rho S_\rho$ is the set of all elements x such that x belongs to every S_ρ. For example, if S_n, $n \in N$, is the set of continuous functions $x(t)$ defined on $[0, 1]$ such that $|x(t)| < 1 + 1/n$ then $\cap_n S_n$ is the set of continuous function on $[0,1]$ such that $|x(t)| < 1$.

Relations

We are all familiar with the connection between, say, the circle of unit radius, centre the origin, drawn in the Cartesian plane and its 'equation' $x^2 + y^2 = 1$. But we have seen that we can consider the Cartesian plane as the product of the real line with itself, and in this sense the curve represents a subset of $R \times R$. We might say that the first and second members of the ordered pairs (x, y) lying on the curve are related.

It is the abstraction of this idea to general sets that leads to the concept of relation in mathematics. If S is a subset of

$$X \times X = \{(x,y) | x \in X, y \in X\}$$

then we say that S is a *relation* on X. This means to say that for every $(x, y) \in X \times X$ the statement $(x, y) \in S$ is meaningful and either true or false. To emphasize the fact that x and y are, in this sense, related, we write xSy if $(x, y) \in S$. For example, let X be the set of all triangles and take S to be the relation 'are congruent'; then $(x, y) \in S$ or xSy if and only if the triangles x and y are congruent. As another example let X be the set of natural numbers and take S to be the relation 'divides'; then $(x, y) \in S$ or xSy if and only if x divides y.

We now introduce some terminology. A relation S on X is said to be

(i) *reflexive* if $xSx \ \forall x \in X$,[†]
(ii) *symmetric* if xSy implies ySx,
(iii) *antisymmetric* if xSy and ySx imply $y = x$,
(iv) *transitive* if xSy and ySz imply xSz.

Notice that in the examples given above 'are congruent' is reflexive, symmetric and transitive while 'divides' is reflexive, antisymmetric and transitive. Relations having these particular combinations of properties are of sufficient importance to be given special names. A relation such as 'are congruent' which is reflexive, symmetric and transitive is called an *equivalence relation*. The importance of this type of relation is that it is closely associated with the splitting-up or division of a set into disjoint subsets. Suppose S is an equivalence relation: for each $x \in X$ the set

$$S_x = \{y \in X \mid xSy\}$$

is called the *equivalence class* of x. Furthermore the family S_x of equivalence classes is pairwise disjoint, that is to say either $S_x = S_y$ or $S_x \cap S_y = \emptyset$. To see this, suppose that $S_x \cap S_y \neq \emptyset$: let $z \in S_x \cap S_y$, then xSz and ySz, and hence ySx. If $u \in S_x$ then xSu, and from ySx we have ySu so $u \in S_y$, and we conclude that $S_x \subseteq S_y$. By symmetry $S_y \subseteq S_x$ and thus we must have $S_x = S_y$ (Example 1.2.1(d)). Since S is reflexive, $x \in S_x$ for all $x \in X$ and it follows that the union $\cup_x S_x$ of the family of equivalence classes of S is X (Example 1.2.1(g)). A pairwise disjoint family of sets whose union is X is called a *partition* of X and we have just shown that the family of equivalence classes of an equivalence relation on X is a partition of X.

An equivalence relation is often written as $x \sim y$ instead of xSy.

Example 1.2.2

(a) An everyday use of the concept of equivalence—but one which is not perhaps recognized as such—is the relation of equality for fractions. The fractions $\frac{3}{4}$ and $\frac{9}{12}$ are clearly not identical, but we treat them as

---

[†] The symbol $\forall$ is an accepted abbreviation for the phrase 'for all' and we shall use it extensively.

being equal. What we are really saying is that a fraction is a representative of an equivalence class of fractions and to this equivalence class we give the name rational number. The explicit relation S in this case is 'a/b and c/d are equal if $ad = bc$ in the sense of integers', where X is taken to be the set of all fractions. Note that the relation 'equals' is itself an equivalence relation on Z.

(b) If X is the set of all points in the (Cartesian) plane then a partition of X might consist of those sets S_x which contain points having the same first coordinate; S_x is then 'a vertical line' and all such lines fill the plane. The equivalence relation is '$((x, y), (u, v)) \in S$ if $x = u$'.

(c) Modular arithmetic provides another example of partitioning. If a, m, r are integers with $m > 1$ then we may write $a \equiv r \pmod{m}$ meaning $mq + r = a$, where q is the quotient and r the remainder. If, for example, $m = 5$, then we obtain five pairwise disjoint subsets of the set of integers Z according to the remainders $0, 1, 2, 3, 4$, viz.,

$$Z_0 = \{\ldots, -10, -5, 0, 5, 10, \ldots\},$$
$$Z_1 = \{\ldots, -9, -4, 1, 6, 11, \ldots\},$$
$$Z_2 = \{\ldots, -8, -3, 2, 7, 12, \ldots\},$$
$$Z_3 = \{\ldots, -7, -2, 3, 8, 13, \ldots\},$$
$$Z_4 = \{\ldots, -6, -1, 4, 9, 14, \ldots\}.$$

The other important type of relation (exemplified by our earlier example, 'divides') is reflexive, antisymmetric and transitive and is called *a partial ordering*. The archetypal example of a partial ordering is the relation 'less than or equal to', symbolized by $\leq$ and defined on N, Z, Q or R: for this reason the symbol $x \leq y$ is used instead of xSy when x, y are elements of any set X and S is a partial ordering.

Let $\leq$ be a partial ordering relation on a set X and let A be a subset of X. If there exists $x \in X$ such that $a \leq x$ for all $a \in A$, then we say x is *an upper bound for* A: an upper bound y for A is called a *supremum* (or least upper bound, l.u.b.) if $y \leq x$ for all upper bounds x for A. The terms *lower bound* and *infimum* (or greatest lower bound, g.l.b.) are defined in an analogous way. When it exists, the supremum (or infimum) of A is unique. We shall subsequently use the abbreviations sup or inf for supremum or infimum respectively. If a set A has both an upper and lower bound we say A is *bounded*; otherwise it is unbounded.

Example 1.2.3

(a) We have already noted that the relation 'divides' is a partial ordering on N. Suppose $A \subset N$ is the subset $\{4, 6\}$: an upper bound for A is any integer divisible by 4 and 6; clearly 12 (the least common multiple) is the supremum. Similarly a lower bound is any number which divides 4 and 6 and the infimum (greatest common divisor) is 2.

(b) The set of real numbers is partially ordered when $\leqslant$ takes its usual meaning, but in addition, for any $x, y \in R$, it is *always* meaningful to write $x \leqslant y$. We can state this another way and say that, for any two elements x, y, either $x \leqslant y$ or $y \leqslant x$. A set having this additional property is said to be *totally* (or linearly) *ordered*. The set R with the relation $<$ is thus totally ordered.

(c) Here are two examples of sets which are partially but not totally ordered.
  (i) If $\leqslant$ is the relation of set inclusion then it is easy to verify that the subsets of a universal set U are partially ordered. However, if two subsets are disjoint it is simply not meaningful to apply the order relation, so the ordering is not total.
  (ii) Let (x_1, x_2, x_3), (y_1, y_2, y_3) be two elements (or points) in $R \times R \times R$. Then we may define $\leqslant$ on this set by $(x_1, x_2, x_3) \leqslant (y_1, y_2, y_3)$ if $x_i \leqslant y_i$, $i = 1, 2, 3$. This is clearly a partial ordering, but, for example, we cannot compare the elements $(1, 2, 3)$ and $(2, 1, 4)$, so the ordering is not total.

(d) (i) The (open) interval $(0, 1)$ of R has 0 as infimum and 1 as supremum.
  (ii) The set $\{1, \frac{1}{2}, \frac{1}{3}, \frac{1}{4} \ldots\}$ has supremum 1 and infimum 0, but notice that whereas 1 is an element of the set, 0 is not. For subsets of the real line, if the supremum is actually an element of the set then we call it the *maximum*, and in the same way we replace infimum by *minimum*. We shall subsequently use the abbreviations max and min, respectively.

  It is a fundamental property of the real numbers that every non-empty subset of R which is bounded above has a supremum: this is sometimes called the *supremum principle* (*see* for example, [2, 10]). Similarly every subset bounded below has an infimum.
  (iii) The subset of rational numbers
  $$A = \{x \in Q \mid 0 < x < \sqrt{2}\}$$
  does not have a supremum in Q.

Functions

We now turn to the extremely important concept of function. A century or so ago the term function meant no more than we mean now by formula: for example, $y = x^2$ would have been recognized as a function, whereas $y = |x|$ might have been a doubtful candidate.

The development of mathematics required that the idea of function be generalized and at the same time rigorously defined. The evolution of the idea of function, culminating in the modern definition, is discussed, for example, in [3].

Let X and Y be two sets which are not necessarily distinct. A *function* from X to Y is a set $F \in X \times Y$ such that if $(x, y) \in F$ and $(x, z) \in F$ then $y = z$. A function then is a special type of relation such that whenever the pairs (x, y) and (x, z) have the same first element, they must also have the same second element.

We now define

(i) the *domain* of F, $\mathscr{D}(F)$, is the set $\mathscr{D}(F) = \{x \in X \mid (x, y) \in F \text{ for some } y \in Y\}$;
(ii) the *range* of F, $\mathscr{R}(F)$,[†] is the set $\mathscr{R}(F) = \{y \in Y \mid (x, y) \in F \text{ for some } x \in X\}$.

Since the second element of the pair $(x, y) \in F$ is uniquely known once x is known we can replace y by the more suggestive symbol $F(x)$, thus reconciling the set definition with the more conventional representation, $y = F(x)$, of a function. In this context the element $F(x) \in Y$ is called the *value of F at x*, or the *image of x under F*.

This view of a function is more akin to the idea of F as a *transformation* or *mapping* which carries (the input) x into (the output) $F(x)$. We often represent this by a diagram such as that of Figure 1.2(a). Notice that the modern definition makes a very clear distinction between the function (or set) F and the value of F at x, $F(x)$. The

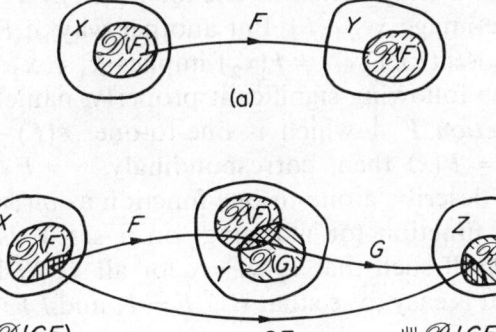

Fig. 1.2

[†] The use of script letters to denote these sets is convenient for our later use.

names function, transformation, mapping, functional, operator are all used to represent function, but each has acquired through use a connotation of its own which tends to promote different usage in different contexts: we shall use all these names in one place or another through the present book. It is also not universal that the value of F at x should be written $F(x)$ with x enclosed in parentheses: in fact on most occasions we shall write Fx, omitting the parentheses. We shall use capital letters to denote mappings except when the range is a subset of the real or complex numbers, when, in accordance with historical convention, we shall use lower case letters.

We shall use the terminology $F: X \to Y$ to mean, 'F is a function mapping a subset of X, namely $\mathscr{D}(F)$ to a subset of Y, namely $\mathscr{R}(F)$'.[†] We say two functions F, H are *equal* if $\mathscr{D}(F) = \mathscr{D}(H)$ and for every $x \in \mathscr{D}(F)$, $F(x) = H(x)$. If $\mathscr{R}(F)$ is a proper subset of Y then we say, 'F is a mapping *into* Y', whereas if $\mathscr{R}(F) = Y$ we say, 'F is a mapping *onto* Y'. If $B \subset Y$, the set $\{x \mid F(x) \in B\}$ is called the *pre-image* of B under F.

Let F be a mapping of a set X into a set Y and G be a mapping of Y into a set Z. We may define a new mapping GF (or $G \circ F$) of X into Z by $GF(x) = G(F(x))$. The mapping GF is called the *composition* or product of G and F (Figure 1.2(b)). It is important to notice that

(i) GF exists only if $\mathscr{R}(F) \cap \mathscr{D}(G) \neq \emptyset$;
(ii) $GF \neq FG$.

In general a point $y \in \mathscr{R}(F)$ will have several pre-images in $\mathscr{D}(F)$. We define a *one-to-one function* as one for which every point $y \in \mathscr{R}(F)$ has a *single* pre-image $x \in \mathscr{D}(F)$. Put another way, if F is one-to-one then, for $x_1, x_2 \in \mathscr{D}(F)$, $F(x_1) \neq F(x_2)$ implies $x_1 \neq x_2$. A one-to-one function has the following significant property, namely: there exists *the inverse function* F^{-1} which is one-to-one $\mathscr{R}(F) \to \mathscr{D}(F)$ and is such that if $y = F(x)$ then, correspondingly, $x = F^{-1}(y)$. For this reason we also describe a one-to-one function as *invertible*.

The *identity* function (or mapping) on a set X is the mapping I_X of X onto itself such that $I_X(x) = x$ for all $x \in X$. If $F: X \to Y$ is invertible then it is easy to see that $F^{-1}F = I_X$ and $FF^{-1} = I_Y$.[‡]

[†] Some authors write $F: X \to Y$ only if $\mathscr{D}(F) = X$ and reserve the term *partial mapping* for our usage.

[‡] It is possible to define a left inverse F_l^{-1} of F such that $F_l^{-1}F = I_X$ but $FF_l^{-1} \neq I_Y$, and similarly a right inverse, but we shall not have need for these concepts in this book (*see*, for example, [29]).

Example 1.2.4

(a) By the graph of a function $f: R \to R$ we shall here understand the usual (Cartesian) plot of the ordered pairs $(x, f(x))$.
  (i) The graph of a function cannot be cut more than once by a vertical line. For example, a semicircle with diameter on the horizontal or x-axis (excluding the end points of the arc) may represent the graph of a function, whereas a complete circle can only represent a relation.
  (ii) The graph of a one-to-one function cannot be cut more than once by a horizontal line: however, a function can be rendered one-to-one by suitably restricting the domain. For example, the function 'sin' is clearly not one-to-one; however if we restrict x to the interval $[-\pi/2, \pi/2]$ then $\sin: [-\pi/2, \pi/2] \to [-1, 1]$ is one-to-one, and we may define an inverse, $\sin^{-1}: [-1, 1] \to [-\pi/2, \pi/2]$, which we conventionally call the principal value (or branch) of $\sin^{-1}$. Similarly, the function $f(x) = x^2$, $x \in (-\infty, \infty)$ is not one-to-one. However, we may define the one-to-one functions $f_1(x) = x^2$, $\mathscr{D}(f_1) = [0, \infty)$ and $f_2(x) = x^2$, $\mathscr{D}(f_2) = (-\infty, 0]$ and we then have $f_1^{-1}(x) = \sqrt{x}$, $f_2^{-1}(x) = -\sqrt{x}$.

(b) Let $f: R \to R$ be such that $f(x) = \sqrt{x}$ with $\mathscr{D}(f) = \{x \mid x \geq 0\}$ and $g: R \to R$ be such that $g(x) = 1 - x^2$ with $\mathscr{D}(g) = R$.
  (i) We have
  $$\mathscr{R}(f) \cap \mathscr{D}(g) = [0, \infty) \cap (-\infty, \infty) = [0, \infty) \neq \emptyset.$$
  so that gf exists; furthermore (Figure 1.2(b)),
  $$\mathscr{R}(gf) = \{z \in Z \mid z = g(x), x \in \mathscr{R}(f) \cap \mathscr{D}(g)\} = (-\infty, 1]$$
  and
  $$\mathscr{D}(gf) = \{x \in X \mid f(x) \in \mathscr{R}(f) \cap \mathscr{D}(g)\} = [0, \infty).$$
  The formula for gf is, of course, $gf(x) = 1 - x$. (Notice that this formula is meaningful for values of x outside $[0, \infty)$.)
  (ii) We have
  $$\mathscr{R}(g) \cap \mathscr{D}(f) = (-\infty, 1) \cap [0, \infty) = [0, 1] \neq \emptyset,$$
  so that fg exists and $\mathscr{R}(fg) = [0, 1]$, $\mathscr{D}(fg) = [-1, 1]$. The formula for fg is $fg(x) = \sqrt{1 - x^2}$.

(c) A (3×2) real matrix $[f_{ij}]$ is, as we shall see later, a transformation of $R \times R$ into $R \times R \times R$: if $[g_{ij}]$ is a real (2×1) matrix then we can define the composition (or matrix product) $[f_{ij}][g_{ij}]: R \to R \times R \times R$, but $[g_{ij}][f_{ij}]$ is not defined.

(d) Consider the set of real continuously differentiable functions $x(t)$ defined on $[0, 1]$ and let F be the operator of differentiation on this set, viz., $Fx = dx/dt$.

(i) F is not one-to-one since two functions which differ by a constant have the same image and thus F^{-1} does not exist.

(ii) Suppose we now take

$$\mathscr{D}(F) = \{x(t) | x \text{ continuously differentiable, } x(0) = 0\},$$

then $F: \mathscr{D}(F) \to \mathscr{R}(F)$ is one-to-one and

$$F^{-1}x = \int_0^t x(\tau) d\tau.$$

This example illustrates a detail of notation which we shall use throughout the book. It is convenient to use the symbols x, y, z, etc., for points of a set, but when these points are themselves functions of, say, a real variable, we need another symbol to represent that variable. For this purpose we shall usually use t, s, τ or σ or, if the variable is, for example, a point in $R \times R \times R$, the 3-tuple (t_1, t_2, t_3). There is no connotation of 'time' with the variable t, indeed it will often represent a 'space-like' variable in a physical sense.

(e) If $F: X \to Y$, $G: Y \to Z$ have inverses F^{-1}, G^{-1}, respectively, and the composition $GF: X \to Z$ is defined, it is not difficult to see that

$$(GF)^{-1} = F^{-1}G^{-1}.$$

(f) (The Kronecker delta) Let $X = N \times N$ and $Y = \{0, 1\}$. The function

$$\delta_i^j = \begin{cases} 1, & i = j, \\ 0, & i \neq j, \end{cases}$$

is called the *Kronecker delta* (or function). We shall find it convenient to make use of this function at many points in the book.

(g) (i) (Sequences) A function S whose domain is the set of natural numbers N and whose range is a subset of a set X is called a *sequence in X*. If $n \in N$ we shall write the value of S at n, $S(n)$, as $x^{(n)}$ and denote the range by $\{x^{(n)}\}$:[†] however, in usage we often speak loosely of 'the sequence $\{x^{(n)}\}$'.

(ii) (Subsequences) Let n be a one-to-one function on N and let $n(k) = n^{(k)}$: the composition $Sn: N \to X$ is called a *subsequence of the sequence S*, and the value of Sn at k is denoted by $(x^{(n)})^{(k)}$.

(h) Countability

(i) A set X is said to be *finite* if there exists a one-to-one mapping of X

[†] $\{x^{(n)}\}$ is a subset of X which is said to be *indexed by n*: subsets may be indexed by other infinite sets than N. We shall often use $\{\rho\}$ to denote a finite ordered subset of N.

onto the subset $\{1, 2, 3, \ldots, k\}$ of the natural numbers. We write $n(X)$ to denote the number of elements in X. Thus, for example, if X and Y are finite it is not difficult to see that

$$n(X \cup Y) = n(X) + n(Y) - n(X \cap Y)$$

(ii) A set which is not finite is said to be *infinite*. A set X is said to be *countably infinite* if there exists a one-to-one mapping of X onto (the whole of) N. A countably infinite set can be *enumerated* (or indexed) by N, and we then write $X = \{x_k | k = 1, 2, 3, \ldots\}$.

A set which is finite or countably infinite is said to be *countable* (or denumerable) otherwise *uncountable* (or non-denumerable). It is a celebrated result that the real numbers are uncountable; the rationals on the other hand are countable (*see*, for example, [2, 10]).

(iii) Two sets are said to have *the same cardinal number* if there exists a one-to-one mapping of one set onto the other. For example the sets $\{1, 2, 3, 4, \ldots\}$ and $\{2, 4, 9, 16, \ldots\}$ have the same cardinal number since the function $F(n) = n^2$ gives a one-to-one correspondence between the sets. Similarly the function $F(m, n) = 2^{m-1}(2n - 1)$ is a one-to-one mapping of $N \times N$ onto N, and hence the set of points in the plane having integer coordinates is countable. We can also see that the set of points in the plane having rational coordinates is likewise countable.

2 Algebraic theory of vector spaces and linear transformations

2.1 Introduction

The word 'space' in mathematics is usually used in the sense of structure. By postulating a set of axioms which are to be obeyed by the elements of the 'space' we build a mathematical structure which is self-consistent. In general, the fewer the axioms, the less structure has the space; as further axioms are added, the structure is enriched and the number of deduced properties (theorems) increases.

The elements of the space are, in a sense, ghosts, since they exist merely by virtue of the fact that they obey the axioms. If we wish to think of the elements in a practical case as concrete entities, such as, for example, force, velocity, functions of a real variable, matrices or random variables, then the sole criterion we must consider is whether such quantities (with our usage of them) satisfy the axioms for the space. This approach is necessarily rather abstract and it is probably true to say that an appreciation of any abstract mathematical structure is only gained through experience with concrete examples.

The particular mathematical structure embodied in the axioms for a vector space is one which we need to understand. From the point of view of applied mathematics this type of space has proved to be of considerable value: roughly speaking, it is a natural extension of the simple addition of numbers. The name 'vector' is used by analogy with the ordinary (directed) vectors of three-dimensional space. Any point in space can be located by a 'sum' of vectors; in fact we need only three non-collinear vectors in order to be able to specify any point as a sum provided we are also allowed to stretch (or shrink) vectors in length.

This idea is capable of extension to spaces having 'n dimensions' and indeed to spaces of 'infinite dimension'. In some sense we can, for example, try to represent a function in which we are interested as a 'vector' sum of a chosen set of reference or coordinate functions. The reader will know that a periodic function can be represented, in the sense of Fourier series, as a weighted sum of sines and cosines. It turns out that many of the quantities in which we are interested when we analyse a physical problem mathematically can be readily interpreted as vectors. If we already have a generalized vector space structure we can use any of its manifold properties immediately in the particular concrete representation.

This chapter deals with the algebraic theory of vector spaces. The use of the qualification 'algebraic' refers to the fact that we shall, in adding vectors, restrict ourselves to a finite number. To be able to assign a meaning to an infinite sum of vectors, we need, in effect, to add (to the vector space axioms) further axioms which allow us to measure length and hence nearness or convergence in the space. This is postponed until Chapter 3. By making this distinction it will be clearer to the reader which results do and which do not depend on the introduction of a 'length' into the vector space.

Once we have defined entities which we call vectors, we can naturally go on to consider mappings on vector spaces. Because of the structure of the vector space, we are led to consider a particular type of mapping called a linear transformation. This has the desirable property that it preserves the basic operations of the vector space so that, for example, the image of the sum of two vectors is the sum of their images. While linear transformations arise naturally in this way, they would not be of more than passing interest to the applied mathematician were it not that many physical systems can be described in terms of equations involving linear transformations. The omnipresent operations of differentiation and integration, for example, are linear transformations. Typically, linear models for physical systems are associated with small disturbances of the system from some equilibrium state. However, even when the describing equations are non-linear, the linear transformation plays a role as a local approximation.

In the important special case of vector spaces of finite dimension, we shall see that every linear transformation, whatever its origin, has a concrete representation as a square or rectangular matrix. Hence much of the later part of the chapter is a setting of results of matrix algebra within the context of vector space theory. The emphasis is on the distinction between results which are independent of whatever reference or coordinate system is used for the space and those that are not.

2.2 The definition of a vector space

Before we give a formal definition of a vector space, it is necessary to say something about the concept of the associated 'scalars'. In an abstract setting the only requirement on these scalars is that they be elements of an algebraic field, but in this book (and in all applications) the scalars will always be the set of real or of complex numbers. As a consequence, the reader does not need to understand the concept of a field [12], since the necessary properties are embodied in the familiar rules of real or complex arithmetic. If we specifically wish to indicate the scalar field we use the terms 'real vector space' or 'complex vector space'.

While the scalars are restricted to being real or complex numbers, the entities we call vectors can assume a wide variety of concrete representations. It cannot be too strongly emphasized that the question of whether the elements of a given set are or are not vectors is completely settled by testing whether these elements satisfy the axioms for a vector space when associated with suitable scalars.

Definition 2.2.1 *A vector space $\mathscr{V}$ is a set of elements called vectors with an operation called* addition, *and an operation called* scalar multiplication *which satisfy the following axioms.*

(a) *Addition axioms: To every pair of vectors $x, y \in \mathscr{V}$, there corresponds a unique vector $x + y \in \mathscr{V}$, the sum of x and y, such that*
  (i) $x + y = y + x$ *(commutative law)*;
  (ii) $(x + y) + z = x + (y + z)$ *(associative law)*;
  (iii) *there exists a unique zero vector $\theta \in \mathscr{V}$ such that $x + \theta = x$ $\forall x \in \mathscr{V}$ (identity element for addition)*;
  (iv) *for every vector x there exists a unique vector $(-x) \in \mathscr{V}$ such that $x + (-x) = \theta$ (additive inverse).*
  The vector $x + (-y)$ is normally written $x - y$.
(b) *Scalar multiplication axioms: To every scalar α and every vector $x \in \mathscr{V}$ there corresponds a unique vector $\alpha x \in \mathscr{V}$ such that*
  (v) $\alpha(\beta x) = (\alpha\beta)x$ *for every scalar β*;
  (vi) $1x = x, 0x = \theta$ $\forall x \in \mathscr{V}$;
  (vii) $\alpha(x + y) = \alpha x + \alpha y,$
  (viii) $(\alpha + \beta)x = \alpha x + \beta x$ } *(distributive laws).*
  One can easily show that the vector $-1x$ is the vector $(-x)$ of axiom (iv).

Throughout the book we will adhere to the convention that scalars are represented by lower case Greek letters: vectors will normally be represented by lower case italic letters, but not invariably so—the zero vector θ is a case in point.

As stated, the axioms are not independent, but they provide a convenient list against which to test possible candidates for vector spaces.

A vector space may contain only the zero (or null) element θ; it is called the *zero space* and will be denoted by $\mathcal{O}$. This is not the same as the set-theoretic notion of an empty set as one containing no elements.

Example 2.2.1 The ordinary vectors of three-dimensional space are vectors in the sense just described, if we interpret addition of vectors as addition of components and scalar multiplication as multiplication of each component by the same scalar. If $x = (\xi_1, \xi_2, \xi_3)$, $y = (\eta_1, \eta_2, \eta_3)$, then

$$x + y = (\xi_1 + \eta_1, \xi_2 + \eta_2, \xi_3 + \eta_3)$$

and

$$\alpha x = (\alpha \xi_1, \alpha \xi_2, \alpha \xi_3).$$

Addition corresponds to the parallelogram law and scalar multiplication to a change of magnitude of the vector. There is no operation corresponding to vector multiplication.

The above suggests the consideration of ordered n-tuples of numbers $(\xi_1, \xi_2, \ldots, \xi_n)$ as possible elements of a vector space: let $x = (\xi_1, \xi_2, \ldots, \xi_n)$, $y = (\eta_1, \eta_2, \ldots, \eta_n)$ and define

$$x + y = (\xi_1 + \eta_1, \xi_2 + \eta_2, \ldots, \xi_n + \eta_n),$$
$$\alpha x = (\alpha \xi_1, \alpha \xi_2, \ldots, \alpha \xi_n).$$

This concrete definition of addition and scalar multiplication on ordered n-tuples satisfies the axioms for a vector space, as is easily verified. The zero vector is $(0, 0, \ldots, 0)$, while the vector $(-x)$ is $(-\xi_1, -\xi_2, \ldots, -\xi_n)$: the scalar ξ_k is called the kth component or coordinate of x. This vector space is generally denoted by $\mathcal{R}_n$ if the scalars are real and $\mathcal{C}_n$ if the scalars are complex. In a sense which will be described later, $\mathcal{R}_n$ is a generalization of familiar three-dimensional space to an n-dimensional space.

Notice that in the statement

$$x + y = (\xi_1 + \eta_1, \xi_2 + \eta_2, \ldots, \xi_n + \eta_n)$$

the $+$ sign on the right-hand side is the ordinary addition of numbers: the $+$ sign on the left-hand side denotes the postulated operation of addition of vectors, and as such is *defined* by the right-hand side.

This is a typical situation; if we wish to construct a particular vector space, we need

(a) to specify a concrete form for its elements;
(b) to define in a concrete way the operations of addition and scalar multiplication.

Example 2.2.2 If the restriction that n is finite in Example 2.2.1 is dropped we may admit as vectors infinite sequences of scalars of the form

$$x = (\xi_1, \xi_2, \ldots, \xi_k, \ldots),$$

where again addition and scalar multiplication are defined component-wise. By imposing various restrictions on the components ξ_k a number of vector spaces can be defined, for example:

(a) the set of all sequences for which $\sum_k \xi_k$ converges;
(b) the set of all sequences in which only a finite number of terms is non-zero;
(c) the set of all bounded sequences (a sequence $\{\xi_k\}$ is said to be bounded if $|\xi_k| < \infty$, for all k).

Example 2.2.3 Let $[a, b]$ be an interval of the real line, and consider the set of all real-valued continuous functions defined on $[a, b]$. If $x(t)$, $y(t)$ are two functions defined on $[a, b]$, then if we take

$$\left. \begin{array}{l} (x+y)(t) = x(t) + y(t), \\ (\alpha x)(t) = \alpha x(t) \end{array} \right\} \forall\, t \in [a, b],$$

we have defined two continuous functions $x + y$ and αx which are the sum of x and y and the scalar multiple of x respectively. With these definitions the set of real-valued continuous functions on $[a, b]$ is a vector space. The zero element is the function having the value zero for all $t \in [a, b]$. Other vector spaces can be formed in this way by placing suitable conditions on the member functions; for example

(a) the set of all continuous functions which have continuous derivatives of order $\leqslant m$ on $[a, b]$ is a vector space;
(b) the set of functions which are integrable on $[a, b]$, that is, those functions $x(t)$ for which

$$\int_a^b x(t)\,dt$$

is finite, is a vector space.

Notice the use of x, y as function names instead of the more usual f, g, h, etc. The elements of a function vector space are the functions x, y themselves and not their values. It might be advisable for the reader to look again at that part of Section 1.2 dealing with functions (p. 10 et seq.).

Example 2.2.4 The set of all polynomial functions defined on the interval $[a, b]$ of degree $\leqslant (m-1)$ is a vector space if we define addition and scalar multiplication as in Example 2.2.3. We may characterize every polynomial $p(t) = a_0 + a_1 t + a_2 t^2 + \ldots + a_{m-1} t^{m-1}$ by its coefficient m-tuple. Addition and scalar multiplication of polynomials can then be transmuted into addition and scalar multiplication of coefficient m-tuples which are elements of the vector space of Example 2.2.1. In this sense the vector space of polynomials of order $\leqslant (m-1)$ is indistinguishable from the vector space of m-tuples: we say the two spaces are *isomorphic*.

Exercises 2.2

1. Which of the following subsets of the set of all triples of real numbers (ξ_1, ξ_2, ξ_3) form a vector space?
    (i) $\xi_1 = 2 - \xi_2; \xi_2, \xi_3$ arbitrary;
    (ii) $\xi_3 = 2\xi_2 - \xi_1; \xi_1, \xi_2$ arbitrary;
    (iii) $\xi_3 = |\xi_1|; \xi_1, \xi_2$ arbitrary;
    (iv) $\xi_1, \xi_2 \geqslant 0, \xi_3$ arbitrary.

2. Show that the set of all $(m \times n)$ matrices can be considered to form a vector space.

3. If $\mathscr{V}, \mathscr{U}$ are two vector spaces over the same field of scalars their *Cartesian product*, denoted $\mathscr{V} \times \mathscr{U}$, consists of the set of ordered pairs (v, u) with $v \in \mathscr{V}, u \in \mathscr{U}$. Show that $\mathscr{V} \times \mathscr{U}$ is a vector space if addition and scalar multiplication are defined by

    $$(v^1, u^1) + (v^2, u^2) = (v^1 + v^2, u^1 + u^2),$$
    $$\alpha(v, u) = (\alpha v, \alpha u).$$

    In this sense show that $\mathscr{R}_n$ (Example 2.2.1) is the n-fold Cartesian product of $\mathscr{R}$ with itself.

2.3 Subspaces

In ordinary three-dimensional space all vectors lying in a fixed plane through the origin can be added and scalarly multiplied to yield vectors which also lie in the plane. For example, if we choose one of the coordinate planes then we are dealing with vectors of the type $(0, \xi_2, \xi_3)$ say, and the above assertion is obviously true. In this way we appear to have set up another vector space using a subset of the vectors belonging to the parent space. This idea is formalized in the concept of a subspace of a vector space $\mathscr{V}$.

Definition 2.3.1 *A non-empty subset $\mathscr{W}$ of a vector space $\mathscr{V}$ is called a subspace of $\mathscr{V}$ if $\mathscr{W}$ is itself a vector space under the rules for addition and scalar multiplication as defined for $\mathscr{V}$.*

In this definition the last phrase is essential. Every vector space has the following trivial subspaces:

1. the zero vector space;
2. the space $\mathscr{V}$ itself.

These two subspaces are called *improper subspaces* of $\mathscr{V}$; all other subspaces are *proper subspaces*. Every subspace of $\mathscr{V}$ contains the zero vector.

Example 2.3.1 In the vector space of n-tuples, the subset of n-tuples with kth element zero is a subspace: more generally the solutions $(\xi_1, \xi_2, \ldots, \xi_n)$ of an equation such as

$$\alpha_1 \xi_1 + \alpha_2 \xi_2 + \ldots + \alpha_n \xi_n = 0$$

form a subspace of the space of n-tuples.

Example 2.3.2 The set of real-valued continuous functions defined on $[a,b]$ and which vanish at a fixed point $c \in [a,b]$ is a subspace of the vector space of real-valued continuous functions on $[a,b]$ (Example 2.2.3). In turn, the set of all continuous functions defined on $[a,b]$ is itself a subspace of the vector space of all functions defined on the same domain.

A useful and readily applied criterion for a subset to be a subspace is given by

Theorem 2.3.1 *A subset $\mathscr{W}$ of a vector space $\mathscr{V}$ is a subspace of $\mathscr{V}$ if, and only if, $\alpha x + \beta y \in \mathscr{W}$ for all α, β and all $x, y \in \mathscr{W}$.*

The proof is easy and is left to the reader.

Example 2.3.3 The subset of n-tuples of the form $(\beta, \xi_2, \xi_3, \ldots, \xi_n)$ with β a fixed non-zero number is not a subspace since the sum of two such vectors does not have first element β.

Example 2.3.4 The set of polynomials defined on $[a,b]$ of degree precisely equal to $n < (m-1)$ is not a subspace of the vector space of polynomials of degree $\leqslant (m-1)$ (Example 2.2.4) since the difference of two polynomials of degree n may not be of degree n: furthermore, this subset does not contain the 'zero' polynomial.

A convenient way of generating subspaces of a vector space $\mathscr{V}$ is as follows. Let $\{x^i\}$ be a finite or infinite set of vectors in $\mathscr{V}$: the vector

$$x = \sum_\rho \lambda_\rho x^\rho,$$

where ρ is an element of any finite ordered subset of the natural numbers, is called a finite linear combination of the vectors $\{x^i\}$. It is easily verified that the set of all finite linear combinations of the set $\{x^i\}$ is a subspace of $\mathscr{V}$. We give this subspace a special name.

Definition 2.3.2 *Let S be an arbitrary, non-void subset of $\mathscr{V}$. The set of all finite linear combinations of vectors in S is called* the subspace spanned by (*or* generated by) *S and is denoted* $[S]$.

The restriction to finite sums is necessary because we have, at this stage, no way of assigning a meaning to an infinite sum of vectors: this difficulty is overcome in Chapter 3.

Example 2.3.5

(i) Let x, y be two vectors in $\mathscr{V}$. Then the sets $\{x, y\}, \{x + y, x - y\}, \{x, 3x, 2x + y\}$ all span the same subspace.

(ii) Let S be the infinite set consisting of the powers of t for $t \in [0, 1]$, viz., $\{1, t, t^2, \ldots, t^n, \ldots\}$; then $[S]$ is the vector space of all finite polynomials in t defined on $[0, 1]$.

Subspaces of a vector space $\mathscr{V}$ can be combined to give further subspaces of $\mathscr{V}$ in two important ways.

Definition 2.3.3 *Let $\mathscr{U}, \mathscr{W}$ be two subspaces of a vector space $\mathscr{V}$, then*

(a) *the* sum *of $\mathscr{U}$ and $\mathscr{W}$, denoted by $\mathscr{U} + \mathscr{W}$, consists of all vectors of the form $u + w$ with $u \in \mathscr{U}$ and $w \in \mathscr{W}$;*
(b) *the* intersection *of $\mathscr{U}$ and $\mathscr{W}$, denoted by $\mathscr{U} \cap \mathscr{W}$, consists of all vectors which are in $\mathscr{U}$ and $\mathscr{W}$. (This coincides precisely with the set-theoretic intersection of the sets $\mathscr{U}$ and $\mathscr{W}$.)*

Theorem 2.3.2 *If $\mathscr{U}$ and $\mathscr{W}$ are subspaces of the vector space $\mathscr{V}$, then the sum $\mathscr{U} + \mathscr{W}$ and the intersection $\mathscr{U} \cap \mathscr{W}$ are also subspaces of $\mathscr{V}$. In general, the union $\mathscr{U} \cup \mathscr{W}$ is not a subspace of $\mathscr{V}$.*

Proof (i) Let $x^1, x^2 \in \mathscr{U} + \mathscr{W}$; then $x^1 = u^1 + w^1, x^2 = u^2 + w^2$ for some $u^1, u^2 \in \mathscr{U}$ and some $w^1, w^2 \in \mathscr{W}$.
Now

$$\begin{aligned} \alpha x^1 + \beta x^2 &= \alpha(u^1 + w^1) + \beta(u^2 + w^2) \\ &= (\alpha u^1 + \beta u^2) + (\alpha w^1 + \beta w^2) \\ &= u^3 + w^3 \end{aligned}$$

with $u^3 \in \mathscr{U}$, $w^3 \in \mathscr{W}$; hence $\alpha x^1 + \beta x^2 \in \mathscr{U} + \mathscr{W}$ and, by Theorem 2.3.1, $\mathscr{U} + \mathscr{W}$ is a subspace of $\mathscr{V}$.

(ii) Let $x^1, x^2 \in \mathscr{U} \cap \mathscr{W}$; then $x^1, x^2 \in \mathscr{U}$ and $\alpha x^1 + \beta x^2 \in \mathscr{U}$; similarly $\alpha x^1 + \beta x^2 \in \mathscr{W}$. Hence $\alpha x^1 + \beta x^2 \in \mathscr{U} \cap \mathscr{W}$ and $\mathscr{U} \cap \mathscr{W}$ is a subspace of $\mathscr{V}$.

(iii) For the last part of the theorem it suffices to produce an example of two subspaces whose union is not a subspace. Let $\mathscr{V} = \mathscr{R}_2$, the space of real 2-tuples, and take $\mathscr{U} = [(\xi_1, 0)]$, $\mathscr{W} = [(0, \xi_2)]$: $\mathscr{U}$ and $\mathscr{W}$ are clearly subspaces, but the sum $\alpha(\xi_1, 0) + \beta(0, \xi_2) \notin \mathscr{U} \cup \mathscr{W}$. However, one can see that the set of vectors $\mathscr{U} \cup \mathscr{W}$ is contained in the subspace $\mathscr{U} + \mathscr{W}$; that is

$$\mathscr{U} \cup \mathscr{W} \subset \mathscr{U} + \mathscr{W}$$

with equality only if $\mathscr{U} \subset \mathscr{W}$ or $\mathscr{W} \subset \mathscr{U}$. ∎

If the subspaces $\mathscr{U}$ and $\mathscr{W}$ have only the zero vector in common, that is

$$\mathscr{U} \cap \mathscr{W} = \mathcal{O},$$

then they are sometimes said to be disjoint. However, this usage of disjoint differs from the set-theoretic notion of disjoint, meaning that the intersection is empty.

If the subspaces $\mathscr{U}$ and $\mathscr{W}$ are such that $\mathscr{V} = \mathscr{U} + \mathscr{W}$, then we say that $\mathscr{V}$ is the sum of $\mathscr{U}$ and $\mathscr{W}$: if, further, $\mathscr{U} \cap \mathscr{W} = \mathcal{O}$, then $v \in \mathscr{V}$ can be *uniquely* expressed in the form $v = u + w$ with $u \in \mathscr{U}$, $w \in \mathscr{W}$. In this case we say that $\mathscr{V}$ is the direct sum of $\mathscr{U}$ and $\mathscr{W}$ and write $\mathscr{V} = \mathscr{U} \oplus \mathscr{W}$.

Definition 2.3.4 *Let $\mathscr{U}$, $\mathscr{W}$ be subspaces of a vector space $\mathscr{V}$. $\mathscr{V}$ is said to be the* direct sum *of $\mathscr{U}$, $\mathscr{W}$, written*

$$\mathscr{V} = \mathscr{U} \oplus \mathscr{W},$$

if, and only if,

  (i) $\mathscr{V} = \mathscr{U} + \mathscr{W}$;
  (ii) $\mathscr{U} \cap \mathscr{W} = \mathcal{O}$.

In this case $\mathscr{W}$ is called a direct complement *of $\mathscr{U}$ in $\mathscr{V}$ and written $\mathscr{U}'$: likewise $\mathscr{U}$ is a direct complement $\mathscr{W}'$ of $\mathscr{W}$ in $\mathscr{V}$.*

The use of 'a' rather than 'the' direct complement is deliberate, since a subspace $\mathscr{U}$ may have many direct complements in $\mathscr{V}$.

Example 2.3.6 Let
$$S_1 = \{(\xi_1, 0, 0), (0, \xi_2, 0)\}, \quad S_2 = \{(0, 0, \xi_3)\},$$
then $\mathcal{U} = [S_1]$, $\mathcal{U}' = [S_2]$ is a direct sum decomposition of $\mathcal{R}_3$. But we can just as well take $\mathcal{U}' = [S_3]$ with $S_3 = \{(\alpha \xi_3, \beta \xi_3, \xi_3)\}$ for any fixed α, β.

The subspace $\mathcal{U}$ is geometrically the (ξ_1, ξ_2) plane, while $\mathcal{U}'$ is any line through the origin not lying in the (ξ_1, ξ_2) plane (Fig. 2.1).

Geometrically a subspace of $\mathcal{R}_3$ is a plane or line passing through the origin. This appears to be a defect of the idea of a subspace, since we may often be interested in a subset of vectors which does not contain the origin but nevertheless has properties analogous to those for a subspace. In $\mathcal{R}_3$ we are now thinking of a line or plane which does *not* pass through the origin.

This difficulty is overcome by the translation of a subspace 'parallel to itself': this gives a subset of vectors which we call a coset (or linear variety).

Definition 2.3.5 *Let $\mathcal{W}$ be a subspace of a vector space $\mathcal{V}$ and let x be an arbitrary vector in $\mathcal{V}$: the $\mathcal{W}$-coset of x is defined by $x + \mathcal{W} = \{x + w | w \in \mathcal{W}\}$.*

Clearly, if z and y are in $x + \mathcal{W}$ then $z - y \in \mathcal{W}$. Although $\mathcal{W}$ is unique, any vector y for which $y - x \in \mathcal{W}$ may replace x in the definition (see Fig. 2.2). The collection of $\mathcal{W}$-cosets may be added and scalarly multiplied according to the rules
$$(x^1 + \mathcal{W}) + (x^2 + \mathcal{W}) = (x^1 + x^2) + \mathcal{W}$$
and
$$\alpha(x^1 + \mathcal{W}) = \alpha x^1 + \mathcal{W}$$
for any $x^1, x^2 \in \mathcal{V}$. With these rules, the collection of $\mathcal{W}$-cosets constitutes a vector space denoted by $\mathcal{V}/\mathcal{W}$ and called the quotient (or difference) space *of $\mathcal{V}$ with respect to $\mathcal{W}$*.

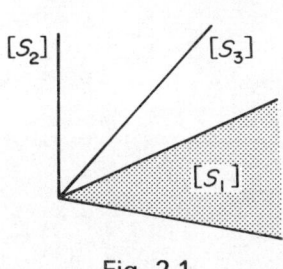

Fig. 2.1

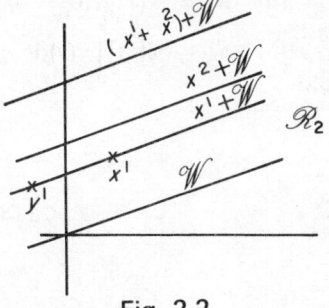

Fig. 2.2

Example 2.3.7 It was pointed out in Example 2.3.3 that the set of n-tuples of the form $(\beta, \xi_2, \xi_3, \ldots, \xi_n)$ is not a subspace; however, this set of vectors is the coset $x + \mathscr{W}$, where $x = (\beta, 0, 0, 0, \ldots, 0)$ and $\mathscr{W}$ is the subspace of n-tuples having first element zero. This coset is equally well defined by $x^1 + \mathscr{W}$, where $x^1 = (\beta, \alpha_2, \alpha_3, \ldots, \alpha_n)$ for any set of scalars $\{\alpha_j\}$. If $y = (\gamma, 0, 0, 0, \ldots, 0)$ then the sum of the cosets $x + \mathscr{W}$ and $y + \mathscr{W}$, namely $(x + y) + \mathscr{W}$, contains all vectors of the form $(\beta + \gamma, \xi_2, \xi_3, \ldots, \xi_n)$.

Certain geometric figures such as spheres, ellipsoids and tetrahedra have the property of being convex, which is to say that any two points in the figures can be joined by a straight line which is itself contained entirely within the figure. Convexity is an important property of certain subsets of a vector space and it may be defined in a purely abstract sense by using the concept of a *line segment* in a vector space. If x, y are two points of a vector space then the line segment joining them is the set of elements $\{\lambda x + (1 - \lambda)y \,|\, 0 \leqslant \lambda \leqslant 1\}$.

Definition 2.3.6 *A set, S, in a vector space is* convex *if the line segment joining any two points in S is contained in S.*

It follows immediately from the definition that every subspace of a vector space is convex and so are cosets of a subspace. It is readily shown that the intersection of convex sets is convex. We may be interested in sets which are not themselves convex, but may be embedded in a convex set: the smallest[†] convex set containing an arbitrary set S is called the *convex hull of S* and is denoted by $\mathrm{Co}(S)$. It is the intersection of all convex sets containing S (Fig. 2.3).

Example 2.3.8 Consider the vector space of continuous functions $x(t)$ defined on $[0, 1]$; the following are examples of sets which are convex but are not subspaces.

(i) $S_1 = \{x(t) \,|\, 0 \leqslant x(t) \leqslant 10\}$;
(ii) $S_2 = \{x(t) \,|\, x(0) = 0, x(1) = 1$ and x is strictly increasing on $[0, 1]\}$;
(iii) $S_3 = \{x(t) \,|\, \int_0^t x(s)\,ds < t\}$.

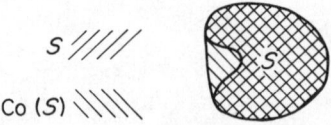

Fig. 2.3

[†] Smallest in this case means 'has no proper subset containing S'.

Exercises 2.3

1. Consider the vector space x of all real-valued continuous functions $x(t)$ defined on $(-\infty, \infty)$. Which of the following conditions define a subspace?
   - (i) $x(1) = 0$
   - (ii) $x(1) + x(-1) = 0$
   - (iii) $x(1) + x(2) = 2$
   - (iv) $\int_{-1}^{1} x(t)\,dt = 0$
   - (v) x is odd, $x(-t) = -x(t)$
   - (vi) x is periodic with period 2π.

2. If $\mathscr{U}, \mathscr{W}$ are subspaces of $\mathscr{V}$, show that $[\mathscr{U} \cap \mathscr{W}] = \mathscr{U} + \mathscr{W}$.

3. Show that the set of all absolutely convergent sequences of real numbers can define a vector space and show that the subset of sequences which converge absolutely to zero is a subspace.

4. Let $\mathscr{U}$ be the subspace of $\mathscr{R}_3$ consisting of the vectors $(0, \xi_2, \xi_3)$ and let $\mathscr{W}$ be the subspace spanned by $(2, 1, 0)$ and $(1, 2, 3)$. Which vectors are in $\mathscr{U} \cap \mathscr{W}$ and which in $\mathscr{U} + \mathscr{W}$?

5. (a) Show that
   - (i) the intersection
   - (ii) the sum

   of two convex sets in $\mathscr{V}$ is convex in $\mathscr{V}$.

   (b) Show that $\mathrm{Co}(S)$, $S \subset \mathscr{V}$, always exists. (Hint: extend (a)(i) to an arbitrary collection of sets in $\mathscr{V}$.)

   (c) Show that the convex hull of the set of points lying on the circumference of a circle in $\mathscr{R}_2$ consists of the circumference plus the interior of the circle.

   (d) Show that the coset $x + \mathscr{W}$ is convex (Definition 2.3.5).

6. Let $\{x^i\}, i = 1, \ldots, n$, be a set of vectors in $\mathscr{V}$ and let S be the set of vectors of the form
   $$\lambda_1 x^1 + \lambda_2 x^2 + \ldots + \lambda_n x^n$$
   with $\lambda_i \geq 0$ and $\sum_{i=1}^{n} \lambda_i = 1$. Show that S is the convex hull of the set $\{x^i\}$. Illustrate in $\mathscr{R}_3$.

   The vectors in S are called *convex combinations* of the $\{x^i\}$: draw an analogy between a convex combination and the centre of mass of a set of particles.

7. A subset C of a real vector space $\mathscr{V}$ is called a *cone* (with vertex at the origin) if, for $x \in C$, $\alpha x \in C$ for all $\alpha > 0$.

   (a) A cone which is a convex set is called a *convex cone*. Show that C is a convex cone if and only if $x + y \in C$ whenever $x, y \in C$.

(b) Define the relation $x \geqslant y$ to mean $x - y \in C$, where C is a convex cone. Show that $\geqslant$ is a partial ordering (see p. 9). Show also that if $x, y, z \in \mathscr{V}$ then $x \geqslant y$ implies $x + z \geqslant y + z$ and $\alpha x \geqslant \alpha y$, $\alpha \geqslant 0$.

(c) Show that the set of vectors $x \geqslant 0$ in $\mathscr{R}_n$ is a convex cone (the *positive orthant*).

2.4 Linear independence, basis and dimension

We have seen that a vector space can be generated as the set of all linear combinations of the elements of a set S. We now turn our attention to the question of redundancy in a spanning set S; that is, we ask whether some members of the set S could be removed without materially changing the subspace generated by S. If we can effect the removal of such redundant elements then we would be left with a *minimal spanning set* $S_m \subset S$ in the sense that S_m contains the minimum number of elements such that $[S_m] = [S]$. This minimum number of spanning vectors is called the *dimension* of the space $[S]$ and is an essential characteristic of the space.

We now proceed to frame the foregoing ideas in a formal manner, beginning with the concept of linear dependence.

Definition 2.4.1a *A vector is* linearly dependent *on a set, S, of vectors if x is in $[S]$. Thus, if $S = \{x^\rho\}$, x is linearly dependent on S if x can be written as the finite linear combination*

$$x = \sum_\rho \lambda_\rho x^\rho.$$

Definition 2.4.1b *A subset, S, of a vector space $\mathscr{V}$ is* linearly dependent *if there exists a finite subset $\{x^\rho\}$ of S and scalars $\{\lambda_\rho\}$, not all zero, such that*

$$\sum_\rho \lambda_\rho x^\rho = \theta.$$

Definition 2.4.1c *A subset of a vector space is* linearly independent *if it is not linearly dependent; that is, if the expression*

$$\sum_\rho \lambda_\rho x^\rho = \theta$$

implies $\lambda_\rho = 0$ for all ρ.

Example 2.4.1

(a) In the space $\mathscr{R}_n$ of real n-tuples let e^i denote the vector $(0, 0, 0, \ldots, 1, \ldots, 0)$

with unity in the ith place. The subset $\{e^i\}$, $i = 1$ to n, is linearly independent. Adjoining any arbitrary n-tuple $\alpha = (\alpha_1, \alpha_2, \ldots, \alpha_n)$ to this set will make the enlarged subset linearly dependent since

$$\alpha - \sum_{i=1}^{n} \alpha_i e^i = \theta.$$

(b) The infinite set of functions $\{t^n\}$, $t \in [0, 1]$, is linearly independent: for, if not, it would imply that scalars $\{\lambda_\rho\}$ exist which are not all zero, such that the finite sum

$$\sum_\rho \lambda_\rho t^\rho = 0 \quad \forall\, t \in [0, 1].^\dagger$$

An important consequence of linear independence of the set of vectors $S = \{x^\rho\}$ is the following: every vector $x \in [S]$ can be expressed *uniquely* in the form

$$x = \sum_\rho \alpha_\rho x^\rho.$$

Indeed, suppose x can also be written

$$x = \sum_\nu \beta_\nu x^\nu$$

then we can arrange the vectors $\{x^\rho\}$, $\{x^\nu\}$ into a single finite indexed set $\{x^\mu\}$ and write

$$x = \sum_\mu \alpha_\mu x^\mu = \sum_\mu \beta_\mu x^\mu,$$

where some of the α_μ, β_μ may be zero. But this implies that

$$\sum_\mu (\alpha_\mu - \beta_\mu) x^\mu = \theta$$

and hence, since the x_μ are linearly independent, that $\alpha_\mu = \beta_\mu$ for all μ.

The question, raised at the beginning of this section, of redundancy in spanning sets is thus resolved in terms of linear dependence. Given a set of vectors S, a minimal spanning set for $[S]$ is a subset $S_m \subseteq S$ whose vectors are linearly independent.

Suppose now that the set S spans the whole vector space $\mathscr{V}$, that is,

$$[S] = [S_m] = \mathscr{V};$$

in this case we call the set S_m a basis for $\mathscr{V}$.

---

† Where confusion might arise between the ρth element of a set and the ρth power, the context will generally make the meaning clear.

Definition 2.4.2 *A set, S, of linearly independent vectors is an* algebraic (*or* Hamel) *basis for the space $\mathscr{V}$ if $[S] = \mathscr{V}$.*

A vector space $\mathscr{V}$ does not have a unique basis; however, the number of vectors in any basis for $\mathscr{V}$ is unique and the number is called the dimension of $\mathscr{V}$.

Definition 2.4.3 *The* dimension, dim $\mathscr{V}$, *of a vector space $\mathscr{V}$ is the number of elements in any basis.*

A vector space having a finite basis is said to be *finite-dimensional*; otherwise *infinite-dimensional* and in this case 'number' in Definition 2.4.3 is to be understood as cardinal number. We shall prove that every basis of a finite-dimensional vector space contains the same number of vectors as a corollary of the following *exchange theorem*.

Theorem 2.4.1 *Let $X = \{x^1, ..., x^r\}$ be a finite spanning set for $\mathscr{V}$; let $Y = \{y^1, ..., y^s\}$ be any linearly independent set in $\mathscr{V}$. Then $s \leqslant r$ and a spanning set for $\mathscr{V}$ can be constructed which consists of the elements of Y together with $r - s$ elements of X.*

Proof (i) We show first that $s \leqslant r$. We shall suppose the converse, namely that $s > r$, and obtain a contradiction. Accordingly, let $y^1, y^2, ..., y^{r+1}$ be $(r+1)$ linearly independent vectors.

Since X is a spanning set, the vector y^1 is equal to a linear combination of the x^i; that is, the set $\{y^1, x^1, x^2, ..., x^r\}$ is linearly dependent. Hence, at least one of the x^i is linearly dependent on the remaining vectors in the set. We remove the first x^i which is a linear combination of the preceding vectors: let this be x^j, then the remaining set

$$\{y^1, x^1, ..., x^{j-1}, x^{j+1}, ..., x^r\}$$

is a spanning set for $\mathscr{V}$. Hence the set of vectors

$$\{y^1, y^2, x^1, x^2, ..., x^{j-1}, x^{j+1}, ..., x^r\}$$

is linearly dependent. Again we remove the first vector x^k which is a linear combination of its predecessors, giving the spanning set

$$\{y^1, y^2, x^1, x^2, ..., x^{j-1}, x^{j+1}, ..., x^{k-1}, x^{k+1}, ..., x^r\}.$$

Continuing in this way, adding one y vector at each stage and removing one x vector, we arrive at the spanning set

$$\{y^1, y^2, ..., y^r\}.$$

But $y^{r+1} \in \mathscr{V}$ and hence is expressible as a linear combination of the spanning set $\{y^1, \ldots, y^r\}$: this contradicts the hypothesis that the $y^i, i = 1, \ldots, r + 1$, are independent, and we conclude that $s \leqslant r$.

(ii) If $s = r$ then $\{y^1, y^2, \ldots, y^r\}$ is a spanning set. If $s < r$ we have shown that after r exchanges the set

$$\{y^1, y^2, \ldots, y^s, x^1, x^2, \ldots, x^{r-s}\}$$

is a spanning set where the remaining elements of the set $\{x^i\}$ have been re-indexed. ∎

Corollary *Every basis of a finite-dimensional vector space contains the same number of vectors.*

Proof Let X be a finite basis of $\mathscr{V}$ containing n elements and let X' be another basis of $\mathscr{V}$ containing n' elements. Theorem 2.4.1 shows that the linearly independent set X' contains at most as many elements, n, as X, that is, $n' \leqslant n$. Interchanging the roles of X and X' we find $n \leqslant n'$, hence $n' = n$. ∎

A proof that the cardinal number of elements in any basis for an infinite-dimensional space is the same is beyond the scope of this text [25]. In any case, the algebraic bases of most commonly used infinite-dimensional spaces are too large and unwieldy to be of practical value. To use such spaces we need to develop a theory of convergent sequences of vectors: this theory is developed in Chapter 3. As a simple example of the difficulties associated with algebraic bases, consider the following: the countable set of vectors $e^1 = (1, 0, 0, \ldots), e^2 = (0, 1, 0, \ldots), e^3 = (0, 0, 1, \ldots), \ldots$ is clearly a linearly independent subset of the vector space of all infinite sequences of real scalars (cf., Example 2.2.2). However, this set is not an algebraic basis since it does not span the space. It is a basis for the vector space of all sequences in which only a finite number of terms is non-zero (Example 2.2.2(b)).

A concept which is often useful when dealing with infinite-dimensional spaces is that of *codimension*. Let $\mathscr{V} = \mathscr{U} \oplus \mathscr{U}'$, where $\mathscr{V}$ is infinite-dimensional: if $\mathscr{U}$ is of finite dimension r then $\mathscr{U}'$, the direct complement, is said to have codimension r. In a sense codimension measures how close the subspace $\mathscr{U}'$ is to being the whole space $\mathscr{V}$. Translated subspaces (i.e., cosets) of codimension 1 are particularly important and are called *hyperplanes* (by analogy with a direct decomposition of $\mathscr{R}_3$ into a plane and a line not lying in the plane).

For example, let $\mathscr{V}$ be the vector space of continuous functions $x(t)$ defined on $[-1, 1]$ and let $\mathscr{U}'$ be the subspace of functions $\tilde{x}(t)$ such that $\tilde{x}(0) = 0$: then $\mathscr{U}'$ has codimension 1 since every $x(t)$ can be written as

$$x(t) = \alpha 1(t) + \tilde{x}(t),$$

where $1(t)$, the function having constant value 1 on $[-1, 1]$, is clearly a basis for $\mathscr{U}$.

Example 2.4.2 The set of n vectors $\{e^i\}$, $i = 1, \ldots, n$ is a basis for $\mathscr{R}_n$ or $\mathscr{C}_n$: it is called *the standard basis*. The set of n functions $\{1, t, t^2, \ldots, t^{n-1}\}$, $t \in [0, 1]$, is a basis for the n-dimensional space whose elements are polynomials of degree $\leqslant (n-1)$. As pointed out in Example 2.2.4 these two spaces are really just two different concrete representations of the n-dimensional (real or complex) vector space which we may denote by $\mathscr{V}_n$. The introduction of a particular basis is always made with some concrete representation in mind. In addition, within this concrete representation we may in effect change our coordinate system by changing the basis vectors.

For example, in $\mathscr{R}_3$ we may use the basis $\{(1, 0, 0), (0, 1, 0), (0, 0, 1)\}$ or equally well the basis $\{(1, 1, 0), (0, 1, 1), (1, 0, 1)\}$. Similarly, in the space of polynomials of degree $\leqslant 2$, we may use the basis $\{1, t, t^2\}$ or $\{1 + t, 1 - t, 1 + t + t^2\}$.

Any particular point or vector $x \in \mathscr{V}_n$ will be represented relative to the basis $\{x^i\}$ by the scalars $\{\xi_i\}$ in the unique linear combination

$$x = \sum_{i=1}^{n} \xi_i x^i.$$

This brings us to the following definition.

Definition 2.4.4 *Let $\{x^\rho\}$ be a basis for the vector space $\mathscr{V}$: for any $x \in \mathscr{V}$ the unique scalars ξ_ρ in the finite linear combination*

$$x = \sum_\rho \xi_\rho x^\rho$$

are called the coordinates *(or* components*) of x relative to the basis $\{x^\rho\}$.*

The coordinates may conveniently be displayed as the sequence $(\xi_1, \xi_2, \ldots, \xi_\rho, \ldots)$, the *coordinate vector* of x relative to $\{x^\rho\}$. It is tempting to denote the coordinates of x by $(x_1, x_2, \ldots)$ but this leads to confusion between vector and scalar quantities. We maintain our convention of using Greek letters to denote scalars, but wherever possible use the subscripted Greek letter which corresponds (in the alphabetic sense) to the italic letter used to denote the vector.

Many results pertaining to vector spaces can be deduced without reference to a basis: this type of basis-free argument is then quite general. However, in actual calculations it is always necessary to introduce a basis or coordinate system at some point, and from that point on we naturally find ourselves performing manipulations with the coordinates of a vector rather than the vector entity itself.

Example 2.4.3 A change of basis will cause the coordinates of a fixed vector to change. Thus, in $\mathscr{R}_3$ let P have coordinate vector (α, β, γ) relative to the basis $\{(1,0,0), (0,1,0), (0,0,1)\}$: relative to the basis $\{(1,1,0), (0,1,1), (1,0,1)\}$ what is the coordinate vector of P? Let us represent these two basis sets by $\{e^i\}, \{b^i\}, i = 1, 2, 3$. Then we see that

$$b^1 = e^1 + e^2,$$
$$b^2 = e^2 + e^3,$$
$$b^3 = e^1 + e^3$$

and conversely that

$$e^1 = \tfrac{1}{2}(b^1 - b^2 + b^3),$$
$$e^2 = \tfrac{1}{2}(b^1 + b^2 - b^3),$$
$$e^3 = \tfrac{1}{2}(-b^1 + b^2 + b^3).$$

Now

$$P = \alpha e^1 + \beta e^2 + \gamma e^3$$
$$= \frac{\alpha}{2}(b^1 - b^2 + b^3) + \frac{\beta}{2}(b^1 + b^2 - b^3) + \frac{\gamma}{2}(-b^1 + b^2 + b^3)$$
$$= \frac{1}{2}(\alpha + \beta - \gamma)b^1 + \frac{1}{2}(-\alpha + \beta + \gamma)b^2 + \frac{1}{2}(\alpha + \beta + \gamma)b^3$$

and hence the coordinate vector of P relative to $\{b^i\}$ is

$$\frac{1}{2}((\alpha + \beta - \gamma), (-\alpha + \beta + \gamma), (\alpha - \beta + \gamma)).$$

Similarly (Example 2.4.2), the polynomial $\alpha_0 + \alpha_1 t + \alpha_2 t^2$ has the coordinate vector

$$\left[\left(\frac{\alpha_0}{2} + \frac{\alpha_1}{2} - \alpha_2\right), \left(\frac{\alpha_0}{2} - \frac{\alpha_1}{2}\right), \alpha_2\right]$$

relative to the basis $\{1 + t, 1 - t, 1 + t + t^2\}$.

Change of basis is dealt with in a general way in Section 2.8.

Example 2.4.4 Consider the linear differential equation

$$\frac{d^2x}{dt^2} + 2b\frac{dx}{dt} + cx = 0.$$

The sum of two solutions $x^1(t)$, $x^2(t)$ is itself a solution, as is the scalar multiple of a solution. The set of all solutions is thus a vector space sometimes called the *solution space*. It is known that a second-order differential equation has at most two independent solutions, so we infer that this space is two-dimensional and we may take as basis any two independent solutions $x^1(t)$, $x^2(t)$ of the equation.

The complete solution of the inhomogeneous equation

$$\frac{d^2x}{dt^2} + 2b\frac{dx}{dt} + cx = f(t)$$

is the coset $\bar{x}(t) + [S]$, where $\bar{x}(t)$ is any solution of the inhomogeneous equation and $S = \{x^1(t), x^2(t)\}$.

In Example 2.2.4 it was pointed out that the vector space of polynomials of order $\leq (m-1)$ is really indistinguishable from the vector space of m-tuples. Having the concept of dimension, we can make a general statement embodying this observation (Exercise 2.5,9)

Theorem 2.4.2 *Let $\mathscr{V}$ be a finite-dimensional vector space of dimension n. If $\mathscr{V}$ is real it is* isomorphic to $\mathscr{R}_n$ *and if it is complex it is* isomorphic to $\mathscr{C}_n$.

This means that there is really only one (real or complex) n-dimensional vector space, although it may be given different concrete representations by a choice of basis containing n elements.

A somewhat similar result can be stated for infinite-dimensional spaces but its usefulness is limited in so far as algebraic basis sets, as has been pointed out, are not themselves very useful in representing such spaces.

Exercises 2.4

1. Find minimal spanning sets for $[S]$ when S contains the vectors
   (i) $(1, 0, 2, 0), (0, 2, 0, 1), (1, 2, 2, 1), (-1, 0, 3, 0)$;
   (ii) $(0, 1, 2, 3), (3, 0, 1, 2), (2, 3, 0, 1), (1, 2, 3, 0)$;
   (iii) $1 + 2t, 1 - t + t^2, 4 + t^2, 1 - t^3, 4 + 2t + t^2 + t^3$.

2. Let $\mathscr{V}$ be a vector space of dimension n. Show that every set of $(n+1)$

vectors in $\mathscr{V}$ is linearly dependent and hence that every system of n homogeneous linear equations in $(n+1)$ or more unknowns always has a non-zero solution.

3. If $\mathscr{V}$ is finite-dimensional and if $\mathscr{U}, \mathscr{W}$ are proper subspaces of $\mathscr{V}$ show that
   (i) $\dim \mathscr{U} < \dim \mathscr{V}$;
   (ii) $\dim (\mathscr{U} + \mathscr{W}) = \dim \mathscr{U} + \dim \mathscr{W} - \dim (\mathscr{U} \cap \mathscr{W})$;
   (iii) $\dim (\mathscr{U} \oplus \mathscr{W}) = \dim \mathscr{U} + \dim \mathscr{W}$.
   (Hint: for (ii) assume bases for $\mathscr{U}, \mathscr{W}$ which contain a basis for $\mathscr{U} \cap \mathscr{W}$.)

4. For those triples in Exercise 2.2,1 which form subspaces find a suitable basis for a direct complement in $\mathscr{R}_3$.

5. What is the dimension of the vector space of Exercise 2.2,2? Suggest a suitable basis set.

6. If x is a non-zero vector in $\mathscr{V}$ show that there is a hyperplane through θ in $\mathscr{V}$ which does not contain x. If $\mathscr{W}$ is a subspace of $\mathscr{V}$ and $x \notin \mathscr{W}$ show that there is a hyperplane through θ which contains $\mathscr{W}$ but not x.

7. Find the codimension of the following subspaces of the space of continuous functions $x(t)$ on $[0, 1]$: the subspace such that
   (i) $x(\tfrac{1}{2}) = 0, x(\tfrac{1}{3}) = x(\tfrac{2}{3})$;
   (ii) $\int_0^1 (t - \tfrac{1}{2})^n x(t) dt = 0, n = 0, 1, 2$.

2.5 Linear transformations

A linear transformation is a mapping or function whose domain is a vector space and whose range is contained in a vector space; furthermore, the function is of such a type that it preserves the vector space operations of addition and scalar multiplication. We shall generally use an italic capital such as T to represent a linear transformation: the image $T(x)$ of the vector x is often written simply Tx, omitting the parentheses.

Definition 2.5.1 *Let $\mathscr{U}, \mathscr{V}$ be two vector spaces with the same system of scalars. Then a function (or mapping)*

$$T: \mathscr{V} \to \mathscr{U}$$

is called a linear transformation *of $\mathscr{V}$ into $\mathscr{U}$ if*

(i) $T(x + y) = Tx + Ty \quad \forall x, y \in \mathscr{V}$;
(ii) $T(\alpha x) = \alpha Tx \quad \forall x \in \mathscr{V}$ *and all scalars α.*

We may note the following:

(a) in (i), (ii) addition and scalar multiplication on the left are operations in $\mathscr{V}$, those on the right are operations in $\mathscr{U}$;

(b) conditions (i), (ii) are equivalent to
$$T(\alpha x + \beta y) = \alpha T x + \beta T y;$$

(c) $\mathscr{V}$ and $\mathscr{U}$ may be the same space;

(d) $T(\theta) = \theta$, $T(-x) = -Tx$;

(e) two linear transformations T_1, T_2 are equal if, and only if, they have the same domain $\mathscr{D}$ and $T_1 x = T_2 x \quad \forall x \in \mathscr{D}$.

The appropriateness of conditions (i) and (ii) in Definition 2.5.1 is clear so far as the operations on vector spaces are concerned, but the notion of linear transformation would not assume the importance it has were it not for the fact that many of the operators met with in applied mathematics fall within the definition. And even for those that do not, it is possible (as described in Chapter 4) to define a linear transformation which, in a local sense, approximates the given operator.

If T_1, T_2 are two linear transformations $\mathscr{V} \to \mathscr{U}$ then they can be added to yield a new linear transformation $T_1 + T_2$ defined by

$$(T_1 + T_2)x = T_1 x + T_2 x.$$

The scalar multiple αT of $T: \mathscr{V} \to \mathscr{U}$ is defined by

$$(\alpha T)x = \alpha T x.$$

With these operations we see that the set of linear transformations $\{T: \mathscr{V} \to \mathscr{U}\}$ is itself a vector space whose elements are linear transformations. The zero vector of this space is the zero transformation 0 defined by $0x = \theta \quad \forall x \in \mathscr{V}$.

Example 2.5.1 Let $\mathscr{V}, \mathscr{U}$ be the finite-dimensional spaces $\mathscr{V}_n, \mathscr{V}_m$ respectively. The linear transformation $T: \mathscr{V}_n \to \mathscr{V}_m$ then maps n-tuples to m-tuples, viz.,

$$T(\xi_1, \xi_2, \ldots, \xi_n) = (\eta_1, \eta_2, \ldots, \eta_m).$$

If T is to be a mapping satisfying (i) and (ii) of Definition 2.5.1 then the only concrete form the transformation can take is the following:

$$\eta_i = \sum_{j=1}^{n} t_{ij} \xi_j, \quad i = 1, \ldots, m,$$

where the $m \times n$ set of scalars t_{ij} is a representation of the linear transformation T.

The scalars t_{ij} can be arranged as an $(m \times n)$ rectangular array

$$\begin{bmatrix} t_{11} & t_{12} & t_{13} & \cdots & t_{1n} \\ t_{21} & t_{22} & t_{23} & \cdots & t_{2n} \\ t_{31} & t_{32} & t_{33} & \cdots & t_{3n} \\ \vdots & \vdots & \vdots & & \vdots \\ t_{m1} & t_{m2} & t_{m3} & \cdots & t_{mn} \end{bmatrix}$$

or, written compactly, $[t_{ij}]$, called a matrix, and we can then proceed to carry out arithmetic manipulations employing the matrix, which are consistent with the properties of a linear transformation. In this sense, every linear transformation $T: \mathscr{V}_n \to \mathscr{V}_m$ has a concrete representation as a rectangular $(m \times n)$ matrix. The connection between linear transformations on finite-dimensional spaces and matrices is taken up in detail in Section 2.8.

Example 2.5.2

(a) Consider the vector space of all polynomials $p(t)$ with real coefficients defined on $[0, 1]$. The operator D, defined by

$$Dp = \frac{dp}{dt},$$

is a linear transformation of this space into itself.

(b) The vector space of all polynomials defined on $[0, 1]$ and of degree $\leq (m - 1)$ is isomorphic to $\mathscr{V}_m$. Let D be the operator d/dt; then, as in (a), D is a linear transformation of this space into itself. But in this case D is $\mathscr{V}_m \to \mathscr{V}_m$ and therefore must have an $(m \times m)$ matrix representation. We can find one particular matrix representation by introducing a basis for the space, say $\{1, t, t^2, \ldots, t^{m-1}\}$. Then a typical vector is

$$p(t) = \sum_{j=0}^{m-1} \alpha_j t^j$$

and

$$Dp = \frac{dp}{dt} = \sum_{j=0}^{m-1} j\alpha_j t^{j-1},$$

so that in terms of coordinates the operator D has the effect $D(\alpha_0, \alpha_1, \ldots, \alpha_j, \ldots, \alpha_{m-1}) = (\alpha_1, 2\alpha_2, \ldots, (j+1)\alpha_{j+1}, \ldots, (m-1)\alpha_{m-1}, 0)$ and the matrix of D is clearly,

$$\begin{bmatrix} 0 & 1 & 0 & 0 & 0 & \cdots & 0 \\ 0 & 0 & 2 & 0 & 0 & \cdots & 0 \\ \vdots & \vdots & \vdots & \vdots & \vdots & & \vdots \\ 0 & 0 & 0 & 0 & j+1 & \cdots & 0 \\ \vdots & \vdots & \vdots & \vdots & \vdots & & \vdots \\ 0 & 0 & 0 & 0 & 0 & \cdots & m-1 \\ 0 & 0 & 0 & 0 & 0 & \cdots & 0 \end{bmatrix}$$

Example 2.5.3 The operation I defined by

$$I(x) = \int_0^1 x(t) \mathrm{d}t$$

is clearly a linear transformation of the (infinite-dimensional) space of integrable functions defined on $[0, 1]$ into the (one-dimensional) space of real numbers.

Example 2.5.4 Suppose L is a general linear second-order differential operator. The domain of L will contain, for example, those functions defined on $[0, 1]$ which are twice differentiable. However, we know that the solution of the operator equation $Lx = y$ is not unique (cf., Example 2.4.4) but can be made unique by specification of boundary conditions on x. For example, we might specify that x satisfies the boundary conditions $x(0) = x(1) = 0$.

In general the domain of a differential operator is taken to include only those functions which satisfy a sufficient set of boundary conditions: in this sense, two differential operators described by the same formula but associated with different sets of boundary conditions, and so having different domains, should be understood as being two distinct operators. Since, by hypothesis, the domain of a linear transformation is a vector space, then the boundary (or initial) conditions must be of the homogeneous type. For example, the functions satisfying $x(0) = x(1) = 0$ are a subspace of the vector space of twice-differentiable functions on $[0, 1]$. We can deal with inhomogeneous boundary conditions by translation, in the same way as we extended subspaces to cosets. Suppose the boundary conditions were $x(0) = a$, $x(1) = 0$, then assuming a solution exists we choose a twice-differentiable function $\tilde{x}(t)$ such that $\tilde{x}(0) = a$, $\tilde{x}(1) = 0$, and form the coset $\tilde{x} + X$, where the functions in X satisfy $x(0) = x(1) = 0$. The operator equation then becomes

$$Lx = y - L\tilde{x}$$
$$= y', \quad x \in X.$$

The range of a linear transformation $\mathscr{R}(T)$, $T: \mathscr{V} \to \mathscr{U}$, is a subspace of $\mathscr{U}$. For, suppose $y^1, y^2 \in \mathscr{R}(T)$, then for some x^1, x^2 in the domain of T, $Tx^1 = y^1$ and $Tx^2 = y^2$. We have

$$\alpha_1 y^1 + \alpha_2 y^2 = \alpha_1 Tx^1 + \alpha_2 Tx^2$$
$$= T(\alpha_1 x^1 + \alpha_2 x^2) = Tx^3,$$

say, and hence the vector $\alpha_1 y^1 + \alpha_2 y^2 \in \mathscr{R}(T)$: this proves that $\mathscr{R}(T)$ is a subspace.

Consider the set of all vectors $x^0 \in \mathscr{V}$ such that $Tx^0 = \theta$, that is, the vectors in $\mathscr{V}$ which are mapped to the origin by T: these vectors

form a subspace of $\mathscr{V}$ since for any two such vectors x^{01} and x^{02},
$$T(\alpha_1 x^{01} + \alpha_2 x^{02}) = \alpha_1 T x^{01} + \alpha_2 T x^{02} = \theta$$
so that $\alpha_1 x^{01} + \alpha_2 x^{02} = x^{03}$, say. This subspace, the *null space* (or kernel) of T, $\mathscr{N}(T)$ is of course a subspace of the domain of T.

Suppose we form a direct sum decomposition of the domain of T,
$$\mathscr{D}(T) = \mathscr{N}(T) \oplus \mathscr{N}'(T),$$
and consider the mapping of the subspace $\mathscr{N}'(T)$ by T. Let $x' \in \mathscr{N}'(T)$, then $y = Tx' \in \mathscr{R}(T)$: indeed, every $y \in \mathscr{R}(T)$ is the image of some $x' \in \mathscr{N}'(T)$, and, furthermore, T is one-to-one $\mathscr{N}'(T) \to \mathscr{R}(T)$ as is easily verified. Hence if $\{x^\rho\}$ is a basis for $\mathscr{N}'(T)$ then $\{Tx^\rho\}$ is a basis for $\mathscr{R}(T)$ and we may conclude that
$$\dim \mathscr{R}(T) = \dim \mathscr{N}'(T) = \operatorname{codim} \mathscr{N}(T).$$

Definition 2.5.2 *If $\rho = \dim \mathscr{R}(T)$ is finite we call this number the rank of T: similarly if $\nu = \dim \mathscr{N}(T)$ is finite we call this number the nullity of T.*

If the domain $\mathscr{D}(T)$ has finite dimension n we may conclude from the discussion above that $n = \rho + \nu$.

Example 2.5.5 Consider the differential equation
$$x'' + x = 1.$$
Let L_1 be the operator
$$\left[\frac{d^2}{dt^2} + 1 \right]$$
with domain those functions defined on $[0, \infty)$ which are twice differentiable. Then (Example 2.4.4) the solution of the operator equation
$$L_1 x = 1.$$
is the coset
$$x = 1 + [S],$$
where $S = \{\cos t, \sin t\}$. The subspace $[S]$ is in fact a representation of the null space $\mathscr{N}(L_1)$ of L_1 which in this case is two-dimensional. An alternative basis set for $\mathscr{N}(L_1)$ would be $\{e^{it}, e^{-it}\}$.

Now let L_2 be the operator
$$\left(\frac{d^2}{dt^2} + 1 \right)$$

with domain those functions which are twice differentiable on $[0, \infty)$ and in addition satisfy the initial conditions $x(0) = 0, x'(0) = 0$: in this case the solution of $L_2 x = 1$ is unique, namely

$$x = 1 - \cos t.$$

The null space of L_2 is the zero vector space; that is, it contains only the function $x(t) = 0$.

We know that if a mapping is one-to-one then we can define an inverse mapping (Chapter 1). A linear transformation T is one-to-one if $\mathcal{N}(T) = \mathcal{O}$, that is to say, if $Tx = \theta$ implies $x = \theta$. Suppose T is many-to-one and $Tx_1 = Tx_2$, then $T(x_1 - x_2) = \theta$, which implies that $\mathcal{N}(T) \neq \mathcal{O}$. Conversely, if $\mathcal{N}(T) = \mathcal{O}$, then $Tx_1 = Tx_2$ implies $x_1 = x_2$ and T is one-to-one. A linear transformation is called *nonsingular* (or *invertible*) if it is one-to-one $\mathcal{D}(T) \to \mathcal{R}(T)$: in this case the inverse $T^{-1} : \mathcal{R}(T) \to \mathcal{D}(T)$ exists and is itself linear, as is easily shown. If T is many-to-one then it is called *singular*.

Example 2.5.6 Let L be the transformation $Lx = dx/dt$ with domain those functions continuously differentiable on $[0, 1]$ and which satisfy the condition $x(0) = 0$. There are no non-zero functions satisfying $dx/dt = 0$ with $x(0) = 0$, so that $\mathcal{N}(L) = \mathcal{O}$ and L^{-1} will exist. Let

$$Lx = \frac{dx}{dt} = y$$

then clearly,

$$x = \int_0^t y(\tau) d\tau$$

and L^{-1} is therefore represented by this integral. However, a more suggestive form for L^{-1} is obtained by defining the function $k(t, \tau)$ of two variables $t \in [0, 1], \tau \in [0, 1]$ by

$$k(t, \tau) = \begin{cases} 1, & 0 \leqslant \tau \leqslant t \leqslant 1, \\ 0, & 0 \leqslant t < \tau \leqslant 1. \end{cases}$$

We may then write

$$x = L^{-1} y = \int_0^1 k(t, \tau) y(\tau) d\tau$$

with $\mathcal{D}(L^{-1}) = \mathcal{R}(L)$, namely the continuous functions on $[0, 1]$. An important class of linear differential operators has inverses which are integral operators of the above type. (Example 3.10.5). The function $k(t, \tau)$ is called the Green's function or kernel (not to be confused with null space).

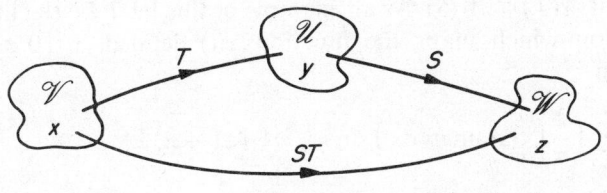

Fig. 2.4

Let T be a linear transformation $\mathscr{V} \to \mathscr{U}$ and let S be a linear transformation $\mathscr{U} \to \mathscr{W}$: then as for composition of functions we may define the *product* (or *composition*, p. 12) of T and S as the linear transformation $R = ST$, meaning $z = Rx = S(Tx)$ (*see* Fig. 2.4).

A particularly important case arises when T, S are both transformations of $\mathscr{V}$ into itself. Then the products ST and TS are always defined, although they are not in general equal.

This means that in the vector space of linear transformations $T: \mathscr{V} \to \mathscr{V}$ we have a further operation of multiplication of vectors and this means that the structure corresponds to an algebra.[†] This algebra has a multiplicative identity represented by the identity transformation I defined by

$$Ix = x \quad \forall x \in \mathscr{V},$$

and when T^{-1} exists we have

$$TT^{-1} = T^{-1}T = I.$$

Multiplication is not commutative ($ST \neq TS$) and there also exist non-zero divisors of zero ($TS = 0$ does not imply $T = 0$ or $S = 0$). The composition $y = T(Tx)$ is conveniently written $y = T^2 x$ and in this way polynomial functions of T can be built up of the form

$$p(T) = \alpha_0 I + \alpha_1 T + \alpha_2 T^2 + \ldots + \alpha_m T^m.$$

Example 2.5.7

(i) On the space of continuously differentiable functions on $[0, 1]$, let $Dx = dx/dt$, $My = ty(t)$, then

(a) MD is the operator $t \dfrac{d}{dt}(\cdot)$;

(b) DM is the operator $\dfrac{d}{dt}(t \cdot)$;

(c) $DM \neq MD$—in fact it is easily shown that $DM - MD = I$.

[†] An algebra is a vector space on which an operation of multiplication is defined. The operation is associative and distributive over addition (*see*, for example, [12]).

(ii) $ST = 0$ if $\mathscr{R}(T) \subset \mathscr{N}(S)$. As an example of this let T be the linear transformation which maps the function $x(t)$ defined in $[0, \pi]$ into the function

$$\sum_{n=1}^{N} \left(\frac{2}{\pi} \int_0^\pi x(s) \sin ns \, ds \right) \sin nt, \qquad t \in [-\pi, \pi],$$

and let S be the linear transformation

$$Sy = \frac{1}{2\pi} \int_{-\pi}^{\pi} y(s) ds :$$

then $ST = 0$.

An important class of linear transformations of $\mathscr{V}$ into itself is connected with a direct sum decomposition of $\mathscr{V}$ (Definition 2.3.4).

Definition 2.5.3 *Let $\mathscr{V}$ have the direct sum decomposition*

$$\mathscr{V} = \mathscr{U} \oplus \mathscr{W}$$

and let $P : \mathscr{V} \to \mathscr{V}$ be the linear transformation with rule

(i) $Pu = u \quad \forall u \in \mathscr{U}$;
(ii) $Pw = 0 \quad \forall w \in \mathscr{W}$.

Then P is called the projection *of $\mathscr{V}$ onto $\mathscr{U}$ along $\mathscr{W}$. Since every $v \in \mathscr{V}$ can be written uniquely as*

$$v = u + w$$

we see that

$$v = Pv + w$$

so that

$$w = (I - P)v,$$

implying that $I - P$ is the projection of $\mathscr{V}$ onto $\mathscr{W}$ along $\mathscr{U}$.
Clearly

$$(I - P)P = P(I - P) = 0.$$

The reason for the name 'projection' is obvious; for any $v \in \mathscr{V}$, Pv is an element of $\mathscr{U}$, so that the vector v is projected into the subspace $\mathscr{U}$. For example, in $\mathscr{R}_3$ if we take $\mathscr{U} = [(1, 0, 0), (0, 1, 0)]$, $\mathscr{W} = [(0, 0, 1)]$ then $P(\xi_1, \xi_2, \xi_3) = (\xi_1, \xi_2, 0)$, which is simply the projection of a vector onto the plane spanned by e^1, e^2. It is obvious that $P(Pu) = Pu$, so that $P^2 = P$: we say P is *idempotent*. We can then pose the question, 'If we have a linear transformation $P: \mathscr{V} \to \mathscr{V}$ such that $P^2 = P$, is P necessarily a projection, and does it imply a direct decomposition of $\mathscr{V}$?' The answers to both these questions are embodied in

Theorem 2.5.1 *If the linear transformation $P: \mathscr{V} \to \mathscr{V}$ is idempotent, then there exist subspaces $\mathscr{U}$ and $\mathscr{W}$ such that $\mathscr{V} = \mathscr{U} \oplus \mathscr{W}$ and P is the projection of $\mathscr{V}$ onto $\mathscr{U}$ along $\mathscr{W}$.*

Proof Let $\mathscr{U} = \mathscr{R}(P)$, i.e., $\mathscr{U} = \{Pv | v \in \mathscr{V}\}$, and $\mathscr{W} = \mathscr{N}(P)$, i.e., $\mathscr{W} = \{v | Pv = 0\}$. We must show that $\mathscr{U}$ and $\mathscr{W}$ span $\mathscr{V}$ and that $\mathscr{U} \cap \mathscr{W} = \mathscr{O}$. First we can trivially write

$$v = Pv + (I - P)v.$$

Now $Pv \in \mathscr{U}$ by definition and

$$P((I - P)v) = P(I - P)v = (P - P^2)v = (P - P)v = \theta,$$

showing that $(I - P)v \in \mathscr{N}(P) = \mathscr{W}$, and hence $\mathscr{V} = \mathscr{U} + \mathscr{W}$. Furthermore, suppose $\tilde{v} \in \mathscr{U} \cap \mathscr{W}$, then

$$P\tilde{v} = \tilde{v} \quad \text{and} \quad P\tilde{v} = \theta,$$

so that $\tilde{v} = \theta$ and $\mathscr{U} \cap \mathscr{W}$ contains only the zero vector. ∎

Eigenvalues of a linear transformation

Given a linear transformation (or a class of linear transformations) it is important to know which properties are intrinsic to the transformation and which are dependent on the particular concrete form adopted by the transformation upon choosing suitable basis sets for the domain and for the range spaces. This search for intrinsic characteristics is particularly important when we consider the class of linear transformations mapping a vector space into itself. Let T be a linear transformation of the vector space $\mathscr{V}$ into itself. Now the simplest of linear transformations is the dilatation $x \to \lambda x$ with operator $T = \lambda I$, where I is the identity transformation. Is it possible that there are vectors $x \in \mathscr{V}$ such that, for these particular vectors, the transformation T simply looks like a dilatation? That is, do there exist vector–scalar pairs $(\lambda; x)$ such that $Tx = \lambda x$? If a finite or denumerably infinite set of such vector–scalar pairs exists and we can find them then the characteristic nature of T is laid bare: this is a point which will be illustrated at several points throughout the present book.

The vectors x are called the *eigenvectors* (proper vectors, characteristic vectors) of T and the scalars λ the *eigenvalues* (proper values, characteristic values): although we speak of the eigenvector x it should be clear that what we are seeking is in effect a one-dimensional subspace of $\mathscr{V}$ since x is indeterminate up to a scalar multiple. The finding of the pairs $(\lambda;x)$, that is, the solution of the operator equation $Tx = \lambda x$, is called the eigenvalue problem for T. The eigenvalue–eigenvector pair is conveniently referred to as an *eigensolution* of T. It should not be assumed that every linear transformation possesses eigenvalues and eigenvectors.

It so happens that many of the problems which arise in modelling physical systems lead directly to eigenvalue problems: for example, in any vibrating system eigenvalues are closely linked with natural vibration frequencies; in control system analysis eigenvalues determine the stability and response of the system; and in quantum physics eigenvalues are connected with the attainable energy levels of the atom.

Any one eigenvalue may have associated with it a number of corresponding eigenvectors. Let λ_ρ be an eigenvalue of T and $\{x^i\}^{(\rho)}$ a set of corresponding eigenvectors together with the vector θ: since for every x^i

$$Tx^i = \lambda_\rho x^i$$

it is readily verified that the x^i span a subspace $\mathscr{V}^{(\rho)}$ which is called the *eigenspace* of T corresponding to λ_ρ. The dimension of the eigenspace is called the geometric multiplicity of λ_ρ (and may be infinite). The eigenspace $\mathscr{V}^{(\rho)}$ is clearly the null space of the transformation $T - \lambda_\rho I$. We shall now prove the following important result.

Theorem 2.5.2 *Eigenvectors belonging to distinct eigenvalues of T are linearly independent.*

Proof Suppose that there exists an eigensolution (λ_μ, x^μ) such that x^μ is a finite linear combination of eigenvectors x^ρ belonging to eigenvalues λ_ρ, $\rho \neq \mu$; that is to say, we postulate that

$$x^\mu = \sum_\rho \xi_\rho x^\rho.$$

Notice that, for any scalar λ,

$$(T - \lambda I)x^\rho = (\lambda_\rho - \lambda)x^\rho$$

so that if we successively apply the operators $T - \lambda_\rho I$ to the above linear relation the right-hand side must vanish and we obtain

$$\left(\prod_\rho (\lambda_\mu - \lambda_\rho)\right) x^\mu = \theta,$$

and since, by hypothesis, $\lambda_\mu \neq \lambda_\rho$ we conclude that $x^\mu = \theta$, the zero vector. This contradiction proves the theorem. ∎

We conclude immediately from the theorem that the eigenspaces $\mathscr{V}^{(\rho)}$ of T are linearly independent, so that we may form the (finite or denumerable) direct sum of the eigenspaces

$$\mathscr{U} = \sum_\rho {}_\oplus \mathscr{V}^{(\rho)}.$$

$\mathscr{U}$ is obviously a subspace of $\mathscr{V}$, but how close $\mathscr{U}$ is to being $\mathscr{V}$ is a very difficult question in general. We shall return to this point in some detail for finite-dimensional spaces in Section 2.8 and for infinite-dimensional spaces in Section 3.9. If the null space of T is not empty, 0 is an eigenvalue of T with corresponding eigenspace $\mathscr{N}(T)$.

Let $(\lambda; x)$ be an eigenvalue–eigenvector pair for T: then $(\lambda^2; x)$ is an eigenvalue–eigenvector pair for T^2 since

$$T^2 x = T(Tx) = T\lambda x = \lambda(Tx) = \lambda^2 x.$$

Extending this result, we see that if $p(\lambda)$ is any (scalar) polynomial of λ then $(p(\lambda); x)$ is an eigensolution for $p(T)$. Hence if the polynomial $p(T)$ of T is such that $p(T) = 0$ (the zero operator) then all the eigenvalues of T satisfy the equation $p(\lambda) = 0$. A polynomial equation having this property, namely that its roots are the eigenvalues of a linear transformation, is called a *characteristic polynomial*.

If T is invertible and $(\lambda; x)$ is an eigensolution then, from

$$(T - \lambda I)x = \theta,$$

we have

$$T^{-1}(T - \lambda I)x = \theta$$

or

$$\left(T^{-1} - \frac{1}{\lambda}I\right)x = \theta,$$

where $\lambda \neq 0$ since $\mathscr{N}(T) = \mathcal{O}$. Hence we see that eigenvectors of T are also eigenvectors of T^{-1} and the corresponding eigenvalues are reciprocal.

Example 2.5.8

(i) Let $T: \mathscr{V}_n \to \mathscr{V}_n$ have the matrix representation (Example 2.5.1)

$$T = \begin{bmatrix} 3 & 1 & 1 \\ 1 & 5 & 1 \\ 1 & 1 & 3 \end{bmatrix}.$$

Then we find that $T(1, 0, -1) = 2(1, 0, -1)$, so that T has the eigensolution $(2; (1, 0, -1))$.

(ii) Let L be the operator $-d^2/dt^2$ with domain those functions $x(t)$ which are twice differentiable in $[0, \pi]$ and satisfy the conditions $x(0) = x(\pi) = 0$. L possesses the denumerable set of eigensolutions $(n^2; \sin nt)$ since, for any integer n,

$$L(\sin nt) = -\frac{d^2}{dt^2}(\sin nt) = n^2 \sin nt$$

and $\sin 0 = \sin n\pi = 0$.

(iii) Let M be the operator

$$\int_0^t x(s)\, ds$$

with domain all polynomials in $[0, \infty)$: the image Mx of any polynomial is of higher degree than x and no eigensolution exists in this space.

(iv) Consider the matrix

$$T = \begin{bmatrix} 0 & -1 \\ 1 & 0 \end{bmatrix}.$$

This matrix represents, relative to the standard basis in $\mathscr{R}_2$, an anticlockwise rotation of vectors through $\pi/2$; for example $T(1, 0) = (0, 1)$. The operator T obviously cannot have an eigensolution in $\mathscr{R}_2$. However, if we consider T to be an operator in $\mathscr{C}_2$ we find that T has the eigensolutions $(i; (-1, i))$ and $(-i; (1, i))$.

Exercises 2.5

1. The transformation $T: \mathscr{R}_2 \longrightarrow \mathscr{R}_2$ given by

$$T(\xi_1, \xi_2) = (\xi_1, -\xi_2)$$

is linear. If we regard the complex numbers as a vector space over the real field through the correspondence $(\xi_1, \xi_2) \leftrightarrow \xi_1 + i\xi_2$, is $T: \mathscr{C}_1 \to \mathscr{C}_1$ linear?

2. (a) Consider the vector space of infinitely differentiable functions on $[0, 1]$: show that differentiation is a linear transformation of this space onto itself which is not one-to-one.
   (b) Consider the vector space of once-differentiable functions on $[0, 1]$: show that the mapping $T(x(t)) = x(t^2)$ of this space into itself is linear and one-to-one but not onto.

3. (a) Show that a linear transformation T on $\mathscr{V}$ is fully defined if the image Tx^i of each basis vector $x^i \in \mathscr{V}$ is known.
   (b) Let $\mathscr{V}_n$ have basis $\{x^i\}$. The transformation T on $\mathscr{V}_n$ is defined by
   $$Tx^j = \sum_{i=1}^{n} x^i, \quad j = 1, \ldots, n.$$
   (i) Find the image of the vector with coordinate n-tuple $(\xi_1, \xi_2, \ldots, \xi_n)$.
   (ii) Find a basis for $\mathscr{R}(T)$.
   (iii) Find a basis for $\mathscr{N}(T)$.

4. A linear transformation $T: \mathscr{V} \to \mathscr{V}$ is *nilpotent* if there exists an integer k, the index of nilpotency, such that, for $m \geq k$, $T^m = 0$, the zero transformation.
   (a) In the space of polynomials of degree $\leq (m - 1)$ show that the differentiation operator is nilpotent with $k = m$.
   (b) If T is nilpotent of index k show that, for any $x \in \mathscr{V}$, the set
   $$S = \{x, Tx, T^2 x, \ldots, T^{k-1} x\}$$
   is linearly independent. Show that if $y \in [S]$ then $Ty \in [S]$. Such a space is said to be a *cyclic subspace* of $\mathscr{V}$ generated by x and T.
   (c) A square matrix is said to be strictly lower triangular if all the entries on and above the main diagonal are zero: show that such a matrix represents a nilpotent transformation. (Hint: try (2×2) and (3×3) first.)
   (d) Show that the matrix of a nilpotent transformation on a cyclic subspace is
   $$\begin{bmatrix} 0 & 0 & 0 & & & & \\ 1 & 0 & 0 & & & & \\ 0 & 1 & 0 & & & & \\ & & & \ddots & & & \\ & & & & 1 & 0 & 0 \\ & & & & 0 & 1 & 0 \end{bmatrix}$$

5. If $T: \mathscr{V} \to \mathscr{U}$ and $S: \mathscr{U} \to \mathscr{W}$, where $\mathscr{V}, \mathscr{U}, \mathscr{W}$ are finite-dimensional, show that
   (i) $\mathscr{R}(ST) \subseteq \mathscr{R}(S)$ and $\rho(ST) \leq \rho(S)$;
   (ii) $\mathscr{N}(ST) \supseteq \mathscr{N}(T)$ and $\nu(ST) \geq \nu(T)$.

6. Let T, S be linear transformations of $\mathscr{V}_n$ to $\mathscr{V}_n$. Show that
   (i) $\rho(T+S) \leq \rho(T) + \rho(S)$;
   (ii) $v(T+S) \geq v(T) + v(S) - n$;
   (iii) $v(T) + v(S) \geq v(ST) \geq \max\{v(T), v(S)\}$;
   (iv) $\rho(T) + \rho(S) - n \leq \rho(ST) \leq \min\{\rho(T), \rho(S)\}$ (Sylvester's law).

7. If P_1 is the projection of $\mathscr{V}$ on $\mathscr{U}^1$ along $\mathscr{W}^1$ and P_2 is the projection of $\mathscr{V}$ on $\mathscr{U}^2$ along $\mathscr{W}^2$ and if $P_1 P_2 = P_2 P_1 = P$, show that P is the projection of $\mathscr{V}$ on $\mathscr{U}^1 \cap \mathscr{U}^2$ along $\mathscr{W}^1 + \mathscr{W}^2$.

8. Let $\mathscr{V}$ be the vector space of real continuous functions on $[0, 1]$. Show that the linear transformation on $\mathscr{V}$ given by

$$Tx(t) = \int_0^t x(s) ds$$

has no eigenvalues.

9. Two vector spaces $\mathscr{V}, \mathscr{W}$ are *isomorphic* if there exists a one-to-one linear transformation of $\mathscr{V}$ onto $\mathscr{W}$, that is to say an *isomorphism*. Prove Theorem 2.4.2 by exhibiting an isomorphism of any n-dimensional space onto the space of n-tuples.

10. Show that if T is $\mathscr{R}_n \to \mathscr{R}_n$ then there exists a polynomial $p(\lambda)$ with real coefficients such that $p(T) = 0$. (Hint: use the result of Exercise 2.4, 5).

11. If T is $\mathscr{V}_n \to \mathscr{V}_n$ then we know that $n = \rho(T) + v(T)$: does this mean that we can write

$$\mathscr{V}_n = \mathscr{R}(T) \oplus \mathscr{N}(T)?$$

12. Verify that $(\pi/4; \sin t)$ is an eigensolution of

$$Tx = \int_0^{\pi/2} \sin t \sin s \, x(s) \, ds$$

13. Let $\mathscr{W}$ be a subspace of $\mathscr{V}$ and let $T: \mathscr{V} \to \mathscr{U}$ be a linear transformation. The mapping $T_\mathscr{W}: \mathscr{W} \to \mathscr{U}$ defined by

$$T_\mathscr{W}(x) = Tx \quad \forall x \in \mathscr{W}$$

is called the *restriction of T to $\mathscr{W}$*.
Show that
   (i) $T_\mathscr{W}$ is linear;
   (ii) $\mathscr{N}(T_\mathscr{W}) = \mathscr{N}(T) \cap \mathscr{W}$;
   (iii) if $\mathscr{V} = \mathscr{W} \oplus \mathscr{N}(T)$ then $T_\mathscr{W}$ is an isomorphism (Exercise 2.5, 9) between $\mathscr{W}$ and $\mathscr{R}(T)$.

14. If T is a linear transformation on $\mathscr{V}_n$ deduce that
    (a) $\mathscr{V}_n \supseteq \mathscr{R}(T) \supseteq \mathscr{R}(T^2) \supseteq \ldots \mathscr{R}(T^p)$,
    (b) $\mathscr{O} \subseteq \mathscr{N}(T) \subseteq \mathscr{N}(T^2) \subseteq \ldots \mathscr{N}(T^p)$.

    Argue further that equality must occur for some $p \leqslant n$ at the same point in both sequences: this number is called the *index* of T and the dimension of $\mathscr{R}(T^p)$, the *implicit rank* of T.

    (c) Show that (cf., Exercise 2.5,11) $\mathscr{V} = \mathscr{R}(T^p) \oplus \mathscr{N}(T^p)$.

2.6 Linear functionals and the dual space

We have seen that the field of real (complex) numbers may be regarded as a one-dimensional vector space over itself; hence it is meaningful to consider linear transformations which map a vector space $\mathscr{V}$ into the vector spaces $\mathscr{R}$ or $\mathscr{C}$ of real or complex numbers. It turns out that such transformations are of particular importance in adding to the structure of vector spaces and they are given the special name of *linear functional*. Since the image of a vector by a linear functional is a real (complex) number, they are, in this sense, similar to the ordinary functions of analysis, and for this reason a lower case italic letter is used to represent them instead of the capital italic letter normally used to denote a linear transformation.

Every linear functional on $\mathscr{V}$ can be described with the aid of an algebraic basis of $\mathscr{V}$. Let $\{x^\rho\}$ be such a basis; then the linear functional l is described by the set of scalars $\{l(x^\rho) = \alpha^\rho\}$ in the sense that if α^ρ is known for each ρ then the effect of l on every element in the space is known.

Definition 2.6.1 *A linear transformation l from a vector space $\mathscr{V}$ into the vector space of real (or complex) scalars is said to be a* linear functional *on $\mathscr{V}$*.

Thus l assigns a scalar $l(x)$ to every vector $x \in \mathscr{V}$ and

$$l(x+y) = l(x) + l(y), \qquad l(\lambda x) = \lambda l(x)$$

$\forall x, y \in \mathscr{V}$ and every scalar λ: it follows that $l(\theta) = 0$. The zero functional assigns the value zero to every vector in $\mathscr{V}$.

Example 2.6.1

(i) In $\mathscr{R}_3$ every vector x is represented, relative to a basis, by the ordered triple (ξ_1, ξ_2, ξ_3): for any real $\alpha_1, \alpha_2, \alpha_3$ the rule

$$l(x) = \alpha_1 \xi_1 + \alpha_2 \xi_2 + \alpha_3 \xi_3$$

is a linear functional on $\mathscr{R}_3$.

(ii) Let $\mathscr{V}$ be the space of continuous functions $x(t)$ defined on $[0, 1]$; then

$$l_1(x) = x(\tfrac{1}{2}), \qquad l_2(x) = \int_0^1 x(t)\,dt$$

are examples of linear functionals on $\mathscr{V}$.

(iii) Again, let $\mathscr{V}$ be the space of continuous functions $x(t)$ defined on $[0, 1]$ and let $y(t)$ be a particular element of $\mathscr{V}$: then we may define a linear functional $l_{y(t)}(x)$ by

$$l_{y(t)}(x) = \int_0^1 y(t) x(t)\,dt \qquad \forall x \in \mathscr{V},$$

thus associating a linear functional with a particular function $y(t)$.

We have seen that the linear transformations $T : \mathscr{V} \to \mathscr{U}$ themselves form a vector space. Hence the linear functionals on $\mathscr{V}$ form a vector space which is called the *algebraic conjugate* or *algebraic dual* of $\mathscr{V}$ and is denoted by $\mathscr{V}^*$.

There are advantages to be gained by adopting a notation for linear functionals which explicitly recognizes that the linear functionals are elements of the vector space $\mathscr{V}^*$. We replace l by the symbol x^* so that the value assigned to $x \in \mathscr{V}$ by the linear functional $x^* \in \mathscr{V}^*$ is $x^*(x)$. Using this notation the zero functional is denoted by θ^*.

We have seen (Example 2.5.1) that linear transformations $T : \mathscr{V}_n \to \mathscr{V}_m$ are represented by matrix arrays relative to chosen bases in $\mathscr{V}_n$ and $\mathscr{V}_m$. Accordingly we anticipate that $x^* : \mathscr{R}_n \to \mathscr{R}$ will be representable (relative to a basis in $\mathscr{R}_n$) by a $(1 \times n)$ matrix array which is, in effect, indistinguishable from an n-tuple of real numbers. Hence we would suspect that, in the finite-dimensional case, the dual space $\mathscr{V}^*$ is isomorphic to the parent space $\mathscr{V}$; this proves to be the case (the argument is identical for complex spaces). Let $\{x^1, x^2, \ldots, x^n\}$ be a basis for $\mathscr{V}_n$ and let x^* be a linear functional on $\mathscr{V}_n$; then the set of scalars $\{\xi^j\}$ defined by

$$x^*(x^j) = \xi^j, \qquad j = 1, \ldots, n,$$

serves to give a representation of the linear functional x^*. We have, for any $x \in \mathscr{V}_n$,

$$x = \sum_{j=1}^{n} \xi_j x^j$$

and hence

$$x^*(x) = \sum_{j=1}^{n} \xi_j x^*(x^j) = \sum_{j=1}^{n} \xi_j \xi^j$$

which defines x^* for every $x \in \mathscr{V}_n$. There is, therefore, a one-to-one correspondence between the n-tuples $(\xi^1, \xi^2, \ldots, \xi^n)$ and the elements x^* of $\mathscr{V}^*$ and it follows that $\mathscr{V}^*$ is isomorphic to $\mathscr{V}_n$. Given a basis $\{x^i\}$ for $\mathscr{V}_n$ we can generate a natural basis $\{x_i^*\}$ for $\mathscr{V}^*$ (also containing n elements) in the following way. Define x_j^* such that

$$x_j^*(x^i) = \delta_j^i, \qquad i, j = 1, \ldots, n,$$

where δ_j^i is the Kronecker delta defined by

$$\delta_j^i = \begin{cases} 1 & \text{if } i = j, \\ 0 & \text{if } i \neq j. \end{cases}$$

Then we have the following result:

Theorem 2.6.1 *Let $\{x^1, x^2, \ldots, x^n\}$ be a basis for $\mathscr{V}_n$ and let x_j^* be the linear functional on $\mathscr{V}_n$ defined by $x_j^*(x^i) = \delta_j^i, j = 1, \ldots, n$. Then $\{x_1^*, x_2^*, \ldots, x_n^*\}$ is a basis for $\mathscr{V}^*$: it is called the* dual basis *of $\{x^i\}$.*

Proof We prove first that the x_j^* are linearly independent and then that they span $\mathscr{V}^*$. Suppose the x_j^* are linearly dependent; then numbers $\lambda^j \neq 0$ exist such that

$$\sum_{j=1}^{n} \lambda^j x_j^* = \theta^*,$$

the zero functional. Operating with both sides of this equation on the vectors $x^i, i = 1, \ldots, n$, gives

$$\left(\sum_{j=1}^{n} \lambda^j x_j^* \right)(x^i) = \sum_{j=1}^{n} \lambda^j x_j^*(x^i) = \sum_{j=1}^{n} \lambda^j \delta_j^i = \lambda^i = \theta^*(x^i) = 0,$$

which implies a contradiction of linear dependence. Let $x^* \in \mathscr{V}^*$ and let the scalar ξ^j be defined by $\xi^j = x^*(x^j)$: we have

$$x^*(x^j) = \xi^j = \sum_{i=1}^{n} \xi^i \delta_i^j = \sum_{i=1}^{n} \xi^i x_i^*(x^j) = \left(\sum_{i=1}^{n} \xi^i x_i^* \right)(x_j),$$

and since the values of x^* and

$$\sum_{i=1}^{n} \xi^i x_i^*$$

coincide on the basis $\{x^j\}$ for $\mathscr{V}_n$, they coincide on the whole of $\mathscr{V}_n$ and we may write

$$x^* = \sum_{i=1}^{n} \xi^i x_i^*. \quad \blacksquare$$

The dual basis has the alternative names *conjugate basis* and *reciprocal basis*.

Since $\mathscr{V}^*$ is isomorphic to $\mathscr{V}_n$ we have simply one n-dimensional space having two basis sets which are related in a particular way to one another. Furthermore, the distinction between vector and linear functional is no longer explicit and it is more convenient to replace the functional notation $x^*(x)$ by the bracket notation $[x^*, x]$ and then to omit the asterisk superscript and replace x^* by another vector symbol, say y. The scalar assigned to the vector x by the linear functional (vector) y is thus denoted $[y, x]$. Similarly, we may omit the asterisk superscript on the basis vectors $\{x_i^*\}$ so that we henceforth denote parent and dual basis sets by $\{x^i\}$ and $\{x_i\}$, respectively. In bracket notation the dual basis $\{x_i\}$ is defined by

$$[x_j, x^i] = \delta_j^i.$$

The use of subscript and superscript is then the distinguishing feature of these two dual basis sets and we can express any vector $x \in \mathscr{V}_n$ in the alternative forms

$$x = \sum_{i=1}^{n} \xi_i x^i = \sum_{i=1}^{n} \xi^i x_i.$$

The ξ_i are called the *covariant* coordinates and the ξ^i the *contravariant* coordinates of x.

Let $y \in \mathscr{V}_n$ have the representations

$$y = \sum_{i=1}^{n} \eta_i x^i = \sum_{i=1}^{n} \eta^i x_i$$

relative to the same dual basis pair. Then the scalar assigned to x by y is given by

$$y(x) = [y, x] = \sum_{i=1}^{n} \sum_{j=1}^{n} \xi_i \eta^j [x^i, x_j] = \sum_{i=1}^{n} \xi_i \eta^i$$

where we have, in a natural way, expressed x (a vector) in terms of the basis $\{x^i\}$ and y (a linear functional) in terms of the basis $\{x_i\}$. This result, namely that

$$[y, x] = \sum_{i=1}^{n} \xi_i \eta^i,$$

justifies the introduction of the dual basis since it gives a convenient expression for the value of $y(x)$ in terms of the (covariant) coordinates of x and (contravariant) coordinates of y.

The bracket notation for the value of the linear functional $y(x)$ suggests that we might also think of x as a linear functional operating on y, namely $x(y)$. The roles of $\mathscr{V}_n$ and $\mathscr{V}^*$ are reversed so that we now think of $\mathscr{V}_n$ as the dual $(\mathscr{V}^*)^*$ of $\mathscr{V}^*$. The isomorphism between $\mathscr{V}_n$ and $\mathscr{V}^*$ extended to an isomorphism between $(\mathscr{V}^*)^*$ and $\mathscr{V}^*$ shows that all these spaces are isomorphic to $\mathscr{V}_n$ and that therefore the dual basis to $\mathscr{V}^*$ is again the parent basis of $\mathscr{V}_n$. The space $\mathscr{V}_n$ is said to be *algebraically reflexive*. It follows that, in the bracket notation, the order of elements is unimportant, that is,

$$[y, x] = [x, y],$$

so that we need not consider which element is vector and which linear functional.

The result that $(\mathscr{V}^*)^* = \mathscr{V}_n$ is crucially dependent on the finite-dimensionality of $\mathscr{V}_n$ and cannot be extended to infinite-dimensional spaces. For an infinite-dimensional space $\mathscr{V}$, a dual space $\mathscr{V}^*$ can be defined and the second dual space $(\mathscr{V}^*)^*$ can also be defined. But it is not true that $(\mathscr{V}^*)^* = \mathscr{V}$ and in general all that can be asserted is that $\mathscr{V} \subset (\mathscr{V}^*)^*$. A proof of these facts is beyond the scope of this text. In any case, as has already been pointed out, a useful development of infinite-dimensional spaces including the construction of linear functionals on such spaces requires the introduction of a concept of length (or norm) into the vector space. We shall find in Chapter 3 that it is possible to define a notion of reflexivity for infinite-dimensional spaces; the concept considered here is based purely on algebraic considerations and is hence properly referred to as algebraic reflexivity.

Example 2.6.2 Suppose $\mathscr{V}$ is infinite-dimensional with a Hamel basis $\{x^\rho\}$: define the linear functionals $\{x_\rho\}$ by $x_\rho(x^\sigma) = \delta_\rho^\sigma$. The $\{x_\rho\}$ are linearly independent but they are not a basis for $\mathscr{V}^*$: for example, the linear functional $x^*(x^\rho) = 1$ for all ρ cannot be written as a finite linear combination of functionals x_ρ. In a finite-dimensional space this linear functional would have the coordinates $(1, 1, \ldots, 1)$ with respect to the dual basis: however, in an infinite-dimensional space only a finite number of the coordinates may be non-zero and hence any such coordinate set can represent $x^*\{x^\rho\} = 1$ only on finite subspaces of $\mathscr{V}$.

The actual construction and use of dual basis sets for finite-dimensional spaces is dealt with in Section 2.8.

In Section 2.4 it was mentioned that a subspace of a vector space having codimension 1 is often called a hyperplane. The range space of a linear functional must have dimension 1 and hence (Section 2.5) the co-dimension of the null space is also 1. This means that there is a correspondence between a hyperplane and the null space of a linear functional. If $\mathscr{I} \subset \mathscr{V}$ is a hyperplane then there is a linear functional x^* on $\mathscr{V}$ such that $\mathscr{I} = \mathscr{N}(x^*)$: the equation of the hyperplane is $x^*(x) = 0$. The coset $j + \mathscr{I}$ is the hyperplane through the point j: its equation is clearly given by $x^*(x) = \alpha$ where $\alpha = x^*(j)$. Thus the set of vectors $\{x | x^*(x) = \alpha\}$ is, for some non-zero functional x^*, a hyperplane in $\mathscr{V}$.

Let us extend the discussion of linear functionals which map elements of a vector space to zero. Let S be an arbitrary subset of vectors in $\mathscr{V}$. We form a subset S^0 of $\mathscr{V}^*$, the dual space, consisting of all those linear functionals in $\mathscr{V}^*$ which map every element of S to zero; that is $x^* \in S^0$ if $x^*(x) = 0 \,\forall x \in S$. The subset S^0 is a subspace of $\mathscr{V}^*$ since for $x^*, y^* \in S^0$ and for any $x \in S$ we have

$$(x^* + y^*)(x) = 0, \qquad \lambda x^*(x) = 0.$$

The subspace $S^0 \subset \mathscr{V}^*$ is called the *annihilator* of $S \subset \mathscr{V}$. Clearly we have the special case $\mathscr{V}^0 = \mathscr{O}^*$ (the space consisting of the zero functional alone) and $\mathscr{O}^0 = \mathscr{V}^*$. From these statements it follows that if $[x, x^*] = 0$ for all $x \in \mathscr{V}$ then x^* is the zero functional.

The following theorem presents some useful results pertaining to subspaces and their annihilators:

Theorem 2.6.2 *If $\mathscr{U}$ and $\mathscr{W}$ are subspaces of $\mathscr{V}$ and $\mathscr{U}^0, \mathscr{W}^0$ are their annihilators, then*

(i) $(\mathscr{U} + \mathscr{W})^0 = \mathscr{U}^0 \cap \mathscr{W}^0$;
(ii) $(\mathscr{U} \cap \mathscr{W})^0 = \mathscr{U}^0 + \mathscr{W}^0$;
(iii) *if $\mathscr{W}^0$ is of finite dimension k then $\mathscr{W}$ is of codimension k.*

Proof Suppose $x^* \in (\mathscr{U} + \mathscr{W})^0$, then $x^*(u + w) = 0$ for all $u \in \mathscr{U}$ and

$w \in \mathcal{W}$: in particular, $x^*(u) = x^*(w) = 0$, hence $x^* \in \mathcal{U}^0 \cap \mathcal{W}^0$. It follows that $(\mathcal{U} + \mathcal{W})^0 \subset \mathcal{U}^0 \cap \mathcal{W}^0$. Conversely, if $x^* \in \mathcal{U}^0 \cap \mathcal{W}^0$ then $x^*(u) = x^*(w) = 0$ for all $u \in \mathcal{U}, w \in \mathcal{W}$ and, from the linearity of x^*, $x^*(u + w) = 0$ and $x^* \in (\mathcal{U} + \mathcal{W})^0$. Thus $\mathcal{U}^0 \cap \mathcal{W}^0 \subset (\mathcal{U} + \mathcal{W})^0$.

The two inclusions establish (i) of the theorem. A similar argument establishes (ii).

To prove (iii) let $\mathcal{V} = \mathcal{U} \oplus \mathcal{W}$ and introduce the basis $\{u^i\}$, $i = 1, \ldots, k$, for $\mathcal{U}$ and the (Hamel) basis $\{w^\rho\}$ for $\mathcal{W}$. Consider the linear functionals w^* which are elements of $\mathcal{W}^0$: by definition they vanish on $\mathcal{W}$, that is, $w^*(w^\rho) = 0$ for all ρ, while on $\mathcal{U}$ their value is arbitrary, say, $w^*(u^i) = \alpha_i$. But this describes each w^* on the whole of $\mathcal{V}$; furthermore, the coordinates of w^* are given by the finitely non-zero sequence $(\alpha_1, \alpha_2, \ldots, \alpha_k, 0, 0, \ldots)$ having exactly k non-zero entries. Thus we have described a subspace, namely $\mathcal{W}^0$, of $\mathcal{V}^*$ having finite dimension k. ∎

Corollary *If $\mathcal{U}$ is of finite dimension k and $\mathcal{V} = \mathcal{U} \oplus \mathcal{W}$, then a direct decomposition of $\mathcal{V}^*$ is given by $\mathcal{V}^* = \mathcal{U}^0 \oplus \mathcal{W}^0$, where $\mathcal{W}^0$ has dimension k.*

Proof Since $\mathcal{V} = \mathcal{U} \oplus \mathcal{W}$, then $\mathcal{V}^0 = \mathcal{O}^* = (\mathcal{U} \oplus \mathcal{W})^0$ and using Theorem 2.6.2(i) we have $\mathcal{O}^* = \mathcal{U}^0 \cap \mathcal{W}^0$: hence $\mathcal{U}^0$ and $\mathcal{W}^0$ have only the zero functional in common. Application of Theorem 2.6.2(iii) then shows that $\mathcal{U}^0 \oplus \mathcal{W}^0$ is a direct decomposition of $\mathcal{V}^*$. ∎

The preceding discussion of linear functionals, dual bases and annihilators has been carried through with little reference to the concrete representation of linear functionals. The form of the linear functionals on infinite-dimensional spaces will be taken up in Chapter 3.

The following examples exhibit concrete forms for the linear functional $x^*(x)$ on finite-dimensional spaces and illustrate the advantage of using dual basis sets in the evaluation of linear functionals.

Example 2.6.3 Let $\mathcal{V}_n$ be the space of n-tuples with standard basis $\{e^i\}$: a typical vector x has the coordinate n-tuple $\{\xi_1, \xi_2, \ldots, \xi_n\}$.

The linear functional x^* with coordinate vector $(\xi^1, \xi^2, \ldots, \xi^n)$ relative to a dual basis assigns the value

$$x^*(x) = \sum_{i=1}^{n} \xi^i \xi_i$$

to x. The n-tuples $(1,0,0,\ldots,0)$, $(0,1,0,0,\ldots)$, $\ldots$, $(0,0,\ldots,1)$ clearly satisfy the condition for a dual basis; hence we may conclude that the basis set $\{e^i\}$ is self-dual (or self-reciprocal); that is,

$$e^i(e^j) = [e^i, e^j] = \delta_i^j.$$

For other basis sets in $\mathscr{V}_n$ the concrete form of linear functionals can be deduced via the standard basis. For example, let the vectors $x^1 = (0, 1, 1)$, $x^2 = (1, 0, 1)$, $x^3 = (1, 1, 0)$ be a basis for $\mathscr{R}_3$. Relative to this basis let $x = (1, 2, 3)$, $x^* = (3, 2, 1)$ be a vector and linear functional respectively: then

$$\begin{aligned}x^*(x) &= 3x^1(x) + 2x^2(x) + x^3(x) \\ &= 3(x^1(x^1) + 2x^1(x^2) + 3x^1(x^3)) \\ &\quad + 2(x^2(x^1) + 2x^2(x^2) + 3x^2(x^3)) \\ &\quad + (x^3(x^1) + 2x^3(x^2) + 3x^3(x^3)).\end{aligned}$$

Now $x^1 = e^2 + e^3$, $x^2 = e^1 + e^3$, $x^3 = e^1 + e^2$, so that

$$x^1(x^1) = 2, \qquad x^1(x^2) = 1, \qquad x^1(x^3) = 1, \qquad \text{etc.,}$$

giving, finally,

$$x^*(x) = 3(2 + 2 + 3) + 2(1 + 4 + 3) + (1 + 2 + 6) = 46.$$

Now let us express the linear functional relative to a dual basis. A dual basis consists of the set $x_1 = \frac{1}{2}(-1, 1, 1)$, $x_2 = \frac{1}{2}(1, -1, 1)$, $x_3 = \frac{1}{2}(1, 1, -1)$ as is readily verified. Relative to this basis x^* has the coordinate n-tuple $(9, 8, 7)$, since

$$\begin{aligned}e^2 + e^3 &= 2x_1 + x_2 + x_3, \\ e^1 + e^3 &= 2x_2 + x_1 + x_3, \\ e^1 + e^2 &= 2x_3 + x_1 + x_2.\end{aligned}$$

Because we have expressed x relative to the original basis and x^* relative to the dual basis, we simply have

$$x^*(x) = (1 \times 9) + (2 \times 8) + (3 \times 7) = 46.$$

This example demonstrates the convenience of representing vectors and linear functionals relative to dual basis sets. (In practice the conversion of x^* to the dual basis would not normally arise—see Example 2.8.2.)

Example 2.6.4

(a) Let $\mathscr{P}_3$ be the three-dimensional space of polynomials on $[0,1]$ of degree equal to or less than two with basis $\{1, t, t^2\}$. If $x(t)$ is a typical vector in this space then a linear functional on $\mathscr{P}_3$ may take the concrete form

$$x^*(x) = \int_0^1 x^*(t)x(t)dt,$$

where $x^*(t) = \alpha_0 + \alpha_1 t + \alpha_2 t^2$. Let $x_i(t)$, $i = 0, 1, 2$, be the dual basis to $\{1, t, t^2\}$ in $\mathscr{P}_3$, that is

$$\int_0^1 x_i(t) t^j dt = \delta_i^j, \quad i, j = 0, 1, 2.$$

We readily find, by solving three sets of three simultaneous equations, that

$$x_0(t) = 3(3 - 12t + 10t^2),$$
$$x_1(t) = -12(3 - 16t + 15t^2),$$
$$x_2(t) = 30(1 - 6t + 6t^2).$$

If, for example, $x(t) = 1 + 2t + 3t^2$ and

$$x^*(t) = 3x_0(t) + 2x_1(t) + x_2(t)$$
$$= -15 + 96t - 90t^2,$$

then

$$x^*(x) = \int_0^1 x^*(t) x(t) dt = (1 \times 3) + (2 \times 2) + (3 \times 1) = 10.$$

(b) Suppose we treat $\mathscr{P}_3$ as a subspace of the space $\mathscr{P}_5$ of polynomials on $[0, 1]$ of degree ≤ 4. Let $\mathscr{Q} \subset \mathscr{P}_5$ have basis $\{t^3, t^4\}$, then clearly $\mathscr{P}_5 = \mathscr{P}_3 \oplus \mathscr{Q}$. According to the corollary to Theorem 2.6.2 we should have a decomposition of $\mathscr{P}_5^*$ into $\mathscr{P}_5^* = \mathscr{P}_3^0 \oplus \mathscr{Q}^0$, in which dim $\mathscr{P}_3^0 = 2$ and dim $\mathscr{Q}^0 = 3$. What are suitable basis sets for $\mathscr{P}_3^0$ and $\mathscr{Q}^0$? Let $y_i(t), i = 0, \ldots, 4$, be the dual basis to $\{1, t, t^2, t^3, t^4\}$ in $\mathscr{P}_5$, that is

$$\int_0^1 y_i(t) t^j dt = \delta_i^j, \quad i, j = 0, \ldots, 4.$$

Then it is readily seen that $\{y_0, y_1, y_2\}$ is a basis for $\mathscr{Q}^0$ and $\{y_3, y_4\}$ is a basis for $\mathscr{P}_3^0$. It is important to notice that the basis of $\mathscr{Q}^0$ is not the same as the dual basis of $\mathscr{P}_3$ when considered as a complete vector space (i.e., the set formerly written $\{x_0, x_1, x_2\}$): the elements $y_i(t)$ will generally contain terms in t^3 and t^4.

(c) The decomposition $\mathscr{P}_5^* = \mathscr{P}_3^0 \oplus \mathscr{Q}^0$ is not, of course, the only direct decomposition of $\mathscr{P}_5^*$ and is often not the most convenient. We can equally well write

$$\mathscr{P}_5^* = \mathscr{P}_3^* \oplus \mathscr{Q}^*,$$

where the basis of $\mathscr{P}_3^*$ is the set $\{x_0, x_1, x_2\}$ of (a) containing polynomials of degree ≤ 2, while $\mathscr{Q}^*$ has the basis (containing polynomials of degree 3 and 4) dual to the basis $\{t^3, t^4\}$ of $\mathscr{Q}$. It will be clear, however, that the complete basis sets so constructed for $\mathscr{P}_5$ and $\mathscr{P}_5^*$ are not dual basis sets.

Indeed, more generally any basis $\{z_1, z_2\}$ could be chosen for the direct complement of $\mathscr{P}_3^*$, provided
(i) its elements are not linear combinations of $\{x_0, x_1, x_2\}$;
(ii) $\{x_0, x_1, x_2, z_1, z_2\}$ spans $\mathscr{P}_5^*$.

Example 2.6.5 In Definition 2.5.3 a projection operator P was introduced having the property that, if $\mathscr{V} = \mathscr{U} \oplus \mathscr{W}$ then $Pu = u \ \forall \ u \in \mathscr{U}$ and $Pw = 0$ $\forall \ w \in \mathscr{W}$.

If $\mathscr{U}$ is finite-dimensional then we can use the dual basis sets in $\mathscr{V}$ and $\mathscr{V}^*$ to give a concrete representation of the projection operator P. From the corollary of Theorem 2.6.2 we know that $\mathscr{V}^*$ has the direct sum decomposition $\mathscr{V}^* = \mathscr{W}^0 \oplus \mathscr{U}^0$, where dim $\mathscr{W}^0 =$ dim $\mathscr{U} = k$, say. Let $\{u^i\}$, $i = 1, \ldots, k$ be a basis for $\mathscr{U}$ and $\{w_i\}$, $i = 1, \ldots, k$, a basis for $\mathscr{W}^0$. Any vector $v \in \mathscr{V}$ has the unique decomposition

$$v = \sum_{i=1}^{k} \alpha_i u^i + w$$

for some set of coordinates α_i, $i = 1, \ldots, k$. Applying the linear functional w_j to this equality we obtain

$$[w_j, v] = \sum_{i=1}^{k} \alpha_i [w_j, u^i] + [w_j, w].$$

Now $w_j \in \mathscr{W}^0$ so that $[w_j, w] = 0$ and, since w_j is a vector in the complete dual basis of $\mathscr{V}^*$, $[w_j, u^i] = \delta_j^i$ and we obtain,

$$\alpha_j = [w_j, v].$$

But

$$\sum_{i=1}^{k} \alpha_i u^i = \sum_{i=1}^{k} [w_i, v] u^i$$

is the projection of v onto $\mathscr{U}$; that is,

$$Pv = \sum_{i=1}^{k} [w_i, v] u^i.$$

(We immediately see that $Pu = u$ and $Pw = 0$.) Hence as soon as the linear functional $[\cdot, \cdot]$ is given a concrete representation then we have a concrete representation for P relative to the basis $\{u^i\}$ of $\mathscr{U}$.

Exercises 2.6

1. Show that the following are linear functionals:
   (i) on the space of continuously differentiable functions,
   $$l(x) = x'(t_0);$$

(ii) on the space of all sequences of real numbers for which $\sum_k \xi_k$ converges (cf., Example 2.2.2(a)),
$$l(x) = \sum_k \xi_k;$$
(iii) on the space of all real $(n \times n)$ matrices $[t_{ij}]$ (cf., Example 2.5.1) the *trace* of $[t_{ij}]$,
$$\operatorname{tr}[t_{ij}] = t_{11} + t_{22} + \ldots + t_{nn}.$$

2. Let $\mathscr{P}_3$ be the vector space of all polynomials $p(t)$ with real coefficients of degree two or less. Let t_1, t_2, t_3 be distinct scalars and let
$$l_i(p) = p(t_i).$$
  (i) Show that the l_i are linearly independent and hence form a basis for $\mathscr{P}_3^*$.
  (ii) Find the dual basis $\{p^i\}$, $i = 1, 2, 3$, for $\mathscr{P}_3$ and hence show that every $p \in \mathscr{P}_3$ can be written
$$p = p(t_1)p^1 + p(t_2)p^2 + p(t_3)p^3.$$
A basis of this type is called an *interpolation basis*.

3. (a) If $\mathscr{W} \subset \mathscr{V}_n$ and l is a linear functional on $\mathscr{W}$ show that there exists a linear functional l_E on $\mathscr{V}_n$ such that $l_E(x) = l(x) \forall x \in \mathscr{W}$. ($l_E$ is called an *extension* of l.)
   (b) Let H be a hyperplane in $\mathscr{V}$. If H does not contain the origin show that there is a unique linear functional $x_1^* \in \mathscr{V}^*$ such that $H = \{x \mid x_1^*(x) = 1\}$.

4. Using the fact that $\mathscr{V}_n$ is algebraically reflexive, show that the subset S of vectors in $\mathscr{V}_n$ annihilated by the linear functionals in S^0 is a subspace of $\mathscr{V}_n$.

5. Show that a set of m real homogeneous linear equations in n unknowns can be written as
$$l_i(x) = 0,$$
where $x \in \mathscr{R}_n$ and $\{l_i\}$, $i = 1, \ldots, m$, is a set of linear functionals on $\mathscr{R}_n$. Hence show that the solution space is the subspace of $\mathscr{R}_n$ annihilated by the functionals l_i, $i = 1, \ldots, m$.

6. In $\mathscr{C}_3$ find the projection of the vector $(1 + i, 1 - i, 2)$ onto the subspace spanned by $(1, 0, -1), (1, 1, 1)$ along the subspace spanned by $(1, 1, 0)$.

2.7 The algebraic dual of a linear transformation

We shall now show that every linear transformation $T: \mathscr{V} \to \mathscr{U}$ generates, in a natural way, a linear transformation from $\mathscr{U}^* \to \mathscr{V}^*$

denoted by T' and called the (algebraic) *dual* or *conjugate of T*.

It turns out that properties like the existence and uniqueness of solutions of an operator equation $Tx = y$ depend on the nature of the range and null space of the dual of T (Exercise 2.7.3). Furthermore, in many instances we come across problems whose solution can be found either from a direct formulation $Tx = y$ or a dual formulation $T'z = w$; these two problems are not always of the same order of difficulty, so that it may be easier or more convenient to find a solution to the dual problem. Also, we shall see later (Section 4.4) that it may be possible, by solving both formulations in an approximate way, to provide upper and lower bounds on the exact solution. Hence the dual transformation plays an important role, not only in characterizing the solution of operator equations, but in a very practical way in methods of approximation.

We begin with the formal definition and then go on to explain and justify it (see Fig. 2.5).

Definition 2.7.1 *Let T be a linear transformation from $\mathscr{V}$ to $\mathscr{U}$. The algebraic dual T' of T is the linear mapping from $\mathscr{U}^*$ to $\mathscr{V}^*$ defined by $[T'x^*, x]_{\mathscr{V}} = [x^*, Tx]_{\mathscr{U}}$ for all $x \in \mathscr{V}$ and $x^* \in \mathscr{U}^*$*

First we verify that T' is linear. By the linear nature of the functional and for $x^*, y^* \in \mathscr{U}^*$,

$$[x^* + y^*, Tx]_{\mathscr{U}} = [x^*, Tx]_{\mathscr{U}} + [y^*, Tx]_{\mathscr{U}}$$
$$= [T'x^*, x]_{\mathscr{V}} + [T'y^*, x]_{\mathscr{V}}$$
$$= [T'x^* + T'y^*, x]_{\mathscr{V}}.$$

But by definition the left-hand side equals $[T'(x^* + y^*), x]_{\mathscr{V}}$ so that

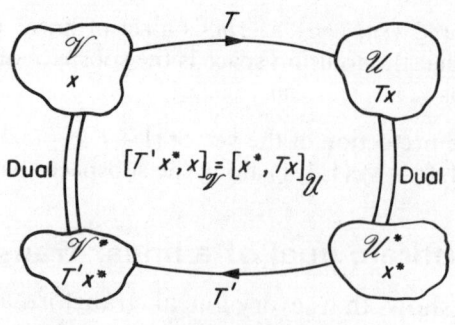

Fig. 2.5

THE ALGEBRAIC DUAL OF A LINEAR TRANSFORMATION

we have

$$[T'(x^* + y^*) - (T'x^* + T'y^*), x] = 0 \quad \forall x \in \mathscr{V},$$

which implies that $T'(x^* + y^*) - (T'x^* + T'y^*) = \theta^*$, the zero functional, and linearity is proved.

Now let us attempt to understand the meaning behind the definition. Let x^* be any vector in $\mathscr{U}^*$ and consider the set of scalars $x^*(Tx) = [x^*, Tx]_{\mathscr{U}}$ generated as x ranges over $\mathscr{V}$. This set of scalars defines a linear functional on $\mathscr{V}$, say $z^* \in \mathscr{V}^*$ such that $z^*(x) = x^*(Tx)$. Hence, having begun with an element x^* of $\mathscr{U}^*$, we have, using T, obtained a member z^* of $\mathscr{V}^*$; that is, we have some kind of mapping $T': \mathscr{U}^* \to \mathscr{V}^*$ given by

$$[x^*, Tx]_{\mathscr{U}} = [z^*, x]_{\mathscr{V}} = [T'x^*, x]_{\mathscr{V}}.$$

Example 2.7.1 Let $L: \mathscr{V} \to \mathscr{U}$ be the operator $L \equiv d/dt$ with domain all polynomials $p(t)$ of degree $\leq (m-1)$ defined on $[0, 1]$ and which satisfy the condition $p(0) = 0$. Let the linear functional on $\mathscr{V}$ and $\mathscr{U}$ have the concrete representation

$$p^*(p) = \int_0^1 p^*(t) p(t) \, dt$$

(cf., Example 2.6.4, where $p^*(t)$ is a polynomial of degree $\leq (m-1)$ in $[0, 1]$). Using the definition of the dual, we have

$$[p^*, Lp] = \int_0^1 p^* \frac{dp}{dt} dt$$

$$= (p^* p)_0^1 - \int_0^1 \frac{dp^*}{dt} p \, dt$$

$$= [L'p^*, p].$$

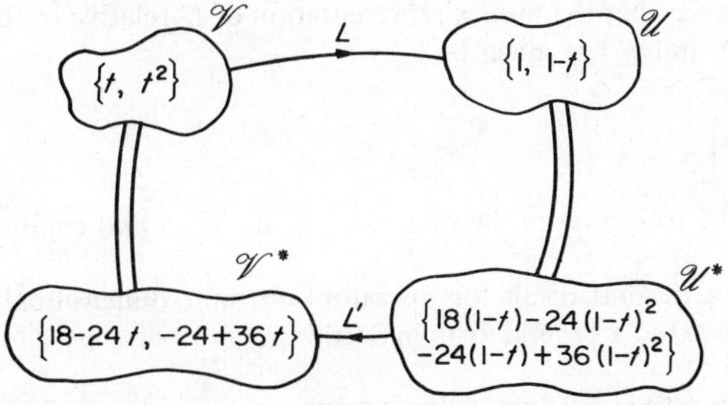

Fig. 2.6

The right-hand side represents a linear functional on $\mathscr{V}$ provided the term $(p^*p)_0^1$ vanishes. Since $p(0) = 0$ we can make this term zero if we insist that $p^*(1) = 0$. Hence we see that the dual of L is the operator $L' \equiv -d/dt$ with domain those polynomials on $[0, 1]$ which vanish at $t = 1$.

Let us choose as a basis for the domain of L the set $S = \{t, t^2\}$: all polynomials in $[S]$ clearly vanish at $t = 0$. The range of L consists of polynomials of degree ≤ 1 and for the range we choose the basis $\{1, 1 - t\}$. (The basis $\{1, t\}$ is a more natural choice, but does not serve in this particular example to illustrate one of the salient features of the dual transformation, L'.)

Since $d(t)/dt = 1$ and $d(t^2)/dt = 2t = 2 - 2(1 - t)$, we see that the matrix representation of $L \equiv d/dt$ relative to the chosen basis sets in $\mathscr{V}$ and $\mathscr{U}$ is

$$L = \begin{bmatrix} 1 & 2 \\ 0 & -2 \end{bmatrix}.$$

Since we know that the dual transformation L' has as domain those polynomials that vanish at $t = 1$, we choose a dual basis for $\mathscr{U}^*$ incorporating the functions $(1 - t)$, $(1 - t)^2$; it is readily verified that the dual basis is $\{(18(1 - t) - 24(1 - t)^2), (-24(1 - t) + 36(1 - t)^2)\}$. Similarly, using the functions $1, t$ we find that a dual basis for $\mathscr{V}^*$ is $\{(18 - 24t), (-24 + 36t)\}$ (see Fig. 2.6). Let us now write down the matrix representation of $L' \equiv -d/dt$ relative to the basis sets in $\mathscr{U}^*$ and $\mathscr{V}^*$. We have

$$-\frac{d}{dt}[18(1 - t) - 24(1 - t)^2] = 18 - 48(1 - t) = -30 + 48t$$

and

$$-\frac{d}{dt}[-24(1 - t) + 36(1 - t)^2] = -24 + 72(1 - t) = 48 - 72t.$$

But

$$-30 + 48t = (18 - 24t) + 2(-24 + 36t)$$

and

$$48 - 72t = -2(-24 + 36t),$$

so that we see that the matrix representation of L' relative to the dual basis sets in $\mathscr{U}^*$ and $\mathscr{V}^*$ is given by

$$L' = \begin{bmatrix} 1 & 0 \\ 2 & -2 \end{bmatrix},$$

which is the transpose of the matrix L (i.e., the rows and columns are interchanged).

This is a general result for operators on finite-dimensional spaces and will be proved in a general form in Section 2.8.

Consider the product transformation ST, where $T: \mathscr{V} \to \mathscr{U}$ and

$S: \mathscr{U} \to \mathscr{W}$, and let $S': \mathscr{W}^* \to \mathscr{U}^*, T': \mathscr{U}^* \to \mathscr{V}^*$ be their respective duals. What is the dual of the product ST? We have, for all $x^* \in \mathscr{W}^*$, $x \in \mathscr{V}$,

$$[x^*, STx] = [x^*, S(Tx)] = [S'x^*, Tx]$$
$$= [T'S'x^*, x] = [(ST)'x^*, x],$$

showing that

$$(ST)' = T'S':$$

note the reversal of order.

Exercises 2.7

1. If $T: \mathscr{V} \to \mathscr{U}$ is a linear transformation and $T': \mathscr{U}^* \to \mathscr{V}^*$ its dual, verify that
   (i) $I' = I$ (I is the identity operator);
   (ii) $(\alpha T)' = \alpha T'$, $\alpha \in \mathscr{R}$ or $\mathscr{C}$;
   (iii) $(T + S)' = T' + S'$;
   (iv) $(T')^{-1} = (T^{-1})'$ provided T is invertible;
   (v) $(T')' = T$ provided $\mathscr{V}, \mathscr{U}$ are finite-dimensional.

2. If $\mathscr{V}_n = \mathscr{U} \oplus \mathscr{W}$ and P is the projection of $\mathscr{V}$ onto $\mathscr{U}$ along $\mathscr{W}$, show that P' is the projection of $\mathscr{V}_n^*$ onto $\mathscr{W}^0$ along $\mathscr{U}^0$.

3. Show that, for $T: \mathscr{V}_m \to \mathscr{U}_n$,
   (i) $\mathscr{N}(T') \subset \mathscr{R}^0(T)$,
   (ii) $\mathscr{R}^0(T) \subset \mathscr{N}(T')$
   and hence that $\mathscr{R}^0(T) = \mathscr{N}(T')$. Thence
   (a) deduce that $\rho(T) = \rho(T')$;
   (b) deduce that the equation $Tx = y$ has a solution if and only if $[y, z] = 0$ for all $z \in \mathscr{N}(T')$;
   (c) verify that the null space of the operator

$$Tx = x(t) - 4 \int_0^1 (3ts - 5t^2s^2) x(s) \, ds$$

   is of dimension 1 with basis vector $(t - t^2)$. Taking

$$[x, y] = \int_0^1 x(t) y(t) \, dt,$$

   deduce that the equation $Tx = y$ has no solution when $y(t) = 1$. Find a function $y(t)$ for which a solution will exist.

2.8 Linear transformations on finite-dimensional spaces

If we restrict ourselves to finite-dimensional spaces we can develop many useful results pertaining to the structure of these spaces, and more particularly to the nature or character of linear transformations from one finite-dimensional space to another. These results can be developed from the algebraic vector space ideas already developed in this chapter.

In contrast, further development in the theory of linear transformations on infinite-dimensional spaces is crucially dependent on the introduction of a measure of distance (or length) into the vector space structure so that questions of convergence of infinite sequences can be discussed (Chapter 3). To be sure, the introduction of 'distance' into finite-dimensional spaces certainly adds to the fund of useful results, but it is not so vital.

In this section we exploit fully Theorem 2.4.2, which states that every vector space of (finite) dimension n is isomorphic to $\mathscr{R}_n$ (or $\mathscr{C}_n$). This allows us to carry the discussion through entirely in terms of vector spaces of real (complex) n-tuples and the linear transformations on these concrete spaces. Example 2.5.1 shows that every linear transformation $T: \mathscr{V}_n \to \mathscr{V}_m$ can be represented by a rectangular ($m \times n$) matrix array $[t_{ij}]$ relative to a pair of basis sets in $\mathscr{V}_n$ and $\mathscr{V}_m$. Hence the subject matter of this section concerns matrices and their properties considered within the context of vector spaces.

It is most important to appreciate that the representation of a linear transformation $T: \mathscr{V}_n \to \mathscr{V}_m$ by a matrix is totally dependent on the particular basis sets chosen for the domain $\mathscr{V}_n$ and codomain $\mathscr{V}_m$; that is to say, a single transformation T will assume many matrix representations. It then becomes important to ask, first, how different matrix representations of T are related, and, second, whether any properties of T are intrinsic to T in the sense that they are independent of the choice of basis sets. Associated with these questions is the determination of a matrix representation of T which is of the simplest possible form—a so-called *canonical representation*.

The coordinate set of a vector x in the vector space of n-tuples relative to a basis is $(\xi_1, \ldots, \xi_n)$. It is convenient in the context of matrices to represent the coordinate set as a $(1 \times n)$ or row matrix written $\lfloor \xi_j \rfloor$, or alternatively as a $(n \times 1)$ or column matrix written $\{\xi_j\}$. In what follows, either form may be used as a matter of convenience. The mapping $y = Tx$ (Example 2.5.1) which takes the concrete form

$$\eta_i = \sum_{j=1}^{n} t_{ij}\xi_j$$

can be compactly written as

$$\{\eta_i\} = [t_{ij}]\{\xi_j\}.$$

The computation is effected by what is generally referred to as 'row-by-column multiplication'; that is, the ith element of $\{\eta_i\}$ is obtained by multiplying the jth element of the ith row of $[t_{ij}]$ by the corresponding element of the vector, namely ξ_j, and summing over j.

Let the vector x be one of the basis vectors, say the kth; then its coordinate set is $(0, 0, \ldots, 1, 0, 0, 0)$ with unity in the kth place. The image of x by T then has the coordinate set $(t_{1k}, t_{2k}, \ldots, t_{nk})$; that is to say, the columns of $[t_{ij}]$ considered as vectors (coordinate sets) are the images of the basis vectors relative to the basis sets in $\mathscr{V}_m$ and $\mathscr{V}_n$. This leads to an alternative and, in many ways, more fruitful interpretation of the mapping $y = Tx$ than in terms of 'row-by-column multiplication'. If we write the columns of $[t_{ij}]$ as vectors $\{t_i\}_j$[†] then the mapping

$$\eta_i = \sum_{j=1}^{n} t_{ij}\xi_j$$

can be written

$$\{\eta_i\} = \sum_{j=1}^{n} \{t_i\}_j \xi_j,$$

showing that every image vector $\{\eta_i\}$ is a linear combination of the columns of $[t_{ij}]$. But this means that the vectors $\{t_i\}_j$ span the range space of T. Hence the rank of T (Definition 2.5.2), which is to say the dimension of the range space of T, is equal to the number of linearly independent columns of $[t_{ij}]$. Correspondingly, the null space $\mathscr{N}(T)$ of T is spanned by the vectors satisfying

$$\sum_{j=1}^{n} \{t_i\}_j \xi_j = 0$$

or, if we represent the rows of $[t_{ij}]$ by $\lfloor t_j \rfloor_i$,

$$\lfloor t_j \rfloor_i \xi_j = 0, \quad i = 1, \ldots, m.$$

If we consider the $\lfloor t_j \rfloor_i$ to be linear functionals in $\mathscr{V}_n^*$ then we see that the (vectors) $\lfloor t_j \rfloor_i$ span $\mathscr{N}^0 \subset \mathscr{V}_n^*$, the annihilator of $\mathscr{N}(T) \subset \mathscr{V}_n$.

[†] There should be no confusion with the notation for displaying the elements of a set.

But Theorem 2.6.2 tells us that dim $\mathcal{N}^0 = \text{codim } \mathcal{N} = n - \nu = \rho$; that is, the number of linearly independent rows of $[t_{ij}]$ equals the rank of $[t_{ij}]$. We may summarize these results in

Theorem 2.8.1 *The rank of a linear transformation T is equal to the number of linearly independent columns or the number of linearly independent rows of a matrix representation for T. Rank is an intrinsic property of T.*

This result is particularly important in studying the existence and uniqueness of solutions of linear algebraic simultaneous equations.

Consider a system of m linear algebraic equations in n unknowns, $\xi_1, \xi_2, \ldots, \xi_n$;

$$\sum_{j=1}^{n} t_{ij}\xi_j = b_i, \qquad i = 1, \ldots, m,$$

which may be written in matrix form as

$$[t_{ij}]\{\xi_j\} = \{b_i\}.$$

The system of equations represents, relative to a pair of bases in $\mathscr{V}_n$ and $\mathscr{V}_m$, the operator equation $Tx = b$. Given T and the vector b the 'solution' of the equation is the vector x. As with any operator equation we must concern ourselves with two basic questions:

1. Does a solution x exist?
2. If it does, is the solution x unique?

It should be clear to the reader that

(a) a solution will exist if, and only if, $b \in \mathscr{R}(T)$, the range space of T;
(b) the solution will be unique only if the null space $\mathscr{N}(T)$ of T is the zero space.

For suppose $\mathscr{N}(T)$ is not the zero space and $x_0 \in \mathscr{N}(T)$; if x_p satisfies $Tx_p = b$ then so does $x_p + x_0$, and hence is also a solution. In fact, we can see immediately that if x_p is any vector satisfying $Tx_p = b$ then the coset $x_p + \mathscr{N}(T)$ contains every solution to $Tx = b$. That is to say, if the nullity of T is ν then the most general solution contains ν undetermined (scalar) multipliers. We can interpret these conditions more precisely in terms of a matrix representation $[t_{ij}]$ of T. We have seen that the range space of T is spanned by the columns $\{t_i\}_j$ of $[t_{ij}]$, hence it follows that a solution will exist if and only if $\{b_i\}$ is a linear combination of the columns of $[t_{ij}]$. Furthermore the nullity of T is given by $v(T) = n - \rho(T)$, where, of course, $\rho(T)$ is the number of linearly independent columns (or rows) of $[t_{ij}]$. If we form the partitioned matrix $[[t_{ij}] \vdots \{b_i\}]$ (p. 69), the 'augmented' matrix, we can see that $\{b_i\}$ can only be a linear combination of the $\{t_i\}_j$ if the (column) rank of the augmented matrix is equal to the rank of $[t_{ij}]$; furthermore the solution will contain $(n - \rho)$ arbitrary multipliers.

The questions of existence and uniqueness are thus answered if we can determine the rank of $[t_{ij}]$ and the rank of the augmented matrix $[[t_{ij}] \vdots \{b_i\}]$. We shall return to this point shortly when we deal with matrix equivalence.

We now turn to an interpretation of the basic operations involving linear transformations in matrix terms. The reader will no doubt be familiar with the matrix operations of matrix addition and matrix multiplication. The purpose of the following paragraphs is to fit these computational schemes into the context of linear transformations.

Let T, S be two linear transformations from $\mathscr{V}_n \to \mathscr{V}_m$ with $(m \times n)$ matrix representations $[t_{ij}], [s_{ij}]$ relative to the same pair of basis sets in $\mathscr{V}_m, \mathscr{V}_n$. The sum of T and S is clearly the matrix $[t_{ij} + s_{ij}]$, being formed simply by adding corresponding elements: the scalar multiple αT is the matrix $[\alpha t_{ij}]$ in which every element is multiplied by α. The zero operator is a matrix with all elements zero while the identity operator $I: \mathscr{V}_n \to \mathscr{V}_n$ has an $(n \times n)$ matrix with units on the principal diagonal and zeros elsewhere. We now consider the product or composition of $T: \mathscr{V}_n \to \mathscr{V}_m$ and $S: \mathscr{V}_m \to \mathscr{V}_l$ having $(m \times n)$ and $(l \times m)$ matrix representations $[t_{ij}], [s_{ij}]$, respectively (*see* p. 41). First we map $x \to y$ with T to give

$$\eta_i = \sum_{j=1}^{n} t_{ij} \xi_j, \qquad i = 1, \ldots, m,$$

and then we map $y \to z$ with S, giving

$$\zeta_k = \sum_{i=1}^{m} s_{ki} \eta_i, \qquad k = 1, \ldots, l,$$

$$= \sum_{i=1}^{m} \sum_{j=1}^{n} s_{ki} t_{ij} \xi_j,$$

which, upon changing the order of summation, becomes

$$\zeta_k = \sum_{j=1}^{n} \left(\sum_{i=1}^{m} s_{ki} t_{ij} \right) \xi_j.$$

But this represents the composite mapping $ST: \mathscr{V}_n \to \mathscr{V}_l$, hence we see that if the $(l \times n)$ matrix representation of ST is $[p_{ij}]$ then

$$p_{ij} = \sum_{r=1}^{m} s_{ir} t_{rj}, \quad i = 1, \ldots, l, \quad j = 1, \ldots, n;$$

that is to say, the ijth element of $[p_{ij}]$ is given by a 'row-by-column multiplication' of the jth column of $[t_{ij}]$ by the ith row of $[s_{ij}]$. Henceforth we shall denote this computational rule in the compact form

$$[p_{ij}] = [s_{ij}][t_{ij}],$$

which in effect defines 'matrix multiplication'.

Let $T: \mathscr{V}_n \to \mathscr{V}_n$ be a one-to-one transformation with $(n \times n)$ matrix $[t_{ij}]$ relative to a chosen basis. We know that T has a unique inverse $T^{-1}: \mathscr{V}_n \to \mathscr{V}_n$ with $(n \times n)$ matrix $[\hat{t}_{ij}]$, say, relative to the same basis. Since $T^{-1} T = TT^{-1} = I$, we have, from the preceding rule for a matrix product,

$$\sum_{k=1}^{n} \hat{t}_{ik} t_{kj} = \sum_{k=1}^{n} t_{ik} \hat{t}_{kj} = \delta_j^k,$$

where δ_j^k is the Kronecker delta. Written as a matrix multiplication this reads

$$[t_{ij}][t_{ij}]^{-1} = [t_{ij}]^{-1}[t_{ij}] = [\delta_i^j].$$

The dual or conjugate T' of the linear transformation T was defined in Section 2.7. Given a matrix representation for $T: \mathscr{V}_n \to \mathscr{V}_m$ relative to a pair of basis sets for $\mathscr{V}_n$ and $\mathscr{V}_m$, what is the matrix representation for $T': \mathscr{V}_m^* \to \mathscr{V}_n^*$ relative to dual basis sets for $\mathscr{V}_m^*$ and $\mathscr{V}_n^*$? Recall that the value of the linear functional $x^* = (\xi^1, \xi^2, \ldots, \xi^n) \in \mathscr{V}_n^*$ on the vector $x = (\xi_1, \xi_2, \ldots, \xi_n) \in \mathscr{V}_n$ is given by

$$x^*(x) = [x^*, x] = \sum_{k=1}^{n} \xi^k \xi_k$$

when x, x^* are referred to dual basis sets. Using this fact we have

$$[y^*, Tx] = \sum_{j=1}^{m} \left(\eta^j \sum_{k=1}^{n} t_{jk} \xi_k \right)$$
$$= \sum_{k=1}^{n} \xi_k \left(\sum_{j=1}^{m} t_{jk} \eta_j \right)$$
$$= [T'y^*, x].$$

We conclude that, since $x^* = T'y^*$,

$$\xi^k = \sum_{j=1}^{m} t_{jk}\eta^j,$$

so that if we take the $(n \times m)$ matrix representation of T' to be $[t'_{ij}]$ we have $t'_{ij} = t_{ji}$. The $(n \times m)$ matrix $[t'_{ij}]$ is thus obtained from the $(m \times n)$ matrix $[t_{ij}]$ simply by interchanging rows and columns: we say that $[t'_{ij}] = [t_{ji}]$ is the *transpose* of $[t_{ij}]$, and this is written

$$[t'_{ij}] = [t_{ji}] = [t_{ij}]^\mathrm{T}$$

(cf., Example 2.7.1).

Partitioning

It is sometimes convenient to write a matrix in a block or partitioned form

$$[t_{ij}]_{(m \times n)} = \left[\begin{array}{c|c} [t_{ij}]_{11} & [t_{ij}]_{12} \\ {\scriptstyle(m_1 \times n_1)} & {\scriptstyle(m_1 \times n_2)} \\ \hline [t_{ij}]_{21} & [t_{ij}]_{22} \\ {\scriptstyle(m_2 \times n_1)} & {\scriptstyle(m_2 \times n_2)} \end{array}\right],$$

where the order of the submatrices is indicated and where $m_1 + m_2 = m$, $n_1 + n_2 = n$. The matrix can then be formally manipulated as a (2×2) matrix with matrix elements. For example, the mapping $[t_{ij}]\{\xi_j\}$ is written

$$\left[\begin{array}{c|c} [t_{ij}]_{11} & [t_{ij}]_{12} \\ \hline [t_{ij}]_{21} & [t_{ij}]_{22} \end{array}\right] \left[\begin{array}{c} \{\xi_j\}_1 \\ \hline \{\xi_j\}_2 \end{array}\right] = \left[\begin{array}{c} \{[t_{ij}]_{11}\{\xi_j\}_1 + [t_{ij}]_{12}\{\xi_j\}_2\} \\ \hline \{[t_{ij}]_{21}\{\xi_j\}_1 + [t_{ij}]_{22}\{\xi_j\}_2\} \end{array}\right],$$

where a conformable partition of the vector $\{\xi_j\}$ has been adopted. Partitioning of a matrix has a simple interpretation as the restriction of the associated linear transformation to certain subspaces of $\mathscr{V}_n$ (see Exercise 2.8,2(c)).

Partitioned matrices may be formally multiplied provided their partition orders are such as to make the various submatrix multiplications conformable (Exercise 2.8,2(b)).

Change of basis

A single linear transformation $T: \mathscr{V}_n \to \mathscr{V}_m$ has many matrix representations depending simply on the choice of basis sets in $\mathscr{V}_n$

and $\mathscr{V}_m$, and we now wish to consider how these different representations are related.

Before doing this, let us look at the simpler problem of finding how the coordinates of a fixed vector change upon changing the basis. Let the vectors $\{x^i\}, i = 1, \ldots, n$, be a basis for $\mathscr{V}_n$. The coordinates of the (fixed) vector x relative to this basis are the (unique) scalars $(\xi_1, \xi_2, \ldots, \xi_n)$ in the representation

$$x = \sum_{i=1}^{n} \xi_i x^i.$$

Suppose we wish to refer x to the new basis $\{\hat{x}^i\}, i = 1, \ldots, n$, for $\mathscr{V}_n$ and let the corresponding coordinates be $(\hat{\xi}_1, \hat{\xi}_2, \ldots, \hat{\xi}_n)$.

To begin with we must be able to relate the 'new' basis $\{\hat{x}^i\}$ to the 'old' basis $\{x^i\}$. Now every vector in $\mathscr{V}_n$ must be representable in terms of the basis $\{\hat{x}^i\}$; that is to say, there must exist n^2 scalars l_{ij} such that

$$x^j = \sum_{i=1}^{n} l_{ij} \hat{x}^i, \qquad j = 1, \ldots, n.$$

We then have

$$x = \sum_{j=1}^{n} \xi_j x^j$$

$$= \sum_{j=1}^{n} \xi_j \left(\sum_{i=1}^{n} l_{ij} \hat{x}^i \right)$$

$$= \sum_{i=1}^{n} \left(\sum_{j=1}^{n} l_{ij} \xi_j \right) \hat{x}^i$$

$$= \sum_{i=1}^{n} \hat{\xi}_i \hat{x}^i$$

so that since the representation of a vector is unique we conclude

$$\hat{\xi}_i = \sum_{j=1}^{n} l_{ij} \xi_j.$$

If we assemble the scalars l_{ij} into the matrix $[l_{ij}]$, we may write

$$\{\hat{\xi}_i\} = [l_{ij}]\{\xi_j\}.$$

Similarly, beginning with

$$\hat{x}^i = \sum_{j=1}^{n} k_{ji} x^j$$

LINEAR TRANSFORMATIONS ON FINITE-DIMENSIONAL SPACES

we find that

$$\{\xi_i\} = [k_{ij}]\{\hat{\xi}_j\}.$$

While the n^2 scalars l_{ij} are freely available to specify the change in basis, they must satisfy the condition that the basis vectors are independent: this implies that the rank of the matrix $[l_{ij}]$ must be n or equivalently that the columns $\{l_i\}_j$ considered as n-vectors are linearly independent. The matrix $[l_{ij}]$ will have an inverse and we recognize immediately that the inverse of $[l_{ij}]$ is the matrix $[k_{ij}]$; that is

$$\sum_{k=1}^{n} k_{ik} l_{kj} = \delta^i_j.$$

What is the structure of $[l_{ij}]$? Suppose we take $\{\xi_i\}$ to be the coordinate set of the basis vector x^k, that is

$$\{\xi_i\} = (0, 0, \ldots, \overset{k\text{th place}}{1}, 0, \ldots, 0),$$

then

$$\{\hat{\xi}_i\} = \{l_i\}_k,$$

the kth column of $[l_{ij}]$. Hence the columns of $[l_{ij}]$ are the coordinate sets of the 'old' basis vectors relative to the 'new'. Conversely the columns of $[k_{ij}]$ are the coordinate sets of the 'new' basis vectors relative to the 'old'.

Example 2.8.1 Let us return to an earlier example, namely Example 2.4.3, and rework it using the matrices $[l_{ij}]$, $[k_{ij}]$. In that example we had (with a slight change in notation) the two basis sets

'old' basis, $\{x^i\} = \{(1, 0, 0); (0, 1, 0); (0, 0, 1)\}$;
'new' basis, $\{\hat{x}^i\} = \{(1, 1, 0); (0, 1, 1); (1, 0, 1)\}.$

Note that the three 3-tuples representing $\{\hat{x}^i\}$ are the coordinates of the $\{\hat{x}^i\}$ (the 'new' basis) relative to the $\{x^i\}$ (the 'old' basis). The matrices $[k_{ij}]$, $[l_{ij}]$ may be inferred from the result of Example 2.4.3 to be given by

$$[k_{ij}] = \begin{bmatrix} 1 & 0 & 1 \\ 1 & 1 & 0 \\ 0 & 1 & 1 \end{bmatrix}, \quad [l_{ij}] = \tfrac{1}{2}\begin{bmatrix} 1 & 1 & -1 \\ -1 & 1 & 1 \\ 1 & -1 & 1 \end{bmatrix}$$

and it is readily verified that these are inverse matrices. Notice that the columns of $[k_{ij}]$ are the coordinates of the 'new' basis relative to the 'old'. In practice, one constructs the matrix $[k_{ij}]$ in exactly this way and then inverts it numerically to obtain $[l_{ij}]$ for use in $\{\hat{\xi}_i\} = [l_{ij}]\{\xi_j\}$.

72   ALGEBRAIC THEORY OF VECTOR SPACES

For the general vector $\{\xi_1, \xi_2, \xi_3\}$ we see that its coordinates relative to the new basis are given by (cf., Example 2.4.3)

$$\begin{bmatrix} \hat{\xi}_1 \\ \hat{\xi}_2 \\ \hat{\xi}_3 \end{bmatrix} = \tfrac{1}{2} \begin{bmatrix} 1 & 1 & -1 \\ -1 & 1 & 1 \\ 1 & -1 & 1 \end{bmatrix} \begin{bmatrix} \xi_1 \\ \xi_2 \\ \xi_3 \end{bmatrix} = \begin{bmatrix} \tfrac{1}{2}(\xi_1 + \xi_2 - \xi_3) \\ \tfrac{1}{2}(-\xi_1 + \xi_2 + \xi_3) \\ \tfrac{1}{2}(\xi_1 - \xi_2 + \xi_3) \end{bmatrix}.$$

The process of changing basis is hence fairly automatic, and depends for its implementation only on a numerical method for obtaining the inverse of a matrix (*see* Exercises 2.8,2(d) and 2.8,11(c)).

Example 2.8.2 We saw in Section 2.6 that the vector space $\mathscr{V}_n$ and its dual $\mathscr{V}_n^*$ are isomorphic. Hence we could consider the dual basis sets $\{x_i\}$ and $\{x^i\}$ to be 'old' and 'new' bases respectively for the single vector space $\mathscr{V}_n$. We then have

$$x_j = \sum_{i=1}^{n} g_{ij} x^i$$

for some set of n^2 scalars g_{ij} (the g_{ij} are analogous to the l_{ij} used earlier) and, using the fundamental property of dual basis sets, we find

$$[x_k, x_j] = \sum_{i=1}^{n} g_{ij} [x_k, x^i]$$

$$= g_{kj}.$$

Consequently, if $\{\xi_i\}, \{\xi^i\}$ are the covariant and contravariant coordinates of a fixed vector $x \in \mathscr{V}_n$,

$$\{\xi_i\} = [g_{ij}] \{\xi^j\}$$

and conversely

$$\{\xi^i\} = [g_{ij}]^{-1} \{\xi_j\}.$$

The symmetric matrix $[g_{ij}]$ in which the elements are defined by $g_{ij} = [x_i, x_j]$ is called the *fundamental matrix*. Its inverse $[g_{ij}]^{-1}$ is more conveniently written $[g^{ij}]$, where clearly we have

$$g^{ij} = [x^i, x^j].$$

The relations between covariant and contravariant coordinates can then be written

$$\{\xi_i\} = [g_{ij}] \{\xi^j\}, \quad \{\xi^i\} = [g^{ij}] \{\xi_j\},$$

so that the fundamental matrices provide a means for 'raising and lowering' the coordinate indices of a fixed vector.

This means that by using the fundamental matrix we can calculate the value of linear functionals without explicit introduction of a dual basis.

LINEAR TRANSFORMATIONS ON FINITE-DIMENSIONAL SPACES

For example, let
$$x = \sum_{i=1}^{n} \xi_i x^i, \qquad y = \sum_{i=1}^{n} \eta_i x^i,$$
then
$$y(x) = [y, x] = \sum_{i=1}^{n} \eta^i \xi_i$$
$$= \lfloor \eta^i \rfloor \{\xi_i\}$$
$$= \lfloor \eta_j \rfloor [g^{ij}]^T \{\xi_i\}$$
$$= \lfloor \eta_j \rfloor [g^{ij}] \{\xi_i\}.$$

The elements of the fundamental matrix $[g_{ij}]$ are defined as the functionals $[x_i, x_j]$ but we cannot, of course, calculate these without resort to the use of coordinates. Now a basis will normally be specified as sets of coordinates relative to the standard basis. Let the basis vectors x^i, x^j have coordinate sets $\{\xi_k\}^i$, $\{\xi_k\}^j$: then, using the fact that the standard basis is self-dual,
$$[x^i, x^j] = \lfloor \xi_k \rfloor^i \{\xi_k\}^j,$$
so that the elements of $[g^{ij}]$ are readily computed. The matrix $[g_{ij}]$ can, if required, be found by inversion.

Example 2.8.3 Let us apply these ideas to a previous example, namely Example 2.6.3. Here the given basis is $(0, 1, 1), (1, 0, 1), (1, 1, 0)$, giving
$$[g^{ij}] = \begin{bmatrix} 2 & 1 & 1 \\ 1 & 2 & 1 \\ 1 & 1 & 2 \end{bmatrix},$$
so that with $x = (1, 2, 3)$ and $x^* = (3, 2, 1)$ we have
$$x^*(x) = [3, 2, 1] \begin{bmatrix} 2 & 1 & 1 \\ 1 & 2 & 1 \\ 1 & 1 & 2 \end{bmatrix} \begin{bmatrix} 1 \\ 2 \\ 3 \end{bmatrix} = [9, 8, 7] \begin{bmatrix} 1 \\ 2 \\ 3 \end{bmatrix} = 46.$$

The inverse of $[g^{ij}]$ is
$$[g_{ij}] = \tfrac{1}{4} \begin{bmatrix} 3 & -1 & -1 \\ -1 & 3 & -1 \\ -1 & -1 & 3 \end{bmatrix}.$$

Associating $[g_{ij}]$ with $[k_{ij}]$ we anticipate that the columns of $[g_{ij}]$ are the coordinate sets of the dual basis relative to the given basis. For example, column one yields
$$\tfrac{1}{4}(3(0, 1, 1) - 1(1, 0, 1) - 1(1, 1, 0)) = \tfrac{1}{2}(-1, 1, 1)$$
as the first dual basis vector (expressed relative to the standard basis) in agreement with the result quoted in Example 2.6.3.

Equivalence

Having considered how the coordinates of a vector change upon changing basis we can now go on to see how the different matrix representations of a linear transformation $T: \mathscr{V}_n \to \mathscr{V}_m$ are related. Since we are dealing with two vector spaces we need to allow for a change of basis in each space, say from $\{x^i\}$ to $\{\hat{x}^i\}$, $i = 1, \ldots, n$, in $\mathscr{V}_n$ and from $\{y^j\}$ to $\{\hat{y}^j\}$, $j = 1, \ldots, m$, in $\mathscr{V}_m$. Let the $(m \times n)$ matrix of T relative to 'old' and 'new' bases be $[t_{ij}]$ and $[\hat{t}_{ij}]$ respectively. We have, relative to the 'old' basis sets

$$\{\eta_i\} = [t_{ij}]\{\xi_j\}.$$

Let the matrices specifying the change of basis in $\mathscr{V}_n, \mathscr{V}_m$ be $[l_{ij}]$, $(n \times n)$ and $[m_{ij}]$, $(m \times m)$, respectively, so that

$$\{\hat{\xi}_i\} = [l_{ij}]\{\xi_j\}, \qquad \{\hat{\eta}_i\} = [m_{ij}]\{\eta_j\}.$$

Now

$$\{\hat{\eta}_i\} = [m_{ij}][t_{ij}]\{\xi_j\}$$
$$= [m_{ij}][t_{ij}][l_{ij}]^{-1}\{\hat{\xi}_j\}$$
$$= [\hat{t}_{ij}]\{\hat{\xi}_j\}$$

by definition. Hence we find that the matrix $[\hat{t}_{ij}]$ of T relative to the 'new' basis sets is related to $[t_{ij}]$ by

$$[\hat{t}_{ij}] = [m_{ij}][t_{ij}][l_{ij}]^{-1} = [m_{ij}][t_{ij}][k_{ij}]:$$

conversely,

$$[t_{ij}] = [m_{ij}]^{-1}[\hat{t}_{ij}][l_{ij}].$$

This brings us to the following definition.

Definition 2.8.1 *Matrices of order $(m \times n)$ which are different representations of a single linear transformation $T: \mathscr{V}_n \to \mathscr{V}_m$ are said to be* equivalent: *if $[t_{ij}]$ is an $(m \times n)$ matrix, $[p_{ij}]$ a non-singular $(n \times n)$ matrix and $[q_{ij}]$ a non-singular $(m \times m)$ matrix, then $[\hat{t}_{ij}]$ is equivalent to $[t_{ij}]$ if*

$$[\hat{t}_{ij}] = [q_{ij}][t_{ij}][p_{ij}].$$

Since equivalent matrices are different representations of a single linear transformation it follows that the rank of a linear transformation is equal to the rank of any one of its (equivalent) matrix representations (cf., Theorem 2.8.1.).

Row reduction, column reduction and canonical form

Let us define the following 'elementary row operations' on the $(m \times n)$ matrix $[t_{ij}]$:

1. interchange the ith and jth rows;
2. multiply the ith row by a scalar α;
3. replace the ith row by the sum of the ith and jth rows.

Each of these elementary operations can be represented as the pre-multiplication of $[t_{ij}]$ by a non-singular $(m \times m)$ matrix $[e_{ij}]$. For each operation above the elements e_{ij} are given by

1. $e_{ji} = e_{ij} = 1, e_{jj} = e_{ii} = 0; e_{lk} = \delta_l^k$ otherwise (this is the unit matrix with ith, jth rows interchanged);
2. $e_{ii} = \alpha \neq 0; e_{lk} = \delta_l^k$ otherwise (this is the unit matrix with the ith unit replaced by α);
3. $e_{ij} = 1; e_{lk} = \delta_l^k$ otherwise (this is the unit matrix with an additional unit in the (i, j)th place).

Each of the $[e_{ij}]$ matrices described above is non-singular since each has m linearly independent columns (Exercise 2.8.8), hence their product is non-singular. If we perform a sequence of row operations $[e_{ij}]^{(r)}$ on $[t_{ij}]$ we are, in effect, generating the equivalent matrix $[m_{ij}][t_{ij}]$, where

$$[m_{ij}] = \prod_r [e_{ij}]^{(r)}.$$

By a systematic application of row operations we can reduce a given matrix $[t_{ij}]$ of rank ρ to the so-called *echelon form* in which

(a) the first ρ rows are non-zero while the remaining $(m - \rho)$ are zero;
(b) the first non-zero element in each non-zero row appears in a column to the right of the first non-zero element of any preceding row.

An example of a matrix in echelon form when $m = 5, n = 8, \rho = 4$ is

$$\begin{bmatrix} * & * & * & * & * & * & * & * \\ 0 & 0 & * & * & * & * & * & * \\ 0 & 0 & 0 & 0 & * & * & * & * \\ 0 & 0 & 0 & 0 & 0 & 0 & 0 & * \\ 0 & 0 & 0 & 0 & 0 & 0 & 0 & 0 \end{bmatrix}$$

Row reduction is a convenient means of establishing the uniqueness and existence of a solution to a set of linear simultaneous equations and for subsequent numerical determination of the solution (Example 2.8.4 below).

What has been said of row reduction can equally be said of column reduction of a matrix leading to a corresponding column echelon form of the matrix: in effect, a column reduction of $[t_{ij}]$ is simply equivalent to a row reduction of the transpose $[t_{ij}]^T$.

By a combination of row and column operations it is clearly possible to reduce an $(m \times n)$ matrix of rank ρ to the canonical partitioned form

$$\left[\begin{array}{c|c} [\delta_i^j]_{(\rho \times \rho)} & [0]_{(\rho \times (n-\rho))} \\ \hline [0]_{(m-\rho) \times \rho} & [0]_{(m-\rho) \times (n-\rho)} \end{array} \right],$$

where the order of the submatrices is indicated.

We may say that there is but one linear transformation $T : \mathscr{V}_n \to \mathscr{V}_m$ of rank ρ, but it has infinitely many $(m \times n)$ matrix representations, the simplest form being the canonical form described above.

Example 2.8.4 Given the system of linear equations

$$[t_{ij}]\{\xi_j\} = \{b_i\}$$

we form the augmented matrix $[[t_{ij}]\,|\,\{b_i\}]$ and row reduce it. In this way we determine whether the rank of $[t_{ij}]$ is equal to the rank of $[[t_{ij}]\,|\,\{b_i\}]$ and if so we have a convenient way of obtaining the solution by back-substitution.

The following numerical example will make the process clear. Let us solve

$$\xi_1 + \xi_2 - \xi_3 + \xi_4 = 1,$$
$$3\xi_1 + \xi_2 - 2\xi_3 + 4\xi_4 = 2,$$
$$2\xi_1 - 3\xi_2 + 10\xi_4 = -1,$$
$$-\xi_1 + 2\xi_2 - \xi_3 + 3\xi_4 = 0.$$

The augmented matrix is

$$\left[\begin{array}{cccc|c} 1 & 1 & -1 & 1 & 1 \\ 3 & 1 & -2 & 4 & 2 \\ 2 & -3 & 0 & 10 & -1 \\ -1 & 2 & -1 & 3 & 0 \end{array} \right]$$

which we row reduce through the following steps:

1. $\begin{bmatrix} 1 & 1 & -1 & 1 & | & 1 \\ 0 & -2 & 1 & 1 & | & -1 \\ 0 & -5 & 2 & 8 & | & -3 \\ 0 & 3 & -2 & 4 & | & 1 \end{bmatrix}$

2. $\begin{bmatrix} 1 & 1 & -1 & 1 & | & 1 \\ 0 & -2 & 1 & 1 & | & -1 \\ 0 & 0 & -1 & 11 & | & -1 \\ 0 & 0 & -1 & 11 & | & -1 \end{bmatrix}$

3. $\begin{bmatrix} 1 & 1 & -1 & 1 & | & 1 \\ 0 & -2 & 1 & 1 & | & -1 \\ 0 & 0 & -1 & 11 & | & -1 \\ 0 & 0 & 0 & 0 & | & 0 \end{bmatrix}$

The ranks of $[t_{ij}]$ and the augmented matrix are equal so that a solution exists—furthermore, the nullity of T is $4-3=1$ and we expect a solution with one arbitrary multiplier. We find a particular solution from the equations

$$\xi_1 + \xi_2 - \xi_3 + \xi_4 = 1,$$
$$-2\xi_2 + \xi_3 + \xi_4 = -1,$$
$$-\xi_3 + 11\xi_4 = -1.$$

Since we are to have an arbitrary multiplier we are at liberty to adopt an arbitrary value for any one of the ξ_i—let us choose $\xi_4 = 0$. Then by back-substitution we find $\xi_3 = 1$, $\xi_2 = 1$, $\xi_1 = 1$, giving the particular solution $(1, 1, 1, 0)$.

The null space of T contains a single basis vector which will obviously be a solution of the equations

$$\xi_1 + \xi_2 - \xi_3 + \xi_4 = 0,$$
$$-2\xi_2 + \xi_3 + \xi_4 = 0,$$
$$-\xi_3 + 11\xi_4 = 0.$$

Since we arbitrarily set $\xi_4 = 0$ in the particular solution, we now take $\xi_4 = 1$, thus giving the solution vector $(4, 6, 11, 1)$. The complete solution is the coset

$$(\xi_1, \xi_2, \xi_3, \xi_4) = (1, 1, 1, 0) + \lambda(4, 6, 11, 1),$$

where λ is an arbitrary multiplier.

It is worth noting that the use of row reduction is superior to the alternative of column reduction since a row reduction is equivalent to a change of basis in $\mathscr{V}_m$ only, so that the (coordinates of) the solution vector are not changed.

A column reduction would have the disadvantage that the solution once found would have to be transformed back to the original basis in $\mathscr{V}_n$.

Similarity

Let us now consider linear transformations which map $\mathscr{V}_n$ into itself: matrix representations of such transformations will be square of order n. As before, in discussing equivalence we are interested in how matrix representations of $T:\mathscr{V}_n \to \mathscr{V}_n$ change when we make changes in basis. This looks superficially like change of basis for $T:\mathscr{V}_n \to \mathscr{V}_m$, but there is one most important difference whose effects are far reaching, namely that since T is a mapping of $\mathscr{V}_n$ into itself we change basis in only one space.

Before going on to consider this in detail it would be as well to recall that the linear transformations of $\mathscr{V}_n$ into itself have a richer structure than the linear transformations $\mathscr{V}_n \to \mathscr{V}_m$ (see Section 2.5). This richness in structure makes the search for a canonical form for square matrices a much more rewarding task than for rectangular matrices. It takes us into the theory of eigenvalues and eigenvectors of T, a matter which we shall take up presently.

Meanwhile, we can formally write down the relation between different matrix representations of T. Let us change the basis in $\mathscr{V}_n$ from $\{x^i\}$ to $\{\hat{x}^i\}$, $i = 1, \ldots, n$, and as before specify the change by the (non-singular) $(n \times n)$ matrix $[l_{ij}]$ so that, in terms of 'old' and 'new' coordinates,

$$\{\hat{\xi}_i\} = [l_{ij}]\{\xi_j\}.$$

By reasoning similar to that leading up to Definition 2.8.1 of equivalence, we find that

$$[\hat{t}_{ij}] = [l_{ij}][t_{ij}][l_{ij}]^{-1}$$

and, conversely,

$$[t_{ij}] = [l_{ij}]^{-1}[\hat{t}_{ij}][l_{ij}].$$

This brings us to:

Definition 2.8.2 *Square matrices of order n which are different representations of a single linear transformation $T:\mathscr{V}_n \to \mathscr{V}_n$ are said to be similar: if $[t_{ij}]$ is a square matrix of order n, and $[l_{ij}]$ a non-singular square matrix of order n then $[\hat{t}_{ij}]$ is similar to $[t_{ij}]$ if*

$$[\hat{t}_{ij}] = [l_{ij}][t_{ij}][l_{ij}]^{-1}.$$

The determinant of a square matrix

The discussion of the theory of eigenvalues and eigenvectors which we shall pursue in the next section is greatly facilitated if we can use the notion of the determinant of a square matrix. The reader will very likely already have encountered determinants, and this section is little more than a brief summary of the basic properties. For a further discussion including a proof of the existence and uniqueness of the determinant function the reader is referred to [14].

The determinant of a square $(n \times n)$ matrix $[t_{ij}]$ is a (real or complex) number which is uniquely determined from the elements of the matrix: we denote this number by $\det [t_{ij}]$ (an alternative notation is $|t_{ij}|$); that is to say, 'det' is a function whose domain is the set of all $(n \times n)$ matrices and whose range is $\mathscr{R}$ or $\mathscr{C}$.

Let $[t_{ij}; k|l]$ denote the matrix obtained from $[t_{ij}]$ by omitting the kth row and lth column: let $M_{kl} = \det [t_{ij}; k|l]$, then M_{kl} is called the *minor* of the element t_{kl} and the signed minor $T_{kl} = (-)^{k+l} M_{kl}$ is called the *cofactor* of t_{kl}. M_{kl} is called a first minor of $[t_{ij}]$: a second minor of $[t_{ij}]$ is obtained by omitting two rows and two columns and so on. The order of a minor refers to the number of rows (or columns) of the corresponding submatrix.

A determinant may be evaluated by a *Laplace expansion* on any row or column, viz.,

$$\det [t_{ij}] = \sum_{k=1}^{n} t_{kl} T_{kl} \quad \text{for any } l$$

$$= \sum_{l=1}^{n} t_{kl} T_{kl} \quad \text{for any } k.$$

The determinant function possesses the following properties (*see*, for example, [11, 14]):

1. If any column (row) consists entirely of zeros the determinant is zero.
2. If each element of a column (row) is multiplied by a scalar λ, the determinant is multiplied by λ.
3. If two columns (rows) are interchanged the determinant changes sign.
4. If a fixed column (row) is the sum of two column (row) vectors the determinant is the sum of the two corresponding determinants.

5. The value of a determinant is unaltered by adding to a column (row) a constant multiple of any other column (row).
6. A Laplace expansion of det $[t_{ij}]$ by alien cofactors vanishes, viz.,

$$\sum_{k=1}^{n} t_{kl} T_{kr} = 0, \quad l \neq r$$

or

$$\sum_{l=1}^{n} t_{kl} T_{rl} = 0, \quad k \neq r.$$

We may note that two important consequences flow directly from the above properties:

(a) Since rows and columns read interchangeably we see that det $[t_{ij}]^T = \det[t_{ij}]$.
(b) If the columns (or rows) of a square matrix are linearly dependent the determinant vanishes: conversely if the columns (or rows) are independent the determinant is non-zero. We saw earlier that if a square matrix of order n was of rank n it was non-singular (or invertible) and we now have an equivalent criterion, namely, a matrix $[t_{ij}]$ is non-singular if, and only if, $\det[t_{ij}] \neq 0$. Indeed, we can go further and assert that if the rank of $[t_{ij}]$ is $\rho \leq n$ then every minor of order $(\rho + 1)$ or greater vanishes while at least one minor of order ρ is non-zero (Exercise 2.8,14).

The determinant of a square matrix is an invariant in the sense that similar matrices (representing one linear transformation relative to different basis sets) have the same determinant. This deduction is based on the following result (Exercise 2.8,15(b)): if

$$[t_{ij}] = [s_{ij}][p_{ij}]$$

then

$$\det[t_{ij}] = \det[s_{ij}] \det[p_{ij}].$$

We see at once that if $\det[t_{ij}] \neq 0$, then $[t_{ij}]^{-1}$ exists and

$$\det[t_{ij}]^{-1} = \frac{1}{\det[t_{ij}]}.$$

Now if matrices $[t_{ij}]$ and $[\hat{t}_{ij}]$ are similar,

$$[\hat{t}_{ij}] = [l_{ij}][t_{ij}][l_{ij}]^{-1}$$

and
$$\det[\hat{t}_{ij}] = \det[l_{ij}]\det[t_{ij}]\det[l_{ij}]^{-1} = \det[t_{ij}].$$

We may therefore talk meaningfully of the 'determinant' of a non-singular linear transformation as being the determinant of any one of its matrix representations.

Eigenvalues and eigenvectors of square matrices

The idea of an eigensolution of a linear transformation was introduced in Section 2.5. We shall now develop the theory of eigenvalues and eigenvectors for linear transformations on finite-dimensional spaces, that is, for square matrices. The development will lead up to a canonical form for similarity.

Let $[t_{ij}]$ be a matrix representation of the linear transformation $T: \mathscr{V}_n \to \mathscr{V}_n$. The eigensolutions $(\lambda; x)$ of T are the solutions of

$$Tx = \lambda x,$$

where λ is an element of the scalar field. We shall assume throughout this section that the scalar field is the field of complex numbers. A direct analogy of this equation in matrix terms would be

$$[t_{ij}]\{\xi_j\} = \tilde{\lambda}\{\xi_i\}.$$

We may write this last equation in the form

$$([t_{ij}] - \tilde{\lambda}[\delta_i^j])\{\xi_j\} = \{0\}$$

and if we are to have non-trivial solutions we know that the rank of the left-hand side matrix must be less than n; that is, the matrix must be singular. We conclude that a necessary and sufficient condition for a non-trivial solution to exist is

$$\det([t_{ij}] - \tilde{\lambda}[\delta_i^j]) = 0,$$

which is called the *characteristic* or *determinantal equation*; written in expanded form, this is

$$\det\begin{bmatrix} t_{11}-\tilde{\lambda} & t_{12} & \cdots\cdots\cdots & t_{1n} \\ t_{21} & t_{22}-\tilde{\lambda} & \cdots\cdots\cdots & t_{2n} \\ \vdots & \vdots & & \vdots \\ t_{n-1,1} & t_{n-1,2} & \cdots\cdots\cdots & t_{n-1,n} \\ t_{n1} & t_{n2} & \cdots\cdots\cdots & t_{nn}-\tilde{\lambda} \end{bmatrix} = 0$$

An expansion of the determinant by row or column reveals that the determinant can be written as a polynomial in $\tilde{\lambda}$ of degree precisely equal to n, say $p(\tilde{\lambda})$, so that we have

$$\det([t_{ij}] - \tilde{\lambda}[\delta_i^j]) = p(\tilde{\lambda}) = 0.$$

The polynomial $p(\tilde{\lambda})$ is called the *characteristic polynomial* of $[t_{ij}]$. The fundamental theorem of algebra [12] tells us that a polynomial of finite degree has at least one root (or zero) in the field of complex numbers (the result is not true if we restrict ourselves to the reals), hence we conclude that the matrix $[t_{ij}]$ possesses at least one eigenvalue. The characteristic polynomial being of degree n, it cannot have more than n roots, hence $[t_{ij}]$ cannot possess more than n eigenvalues.

Let $[\hat{t}_{ij}]$ be another matrix representation of T relative to another basis in $\mathscr{V}_n$. We know that $[\hat{t}_{ij}]$ is similar to $[t_{ij}]$, that is,

$$[\hat{t}_{ij}] = [l_{ij}][t_{ij}][l_{ij}]^{-1}.$$

The characteristic polynomial of $[\hat{t}_{ij}]$ is

$$\det([\hat{t}_{ij}] - \tilde{\mu}[\delta_i^j]) = q(\tilde{\mu}) = 0.$$

But we may write

$$[\hat{t}_{ij}] - \tilde{\mu}[\delta_i^j] = [l_{ij}]([t_{ij}] - \mu[\delta_i^j])[l_{ij}]^{-1},$$

and since we have already shown that the determinants of similar matrices are equal we conclude that $q \equiv p$ and the eigenvalues $\tilde{\mu}$, $\tilde{\lambda}$ of $[\hat{t}_{ij}]$ and $[t_{ij}]$ coincide.

Indeed, we may identify the set of eigenvalues $\{\lambda_k\}$ of the linear transformation T as being coincident with the set of eigenvalues of any of its (similar) matrix representations. The eigenvalues of a linear transformation are invariants and characteristic of the transformation itself. The eigenvectors are of course dependent on the basis used to represent T.

As we have already seen in Section 2.5, if λ_k is an eigenvalue of T then the eigenspace $\mathscr{V}^{(k)}$ is the null space of the transformation $T - \lambda_k I$.

The dimension of the null space is the *geometric multiplicity* of λ_k and is, of course, equal to the nullity of $(T - \lambda_k I)$. Relative to a given basis the eigenvectors $x^{(k)} = \{\xi_i\}^{(k)}$ are the solution n-tuples of the matrix equation

$$([t_{ij}] - \lambda_k[\delta_i^j])\{\xi_j\}^{(k)} = 0.$$

Any eigenvalue of T also possesses an *algebraic multiplicity* which is related to a minimal factorization of the characteristic polynomial. Let $\lambda_1, \lambda_2, \ldots, \lambda_m$ be the roots of $p(\lambda) = 0$; then we may factorize $p(\lambda)$ into

$$p(\lambda) = (\lambda - \lambda_1)^{d_1}(\lambda - \lambda_2)^{d_2} \ldots (\lambda - \lambda_m)^{d_r},$$

where $\sum d_i = n$ [14]. The index d_k of $(\lambda - \lambda_k)$ is the algebraic multiplicity of λ_k. The geometric and algebraic multiplicities of λ_k are not simply connected, although it can be shown that the geometric multiplicity is always less than or equal to the algebraic multiplicity [13]. We shall be most concerned with the geometric multiplicity in the discussion which follows.

It should perhaps be emphasized at this point that the practical computation of eigenvalues and eigenvectors is never based on the direct solution of the characteristic polynomial; many efficient computational schemes are available which are described in innumerable texts on numerical methods (e.g., [18–21]).

It follows from Theorem 2.5.2 that we may form a subspace $\mathcal{U}$ of $\mathcal{V}_n$ as the direct sum of the eigenspaces of T; that is,

$$\mathcal{U} = \sum_{k=1}^{r} {}_\oplus \mathcal{V}^{(k)},$$

where $r \leq n$ is the number of distinct eigenvalues of T. Suppose this direct sum actually spans $\mathcal{V}_n$: then we have a basis for $\mathcal{V}_n$ consisting entirely of eigenvectors of T and we may write, for any $x \in \mathcal{V}_n$,

$$x = x^{(1)} + x^{(2)} + \ldots + x^{(r)},$$

where $x^{(k)} \in \mathcal{V}^{(k)}, k = 1, \ldots, r$. But then we see that

$$Tx = \lambda_1 x^{(1)} + \lambda_2 x^{(3)} + \ldots + \lambda_r x^{(r)},$$

so that in terms of this basis the structure of the transformation T is particularly clear. If we consider $x^{(s)}$ to be the projection (Definition 2.5.3) of x onto $\mathcal{V}^{(s)}$ along

$$\sum_{k \neq s} {}_\oplus \mathcal{V}^{(k)}$$

then we see that the effect of T operating on x is simply to stretch (or shorten) its projection in each eigenspace. In a sense then T is itself a 'direct sum' of simple dilatation operators whose effect is entirely confined to the corresponding eigenspaces. Because the operation of T on elements of the eigenspace $\mathcal{V}^{(k)}$ is to produce image vectors in $\mathcal{V}^{(k)}$, the subspace $\mathcal{V}^{(k)}$ is said to be *invariant* with respect to T. We can give an immediate interpretation of this as a partitioned matrix representation of T relative to a basis of eigenvectors. It should be clear that T will consist of a diagonal partitioned matrix of the form

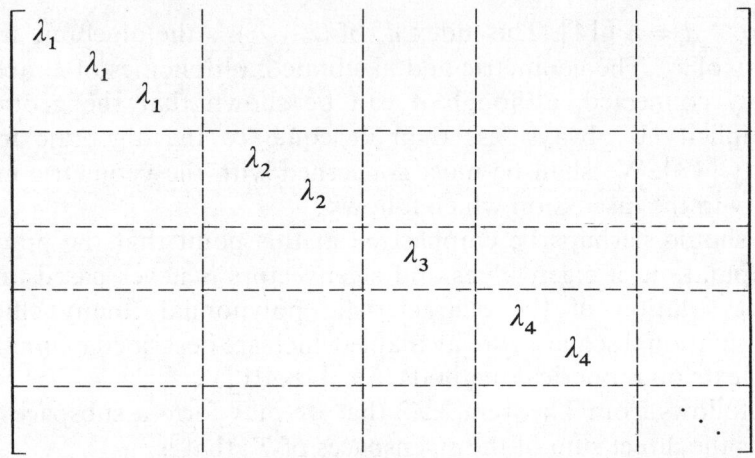

since each projection $x^{(k)}$ of x is itself a linear combination of eigenvectors belonging to λ_k.

We have the following definition:

Definition 2.8.3 *A linear transformation $T: \mathscr{V}_n \to \mathscr{V}_n$ is diagonizable if the eigenvectors of T span $\mathscr{V}_n$. Any matrix representation of T is then similar to a diagonal matrix having the eigenvalues of T as its entries, each eigenvalue appearing as often as its geometric multiplicity. This is a canonical form with respect to similarity for a diagonalizable transformation.*

The last part of the definition may not seem entirely obvious since we have pursued the discussion without explicit reference to change of basis in $\mathscr{V}_n$. Given a representation $[t_{ij}]$ for T, what we are maintaining is that

$$[\lambda_i \delta_i^j] = [\xi_{ij}]^{-1}[t_{ij}][\xi_{ij}],$$

where the columns of $[\xi_{ij}]$ are the eigenvectors $\{\xi_{ij}\}^{(k)}$ of $[t_{ij}]$ (by hypothesis $[\xi_{ij}]$ is non-singular).

Rearranging this equation, we have

$$[t_{ij}][\xi_{ij}] = [\xi_{ij}][\lambda_i \delta_i^j],$$

which is nothing more than a portmanteau expression embodying

$[t_{ij}]\{\xi_j\}^{(k)} = \lambda_k\{\xi_j\}^{(k)}$ for all eigensolutions, as the reader can easily verify.

We have thus partially answered the question initially posed of finding, via similarity, a canonical form for a linear transformation T. It is essential for similarity to a diagonal matrix that the eigenvectors of T span $\mathscr{V}_n$. One important special case when this is certainly true is when all the eigenvalues of T are distinct and the eigenspaces are all one-dimensional. More generally it is easy to see that if there are to be n independent eigenvector's of T then, for each eigenvalue, the geometric and algebraic multiplicities must be equal.

When the eigenvectors of T do not span $\mathscr{V}_n$ the canonical form for $[t_{ij}]$ and its derivation are much more complicated. Here we will only indicate the result and refer the reader elsewhere for the details [11, 13, 14, 31]. Basically the problem resides in adding to the (numerically inadequate) eigenvectors of T a sufficient number of linearly independent vectors so as to form a basis for $\mathscr{V}_n$.

An element of $\mathscr{N}((T - \lambda_k I)^n)$, the null space of $(T - \lambda_k I)^n$, is called a *generalized eigenvector* of exponent n belonging to λ_k. We know from Exercise 2.5,14(b) that the null spaces $\mathscr{N}((T - \lambda_k I)^n)$ form an increasing sequence; if p is the index of $(T - \lambda_k I)$ then we call the space $\mathscr{V}_g^{(k)} = \mathscr{N}((T - \lambda_k I)^p)$ the *generalized eigenspace* of T belonging to λ_k. By generating cyclic subspaces (cf., Exercise 2.5,4(b)) whose direct sum is $\mathscr{V}_g^{(k)}$ we can represent the restriction (Exercise 2.5,13) of $(T - \lambda_k I)$ to $\mathscr{V}_g^{(k)}$ in a canonical form consisting of a matrix having zero everywhere except for (some) units on the subdiagonal (Exercise 2.5,4(d)). Hence a canonical form for T restricted to $\mathscr{V}_g^{(k)}$ consists of the sum of a diagonal matrix all of whose entries are λ_k plus a matrix having (some) units on its subdiagonal (Exercise 2.8,19).

It can be shown that $\mathscr{V}$ can be decomposed as a direct sum of the generalized eigenspaces $\mathscr{V}_g^{(k)}$ [13, 14] with the result that a canonical form for $[t_{ij}]$ is the partitioned matrix

$$[t_{ij}] = \begin{bmatrix} \lambda_1[\delta_i^j] + [\alpha_{ij}]^{(1)} & & \\ & \lambda_2[\delta_i^j] + [\alpha_{ij}]^{(2)} & \\ & & \ddots \end{bmatrix},$$

where the α_{ij} are zero except for (some) units in each subdiagonal. The number and arrangement of the units is unique and the matrix is said to be in *Jordan canonical form*: the determination of this pattern is too complicated to describe here. We see that the presence of roots whose algebraic multiplicity is greater than their geometric multiplicity adds to the 'diagonal' representation of $[t_{ij}]$ a matrix of the type described above.

Example 2.8.5

(a) We shall find the eigensolutions of the matrix

$$[t_{ij}] = \begin{bmatrix} 2 & 1 & 1 \\ -2 & 1 & 3 \\ 3 & 1 & -1 \end{bmatrix}.$$

The characteristic polynomial det $([t_{ij}] - \lambda[\delta_i^j])$ is readily verified to be

$$(\lambda + 2)(\lambda - 1)(\lambda - 3),$$

and this has the distinct roots $\lambda_1 = -2, \lambda_2 = 1, \lambda_3 = 3$. To find the eigenvectors corresponding to λ_1, say, we solve the set of equations

$$([t_{ij}] - \lambda_1[\delta_i^j])\{\xi_j\}^{(1)} = 0,$$

which in this case is

$$\begin{bmatrix} 4 & 1 & 1 \\ -2 & 3 & 3 \\ 3 & 3 & 3 \end{bmatrix} \begin{bmatrix} \xi_1^{(1)} \\ \xi_2^{(1)} \\ \xi_3^{(1)} \end{bmatrix} = \begin{bmatrix} 0 \\ 0 \\ 0 \end{bmatrix}.$$

A row reduction leads to

$$\begin{bmatrix} 4 & 1 & 1 \\ 0 & 1 & 1 \\ 0 & 0 & 0 \end{bmatrix} \begin{bmatrix} \xi_1^{(1)} \\ \xi_2^{(1)} \\ \xi_3^{(1)} \end{bmatrix} = \begin{bmatrix} 0 \\ 0 \\ 0 \end{bmatrix},$$

verifying that the root λ_1 possesses an eigenspace of dimension 1 and giving as a basis vector $(0, 1, -1)$. In like manner, we find the eigenvectors $(3, -5, 2)$, $(5, 1, 4)$ respectively. The matrix $[\xi_{ij}]$ whose columns are these eigenvectors is

$$\begin{bmatrix} 0 & 3 & 5 \\ 1 & -5 & 1 \\ -1 & 2 & 4 \end{bmatrix},$$

and its inverse $[\xi_{ij}]^{-1}$ is readily verified to be (cf., Exercise 2.8,13).

$$\frac{1}{30} \begin{bmatrix} 22 & 2 & -28 \\ 5 & -5 & -5 \\ 3 & 3 & 3 \end{bmatrix}.$$

We may verify, by matrix multiplication, that $[\xi_{ij}]^{-1}[t_{ij}][\xi_{ij}]$ is a diagonal matrix with $\lambda_1, \lambda_2, \lambda_3$ as its entries.

(b) Now consider the matrix
$$[t_{ij}] = \begin{bmatrix} 1 & 0 & \sqrt{2} \\ 0 & 2 & 0 \\ \sqrt{2} & 0 & 0 \end{bmatrix}$$
which has the characteristic polynomial
$$(\lambda - 2)^2(\lambda + 1)$$
and thus has an eigenvalue ($\lambda_2 = 2$) with algebraic multiplicity 2. First, as in (a), we may find without difficulty the eigenvector $(1, 0, -\sqrt{2})$ belonging to $\lambda_1 = -1$. The matrix $([t_{ij}] - \lambda_2[\delta_i^j])$ is
$$\begin{bmatrix} -1 & 0 & \sqrt{2} \\ 0 & 0 & 0 \\ \sqrt{2} & 0 & -2 \end{bmatrix},$$
which may be row reduced to
$$\begin{bmatrix} 1 & 0 & -\sqrt{2} \\ 0 & 0 & 0 \\ 0 & 0 & 0 \end{bmatrix}.$$
The rank of this matrix is 1, so that the nullity is 2; hence the geometric multiplicity of λ_2 is 2, equal to its algebraic multiplicity and $[t_{ij}]$ is diagonalizable. The two linearly independent eigenvectors belonging to λ_2 are easily seen to be $(0, 1, 0)$ and $(\sqrt{2}, 0, 1)$. Again, if we form the matrix $[\xi_{ij}]$ of eigenvectors we find that $[\xi_{ij}]^{-1}[t_{ij}][\xi_{ij}]$ is the diagonal matrix of eigenvalues.

Had we found that the rank of the matrix was 2, rather than 1, then we would have immediately concluded that $[t_{ij}]$ was not diagonalizable.

(c) Let
$$[t_{ij}] = \begin{bmatrix} 0 & 1 \\ -\omega^2 & -2\gamma\omega \end{bmatrix}:$$
the characteristic polynomial is $\lambda^2 + 2\gamma\omega\lambda + \omega^2$ with roots λ_1, $\lambda_2 = \omega(-\gamma \pm \sqrt{\gamma^2 - 1})$. These roots are
1. real and distinct if $\gamma^2 > 1$;
2. real and equal if $\gamma^2 = 1$;
3. complex conjugate if $\gamma^2 < 1$.

The eigenvectors are $(1, \lambda_1)$ and $(1, \lambda_2)$ and these are
(a) real and span $\mathscr{V}_2$ if $\gamma^2 > 1$;
(b) complex conjugate and span $\mathscr{V}_2$ if $\gamma^2 < 1$.

In the case when $\gamma^2 = 1$ the rank of $([t_{ij}] - \lambda_1[\delta_i^j])$ is one and $[t_{ij}]$ is not diagonalizable: the Jordan canonical form is

$$\begin{bmatrix} -\omega & 0 \\ 1 & -\omega \end{bmatrix}.$$

The basis for $\mathscr{V}_2$ in this case is $\{(1, 0), (\omega, -\omega^2)\}$.

The first vector is a solution of $([t_{ij}] + \omega[\delta_i^j])^2 \{\xi_i\}_g = 0$, while the second is simply

$$([t_{ij}] + \omega[\delta_i^j])\{\xi_i\}_g = ([t_{ij}] + \omega[\delta_i^j])\begin{bmatrix} 1 \\ 0 \end{bmatrix}.$$

Eigenvectors and eigenvalues of the dual transformation

If $T' : \mathscr{V}_n^* \to \mathscr{V}_n^*$ is the dual of $T : \mathscr{V}_n \to \mathscr{V}_n$, consider the eigensolutions of $T'y = \lambda' y$.[†] If $[t_{ij}]$ is the matrix representation of T relative to a chosen basis for $\mathscr{V}_n$ then we know that, relative to the dual basis, the matrix representation of T' is the transpose $[t_{ij}]^T$.

The characteristic equation of $[t_{ij}]^T$ is, by definition,

$$\det([t_{ij}]^T - \lambda'[\delta_i^j]) = 0$$

or

$$\det([t_{ij}] - \lambda'[\delta_i^j])^T = 0.$$

But the determinant of the transpose of a matrix is equal to the determinant of the matrix itself, so that we may conclude that the characteristic polynomial for $[t_{ij}]^T$ is (in an indeterminate form containing λ') identical to the characteristic polynomial for $[t_{ij}]$. Hence the eigenvalue sets of $[t_{ij}]$ and $[t_{ij}]^T$ and hence of T and T' are the same.

Let $y^{(j)}$ denote an eigenvector of $[t_{ij}]^T$ and consider the linear functional $y^{(j)}$ operating on the vector $Tx^{(i)}$; we have

$$[y^{(j)}, Tx^{(j)}] = [y^{(j)}, \lambda_i x^{(i)}] = \lambda_i[y^{(j)}, x^{(i)}].$$

But, by the definition of the dual, the left-hand side of the above equation is equal to

$$[T'y^{(j)}, x^{(i)}] = \lambda_j[y^{(j)}, x^{(i)}].$$

We conclude that

$$\lambda_i[y^{(j)}, x^{(i)}] = \lambda_j[y^{(j)}, x^{(i)}]$$

[†] Recall that $\mathscr{V}_n^*$ is isomorphic to $\mathscr{V}_n$ so that T and T' are both in effect mappings of $\mathscr{V}_n$ into itself.

or

$$(\lambda_i - \lambda_j)[y^{(j)}, x^{(i)}] = 0.$$

Hence, provided $i \neq j$, the jth eigenspace of T' is an annihilator of the ith eigenspace of T and vice versa. We know that $\mathscr{V}_n$ can be decomposed as a direct sum of eigenspaces (or generalized eigenspaces) of T. The corollary to Theorem 2.6.2 shows that any direct sum decomposition of $\mathscr{V}_n$ induces a direct sum decomposition of $\mathscr{V}_n^*$ in terms of the corresponding annihilators.

When T is diagonalizable the eigenvectors are a basis for $\mathscr{V}_n$ and it is clear that the eigenvectors of T' are a basis for $\mathscr{V}_n^*$: furthermore, this basis for $\mathscr{V}_n^*$ is reciprocal to that of $\mathscr{V}_n$ if we choose the arbitrary multiplier(s) of $x^{(i)}$ and $y^{(i)}$ so that

$$[y^{(i)}, x^{(i)}] = 1, \quad i = 1, \ldots, n.$$

These results are practically very useful since they give us a way of generating a basis reciprocal to the eigenvector basis of $\mathscr{V}_n$ without the use of matrix inversion. This is particularly important if we are interested only in a subset of the eigensolutions of T and we wish to diagonalize T with respect to the subspace spanned by the corresponding eigenvectors.

These remarks are illustrated in Example 2.9.4.

Example 2.8.6

(a) In Example 2.8.5 we found the eigensolutions of the matrix

$$\begin{bmatrix} 2 & 1 & 1 \\ -2 & 1 & 3 \\ 3 & 1 & -1 \end{bmatrix}$$

to be $(-2; (0, 1, -1))$, $(1; (3, -5, 2))$ and $(3; (5, 1, 4))$. Let us find the eigenvectors $\{\eta_i\}^{(k)}$ of the (dual) transpose

$$\begin{bmatrix} 2 & -2 & 3 \\ 1 & 1 & 1 \\ 1 & 3 & -1 \end{bmatrix}.$$

Notice that these 'eigenvectors' are the coordinate sets of the eigenvectors of T' relative to the basis which is dual to that in which T is expressed. For the eigenvalue -2 we have

$$\begin{bmatrix} 4 & -2 & 3 \\ 1 & 3 & 1 \\ 1 & 3 & 1 \end{bmatrix} \begin{bmatrix} \eta_1^{(1)} \\ \eta_2^{(1)} \\ \eta_3^{(1)} \end{bmatrix} = \begin{bmatrix} 0 \\ 0 \\ 0 \end{bmatrix},$$

which row reduces to

$$\begin{bmatrix} 4 & -2 & 3 \\ 0 & -14 & -1 \\ 0 & 0 & 0 \end{bmatrix} \begin{bmatrix} \eta_1^{(1)} \\ \eta_2^{(1)} \\ \eta_3^{(1)} \end{bmatrix} = \begin{bmatrix} 0 \\ 0 \\ 0 \end{bmatrix},$$

giving the eigenvector $(11, 1, -14)$. It is easily verified that this vector annihilates any vector in the subspace spanned by $(3, -5, 2)$ and $(5, 1, 4)$. In the same way we find the eigenvectors $(1, -1, -1)$ and $(1, 1, 1)$. We can make these eigenvectors a dual basis by insisting that

$$[\{\eta_i\}^{(k)}, \{\xi_{ij}\}^{(k)}] = 1, \quad k = 1, 2, 3,$$

and this gives the set $\{\eta_i\}^{(1)} = \frac{1}{15}(11, 1, -14)$, $\{\eta_i\}^{(2)} = \frac{1}{6}(1, -1, -1)$, $\{\eta_i\}^{(3)} = \frac{1}{10}(1, 1, 1)$.

It will be seen that if we construct a matrix $[\eta_{ij}]$ whose columns are these eigenvectors, then it is identical with the transpose of $[\xi_{ij}]^{-1}$. This result follows, of course, from the fact that row-by-column matrix multiplication means that the (i, j)th entry in the product matrix can, if desired, be interpreted as the value of a linear functional defined by the ith row of the first matrix on a vector defined by the jth column of the second.

(b) In part (c) of Example 2.8.5 we found the eigensolutions of the matrix

$$\begin{bmatrix} 0 & 1 \\ -\omega^2 & -2\gamma\omega \end{bmatrix}$$

to be $(\lambda_1; (1, \lambda_1)), (\lambda_2; (1, \lambda_2))$ when $\gamma \neq 1$.

The eigenvectors of $[t_{ij}]^{\mathrm{T}}$ are $(-\omega^2, \lambda_1)$ and $(-\omega^2, \lambda_2)$; we convert these into a dual basis by dividing by the scalar $[\{\eta_i\}^{(k)}, \{\xi_{ij}\}^{(k)}]$, in each case giving the vectors

$$\frac{1}{\lambda_1^2 - \omega^2}(-\omega^2, \lambda_1), \quad \frac{1}{\lambda_2^2 - \omega^2}(-\omega^2, \lambda_2)$$

and, using the fact that $\omega^2 = \lambda_1 \lambda_2$, these may be rewritten

$$\frac{1}{\lambda_2 - \lambda_1}(\lambda_2, -1), \quad \frac{1}{\lambda_2 - \lambda_1}(-\lambda_1, 1).$$

We end this section with some remarks aimed at emphasizing the importance of the scalar field in deriving a canonical form for similar matrices. If we had restricted ourselves to real vector spaces we could not have used the fact that every characteristic polynomial of degree n, and hence every square matrix of order n, has n roots (when each root is counted with its algebraic multiplicity). Restriction to the real field does yield a canonical form for square matrices [13], but it owes nothing to eigensolutions and is not particularly revealing of the character of the associated linear transformation.

If a linear transformation $T:\mathscr{V}_n \to \mathscr{V}_n$ has, relative to some basis in $\mathscr{V}_n$, a matrix representation having only real entries, then its characteristic polynomial has real coefficients and consequently the eigenvalues are either real or occur in complex conjugate pairs. If $(\lambda_k; \{\xi_i\}^{(k)})$ is an eigensolution of $[t_{ij}]$ then so is $(\bar{\lambda}_k; \{\bar{\xi}_i\}^{(k)})$ for, upon taking the complex conjugate of

$$\sum_{j=1}^n t_{ij}\xi_j^{(k)} = \lambda_k \xi_i^{(k)},$$

we obtain

$$\sum_{j=1}^n t_{ij}\bar{\xi}_j^{(k)} = \bar{\lambda}_k \bar{\xi}_i^{(k)},$$

since the t_{ij} are real.

In many applications a physical system is described in the first instance in terms of real-valued variables and real parameters, so that we very often meet the situation so described. Nevertheless, in order to understand the characteristics of the system as revealed by its eigenvalues and eigenvectors we must clearly treat the problem within the overall context of the complex number field.

Exercises 2.8

1. (a) Verify that the row-by-column matrix multiplication

    $$[t_{ij}]\{\xi_j\} = \begin{bmatrix} 1 & 2 & 4 & 5 \\ 3 & 1 & 0 & 2 \end{bmatrix}\begin{bmatrix} 1 \\ 4 \\ 8 \\ 9 \end{bmatrix}$$

    yields a vector which is a linear combination of independent columns of the matrix $[t_{ij}]$.
   (b) Verify numerically that

    $$\lfloor \eta_i \rfloor [t_{ij}]\{\xi_j\} = \lfloor \xi_i \rfloor [t_{ij}]^T\{\eta_j\},$$

    where $[t_{ij}], \{\xi_j\}$ are given in part (a) and $(\eta_1, \eta_2) = (2, 3)$.
   (c) Let $[s_{ij}]$ be an $(m \times m)$ non-singular matrix and $[t_{ij}]$ an $(m \times n)$ matrix. Use the fact that every vector in the range space of $[t_{ij}]$ is a linear combination of the linearly independent columns of

$[t_{ij}]$ to show that
$$\rho([s_{ij}][t_{ij}]) = \rho([t_{ij}]).$$

(d) An $(n \times n)$ matrix $[t_{ij}]$ which satisfies the properties
   (i) $0 \leqslant t_{ij} \leqslant 1$, all i, j,
   (ii) $\sum_{i=1}^{n} t_{ij} = 1$, all j,

is called a *Markov matrix*. Show that the product of two Markov matrices is a Markov matrix.

2. (a) If
$$[t_{ij}] = \begin{bmatrix} [t_{ij}]_{11} & [t_{ij}]_{12} & [t_{ij}]_{13} \\ [t_{ij}]_{21} & [t_{ij}]_{22} & [t_{ij}]_{23} \end{bmatrix}$$

is a partitioned matrix of order $(3 + 5) \times (2 + 4 + 1)$, write down the orders of the submatrices $[t_{ij}]_{kl}$.

(b) If
$$[t_{ij}] = \begin{bmatrix} [t_{ij}]_{11} & [t_{ij}]_{12} \\ [t_{ij}]_{21} & [t_{ij}]_{22} \end{bmatrix}, \quad [s_{ij}] = \begin{bmatrix} [s_{ij}]_{11} & [s_{ij}]_{12} \\ [s_{ij}]_{21} & [s_{ij}]_{22} \end{bmatrix},$$

where $[t_{ij}]$ is $(m \times n)$ and $[s_{ij}]$ is $(n \times p)$, write down an expression for $[t_{ij}][s_{ij}]$ in terms of the submatrices $[t_{ij}]_{kl}$, $[s_{ij}]_{kl}$. Suggest permissible choices for the orders of the submatrices. Evaluate

$$\begin{bmatrix} 0 & 0 & 1 \\ 0 & -1 & 2 \\ \hline 1 & -1 & 2 \end{bmatrix} \begin{bmatrix} 1 & 0 \\ -1 & 0 \\ \hline 0 & 2 \end{bmatrix}$$

by block multiplication and verify your result by using row-by-column multiplication.

(c) Show that the $(m \times p)$ matrix $[t_{ij}]_{11}, p < n$, is the restriction (Exercise 2.5,13) of the $(m \times n)$ matrix $[t_{ij}] = [[t_{ij}]_{11} \mid [t_{ij}]_{12}]$ to the subspace of $\mathscr{V}_n$ spanned by the first p basis vectors of $\mathscr{V}_n$. Interpret results (ii) and (iii) of Exercise 2.5,13 for this case.

(d) Let
$$[t_{ij}] = \begin{bmatrix} [a_{ij}] & \{b_i\} \\ \lfloor c_j \rfloor & \alpha \end{bmatrix}, \quad [t_{ij}]^{-1} = \begin{bmatrix} [e_{ij}] & \{f_i\} \\ \lfloor g_j \rfloor & \beta \end{bmatrix},$$

where $[t_{ij}]$ is non-singular $(n \times n)$, $[a_{ij}]$, $[e_{ij}]$ are $[(n-1) \times (n-1)]$, $\{b_i\}$, $\{f_i\}$ are $[(n-1) \times 1]$, $\lfloor c_j \rfloor$, $\lfloor g_j \rfloor$ are $[1 \times (n-1)]$ and α, β are scalars.

Show that

$$[a_{ij}][e_{ij}] + \{b_i\}\lfloor g_j\rfloor = [\delta_i^j],$$
$$[a_{ij}]\{f_j\} + \beta\{b_i\} = \{0\},$$
$$\lfloor c_i\rfloor[e_{ij}] + \alpha\lfloor g_j\rfloor = \lfloor 0\rfloor,$$
$$\lfloor c_i\rfloor\{f_i\} + \alpha\beta = 1,$$

and hence that

$$\beta = \frac{1}{(\alpha - \lfloor c_i\rfloor[a_{ij}]^{-1}\{b_j\})},$$
$$\lfloor g_j\rfloor = -\beta\lfloor c_i\rfloor[a_{ij}]^{-1},$$
$$\{f_i\} = -\beta[a_{ij}]^{-1}\{b_j\},$$
$$[e_{ij}] = [a_{ij}]^{-1}([\delta_i^j] - \{b_i\}\lfloor g_j\rfloor).$$

Thus, knowing $[a_{ij}]^{-1}$ we can find $[t_{ij}]^{-1}$: this is the basis of the method 'inversion by bordering'. Beginning with a (principal) (2×2) matrix we border it to (3×3) and find the inverse, then border to (4×4) and so on until the full matrix is inverted. This is not a commonly used computing method but is useful for some types of matrix.

Find the inverse of

$$\begin{bmatrix} 1 & 2 \\ 4 & 6 \end{bmatrix}$$

and thence, using the above method, find the inverse of

$$\begin{bmatrix} 1 & 2 & 3 \\ 4 & 6 & 7 \\ 5 & 8 & 9 \end{bmatrix}.$$

3. A point in $\mathscr{R}_3$ has coordinates $(3, 2, 1)$ relative to the standard basis: find its coordinates relative to the basis $\{(1, 4, 5), (2, 6, 8), (3, 7, 9)\}$. (Hint: use result of (d) in the previous exercise.)

4. A linear transformation $T: \mathscr{R}_2 \to \mathscr{R}_2$ maps the points $(1, 1), (0, 1)$ to the points $(2, 0), (1, -1)$ respectively.
   (i) Choose e^1, e^2 as a basis and find Te^1, Te^2.
   (ii) Find the matrix representation of T relative to $\{e^1, e^2\}$.
   (iii) What is the image of the vector (ξ_1, ξ_2)?
   (iv) Choose basis vectors $(1, 1), (0, 1)$: what is now the matrix representation of T?
   (v) Show that T is non-singular and find from first principles a matrix representation of T^{-1} relative to $\{e^1, e^2\}$.

5. The vector $x \in \mathscr{V}_n$ has covariant coordinates $(\xi_1, \xi_2, \ldots, \xi_n)$ while

$y \in \mathscr{V}_n$ has contravariant coordinates $(\eta^1, \eta^2, \ldots, \eta^n)$. Show that if a change of basis is made so that x and y now have coordinates $(\hat{\xi}_1, \hat{\xi}_2, \ldots, \hat{\xi}_n)$ and $(\hat{\eta}^1, \hat{\eta}^2, \ldots, \hat{\eta}^n)$, respectively, then

$$\lfloor \hat{\eta}^i \rfloor \{\hat{\xi}_i\} = \lfloor \eta^i \rfloor \{\xi_i\}.$$

6. (a) Define the relation $T \sim S$ to mean that the linear transformations T and S have the same rank (Definition 2.5.2): verify that $\sim$ is an equivalence relation (p. 8).
   (b) If $T \sim S$ is (i) $T^2 \sim S^2$, (ii) $TS \sim ST$?

7. The matrix of $T: \mathscr{R}_3 \to \mathscr{R}_2$ is given by

$$\begin{bmatrix} 1 & 2 & 4 \\ 3 & 1 & 0 \end{bmatrix}$$

relative to standard basis sets in $\mathscr{R}_3, \mathscr{R}_2$: find the matrix representation of T relative to the basis sets $\{(1, 1, 0); (0, 1, 1); (1, 0, 1)\}$ for $\mathscr{R}_3$ and $\{(1, 1); (0, 1)\}$ for $\mathscr{R}_2$ (cf., Example 2.8.1).

8. (a) Verify that each of the elementary row operations (p. 75) can be represented as pre-multiplication of $[t_{ij}]$ by the matrices $[e_{ij}]$.
   (b) Verify that each of the $(m \times m)$ matrices $[e_{ij}]$ has m linearly independent columns.
   (c) Calculate the determinant of each type of $[e_{ij}]$ matrix.

9. (a) By row reduction, find the rank of the following matrices

   (i) $\begin{bmatrix} 1 & 2 & 3 & 2 \\ 2 & 3 & 5 & 1 \\ 1 & 3 & 4 & 5 \end{bmatrix}$

   (ii) $\begin{bmatrix} 1 & 2 & 1 & 2 \\ 1 & 3 & 2 & 2 \\ 2 & 4 & 3 & 4 \\ 3 & 7 & 4 & 6 \end{bmatrix}$

   (iii) $\begin{bmatrix} 1 & 2 & -2 & 3 \\ 2 & 5 & -4 & 6 \\ -1 & -3 & 2 & -2 \\ 2 & 4 & -1 & 6 \end{bmatrix}$

   (b) Check your answers to (a) by column reduction.
   (c) Check that each matrix in (a) can, by row and column reduction, be reduced to the canonical form for equivalence (p. 76).

10. Check for consistency and solve, where possible, the following sets of

equations:

(i) $\xi_1 + \xi_2 + \xi_3 + \xi_4 = 0$,
$\xi_1 + \xi_2 + \xi_3 - \xi_4 = 4$,
$\xi_1 + \xi_2 - \xi_3 + \xi_4 = -4$,
$\xi_1 - \xi_2 + \xi_3 + \xi_4 = 2$;

(ii) $2\xi_1 - \xi_2 + 3\xi_3 = 1$,
$3\xi_1 + 2\xi_2 - \xi_3 = 4$,
$\xi_1 - 4\xi_2 + 7\xi_3 = -2$.

11. A matrix in echelon form can, by further row operations, be reduced to a form in which
   (i) the first non-zero entry in each non-zero row is unity;
   (ii) each column which contains the leading non-zero entry of some row has all other entries 0.
   The matrix is then said to be in *row-reduced echelon form*.
   (a) Carry out this further reduction on the matrices in Exercise 2.8,9(a).
   (b) Show that any non-singular ($n \times n$) matrix can be row reduced to the identity matrix.
   (c) The result (b) allows us to find the inverse of a matrix $[t_{ij}]$ by row operations. We form the partitioned matrix

   $$[[t_{ij}] \vdots [\delta_i^j]]$$

   and row reduce $[t_{ij}]$ to the identity: if we simultaneously perform the same operations on the identity matrix, then we arrive at the partitioned matrix

   $$[[\delta_i^j] \vdots [t_{ij}]^{-1}].$$

   Prove this result.
   (d) By the method outlined in (c) find the inverse of

   $$\begin{bmatrix} 3 & 7 & 8 & 15 \\ 2 & 5 & 6 & 11 \\ 2 & 6 & 10 & 19 \\ 4 & 11 & 19 & 38 \end{bmatrix}$$

   (Note: solution of a set of n simultaneous equations in n unknowns by reduction to echelon form is commonly known as *Gaussian elimination*: reduction to row-reduced echelon form is known as the *Gauss–Jordan method* [18, 21].)

12. Relative to the standard basis in $\mathcal{R}_4$ a transformation T has the

representation

$$\begin{bmatrix} 1 & 3 & 1 & 0 \\ 2 & 4 & 2 & 0 \\ 3 & 0 & 4 & 6 \\ 4 & 0 & 2 & 8 \end{bmatrix}.$$

Find the matrix representation relative to the basis $(1, 1, 1, 1)$, $(0, 1, 1, 1)$, $(0, 0, 1, 1)$, $(0, 0, 0, 1)$.

13. (a) Evaluate the following determinants by using the properties of determinants and not solely by a Laplace expansion:

   (i) $\begin{bmatrix} 1 & \omega & \omega^2 \\ \omega & \omega^2 & 1 \\ \omega^2 & 1 & \omega \end{bmatrix}$, $\omega^3 = 1$;

   (ii) $\begin{bmatrix} 1 & 1 & 1 \\ \alpha & \beta & \gamma \\ \alpha^2 & \beta^2 & \gamma^2 \end{bmatrix}$.

   (b) The *adjugate* of the $(n \times n)$ matrix $[t_{ij}]$ is defined to be the transpose of the matrix of cofactors of $[t_{ij}]$, i.e., $\text{Adj } [t_{ij}] = [T_{ij}]^T$. Show that $[t_{ij}] \text{ Adj } [t_{ij}] = \det [t_{ij}] [\delta_i^j]$ and hence that, if $[t_{ij}]^{-1} = [t'_{ij}]$, then

   $$t'_{ij} = \frac{T_{ji}}{\det [t_{ij}]}, \quad \det [t_{ij}] \neq 0.$$

   (c) Find the inverse of the matrix in Exercise 2.8,2(d) using this result. (Note: this is *not* a practical computational method for finding inverses.)

   (d) Let the $(n \times n)$ matrix $[t_{ij}]$ be non-singular. Using (b) show that the ith element of the solution vector of the set of equations,

   $$[t_{ij}]\{\xi_j\} = \{b_i\}$$

   is given by

   $$\xi_i = \left(\sum_{j=1}^{n} b_j T_{ji} \right) / \det [t_{ij}].$$

   Hence show that

   $$\xi_i = \det [t_{ij}(b_i)] / \det [t_{ij}],$$

   where $[t_{ij}(b_i)]$ is the matrix obtained from $[t_{ij}]$ by replacing the ith column by the vector $\{b_i\}$.

Solve the equations
$$10\xi_1 - \xi_2 + 3\xi_3 = 8,$$
$$\xi_1 - 10\xi_2 + 2\xi_3 = 7,$$
$$2\xi_1 - 2\xi_2 + 5\xi_3 = 9$$
using this result.

(Note: this method of solution of equations is called *Cramer's rule*: it is *not* a practical method except when n is small. It is of occasional use as a theoretical result.)

14. Show that the rank of an $(m \times n)$ matrix $[t_{ij}]$ is ρ if and only if every minor of $[t_{ij}]$ of order $(\rho + 1)$ vanishes while at least one minor of order ρ is non-zero.

    (Hint: (i) to prove necessity use the fact that $(\rho + 1)$ columns of $[t_{ij}]$ are linearly dependent; (ii) to prove sufficiency assume that a non-zero minor of order ρ exists.)

15. (a) If $[e_{ij}]$ is an elementary matrix and $[t_{ij}]$ an $(n \times n)$ matrix, show that
    $$\det([e_{ij}][t_{ij}]) = \det[e_{ij}] \det[t_{ij}]$$

    (b) Using (a) and the fact that any square matrix is equivalent to a diagonal matrix show that, for any two square matrices $[t_{ij}]$, $[s_{ij}]$,
    $$\det([s_{ij}][t_{ij}]) = \det([t_{ij}][s_{ij}]) = \det[t_{ij}]\det[s_{ij}].$$

16. (a) Find the eigenvalues and eigenvectors of the matrices

    (i) $\begin{bmatrix} 2 & -1 & -1 \\ 0 & 3 & 2 \\ -1 & 1 & 2 \end{bmatrix}$, (ii) $\begin{bmatrix} 1+i & 0 & 0 \\ 2-2i & 1-i & 0 \\ 2i & 0 & 1 \end{bmatrix}$

    (b) Show directly that each matrix in (a) is similar to a diagonal matrix by finding its representation relative to a basis of eigenvectors.

    (c) Let
    $$p(\lambda) = (-1)^n(\lambda^n + \alpha_1 \lambda^{n-1} + \alpha_2 \lambda^{n-2} + \ldots + \alpha_n)$$
    be the characteristic polynomial of the $(n \times n)$ matrix $[t_{ij}]$. Show that

    (i) $-\alpha_1 = \sum_{i=1}^{n} t_{ii} = \text{tr}[t_{ij}]$ (cf., Exercise 2.6, 1(iii));

    (ii) $(-1)^n \alpha_n = \det[t_{ij}]$.

    Hence show that if $\lambda_1, \lambda_2, \ldots, \lambda_n$ are the eigenvalues of $[t_{ij}]$ (not necessarily distinct) then

(iii) $\operatorname{tr}[t_{ij}] = \lambda_1 + \lambda_2 + \ldots + \lambda_n$;

(iv) $\det[t_{ij}] = \lambda_1 \lambda_2 \ldots \lambda_n$

Verify these results numerically for the matrices of part (a).

17. The Cayley–Hamilton theorem states that 'a matrix satisfies its own characteristic equation'. That is to say that if

$$\lambda^n + \alpha_1 \lambda^{n-1} + \ldots + \alpha_n = 0$$

is the characteristic equation of the linear transformation $T: \mathscr{V}_n \to \mathscr{V}_n$ then (cf., p. 45)

$$T^n + \alpha_1 T^{n-1} + \ldots + \alpha_n I = 0.$$

An elementary proof of this theorem is given, for example, in [16] (*see also* [17]).

Verify the result numerically for the matrix of Exercise 2.8,16(a)(i) and use the theorem to compute the seventh power of this matrix.

18. Show that zero is the only eigenvalue of the linear transformation $Dp = dp/dt$ on the space $\mathscr{P}_m$ of polynomials of degree $\leq (m-1)$ (Example 2.5.2), and that this eigenvalue has algebraic multiplicity m and geometric multiplicity 1. Show that any polynomial p_g of degree $(m-1)$ is a generalized eigenvector of D of exponent m and that the set of vectors $\{p_g, Dp_g, D^2 p_g, \ldots, D^{m-1} p_g\}$ is a basis for the space $\mathscr{P}_m$. Find the matrix of D relative to this basis.

19. (a) Show that the (4×4) Jordan matrix

$$[t_{ij}] = \begin{bmatrix} \tilde{\lambda} & 0 & 0 & 0 \\ 1 & \tilde{\lambda} & 0 & 0 \\ 0 & 1 & \tilde{\lambda} & 0 \\ 0 & 0 & 1 & \tilde{\lambda} \end{bmatrix}$$

has the single eigenvalue $\tilde{\lambda}$ of algebraic multiplicity 4 and the single eigenvector $(0, 0, 0, 1)$. Show that $[t_{ij}] - \tilde{\lambda}[\delta_i^j]$ has index of nilpotency 4 (cf., Exercise 2.5,4(d)). Generate a cyclic basis for $\mathscr{R}_4$ beginning with a generalized eigenvector of $[t_{ij}]$ of exponent 4.

(b) (i) Show that the eigenvalues of

$$\begin{bmatrix} 3 & -1 & 1 & -3 \\ 5 & 0 & -2 & 0 \\ 7 & -2 & 0 & -3 \\ 2 & -1 & 1 & -2 \end{bmatrix}$$

are $\lambda_1 = -2$ with algebraic multiplicity 1 and $\lambda_2 = 1$ with algebraic multiplicity 3.

(ii) Find the eigenvector $\{\xi_i\}^{(1)}$ belonging to λ_1.

(iii) Show that the geometric multiplicity of λ_2 is two and find two independent eigenvectors $\{\xi_i\}_1^{(2)}$, $\{\xi_i\}_2^{(2)}$ belonging to λ_2.

(iv) Find a vector; $\{\xi_i\}_g^{(2)}$ say, such that

$$([t_{ij}] - \lambda_2[\delta_i^j])^2 \{\xi_i\}_g^{(2)} = \{0\}$$

and which is not a linear combination of $\{\xi_i\}_1^{(2)}$, $\{\xi_i\}_2^{(2)}$. (The existence of such a vector is guaranteed by the result of Exercise 2.5,14.)

(v) Show that $\{\xi_i\}^{(1)}$, $\{\xi_i\}_1^{(2)}$, $\{\xi_i\}_g^{(2)}$ and $([t_{ij}] - \lambda_2[\delta_i^j])\{\xi_i\}_g^{(2)}$ are a basis for $\mathscr{R}_4$ and find the representation of $[t_{ij}]$ relative to this basis.

20. (a) Find the eigenvectors of the transpose of the matrix in Exercise 2.8, 16(a)(i) and show that they are a dual basis for $\mathscr{R}_3$.

(b) Show that, if $[t_{ij}]$ is diagonizable, the inhomogeneous equation

$$([t_{ij}] - \lambda[\delta_i^j])\{\xi_j\} = \{b_i\}, \qquad i,j = 1,\ldots,n,$$

has the solution

$$\{\xi_i\} = \sum_{k=1}^{n} \frac{\lfloor\eta_i\rfloor^{(k)}\{b_i\}}{\lambda_k - \lambda} \cdot \frac{\{\xi_i\}^{(k)}}{\lfloor\eta_i\rfloor^{(k)}\{\xi_i\}^{(k)}}$$

provided $\lambda \neq \lambda_k$, where $(\lambda_k; \{\xi_i\}^{(k)})$ are the eigensolutions of $[t_{ij}]$ and $(\lambda_k; \{\eta_i\}^{(k)})$ are the eigensolutions of $[t_{ij}]^T$.

21. Assume that the eigenvalues λ_j of $T: \mathscr{V}_n \to \mathscr{V}_n$ satisfy the inequalities

$$|\lambda_1| > |\lambda_2| \geq |\lambda_3| \geq \ldots \geq |\lambda_n|.$$

If $x^{(k)}$ are the eigenvectors of T then any vector $x \in \mathscr{V}_n$ can be represented by

$$x = \sum_{k=1}^{n} \xi_k x^{(k)}$$

and if we compute $T^r x$ we obtain

$$T^r x = \sum_{k=1}^{n} \xi_k \lambda_k^r x^{(k)} = \lambda_1^r \sum_{k=1}^{n} \xi_k \left(\frac{\lambda_k}{\lambda_1}\right)^r x^{(k)}.$$

For sufficiently large r the summation on the right-hand side will converge to the vector $\xi_1 x^{(1)}$, that is, the eigenvector belonging to λ_1.

This method of obtaining a dominant eigenvalue is called the *power method*. To implement it we choose an arbitrary vector x_0 and compute $x_1 = Tx_0$, $x_2 = Tx_1$ and so on. At each stage we scale the resulting vector so as to make the numerically largest element equal unity. This multiplier is, at each stage, an approximation to λ_1 (show this).

Use this method to find the dominant eigenvalue and corresponding eigenvector of the matrix of Exercise 2.8,16(a)(i).

The power method can be extended to the calculation of non-dominant eigenvalues (*see*, for example, [17, 18, 21] and Exercise 3.8,5(a)). Some modification is required if the dominant roots are a complex conjugate pair [17].

2.9 Application examples

Example 2.9.1 Consider the (indeterminate) pin-jointed bar structure shown in Fig. 2.7. The joint at A is loaded by horizontal and vertical force components f^1, f^2, respectively, with corresponding deflection components d_1, d_2.

Let the extension of each bar be e_i and the force in each bar be t^i (tensile).

From kinematic considerations we see that the vector of extensions $\{e_i\}$ is related to the displacements $\{d_i\}$ by the transformation

$$\begin{bmatrix} e_1 \\ e_2 \\ \vdots \\ e_i \\ \vdots \\ e_n \end{bmatrix} = \begin{bmatrix} \cos\theta_1 & \sin\theta_1 \\ \cos\theta_2 & \sin\theta_2 \\ \vdots & \vdots \\ \cos\theta_i & \sin\theta_i \\ \vdots & \vdots \\ \cos\theta_n & \sin\theta_n \end{bmatrix} \begin{bmatrix} d_1 \\ d_2 \end{bmatrix},$$

say

$$e_i = \sum_{j=1}^{2} l_{ij} d_j, \quad i = 1, \ldots, n, \quad \text{or} \quad \{e_i\} = [l_{ij}]\{d_j\}.$$

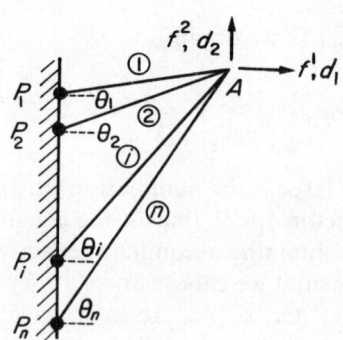

Fig. 2.7

APPLICATION EXAMPLES 101

Similarly, from force equilibrium at the joint A, we have

$$\begin{bmatrix} f^1 \\ f^2 \end{bmatrix} = \begin{bmatrix} \cos\theta_1 & \cos\theta_2 \ldots \cos\theta_i \ldots \cos\theta_n \\ \sin\theta_1 & \sin\theta_2 \ldots \sin\theta_i \ldots \sin\theta_n \end{bmatrix} \begin{bmatrix} t^1 \\ t^2 \\ \vdots \\ t^i \\ \vdots \\ t^n \end{bmatrix},$$

thus

$$f^j = \sum_{i=1}^{n} l_{ij} t^i, \quad j = 1, 2, \quad \text{or} \quad [f^j] = [l_{ij}]^T \{t^i\}.$$

This type of result is general and will apply to any similar situation involving forces and displacements, voltages and currents, etc. That it follows directly from the definition of the dual transformation can be seen as follows. Let $\mathscr{E}_n$ denote the space of 'internal' displacement vectors e and $\mathscr{D}_2^*$ denote the space of 'external' displacement vectors d (*see* Figure 2.8); also let $L: \mathscr{D}_2^* \to \mathscr{E}_n$ denote the transformation between them, viz., $e = Ld$.

The dual space $\mathscr{E}_n^*$ of $\mathscr{E}_n$ may naturally be identified with the space whose vectors t physically represent the forces in the set of bars. If the vectors e, t are referred to dual basis sets in $\mathscr{E}_n$, $\mathscr{E}_n^*$, respectively, then we have

$$t(e) = [t, e] = \sum_{i=1}^{n} t^i e_i.$$

In this case we can give the linear functional $t(e)$ the physical interpretation of 'work', being the sum product of the force components with their corresponding displacement. The definition of the dual, namely that if $e = Ld$ and $f = L't$ then $[e, t] = [d, f]$, is nothing more than a statement that 'the virtual work of the external forces through the external displacements is equal to the virtual work of the internal forces through the corresponding internal displacements'—a result which is usually called the *virtual work identity* in structural analysis.

There is no requirement in the above statement that the forces and displacements should be the actual vectors which would be generated in the framework due to a given loading system (although the statement is of course

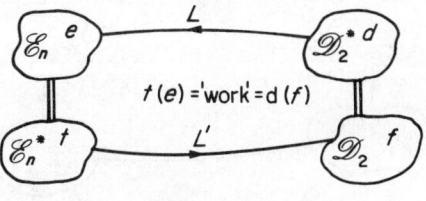

Fig. 2.8

true in this special case). It is for this reason that the adjective 'virtual' is used.

As soon as a 'constitutive equation' is supplied which relates the internal forces and displacements, the problem can be solved by using the virtual work identity. For example, if the rods suffer only small strains we may write

$$t^i = k_i e_i,$$

where k_i depends on the length, sectional area and elastic modulus of rod i.

Using the relations between the external and internal displacements gives

$$f^j = \sum_{i=1}^n l_{ij} t^i = \sum_{i=1}^n l_{ij} k_i e_i = \sum_{i=1}^n l_{ij} k_i \sum_{k=1}^2 l_{ik} d_k = \sum_{k=1}^2 \left(\sum_{i=1}^n l_{ij} k_i l_{ik} \right) d_k$$

upon changing the order of summation. Writing

$$s_{jk} = \sum_{i=1}^n l_{ij} k_i l_{ik}$$

we obtain

$$\{f^j\} = [s_{jk}]\{d_k\},$$

where $[s_{jk}]$ is the (2×2) 'stiffness matrix' for the framework loaded at A. Notice that $[s_{jk}]$ is a symmetrical matrix. If we assemble the k_i into a diagonal matrix $[k_j \delta_i^j]$ we may write

$$[s_{jk}] = [l_{ij}]^T [k_j \delta_i^j][l_{ik}].$$

This example contains the essential ideas of the stiffness method for the analysis of structures. The summation defining the s_{jk} in effect assembles the unconnected bars (with their associated stiffnesses) so that each contributes correctly to the overall stiffness of the structure.

The relation between the matrices $[s_{ij}]$ and $[k_j \delta_i^j]$ in this example is an important one and will arise in other physical problems where changes of basis are made. In fact, we shall now develop a more general form of the foregoing result.

Let $\mathscr{V}_n$, $\mathscr{U}_m$ be n- and m-dimensional vector spaces, respectively, and let $\mathscr{V}_n^*$, $\mathscr{U}_m^*$ be their conjugates. Since $\mathscr{V}_n$, $\mathscr{V}_n^*$ are (isomorphic) vector spaces

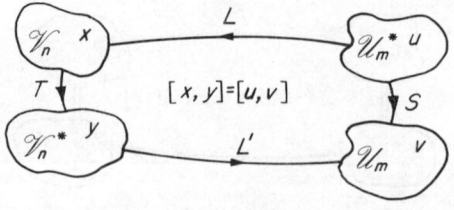

Fig. 2.9

we may define a transformation $T: \mathscr{V}_n \to \mathscr{V}_n^*$ so that for $x \in \mathscr{V}_n, y \in \mathscr{V}_n^*$

$$y = Tx$$

(see Fig. 2.9).

Let L be a transformation $\mathscr{U}_m^* \to \mathscr{V}_n$ and $L': \mathscr{V}_n^* \to \mathscr{U}_m$ its conjugate. Then we can show that the mapping $T: \mathscr{V}_n \to \mathscr{V}_n^*$ implies a mapping $S: \mathscr{U}_m^* \to \mathscr{U}_m$ given by

$$S = L'TL.$$

We begin with $[x, y] = [u, v]$, wherein $x = Lu$ and $v = L'y$. Thus we have

$$[u, v] = [x, Tx] = [Lu, Tx] = [u, L'TLu] = [u, Su],$$

and since this is true for all $u \in \mathscr{U}_m^*$ the result follows.

Example 2.9.2 A uniform beam of length l and bending stiffness EI is acted upon only by shears and moments at its ends $x = 0, l$. The deflection curve $v(x), x \in [0, l]$ is an integral of the differential equation

$$EI \frac{d^4 v}{dx^4} = 0.$$

Writing $\xi = x/l$ we see that the most general integral takes the form

$$v(\xi) = \beta_0 \xi^3 + \beta_1 \xi^2 + \beta_2 \xi + \beta_3$$

and hence is a vector in $\mathscr{R}_4$ with basis $\{\xi^3, \xi^2, \xi, 1\}$, $\xi \in [0, 1]$.

The deflection curves ξ^n have no particular physical significance and in applications it is more convenient to change to a basis set which has as coordinates the vertical displacement and (small) rotation of each end of the beam: the corresponding 'force' components (cf., Example 2.9.1) are then the applied shears and moments.

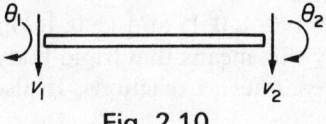

Fig. 2.10

Referring to Fig. 2.10, let v_1, θ_1; v_2, θ_2 be the relevant displacements and rotations of the ends $\xi = 0$, $\xi = 1$ of the beam, respectively. Since $v_1 = v(0)$, $\theta_1 = v'(0)$, etc., we find

$$(l\theta_1, v_1, l\theta_2, v_2) = [l_{ij}](\beta_0, \beta_1, \beta_2, \beta_3),$$

where

$$[l_{ij}] = \begin{bmatrix} 0 & 0 & 1 & 0 \\ 0 & 0 & 0 & 1 \\ 3 & 2 & 1 & 0 \\ 1 & 1 & 1 & 1 \end{bmatrix}.$$

We find (for example, by partitioning—Exercise 2.8,2(d)) that

$$[l_{ij}]^{-1} = \begin{bmatrix} 1 & 2 & 1 & -2 \\ -2 & -3 & -1 & 3 \\ 1 & 0 & 0 & 0 \\ 0 & 1 & 0 & 0 \end{bmatrix},$$

so that the 'new' basis functions are given (in terms of the 'old') by

$$\{\xi(1-\xi)^2, (1+2\xi)(1-\xi)^2, -\xi^2(1-\xi), \xi^2(3-2\xi)\}$$

(*see* Fig. 2.11).

The forces acting on the beam are the shears and moments at its ends, given by

$$S_1 = EIv'''(0), \qquad M_1 = -EIv''(0),$$
$$S_2 = -EIv'''(1), \qquad M_2 = EIv''(1),$$

respectively, the signs being chosen so as to make corresponding 'force' and 'displacement' have the same sense.

By a direct computation of these we find

$$(M_1, lS_1, M_2, lS_2) = [k_{ij}](l\theta_1, v_1, l\theta_2, v_2),$$

where the stiffness matrix

$$[k_{ij}] = \begin{bmatrix} 2 & 3 & 1 & -3 \\ 3 & 6 & 3 & -6 \\ 1 & 3 & 2 & -3 \\ -3 & -6 & -3 & 6 \end{bmatrix} \frac{2EI}{l^2}$$

relates the shears and moments at the beam ends to the corresponding displacements and rotations.

Notice that the vectors $(0, 1, 0, 1)$ and $(1, 0, 1, 1)$ are a basis for the null space of $[k_{ij}]$: physically this means that 'rigid body motions' or 'strainless' motions of the beam generate no reactions. It also means of course that $[k_{ij}]$ is not invertible.

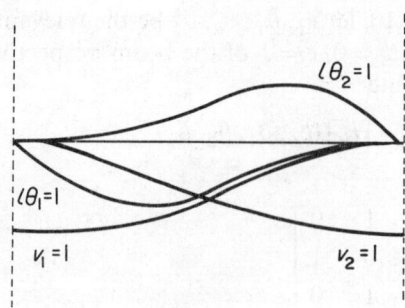

Fig. 2.11

We can apply the general result deduced at the end of Example 2.9.1 to find (a) the form of the 'force systems' corresponding to the basis $\{\xi^3, \xi^2, \xi, 1\}$ in terms of the end shears and moments, and (b) the stiffness matrix relative to this basis.

(a) If we denote the generalized force components corresponding to the displacements $\{\xi^3, \xi^2, \xi, 1\}$ by $(\gamma_0, \gamma_1, \gamma_2, \gamma_3)$ then

$$(\gamma_0, \gamma_1, \gamma_2, \gamma_3) = [l_{ij}]^T (M_1, lS_1, M_2, lS_2)$$

and the 'generalized forces' consist of linear combinations of end moments and shears: γ_3/l represents the total lateral force on the beam and γ_2 the total moment about the left-hand end.

(b) The stiffness matrix $[s_{ij}]$ relative to the basis $\{\xi^3, \xi^2, \xi, 1\}$, that is, the transformation

$$\{\gamma_i\} = [s_{ij}]\{\beta_j\},$$

is given (Example 2.9.1) by

$$[s_{ij}] = [l_{ij}]^T [k_{ij}] [l_{ij}] \begin{bmatrix} 6 & 3 & 0 & 0 \\ 3 & 2 & 0 & 0 \\ 0 & 0 & 0 & 0 \\ 0 & 0 & 0 & 0 \end{bmatrix} \frac{2EI}{l^2}.$$

This confirms that 1 and ξ are 'strainless' displacement modes: the displacement ξ^2, for example, is produced by the end moments and shears associated with $\gamma_0 = 3$, $\gamma_1 = 2$, $\gamma_2 = 0$ and $\gamma_3 = 0$. Note that the last two conditions, namely $\gamma_2 = 0$ and $\gamma_3 = 0$, represent overall force and moment equilibrium of the beam; these are satisfied for all possible displacement shapes.

Example 2.9.3 Many problems in economics or management involving allocation of resources or manufacturing schedules lead to the solution of a set of linear inequalities, the so-called *linear programming problem*.

We begin by defining a vector inequality in $\mathcal{R}_n$: let $x, y \in \mathcal{R}_n$, then we write $x \geqslant y$ if (relative to a fixed basis in $\mathcal{R}_n$) $x_i \geqslant y_i$, $i = 1, \ldots, n$ (see p. 10).

The linear programming problem takes the form: given $T: \mathcal{R}_n \to \mathcal{R}_m$, $b \in \mathcal{R}_m$ and $c^* \in \mathcal{R}_n^*$ determine $x \in \mathcal{R}_n$ such that $Tx \geqslant b$, $x \geqslant 0$ and $c^*(x)$ is a minimum.

The symmetry of Fig. 2.12 suggests the formulation of a dual linear programming problem, namely, find $y^* \in \mathcal{R}_m^*$ such that $T'y^* \leqslant c^*$, $y^* \geqslant 0$ and $b(y^*)$ is a maximum. Choose any $x \in \mathcal{R}_n$, $y^* \in \mathcal{R}_m^*$ such that $x \geqslant 0$, $y^* \geqslant 0$. Since $b \leqslant Tx$ and $y^* \geqslant 0$,

$$b(y^*) \leqslant Tx(y^*) = [Tx, y^*] = [x, T'y^*] = T'y^*(x),$$

and since we require that $T'y^* \leqslant c^*$ we have

$$b(y^*) \leqslant c^*(x).$$

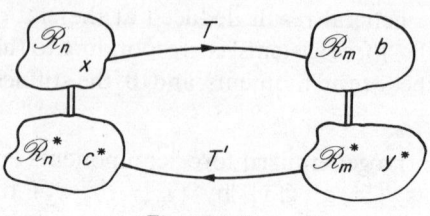

Fig. 2.12

If we maximize the left-hand side for all positive y^* satisfying $T'y^* \leq c^*$, we still have

$$\max_{y^*} b(y^*) \leq c^*(x);$$

similarly, if we minimize the right-hand side for all positive x satisfying $Tx \geq b$,

$$\max_{y^*} b(y^*) \leq \min_{x} c^*(x).$$

It follows that if y_0^* and x_0 are such that

$$b(y_0^*) = c^*(x_0)$$

then both y_0^* and x_0 are the required (optimal) solution vectors. It should be borne in mind that the existence of an optimal solution is not guaranteed: nor is an effective method for finding the optimal solution a straightforward matter. A full treatment of this type of problem can be found in [8, 11], for example.

The importance of the dual transformation cannot be overemphasized in the understanding and solution of this type of problem.

Example 2.9.4 Many problems in linear control theory or in the transient behaviour of mechanical or electrical systems under small disturbances lead to a system of first-order ordinary differential equations of the form

$$\frac{d\xi_i}{dt} = \sum_{j=1}^{n} t_{ij} \xi_j + \zeta_i(t), \qquad i = 1, \ldots, n,$$

where the $\xi_i(t)$ are state variables (or generalized coordinates), the $\zeta_i(t)$ are known functions of time defined for $t > 0$ and the t_{ij} are (system) constants: the system of equations is subject to the initial values $\xi_i(0) = \xi_i^0$, $i = 1, \ldots, n$.

The system of equations is a matrix representation of the operator equation

$$\left(\frac{d}{dt} - T\right) x(t) = z(t), \qquad x(0) = x^0,$$

where $x(t)$ is the state vector, x^0 the initial state and $z(t)$ the forcing function or input.

We shall assume that T possesses n distinct eigenvalues $\lambda_1, \lambda_2, \ldots, \lambda_n$ and we denote the corresponding eigenvectors of $[t_{ij}]$ by $\{\xi_i\}^{(k)}$ and of $[t_{ij}]^T$ by $\{\eta_i\}^{(k)}$. If $\{\mu_i\}$ is the coordinate vector of x relative to the basis

$\{\{\xi_i\}^{(k)}\}$ then

$$x(t) = \sum_{k=1}^{n} \mu_k(t) \{\xi_i\}^{(k)}, \qquad x^0 = \sum_{k=1}^{n} \mu_k^0 \{\xi_i\}^{(k)}.$$

Substituting in the differential equation we obtain

$$\sum_{k=1}^{n} \left(\frac{d}{dt} - [t_{ij}]\right) \mu_k(t) \{\xi_j\}^{(k)} = \{\zeta_i(t)\}$$

or

$$\sum_{k=1}^{n} \left(\frac{d}{dt} - \lambda_k\right) \mu_k(t) \{\xi_j\}^{(k)} = \{\zeta_i(t)\}.$$

We now operate on this vector equation with the linear functional having coordinate vector $\{\eta_i\}^{(l)}$. If we assume that we have arranged (Example 2.8.6) that $[\{\eta_i\}^{(k)}, \{\xi_i\}^{(k)}] = 1, k = 1, \ldots, n$, then, since $\{\eta_i\}^{(l)}$ is the annihilator of the eigenspaces $\mathscr{V}^{(k)}$ of $\mathscr{V}$ for $k \neq l$, we have

$$\frac{d\mu_k}{dt} - \lambda_k \mu_k = [\{\eta_i\}^{(k)}, \{\zeta_i(t)\}], \qquad k = 1, \ldots, n.$$

The initial values for the μ_k are, in the same fashion, given by

$$\mu_k^0 = [\{\eta_i\}^{(k)}, \{\xi_i^0\}].$$

We have now arrived at a system of uncoupled first-order differential equations for the coordinates $\mu_k(t)$. In the kth equation we multiply by the integrating factor $e^{-\lambda_k t}$, giving

$$\frac{d}{dt}(e^{-\lambda_k t} \mu_k(t)) = [\{\eta_i\}^{(k)}, \{\zeta_i(t)\}] e^{-\lambda_k t},$$

and upon integration and using the initial value μ_k^0 to eliminate the constant of integration we find

$$\mu_k(t) = [\{\eta_i\}^{(k)}, \{\xi_i^0\}] e^{\lambda_k t} + \int_0^t [\{\eta_i\}^{(k)}, \{\zeta_i(\tau)\}] e^{\lambda_k(t-\tau)} d\tau.$$

Finally, the solution for the state vector is

$$\{\xi_i(t)\} = \sum_{k=1}^{n} [\{\eta_i\}^{(k)}, \{\xi_i^0\}] e^{\lambda_k t} \{\xi_i\}^{(k)}$$

$$+ \sum_{k=1}^{n} \int_0^t [\{\eta_i\}^{(k)}, \{\zeta_i(\tau)\}] e^{\lambda_k(t-\tau)} d\tau \{\xi_i\}^{(k)}.$$

The above procedure is clearly equivalent to a similarity transformation of $[t_{ij}]$ to the diagonal matrix of its eigenvalues. However, the advantages of using the eigenvectors of the dual are twofold:

1. In a practical case the computation of the two sets of eigenvectors is more satisfactory than inversion of the matrix of eigenvectors of $[t_{ij}]$.

2. There is no need to find *all* the eigenvectors of $[t_{ij}]$ if we are interested only in the behaviour of *some* of the coordinates $\mu_k(t)$. Very often some of the eigenvalues of large modulus will represent transient motions which are very rapidly damped and which we may wish to ignore; in such a case we can still compute the contribution to the state vector from the remaining eigenvalues and eigenvectors.

The type of solution presented here is often called the *spectral solution* since it represents the state vector as a sum of exponential motions each of which is associated with a (time-invariant) eigenvector or mode of the system.

Let us apply the foregoing analysis to the very simple example of a spring–mass–dashpot system (Fig. 2.13) described by the equation

$$m\ddot{d} + c\dot{d} + kd = p(t),$$

where $d(t)$ is the displacement of the mass and $p(t)$ is the forcing function. Let us rewrite the equation as

$$\ddot{d} + 2\gamma\omega\dot{d} + \omega^2 d = q(t),$$

where $\omega^2 = k/m$ is the undamped natural frequency,

$$\gamma = c/2\sqrt{mk} > 0$$

is a relative measure of damping and

$$q(t) = p(t)/m$$

is a normalized forcing function.

We now recast the equation of motion as a first-order system by defining the state variables $\xi_1(t) = d(t)$ and $\xi_2(t) = \dot{d}(t)$; this results in the (2 × 2) system equation

$$\frac{d}{dt}\begin{bmatrix} \xi_1 \\ \xi_2 \end{bmatrix} - \begin{bmatrix} 0 & 1 \\ -\omega^2 & -2\gamma\omega \end{bmatrix}\begin{bmatrix} \xi_1 \\ \xi_2 \end{bmatrix} = \begin{bmatrix} 0 \\ q \end{bmatrix}$$

with initial values $\xi_1^0 = d(0)$ and $\xi_2^0 = \dot{d}(0)$.

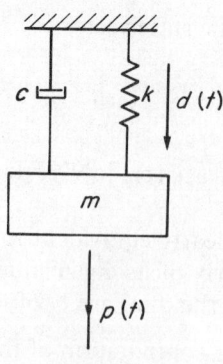

Fig. 2.13

APPLICATION EXAMPLES 109

We now refer to Example 2.8.5(c):

(a) If $\gamma > 1$ we write the roots as λ_1, λ_2 and, using the (first component of the) spectral solution, we find

$$d(t) = \frac{d(0)(\lambda_2 e^{\lambda_1 t} - \lambda_1 e^{\lambda_2 t})}{\lambda_2 - \lambda_1} + \frac{\dot{d}(0)(e^{\lambda_2 t} - e^{\lambda_1 t})}{\lambda_2 - \lambda_1}$$

$$- \frac{1}{\lambda_2 - \lambda_1} \int_0^t q(\tau)(e^{\lambda_1(t-\tau)} - e^{\lambda_2(t-\tau)}) d\tau.$$

(b) If $\gamma < 1$ we write $\lambda_1 = \mu + i\tilde{\omega}$, $\lambda_2 = \mu - i\tilde{\omega}$, giving

$$d(t) = \frac{d(0)}{\tilde{\omega}} e^{\mu t}(\tilde{\omega} \cos \tilde{\omega} t - \mu \sin \tilde{\omega} t) + \frac{\dot{d}(0)}{\tilde{\omega}} e^{\mu t} \sin \tilde{\omega} t$$

$$- \int_0^t q(\tau) e^{\mu(t-\tau)} \sin \tilde{\omega}(t-\tau) d\tau.$$

(c) If $\gamma = 1$ the system matrix has the Jordan form

$$\begin{bmatrix} -\omega & 0 \\ 1 & -\omega \end{bmatrix}.$$

For μ_1, μ_2 this gives the equations

$$\frac{d\mu_1}{dt} + \omega \mu_1 = \frac{q(t)}{\omega},$$

$$\frac{d\mu_2}{dt} + \omega \mu_2 - \mu_1 = -\frac{q(t)}{\omega^2}$$

with solutions

$$\mu_1(t) = (d(0) + \frac{1}{\omega} \dot{d}(0)) e^{-\omega t} - \frac{1}{\omega} \int_0^t q(\tau) e^{-\omega(t-\tau)} d\tau$$

and

$$\mu_2(t) = t d(0) e^{-\omega t} - \frac{\dot{d}(0)}{\omega^2}(1 - \omega t) e^{-\omega t} + \frac{1}{\omega^2} \int_0^t q(\tau) e^{-\omega(t-\tau)} d\tau$$

$$- \frac{1}{\omega} \int_0^t e^{\omega(t-\tau)} \int_0^\tau q(\tau') e^{-\omega(\tau-\tau')} d\tau',$$

giving

$$d(t) = \mu_1(t) + \omega \mu_2(t) = d(0)(1 + \omega t) e^{-\omega t} + \dot{d}(0) t e^{-\omega t}$$

$$- \int_0^t e^{-\omega(t-\tau)} \int_0^\tau q(\tau') e^{-\omega(\tau-\tau')} d\tau'.$$

3 Metric, normed and inner product spaces

3.1 Introduction

In this chapter we introduce the concept of 'distance', and with it a measure of 'nearness', into our mathematical structure. We do this by endowing a set of objects or elements with a metric, a function defined on pairs of elements of the set whose value is a positive real number and which satisfies a set of axioms whose genesis lies in a generalization of the properties enjoyed by the 'distance between points' in ordinary three-dimensional space. In this way we arrive at a metric space.

Perhaps the most important concept in analysis is that of limit: armed with a metric we are able to discuss sequences of elements which converge to a limit in the sense that the 'distance' between successive terms of the sequence and the limit can be made as small as we please. In practice, we do not often know the limit of a sequence—indeed, it may be precisely what we are trying to find—and the foregoing criterion for convergence is virtually useless. Instead, what we would like to say is something like the following: successive terms of a sequence are 'getting nearer' to each other (this we can readily test), hence they are themselves 'getting nearer' to something and that something is the limit. Unfortunately this is not always true and one of the main considerations in our discussion of metric spaces, namely the idea of completeness, springs from this difficulty.

The metric also allows us to consider 'continuity' of mappings of one metric space into another: roughly speaking, a mapping is continuous if the images of a pair of neighbouring points are also neighbours in the sense of being 'near'. Both convergence and continuity are obvious generalizations to metric spaces of corresponding concepts which are no doubt familiar to the reader in the case of functions of a single real variable.

If we graft a suitable metric onto a vector space we obtain a normed vector space, the word norm generally being used in this context, in place of distance or metric. Now we can give a precise meaning to infinite summations, for example, or to other non-finite operations in the space. Retracing the path we followed in Chapter 2 we look at linear transformations with continuity in mind, we consider the dual of a normed vector space, and in many other ways we examine the impact of a 'distance function' on the structure of a vector space.

Distance is not the only feature of ordinary three-dimensional space which is missing from the primeval vector space; there is also angle. Taking a clue from the ordinary dot product of two space vectors as representing a measure (the cosine) of the angle between them, we can define a binary operation between vectors which may properly be referred to as an inner product. This is in contrast to scalar multiplication, which is essentially an external product, involving as it does a vector and an element of the scalar field. As usual, the inner product is introduced on an axiomatic basis; the resulting vector space is called an inner product space and is also a normed space by virtue of the fact that the inner product of a vector with itself yields a natural norm. The connection with angle in the geometric sense becomes somewhat tenuous, but one significant element remains, that of orthogonality or of 'being at right-angles', where in ordinary three-dimensional space the dot product vanishes. Following immediately upon this is the possibility of conceiving of orthogonal or Cartesian axis, or basis, systems—not only for finite-dimensional, but also for infinite-dimensional inner product spaces. Such spaces and the linear transformations on them bear a more than passing resemblance to finite-dimensional spaces; for example, every linear transformation has a representation as an infinite matrix. There also arises the possibility of identifying the values of inner products with those of linear functionals with the result that an inner product space is, in a sense, its own dual. The dual of a linear transformation, being an operator in the same space, then has the possibility of being equal to itself, and it turns out that the class of linear transformations having this property (the so-called symmetrical transformations) possess a rather nice structure and can, like finite matrices, be reduced to a sum (albeit infinite) of simple dilatations.

3.2 Metric spaces

A metric space is a set in which every pair of elements has associated with it a well-defined 'distance'. By utilizing the distance we can define convergence of sequences of elements by saying that a sequence $\{x^{(k)}\}$ of elements converges to the element x if the distance between $x^{(k)}$ and x tends to zero as $k \to \infty$. Thus, by defining 'distance' we are able to discuss convergence of elements in terms of the usual convergence of a sequence of real numbers, namely the sequence of distances between x and $x^{(k)}$.

In the following definition S is any set of elements, not necessarily a vector space.

Definition 3.2.1 *A metric (or distance function) on a set S is a real-valued function $d(x, y)$ defined for all pairs of elements x and y in S and which satisfies the following axioms:*

1. $d(x, y) > 0$; $d(x, y) = 0$, if and only if, $x = y$;
2. $d(x, y) = d(y, x)$ $\quad \forall x, y \in S$;
3. $d(x, z) \leqslant d(x, y) + d(y, z)$ $\quad \forall x, y, z \in S$.

A metric space denoted by (S, d), consists of a set S and a metric d on S.

Axiom 1 is natural and reasonable for any measure of distance: it should be noted that we are actually identifying any two elements x, y for which $d(x, y) = 0$. Axiom 2 (the symmetry condition) states that the distance from x to y is the same as the distance from y to x and is also a rather obvious requirement. Axiom 3 is called the 'triangle inequality' and is not an obvious requirement: suffice it to say that without it the distance function would not have the properties we expect of it. We shall use the triangle inequality very shortly to show that a convergent sequence has a unique limit. It is important to keep in mind that different metrics d_1, d_2, on a single set S define distinct metric spaces $(S, d_1), (S, d_2)$. By a *subspace* of a metric space we mean a metric space whose points form a subset of those of S and whose metric agrees with that defined on S.

Example 3.2.1

(a) Consider the vector space $\mathscr{R}_2$ of the ordered pairs of real numbers. Let $x = (\xi_1, \xi_2)$, $y = (\eta_1, \eta_2)$, then we may define the following metrics on $\mathscr{R}_2$:

$$d_1(x, y) = |\xi_1 - \eta_1| + |\xi_2 - \eta_2|,$$
$$d_2(x, y) = \{(\xi_1 - \eta_1)^2 + (\xi_2 - \eta_2)^2\}^{1/2},$$
$$d_p(x, y) = \{|\xi_1 - \eta_1|^p + |\xi_2 - \eta_2|^p\}^{1/p}, \quad 1 \leqslant p < \infty,$$
$$d_\infty(x, y) = \max\{|\xi_1 - \eta_1|, |\xi_2 - \eta_2|\},$$
$$\tilde{d}(x, y) = \frac{d(x, y)}{1 + d(x, y)},$$

where d is any metric on $\mathcal{R}_2$. These metrics are also immediately applicable to the complex space $\mathscr{C}_2$ if the modulus signs are interpreted as the modulus of a complex number. Each of the metrics obviously satisfies axioms 1 and 2 of Definition 3.2.1 and that axiom 3 is satisfied by d_1, d_p and d_∞ is easy to show: that $\tilde{d}$ satisfies axiom 3 follows from the result of Exercise 3.2,1(b).

The above metrics can be extended to the vector spaces $\mathcal{R}_n$ or $\mathscr{C}_n$ by defining

$$d_p(x, y) = \left\{ \sum_{i=1}^{n} |\xi_i - \eta_i|^p \right\}^{1/p}, \quad 1 \leq p < \infty,$$

and

$$d_\infty(x, y) = \max_i \{|\xi_i - \eta_i|\}$$

That the triangle inequality is satisfied by these metrics follows immediately from Minkowski's inequality, which is proved in Appendix 3.A.

(b) We now consider metrics on the vector spaces of infinite sequences $(\xi_1, \xi_2, \ldots)$ described in Example 2.2.2.

  (i) On the space of sequences for which $\sum_{i=1}^{\infty} |\xi_i| < \infty$,

$$d_1(x, y) = \sum_{i=1}^{\infty} |\xi_i - \eta_i|$$

  is a metric.

  (ii) On the space of all bounded sequences,

$$d_\infty(x, y) = \sup_i \{|\xi_i - \eta_i|\}$$

  is a metric.

  (iii) On the space of infinite sequences,

$$d(x, y) = \sum_{i=1}^{\infty} \frac{1}{2^i} \frac{|\xi_i - \eta_i|}{1 + |\xi_i - \eta_i|}$$

  is a metric.

  (iv) On the space of sequences for which $\sum_{i=1}^{\infty} |\xi_i|^p < \infty$,

$$d_p(x, y) = \left\{ \sum_{i=1}^{\infty} |\xi_i - \eta_i|^p \right\}^{1/p}$$

  is a metric.

(c) (i) Consider the vector space of all continuous functions $x(t)$ defined on $[0, 1]$. We may take

$$d_\infty(x, y) = \sup_{t \in [0,1]} \{|x(t) - y(t)|\}.$$

(ii) The same metric can be used on the set of all bounded functions on $[0, 1]$.

(iii) Corresponding to the d_p metric for $\mathscr{R}_n$ we have, for the vector space of continuous functions on $[0, 1]$, the family of metrics

$$d_p(x, y) = \left\{ \int_0^1 |x(t) - y(t)|^p \, dt \right\}^{1/p}, \qquad 1 \leqslant p < \infty.$$

(iv) Consider the set of functions $x(t)$ which are (absolutely) integrable to the rth power over $[0, 1]$: then the distance function d_p of (iii) with $1 \leqslant p \leqslant r$ is a metric for this set provided we agree to consider the functions $x(t)$, $y(t)$ as being equal if $d_p(x, y) = 0$. This does not necessarily imply that $x(t) = y(t)$ for all $t \in [0, 1]$. The difficulty arises here because we need to interpret integration in the Lebesgue rather than the Riemann sense, a point which is dealt with later (Example 3.3.1(e)) in some detail.

(d) Define the metric on an arbitrary space S by

$$d(x, y) = \begin{cases} 0 & \text{if } x = y, \\ 1 & \text{if } x \neq y. \end{cases}$$

This is called the *discrete* metric on S.

While this is a rather uninteresting metric, it shows that we may make any arbitrary set into a metric space; in addition, it is often useful to employ this metric to produce counter-examples showing that 'obvious' properties of the metric may not hold in all metric spaces.

The existence of a distance between points or elements of a set allows us to give a concrete meaning to convergence of a sequence and the related notion of limit in a metric space.

Definition 3.2.2 *A sequence* $\{x^{(k)}\}$ *of elements of* (S, d) *converges with respect to the metric to an element* $x \in (S, d)$ *if* $\lim_k d(x^{(k)}, x) = 0$. *The element x is called the* limit of the sequence *and we write*

$$\lim_k x^{(k)} = x \qquad \text{or} \qquad x^{(k)} \to x.$$

A sequence $\{x^{(k)}\}$ of elements of a metric space can converge to at most one limit. Suppose, on the contrary, that $x^{(k)} \to x$ and $x^{(k)} \to y$, then, from the triangle inequality,

$$d(x, y) \leqslant d(x, x^{(k)}) + d(y, x^{(k)});$$

but the right-hand side tends to zero as $k \to \infty$; consequently $d(x, y) \to 0$ and $x = y$.

The triangle inequality can also be used to show that the distance function $d(x, y)$ is continuous; that is, if $x^{(k)} \to x$ and $y^{(k)} \to y$ then $d(x^{(k)}, y^{(k)}) \to d(x, y)$ (Exercise 3.2,4).

Example 3.2.2 We shall now discuss the nature of convergence in some of the metric spaces of Example 3.2.1.

(a) In the space $\mathscr{R}_n$ (or $\mathscr{C}_n$) with metric d_p, convergence with respect to the metric is equivalent to componentwise convergence. That is to say, if

$$x^{(k)} = (\xi_1^{(k)}, \xi_2^{(k)}, \ldots, \xi_n^{(k)})$$

tends to the limit $x = (\xi_1, \xi_2, \ldots, \xi_n)$ then $\xi_i^{(k)} \to \xi_i$ for all $i = 1, \ldots, n$. Conversely, if $\xi_i^{(k)} \to \xi_i$ for all i then this implies that $d_p(x^{(k)}, x) \to 0$ (Exercise 3.2,5(a)).

For all finite-dimensional spaces, convergence with respect to a metric is equivalent to componentwise convergence; all metrics on finite-dimensional spaces are equivalent in this sense and the choice of metric is purely a matter of convenience.

(b) In infinite-dimensional spaces, while convergence with respect to the metric implies componentwise convergence, the converse is not necessarily true. That is to say, convergence with respect to the metric is a stronger condition than convergence of components.

For example, convergence of $x^{(k)} = (\xi_1^{(k)}, \xi_2^{(k)}, \ldots, \xi_n^{(k)}, \ldots)$ to $x = (\xi_1, \xi_2, \ldots, \xi_n, \ldots)$, $\sum_{i=1}^{\infty} |\xi_i|^2 < \infty$, with respect to the metric d_2 means that

$$\lim_k \sum_{i=1}^{\infty} (\xi_i^{(k)} - \xi_i)^2 = 0,$$

and this certainly implies that $\xi_i^{(k)} \to \xi_i$ for all $i = 1, 2, \ldots$. However, the converse will not follow, as a simple counter-example will make clear. Let $x^{(k)} = e^k$, the unit vector with unity in the kth place, and let $x = \theta$. For any i, $\xi_i^{(k)} = \delta_i^k$ while $\xi_i = 0$, and we see that it is certainly true that $\xi_i^{(k)} \to \xi_i = 0$: however, $\sum_{i=1}^{\infty} (\xi_i^{(k)} - \xi_i)^2 = 1$ whatever member of the sequence is considered, so that $d_2(x^{(k)}, x) \not\to 0$ and the sequence does not converge.

For the space of bounded sequences with metric $d(x, y) = \sup_i \{|\xi_i - \eta_i|\}$ convergence with respect to the metric is equivalent to uniform convergence of the components. For, if $x^{(k)} \to x$ then

$$d(x^{(k)}, x) = \sup_i \{|\xi_i^{(k)} - \xi_i|\} < \varepsilon$$

for all $k > k_0(\varepsilon)$, say, so that for $k > k_0(\varepsilon)$ and every i,

$$|\xi_i^{(k)} - \xi_i| < \varepsilon.$$

Conversely, suppose $|\xi_i^{(k)} - \xi_i| < \varepsilon$ for $k > k_0(\varepsilon)$ and all i, then $d(x^{(k)}, x) \to 0$. So we conclude that if the components of $x^{(k)}$ converge *uniformly* to the components of x then this implies convergence with respect to the metric. For this reason the metric used here is often called the *uniform metric*.

It is possible to show [23, 28] that simple convergence of components, namely $\xi_i^{(k)} \to \xi_i$ for all i, is equivalent to convergence with respect to the metric $\tilde{d}$ of Example 3.2.1 (b) (iii).

(c) On the space of continuous functions $x(t)$ defined on $[0, 1]$ convergence with respect to the metric

$$d_\infty(x, y) = \max_{t \in [0,1]} \{|x(t) - y(t)|\}$$

is equivalent to uniform convergence of functions in the usually accepted sense. For example, the sequence of functions

$$x^{(k)}(t) = t^k(1-t)$$

converges uniformly to $x(t) = 0$ in $[0, 1]$ since

$$d_\infty(x^{(k)}, x) = \max_t \{t^k(1-t)\}$$

$$= \left(\frac{k}{k+1}\right)^k \frac{1}{1+k} < \frac{1}{1+k},$$

which tends to zero as $k \to \infty$.

Convergence with respect to the metrics d_p of Example 3.2.1 (c)(iii) is often called 'convergence in the mean of order p': this type of convergence is less stringent than uniform convergence. In fact, we can say that, on a bounded interval, uniform convergence implies convergence in the mean of order $p > 1$ and furthermore convergence in the mean of order p implies convergence in the mean of every order less than p; the converse of these statements is *not* true. It is for this reason that the uniform metric is given the symbol d_∞.

The following example illustrates that convergence with respect to d_1 does not imply pointwise convergence. Consider the sequence of functions

$$x^{(k)}(t) = \begin{cases} kt, & t \in [0, 1/k], \\ 1, & t \in [1/k, 1], \end{cases}$$

which converges (pointwise) to the function $x(t) = 1$, except at $t = 0$. We shall show that, with respect to the metric d_1, the $x^{(k)}$ converge to $x(t) = 1$. We have

$$d_1(x^{(k)}, x) = \int_0^1 (x(t) - x^{(k)}(t)) dt$$

$$= \int_0^{1/k} (1-kt)dt = \frac{1}{2k},$$

which tends to zero as $k \to \infty$. A similar result would be true for the metric d_2.

Practically speaking, the metrics d_p measure the distance between two functions $x(t)$ and $y(t)$ in an integral or mean sense; this type of measure is very useful (particularly for $p = 1, 2$) in many applications as we shall see in later chapters.

It can be shown [23] that pointwise convergence of functions cannot be associated with any metric on a non-trivial class of functions.

We can give an alternative characterization of convergence in a metric space by using the concept of a ball.

Definition 3.2.3 *The set $B(x_0, r)$ of all points $x \in (S, d)$ satisfying the inequality $d(x_0, x) < r$ is called an* open ball *with centre $x_0 \in (S, d)$ and radius $r > 0$. If the inequality is replaced by $d(x_0, x) \leq r$ we speak of a* closed ball.

We then have: $x = \lim_k x^{(k)}$ if, and only if, for every ball $B(x, \varepsilon)$ with centre at x there exists $k_0(\varepsilon)$ such that $x^{(k)} \in B(x, \varepsilon)$ for $k > k_0$.

This implies that every sphere with centre at x must contain infinitely many elements of the sequence. The elements of a sequence must have different indices, but they need not be distinct points of (S, d); for example, the sequence $\{1, 1, 1, 1, \ldots\}$ has the limit 1.

A subset of a metric space is called *bounded* if it is contained in some ball of finite radius.

Example 3.2.3 Let us consider the 'shape' of the unit balls associated with some of the metric spaces of Example 3.2.1.

(a) Figure 3.1 shows the unit closed balls $d_p(x, 0) \leq 1$ for $p = 1, 2, \infty$ for the vector space $\mathscr{R}_2$. For these examples the idea of a bounded set conforms with what we expect in geometric terms. However, if we

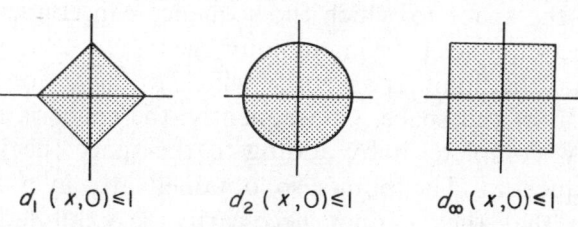

Fig. 3.1

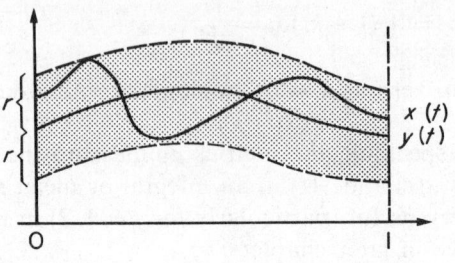

Fig. 3.2

consider the metric $\tilde{d}$ on $\mathcal{R}_2$ then we see that the whole space is bounded since it is contained within the unit ball; in this case boundedness does not really mean very much.

(b) In the vector space of continuous functions defined on $[0, 1]$ with the uniform metric d_∞ the ball of radius r with centre $y(t)$ consists of all functions $x(t)$ for which $|x(t) - y(t)| < r, t \in [0, 1]$ (see Fig. 3.2).

(c) The points in $B(0,1) < 1$ for the metric

$$d_1(x, y) = \int_0^1 |x(t) - y(t)| \, dt$$

are all those functions $x(t)$ for which

$$\int_0^1 |x(t)| \, dt < 1.$$

Complete metric spaces

We have defined convergence in a metric space by saying that $x^{(k)} \to x$ if $d(x^{(k)}, x) \to 0$. The question that immediately springs to mind from a practical point of view is: 'How can we test convergence of a sequence if we do not, *a priori*, know the limit?' We could argue that if successive terms of the sequence are approaching each other (in distance) then they are approaching something, namely the limit. Unfortunately this line of reasoning may not hold because there is no point in the space to which the sequence can converge; loosely speaking, the sequence is trying to converge to a 'hole' in the space in the sense that the 'limit' is an entity which we have not included in the definition of the space. Consequently, the space is incomplete; but we may 'complete' it by adding to the space the limits of all possible sequences. The points so obtained are, in a sense, ideal elements in that they cannot necessarily be exhibited, but have existence only as the limits of sequences.

One of the more interesting aspects of functional analysis is to see what 'ideals' should be added in order to complete various metric spaces: we shall investigate several examples shortly. The importance of using only metric spaces which are complete is related to the practical consideration that many technical problems are solved approximately by computing sequences of functions which, we hope, are better and better approximations to the solution of the problem; it would be most unfortunate if our sequence of approximations were to disappear down a 'hole'!

As a simple example, consider the subspace $(0, 1]$ of the real line with the metric d_1; the members of the sequence $1/k$, $k = 1, 2, \ldots$, certainly approach one another in distance, but the sequence does not converge in the subspace because the obvious limit 0 is not in the space: we can readily 'complete' the space by adjoining or including the point 0.

A less trivial example and one which illustrates 'ideals' is that of the rational numbers. The space of rational numbers is known to be incomplete because sequences exist which may not converge to rational numbers but converge, for example, to an irrational number like $\sqrt{2}$. The space is completed by adjoining all the irrational numbers, so giving the real line. But what is an irrational number, save the limit of a sequence of rational numbers? It is an ideal in the sense that one cannot exhibit $\sqrt{2}$ as a rational number, but only as a sequence of approximations each of which is a rational number. This does not prevent us in any way from using an entity to which we give the symbol $\sqrt{2}$ having the property that $(\sqrt{2})^2 = 2$. The answer to the question, 'Have you seen $\sqrt{2}$?' must be, 'No, but I have seen arbitrarily close approximations to it'. From an engineering or physics point of view we are content with such approximations, and in essence every numerical answer to a practical problem must be considered in this context.

We begin to put these heuristic ideas on a firmer basis by defining a Cauchy sequence.

Definition 3.2.4 *A sequence $\{x^{(k)}\}$ in a metric space (S, d) is said to be a* Cauchy (*or* fundamental) *sequence if $d(x^{(k)}, x^{(l)}) \to 0$ as $k, l \to \infty$. This means that for every $\varepsilon > 0$ there exists N_ε such that $d(x^{(k)}, x^{(l)}) \leq \varepsilon$ for any $k, l \geq N_\varepsilon$.*

Every convergent sequence in a metric space is a Cauchy sequence, but a Cauchy sequence may not be convergent in the space, as we have already discussed.

120 METRIC, NORMED AND INNER PRODUCT SPACES

Definition 3.2.5 *A metric space (S, d) is said to be* **complete** *if every Cauchy sequence in (S, d) has a limit in (S, d).*

Here are two examples of incomplete metric spaces.

Example 3.2.4

(a) Consider the set of all sequences $(\xi_1, \xi_2, \ldots, \xi_k, \ldots)$ in which only a finite number of terms are not equal to zero with the metric

$$d_\infty(x, y) = \max_i |\xi_i - \eta_i|.$$

Define a sequence of elements by

$$x^{(k)} = (1, \tfrac{1}{2}, \tfrac{1}{3}, \ldots, 1/k, 0, 0, \ldots).$$

This is a Cauchy sequence, since

$$d_\infty(x^{(k)}, x^{(l)}) = \max(1/k, 1/l)$$

tends to zero as $k, l \to \infty$. However, the sequence cannot converge to any element having only a finite number of non-zero terms.

(b) Let S be the set of continuous functions on $[0, 1]$ with the metric

$$d_1(x, y) = \int_0^1 |x(t) - y(t)| \, dt.$$

Define a sequence of continuous functions by

$$x^{(k)}(t) = \begin{cases} 0, & 0 \leq t \leq [\tfrac{1}{2} - 1/k], \\ kt - k/2 + 1, & [\tfrac{1}{2} - 1/k] \leq t \leq \tfrac{1}{2}, \\ 1, & t \geq \tfrac{1}{2} \end{cases} \quad k \geq 2;$$

(*see* Fig. 3.3). We have

$$d_1(x^{(k)}, x^{(l)}) = \tfrac{1}{2} |1/k - 1/l|$$

so that $d_1(x^{(k)}, x^{(l)}) \to 0$ as $k, l \to \infty$ and the sequence is Cauchy.

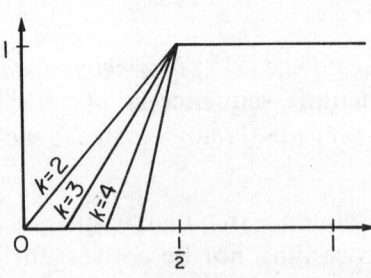

Fig. 3.3

Now suppose there is a continuous limit function $x(t)$, then

$$d_1(x^{(k)}, x) = \int_0^{1/2-1/k} |x(t)|\,dt + \int_{1/2-1/k}^{1/2} |x^{(k)}(t) - x(t)|\,dt$$
$$+ \int_{1/2}^{1} |1 - x(t)|\,dt$$

and

$$\lim_k d_1(x^{(k)}, x) = 0$$

implies

$$x(t) = \begin{cases} 0, & t \in [0, \tfrac{1}{2}) \\ 1, & t \in (\tfrac{1}{2}, 1]; \end{cases}$$

but since no continuous function possesses this property, the sequence $\{x^{(k)}\}$ has no limit in the space of continuous functions with metric d_1.

The stumbling block here appears to be the requirement of continuity, so we abandon it and consider the space of all functions which are bounded on $[0, 1]$ with metric d_1. Unfortunately this metric space is not complete. For example, the sequence of functions

$$x^{(k)}(t) = \begin{cases} \sqrt{k}, & 0 \leqslant t \leqslant 1/k, \\ 1/\sqrt{t}, & 1/k \leqslant t \leqslant 1, \end{cases}$$

is easily seen to be a Cauchy sequence (Exercise 3.2, 7(a)) which does not converge to a bounded integrable function but to the function

$$x(t) = \begin{cases} 0, & t = 0 \\ t^{-1/2}, & 0 < t \leqslant 1. \end{cases}$$

Very well then; let us extend the space of functions considered to include unbounded functions for which an improper integral exists, and interpret the integral in d_1 as an improper integral when necessary. The metric space so obtained is still not complete and it fails to be complete for a very subtle reason connected with the fact that we are implying the use of the Riemann integral in defining d_1. One can construct sequences of functions which have, as limits, functions with an infinite number of finite discontinuities in $[0,1]$. Such limit functions are not Riemann integrable and in order to proceed further we have to invoke the more general Lebesgue integral. We return to this point in Example 3.3.1(e). A simple example of a function which is not Riemann integrable in $[0, 1]$ is

$$x(t) = \begin{cases} 1, & \text{if } t \text{ is rational,} \\ 0, & \text{if } t \text{ is irrational.} \end{cases}$$

As the outlook appears to be rather gloomy, let us gain some reassurance at this stage by exhibiting one or two complete metric spaces.

Example 3.2.5

(a) The vector space $\mathscr{R}$, the real line, with the metric d_1 is complete. We have already seen, in an heuristic way, how the limits of rational sequences lead to the real numbers. This is a celebrated and fundamental result in analysis and will simply be assumed known in all that follows [10]. The completeness of the vector space $\mathscr{C}$ with the usual modulus metric follows from the completeness of $\mathscr{R}$.

(b) We can show the completeness of $\mathscr{V}_n$ by using the completeness of $\mathscr{R}$ (or $\mathscr{C}$). We saw in Example 3.2.2(a) that, for all finite-dimensional spaces, convergence with respect to a metric is equivalent to componentwise convergence. Suppose that the sequence

$$x^{(k)} = \sum_{i=1}^{n} \xi_i^{(k)} x^i$$

is Cauchy in $\mathscr{V}_n$: then $|\xi_i^{(k)} - \xi_i^{(l)}| \to 0$ as $k, l \to \infty$ for every i. Each sequence of coordinates is a Cauchy sequence and hence has a limit in $\mathscr{R}$ or $\mathscr{C}$. Hence the vectors $x^{(k)}$ converge (with respect to the metric) in $\mathscr{V}_n$.

(c) The continuous functions on $[0, 1]$ with metric d_1 were seen in Example 3.2.4(b) to lead to an incomplete metric space. However, if we adopt the uniform metric d_∞ we can use the fact that the uniform limit of a sequence of continuous functions is continuous and that the space is therefore complete. That our expectations are fulfilled is readily verified. Let $\{x^{(k)}\}$ be a Cauchy sequence, so that

$$d_\infty(x^{(k)}, x^{(l)}) = \max_{t \in [0,1]} |x^{(k)}(t) - x^{(l)}(t)| \to 0$$

as $k, l \to \infty$. For every fixed $t \in [0, 1]$ the values of $x^{(k)}(t)$ will form a Cauchy sequence of real or complex numbers. By the completeness of the real or complex numbers $\{x^{(k)}(t)\}$ converges to a number $x(t)$ and this determines a function $x(t)$ defined on $[0, 1]$. If we can show that $x^{(k)}(t) \to x(t)$ uniformly as $k \to \infty$ then the metric space is shown to be complete.

For a given $\varepsilon > 0$, take N so large that $|x^{(k)}(t) - x^{(l)}(t)| < \varepsilon$ for all $k, l > N$ and all $t \in [0, 1]$. Now let $l \to \infty$ for fixed t, giving

$$|x^{(k)}(t) - x(t)| < \varepsilon$$

for all $k > N$ and all $t \in [0, 1]$. This means that $x^{(k)}(t) \to x(t)$ uniformly in [0, 1] and since the uniform limit of a sequence of continuous functions is continuous the required result follows.[†]

It would be as well, at this point, to clarify the relationship between completeness and closure in a metric space. A subset of a metric space is *closed* if the set contains the limit of every convergent sequence in the set. The *closure* of a set S, denoted $\bar{S}$, is the set S together with the limits of all convergent sequences. Closure and closed set are general topological notions (Appendix 3.B) which do not in any way depend on the existence of a metric but, for our purposes, the definition just given will always suffice. Completeness, on the other hand, only has meaning in a metric space and specifically allows us to infer convergence of a sequence if it is Cauchy. The two concepts are obviously closely related in a metric space: indeed, we may assert that if (S, d) is a complete metric space and (M, d) is a subspace of (S, d) then (M, d) is complete if, and only if, M is a closed set in (S, d) [25, 29].

Dense subsets

A famous theorem in analysis due to Weierstrass states that if f is a continuous function on the closed interval [0, 1] then for every number $\varepsilon > 0$ there exists a polynomial function $p(t)$ such that

$$|f(t) - p(t)| < \varepsilon$$

for all $t \in [0, 1]$. The theorem means that we can approximate a continuous function in [0, 1] arbitrarily closely (in modulus) by a polynomial [9].

The polynomials $p(t)$ are obviously a subspace of the metric space of continuous functions with the uniform metric. With this interpretation we can see that the theorem asserts that there is a subset of functions (namely the polynomials) such that the distance between an arbitrary element of the space and an element of the subset can be made as small as we please.

[†] This result is proved in advanced calculus [9]. We give a brief outline of the proof. Continuity of $x^{(k)}(t)$ at t_0 implies $|x^{(k)}(t) - x^{(k)}(t_0)| < \varepsilon$ for $|t - t_0| < \delta$. By uniform convergence we can find an N such that for $k > N, |x^{(k)}(t) - x(t)| < \varepsilon$. Now

$$|x(t) - x(t_0)| \leq |x(t) - x^{(k)}(t)| + |x^{(k)}(t) - x^{(k)}(t_0)| + |x^{(k)}(t_0) - x(t_0)|$$

and each term on the right can be made smaller than ε, hence $|x(t) - x(t_0)| < 3\varepsilon$ and the continuity of $x(t)$ follows.

We extend this idea to a general metric space by the following definition.

Definition 3.2.6 *A subset A of the metric space (S, d) is said to be* dense *(or everywhere dense) in (S, d) if, for every $x \in (S, d)$ and $\varepsilon > 0$, there is a $y \in A$ such that $d(x, y) < \varepsilon$.*

If a set $A^{(2)}$ is dense in $(A^{(1)}, d^{(2)})$, $A^{(1)}$ is dense in $(S, d^{(1)})$ and convergence with respect to $d^{(2)}$ implies convergence with respect to $d^{(1)}$ then $A^{(2)}$ is dense in $(S, d^{(1)})$.

Example 3.2.6

(a) Weierstrass' theorem referred to above asserts that the set of all polynomials on $[0, 1]$ is dense in the space of continuous functions with the uniform metric.

(b) In the real line $\mathscr{R}$ with the metric d_1, the rational numbers are a dense subset. Every real number can be approximated as closely as we desire by a rational number. Similarly, in the vector space $\mathscr{R}_n$ with any metric, the set of vectors with rational coordinates is dense.

This last example illustrates an important idea. The rationals are an incomplete metric space, but if we add the limits of all Cauchy sequences of rationals we obtain the complete metric space of the reals. This suggests that any incomplete metric space might be considered to be a dense subset of a complete metric space. Before we can state this as a general result, there are two points that require clarification. The first concerns the 'equivalence' of Cauchy sequences. We should not add more limits to our incomplete space than is strictly necessary. In particular, we should regard two Cauchy sequences $\{x^{(k)}\}, \{\tilde{x}^{(k)}\}$ as equivalent if $d(x^{(k)}, \tilde{x}^{(k)}) \to 0$ as $k \to \infty$. In this way all Cauchy sequences which converge to the same (generalized) limit are identified (or are members of the same equivalence class).

Second, we want to ensure that the (ideal) limit elements which we add are subject to the same metric as those of the incomplete space. We do this by stipulating that if x, y are the limits of any two equivalence classes of Cauchy sequences $\{x^{(k)}\}, \{y^{(k)}\}$ then

$$d(x, y) = \lim_k d(x^{(k)}, y^{(k)}).$$

The addition of the limits then gives a larger metric space which has the same metric structure as the incomplete space. Metric spaces which are essentially the same and differ only in the 'labelling' of elements are termed *isometric*.

Theorem 3.2.1 *Any incomplete metric space (S, d) can be embedded, as a dense subset, in a complete metric space. The latter is called a* completion *of (S, d). Any two completions are isometric and hence we may speak of* the completion of (S, d).

We shall not give a proof of this theorem [26, 30].

Separable space

The rational numbers are countable, that is, they can be systematically arranged so that they are in one-to-one correspondence with the positive integers: the real numbers, on the other hand, are uncountable. The above Example 3.2.6(b) shows that we may approximate to any element of an uncountable set by an element of a countable set. A generalization of this idea to general metric spaces brings us to

Definition 3.2.7 *A metric space is said to be* separable *if it contains a countable subset which is (everywhere) dense. A separable space may contain more than one set which is dense and countable.*

Example 3.2.7

(a) The space $\mathscr{V}_n$ with any metric is separable since the set of vectors with rational coordinates is countable (cf., Example 3.2.6(b)).

(b) We have seen in Example 3.2.6(a) that the set of polynomials on $[0, 1]$ is dense in the space of continuous functions on $[0, 1]$ with the uniform metric. But for every polynomial in $[0, 1]$ we can find a polynomial with rational coefficients which will approximate it as closely as we desire. The set of polynomials with rational coefficients is countable, hence we conclude that the space of continuous functions on $[0, 1]$ with the uniform metric is separable.

(c) The space of infinite sequences $\{\xi_1, \xi_2, \ldots, \xi_n, \ldots\}$, $\sum_{i=1}^{\infty} \xi_i^2 < \infty$, with the metric d_2 is separable (cf., Example 3.2.1(b)(iv)).

Let A be the subset of sequences of the form $(\alpha_1, \alpha_2, \alpha_3, \ldots, \alpha_n, 0, 0, 0, \ldots)$ having only a finite number of non-zero entries and in which the α_i are rational numbers.

Let $x = (\xi_1, \xi_2, \ldots)$ be an arbitrary element of the space. Since $\sum_{i=1}^{\infty} \xi_i^2 < \infty$ it follows that, for any $\varepsilon > 0$, there exists an integer N

such that
$$\sum_{i=N+1}^{\infty} \xi_i^2 < \frac{\varepsilon^2}{2}.$$

Now choose an element $\tilde{x} = (\alpha_1, \alpha_2, \ldots, \alpha_N, 0, 0, \ldots)$ such that
$$\sum_{i=1}^{N} (\xi_i - \alpha_i)^2 < \frac{\varepsilon^2}{2}.$$

We have
$$(d_2(x, \tilde{x}))^2 = \sum_{i=1}^{N} (\xi_i - \alpha_i)^2 + \sum_{i=N+1}^{\infty} \xi_i^2 < \varepsilon^2,$$
whence
$$d_2(x, \tilde{x}) < \varepsilon.$$

The set of finitely non-zero sequences with rational components is countable and we have shown it to be dense in the space of square summable sequences with the metric d_2. Hence this metric space is separable.

(d) The space of bounded infinite sequences with the metric d_∞ is not separable (cf., Example 3.2.1(b)(ii)). Consider the set S of all elements $x = (\xi_1, \xi_2, \xi_3, \ldots)$ in which ξ_i is either 0 or 1. This set is uncountable, since we may consider the sequence as the binary representation of a number in $[0, 1]$. For any two elements $x, y \in S$, $d_\infty(x, y) = 1$. Now suppose there exists in the space a countable dense subset A. Surround each of the elements of A by an open ball of radius $\varepsilon \leq \frac{1}{2}$. Since S is uncountable it follows that at least one of the balls must contain more than one element of S. But this is not possible since the elements of S are unit distance apart.

In a sense, the statement that a metric space is separable is equivalent to saying that it does not contain an impossibly large number of elements. We can reach every element of the space through a sequence of elements belonging to a countable set which, in terms of cardinality, is the 'smallest' type of infinite set.

We may recall that when we discussed basis in Chapter 2, the point was made that, for infinite-dimensional vector spaces, the Hamel or algebraic basis was of little or no practical significance. The separable (vector) metric space with its dense countable subset is perhaps just the substitute for the algebraic basis that we might hope to find having introduced the concept of distance into our vector spaces. Most of the metric spaces which will be of value to us are separable and we shall discuss the nature of the possible associated dense, countable subsets when we deal with Banach spaces in Section 3.3. Could we go further and ask whether there is a *finite* set of points in a metric space, such that every point of the space is 'near' a point of the finite set? The answer to this question is only a qualified 'yes', since the imposition of this type of requirement on a metric space is so severe that in practical terms it is useful only in very special circumstances.

Total boundedness and compact spaces

A set in a metric space is bounded if it lies within some ball of finite radius. However, as we showed at the end of Example 3.2.3(a), boundedness may not be a very strong condition. We would like to have a property which is analogous to the following property of a closed and bounded interval $[a, b]$ of the real line: for any number $\varepsilon > 0$ we can find a finite set of points $x_n \in [a, b]$ such that, for any $x \in [a, b]$, there exists x_n such that $|x - x_n| < \varepsilon$. For example, the point set $x_n = a + n(b - a)/N, n = 1, \ldots, N - 1$, will do if $(b - a)/N < \varepsilon$. Notice that we cannot do this if the interval is not bounded. The property of the interval $[a, b]$ that we are illustrating here is that of total boundedness. For complete metric spaces we can approach the idea of compactness through that of total boundedness.

Definition 3.2.8 *A subset A of a metric space (S, d) is said to be* totally bounded *if, for every $\varepsilon > 0$, A contains a finite set A_ε called an ε-net, such that for each $x \in (S, d)$ there is a $y \in A_\varepsilon$ such that $d(x, y) < \varepsilon$.*

Notice that the definition specifies 'every $\varepsilon > 0$'; making ε smaller will enlarge the finite ε-net, but it will still exist.

If the metric space (S, d) is totally bounded then it is separable (Exercise 3.2,9(a)). Let us give an example of a subset of a metric space which is bounded but not totally bounded.

Example 3.2.8 Consider the space of infinite sequences with the metric d_2. The set of unit vectors e^k (cf., Example 3.2.2(b)) is bounded, as is clear, but not totally bounded. For $j \neq k$, we have $d_2(e^k, e^j) = \sqrt{2}$. Suppose an ε-net existed for $\varepsilon = \frac{1}{2}$, then this would imply that $A_{1/2}$ contains points which are distant less than $\frac{1}{2}$ from e^k, or equivalently, that $A_{1/2}$ must have points in each of the balls $B(e^k, \frac{1}{2})$. But the balls $B(e^k, \frac{1}{2})$, $B(e^j, \frac{1}{2})$ are disjoint for $k \neq j$ and are at least countably infinite. Therefore a finite ε-net does not exist and the set is not totally bounded. Every totally bounded set is bounded, but as this example shows the converse is not true: total boundedness is a much stronger condition than boundedness. The (countable) set $\{e^k\}$ fails to be totally bounded because its points are 'too widely' spaced: this is in contrast to the interval $[a, b]$ in which the points are uniformly closely crowded (or compact).

Definition 3.2.9 *A metric space (S, d) is said to be* compact *(or* sequentially compact*) if it is complete and totally bounded.*

Compactness can be defined in several different ways and is a topological property. Sequential compactness for a metric space is so called because if a set A is compact in a metric space (S, d) every infinite sequence in A has a convergent subsequence with limit in A (Example 1.2.4(g)). We shall occasionally use the convergent subsequence property to characterize a compact space, but it is not trivial to establish its equivalence to Definition 3.2.9 and we shall not do so here (*see*, for example, [24–26, 29, 30]).

Example 3.2.9

(a) If A is a finite subset of a metric space then A is compact: A is its own ε-net!

(b) A set A of the real line $\mathscr{R}$ with the usual metric is compact if, and only if, it is closed and bounded. (This is a famous result in analysis due to Heine and Borel [25].)

We can extend this result to show that a bounded closed subset of $(\mathscr{R}_n, d_2)$ is compact. First of all, we know that $(\mathscr{R}_n, d_2)$ (and its closed subsets) are complete, hence we need only to show that a bounded subset of $(\mathscr{R}_n, d_2)$ is totally bounded. Let A be contained in the closed unit ball $B(0, 1)$. Let A_ε be the set of n-tuples $(\eta_1, \eta_2, \ldots, \eta_n)$ such that η_j can take only the values $m/N, m \leqslant |N|$ and

$$\sum_{i=1}^{n} \eta_i^2 \leqslant 1.$$

The set A_ε is finite and contained in A. For any $x \in A$ we can find $y \in A_\varepsilon$ such that

$$|\xi_i - \eta_i| \leqslant 1/K$$

(the interval $[-1, 1]$ is compact). Hence

$$(d_2(x, y))^2 \leqslant \sum_{i=1}^{n} \frac{1}{K^2} = \frac{n}{K^2},$$

and if we choose K so that $\sqrt{n}/K < \varepsilon$ then we have an ε-net for $B(0, 1)$

in $\mathscr{R}_n$. The proof is not vitally dependent on the use of the metric d_2 rather than d_p and hence is true for $(\mathscr{R}_n, d_p)$ and by an obvious extension for $(\mathscr{C}_n, d_p)$.

(c) In Example 3.2.5(c) we saw that the space of continuous functions on [0, 1] with the uniform metric is complete. Can we find a totally bounded subset? The closed ball $B(0, 1)$ itself will not do, since it is not totally bounded (Exercise 3.2, 9(c)).

Consider the set of functions $x(t)$ such that

$$|x(t) - x(s)| < L|t - s|:$$

it can be shown (Exercise 3.2, 9(d)) that this set is totally bounded by exhibiting a finite set of piecewise linear functions which form an ε-net (*see* Fig. 3.4). Functions satisfying the above inequality are said to be Lipschitz and the existence of an ε-net shows the set to be compact. A more general result of this type exists called the Arzela–Ascoli theorem [25, 26] which states that compact subsets comprise those continuous functions which are (uniformly) bounded and equicontinuous.[†]

Continuous operators on compact sets

Let T be an operator (mapping, transformation) whose domain $\mathscr{D}$ and range $\mathscr{R}$ belong to metric spaces $(X, d_X), (Y, d_Y)$ respectively.

Definition 3.2.10 *An operator T, mapping a set $\mathscr{D} \subset X$ into a space Y is continuous at the point $x_0 \in \mathscr{D}$ if, for every $\varepsilon > 0$, there exists $\delta > 0$ such that*

$$d_Y(Tx, Tx_0) < \varepsilon$$

whenever

$$d_X(x, x_0) < \delta.$$

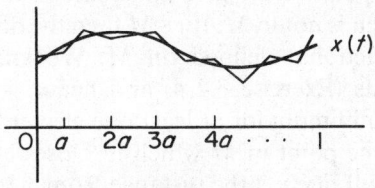

Fig. 3.4

[†] A collection of functions $x(t)$ defined on a closed interval is *equicontinuous* if, for any $\varepsilon > 0$, there exists $\delta > 0$ such that $|t_1 - t_2| < \delta$ implies $|x(t_1) - x(t_2)| < \varepsilon$ for all $x(t)$.

In general δ is dependent on both ε and x_0. T is continuous if it is continuous at each point of its domain. If a δ can be found which will serve for all $x_0 \in \mathscr{D}$ then T is said to be uniformly continuous.

The above definition coincides with the usual definition of continuity for a function of one variable when $X = Y = R$. The definition could be restated in terms of sequences by stating that T is continuous at x_0 if, for any sequence $x^{(k)} \to x_0 \in \mathscr{D}$, $Tx^{(k)} \to Tx_0 \in Y$.

If an ordinary function, f, of one variable is continuous on a closed interval $[a, b]$ of the real line, then the values $\sup[f(x)|x \in [a, b]]$ and $\inf[f(x)|x \in [a, b]]$ are finite and are actually attained at some points in $[a, b]$; that is to say, the maximum and minimum values of $f(x)$ are the extreme points of the (closed) range of f. Besides the continuity of f, this result is vitally dependent on the fact that a closed and bounded interval of the real line is compact. The corresponding result for operators on metric spaces is

Theorem 3.2.2 *If T is continuous, then the image of a compact set is compact.*

Proof Let $y^{(k)} = Tx^{(k)}$ be a sequence in $\mathscr{R}(T) \subset Y$. Since $\mathscr{D}(T)$ is compact, there is a subsequence $(x^{(k)})^{(p)}$ of $x^{(k)}$ which converges to a limit $x \in \mathscr{D}(T)$. Then it follows from the continuity of T that the subsequence $(y^{(k)})^{(p)} = T(x^{(k)})^{(p)}$ of the sequence $y^{(k)}$ converges to $y = Tx$ in $\mathscr{R}(T)$: hence $\mathscr{R}(T)$ is compact. ∎

Corollary *If T is a continuous mapping from a metric space to the real line (i.e., T is a functional) then the image of a compact set is closed and bounded and hence $T(x)$ attains its supremum and infinum.*

Example 3.2.10 Suppose we are given a subset M of a metric space (X, d) and an element x which is not in M. If $y \in M$ then the distance function $d(x, y)$ is, for fixed x, a functional defined on M. We know that the distance function is continuous (Exercise 3.2, 4) and hence if M is a compact set, $d(x, y)$ will attain its minimum for at least one element $y \in M$. That is to say, there exists at least one point in M which is 'closest' to x: let y_0 be such a point, then we may call $d(x, y_0)$ the distance from x to the set M and write this distance as $d(x, M)$.

Without the hypothesis of the compactness of M we cannot assert that a 'closest' vector even exists. We shall return to this problem in Section 3.3.

Exercises 3.2

1. (a) Show that the metrics d_1, d_2 and d_∞ on $\mathscr{R}_2$ (Example 3.2.1(a)) satisfy the inequalities $d_1 \geqslant d_2 \geqslant d_\infty$; illustrate by computing the distances d_1, d_2, d_∞ between the points (2, 3) and (5, 4).

   (b) If $\alpha \geqslant 0, \beta \geqslant 0, \gamma \geqslant 0$ and $\alpha \leqslant \beta + \gamma$ deduce that

   $$\frac{\alpha}{1+\alpha} \leqslant \frac{\beta}{1+\beta} + \frac{\gamma}{1+\gamma}$$

   and use this result to prove that $\tilde{d} = d/(1+d)$ is a metric if d is a metric.

   (c) Verify the metric axioms for the distance functions given in Example 3.2.1(b). (Hint: for (iii) establish a result similar to (b) above.)

   (d) Compute the distances d_1, d_2, d_∞ (Example 3.2.1(c)(i),(iii)) between the continuous functions

   (i) t and t^2 on $[0, 1]$,

   (ii) $e^{i\lambda t}$ and $e^{i\mu t}$ on $[-\pi, \pi]$,

   (iii) $x(t)$ and the zero function on $[0, 1]$, where

   $$x(t) = \begin{cases} \frac{1}{\varepsilon}\left[t - \left(\frac{1}{2} - \varepsilon\right)\right], & t \in \left[\frac{1}{2} - \varepsilon, \frac{1}{2}\right], \\ -\frac{1}{\varepsilon}\left[t - \left(\frac{1}{2} + \varepsilon\right)\right], & t \in \left[\frac{1}{2}, \frac{1}{2} + \varepsilon\right], \\ 0 & \text{elsewhere,} \end{cases}$$

   and where $0 < \varepsilon < \frac{1}{2}$.

   (e) Find the constant functions which are, in the sense of d_1, d_2 and d_∞ of Example 3.2.1(c)(iv), nearest to the step function

   $$x(t) = \begin{cases} 0, & t \in [0, \beta], \\ 1, & t \in [\beta, 1]. \end{cases}$$

   (f) Let S be the set of all infinite sequences of positive integers $n = \{n_1, n_2, n_3, \ldots\}$. Show that (S, d) is a metric space if we define

   $$d(n, m) = \begin{cases} 0, & \text{if } n_i = m_i, i = 1, 2, \ldots, \\ 1/N, & \text{where } N \text{ is the smallest index for which } n_i \neq m_i. \end{cases}$$

2. A real-valued function $\hat{d}(x, y)$ on a set S is a *pseudometric* (or *premetric*) if it satisfies axioms 2 and 3 of Definition 3.2.1, but satisfies a weaker

form of axiom 1 in which 'if, and only if,' is replaced by 'if'. Show that the relation on S defined by $x \sim y$ if, and only if, $\hat{d}(x, y) = 0$ is an equivalence relation. Verify that, in $\mathcal{R}_2$, $\hat{d}(x, y) = |\xi_1 - \eta_1|$ is a pseudometric.

3. Given a finite collection of metric spaces $(S^{(1)}, d^{(1)})$, $(S^{(2)}, d^{(2)})$, ..., $(S^{(n)}, d^{(n)})$ it is possible to define a metric on the product space $S^{(1)} \times S^{(2)} \times \ldots \times S^{(n)}$ in many different ways.

For example if $x = (x_1, x_2, \ldots, x_n)$ and $y = (y_1, y_2, \ldots, y_n)$ are points in $S^{(1)} \times S^{(2)} \times \ldots \times S^{(n)}$ then each of the following defines a metric:

(i) $d(x, y) = \sum_{i=1}^{n} d^{(i)}(x_i, y_i)$;

(ii) $d(x, y) = \max_{i} d^{(i)}(x_i, y_i)$;

(iii) $d(x, y) = \left\{ \sum_{i=1}^{n} (d^{(i)}(x_i, y_i))^2 \right\}^{1/2}$

Verify that each of these satisfies the axioms for a distance function in the case $n = 2$. Illustrate by considering $\mathcal{R}_2$ as $\mathcal{R}_1 \times \mathcal{R}_1$.

4. Prove that a distance function $d(x, y)$ is continuous by showing that if $x^{(k)} \to x$ and $y^{(k)} \to y$ then $d(x^{(k)}, y^{(k)}) \to d(x, y)$.

5. (a) Prove that, in the space $\mathscr{C}_n$ with metric d_p (Example 3.2.2(a)) componentwise convergence implies and is implied by convergence with respect to the metric.
   (b) (Example 3.2.1(c)) Discuss this conjecture: on a bounded interval, convergence in the mean of order p_1 implies convergence in the mean of order p_2 if $p_1 > p_2 \geqslant 1$. Give an example to illustrate that this is *not* true on an unbounded interval (*see* Exercise 3.3,4(e)).

6. (a) On the space of continuous functions on $[0, 1]$ show that convergence with respect to the uniform metric d_∞ (Example 3.2.2(c)) is equivalent to uniform convergence.
   (b) Show that the function sequence
   $$x^{(k)}(t) = k^2 t (1 - t)^k$$
   is pointwise convergent to $x(t) = 0$ on $[0, 1]$ but is not convergent with respect to the uniform metric.

7. (a) Show that the sequence
   $$x^{(k)}(t) = \begin{cases} \sqrt{k}, & 0 \leqslant t \leqslant 1/k, \\ 1/\sqrt{t}, & 1/k \leqslant t \leqslant 1, \end{cases}$$
   is Cauchy in the space of bounded functions on $[0, 1]$ with metric d_1.

(b) Are the following sequences of continuous functions on $[0, 1]$ Cauchy with respect to (i) the distance d_∞, (ii) the distance d_1?

$$x^{(k)}(t) = t^k;$$

$$x^{(k)}(t) = \begin{cases} kt^2, & t \in [0, 1/k], \\ t, & t \in [1/k, 1]. \end{cases}$$

(c) Consider the metric space of bounded real functions on $[0, 1]$ with the uniform metric d_∞ (Example 3.2.1(c)(ii)). Show that this metric space is complete.

(d) Show that if $\{x^{(k)}\}$ is a Cauchy sequence in a metric space (S, d) then the sequence $\{x^{(k)}\}$ is bounded. (Hint: find a finite ball which contains every point of the sequence.)

(e) If $(S^{(1)}, d^{(1)})$, $(S^{(2)}, d^{(2)})$ are complete metric spaces, show that the product space $S^{(1)} \times S^{(2)}$ with distance function $d^{(1)} + d^{(2)}$ is complete (Exercise 3.2.3).

8. Prove that every subset (A, d) of a separable metric space (S, d) is separable. (Hint: assume a countable dense set in (S, d)—label those points of (A, d) which are within distance $1/k$ of this set—show that these are countable and dense in (A, d).)

9. (a) Show that every totally bounded metric space is separable. (Hint: the union of $1/n$-nets, $n = 1, 2, \ldots$, is a countable dense set.)
   (b) Show that in $\mathscr{R}_n$ total boundedness is equivalent to boundedness.
   (c) Consider the metric space of continuous functions on $[0, 1]$ with the uniform metric. Show that the closed unit ball is not totally bounded and hence not compact. (Hint: consider the function set

   $$x_k(t) = \begin{cases} kt, & t \in [0, 1/k], \\ 1, & t \in [1/k, 1], \end{cases}$$

   assume the existence of an ε-net and obtain a contradiction.)
   (d) Show that the set S_L of functions $x(t)$ which satisfy a Lipschitz condition with constant L on $[0, 1]$ is totally bounded (Example 3.2.9(c)) by the following procedure:
      (i) Divide $[0, 1]$ into N equal subintervals

      $$0 = t_0 < t_1 < t_2 < \ldots < t_N = 1, t_i - t_{i-1} = 1/N.$$

      (ii) Define the set of all piecewise linear functions $y(t)$ on $[0, 1]$ (see Fig. 3.4) with $y(t_i) = k/M$ for some integer k and some positive integer M, $|k| \leq M$. Impose the condition $|y(t_i) - y(t_{i-1})| \leq L/N$: this set is finite.
      (iii) Show that there exists $y(t)$ such that

      $$|x(t_i) - y(t_i)| \leq 1/M, \quad x \in S_L,$$

and hence that the set of piecewise linear functions is an ε-net.

10. (a) Show that if $T: X \to Y$ is continuous (Definition 3.2.10) then $\lim_k T(x^{(k)}) = T(\lim_k x^{(k)})$ for every convergent sequence in X. Hence show that

$$\lim_k \int_0^1 x^{(k)}(t) dt = \int_0^1 \lim_k x^{(k)}(t) dt,$$

where the $x^{(k)}$ are the continuous functions on $[0, 1]$ with the uniform metric.

(b) Use the example of the function $x(t) = 1/t$ on $(0, 1]$ to show that the continuous image of an open bounded set need not be bounded.

(c) Let S be the subset of continuous functions on $[0, 1]$ for which $x(0) = 0$, $x(1) = 1$ and $d_\infty(x, 0) \leq 1$. Show that this bounded closed set is not compact by showing that the continuous functional

$$Tx = \int_0^1 x^2(t) dt$$

does not attain its infinum on this set (Corollary of Theorem 3.2.2). (Hint: using $x(t) = t^k$ show that inf $Tx = 0$.)

(d) Show that if $T: X \to Y$ is not uniformly continuous on $\mathscr{D}(T) \subset X$ then there exists $\varepsilon > 0$ and two sequences $\{x^{(k)}\}$, $\{y^{(k)}\}$ in $\mathscr{D}$ such that, for every k, $d_X(x^{(k)}, y^{(k)}) < 1/k$ but $d_Y(Tx^{(k)}, Ty^{(k)}) > \varepsilon$. Use this result to show that every T which is continuous on a compact set is uniformly continuous. (Hint: assume T is continuous but not uniformly continuous and obtain a contradiction.)

3.3 Normed vector spaces

The majority of the examples used in Section 3.2 to illustrate the features of metric spaces were also vector spaces. This is not altogether surprising, since it is natural to try to combine the algebraic structure of vector spaces with the topological structure of metric spaces, thereby generating a class of spaces with a rather rich structure. These spaces, called normed vector spaces, will be of major concern to us in the remainder of the book: they prove to be of great usefulness in the formulation and solution of the multiplicity of operator equations thrown up by constructing models of physical systems.

A metric on a vector space defines the 'distance' $d(x, y)$ between two points x and y. But every vector space contains the zero vector θ, and hence we can always find the distance between a non-zero vector

x and θ, namely $d(x, \theta)$. It is then natural to refer to this as the 'length' or *norm* of the vector x. Is this, then, all we need to convert a vector space into a normed vector space? Not quite. We must choose our distance function so that it depends, not simply on x and y as for a metric space, but on the *difference* between x and y, namely $(x - y)$.

With these preliminaries we can now define the length or *norm* function.

Definition 3.3.1 *A norm (or length function) on a vector space $\mathscr{V}$ is a real-valued function, $\|x\|$, defined for all vectors $x \in \mathscr{V}$ and which satisfies the following axioms*:

1. $\|x\| > 0$; $\|x\| = 0$ *if, and only if,* $x = \theta$;
2. $\|x + y\| \leqslant \|x\| + \|y\|$ $\forall x, y \in \mathscr{V}$;
3. $\|\lambda x\| = |\lambda| \|x\|$, λ an arbitrary scalar.

A normed vector space, *denoted by* $(\mathscr{V}, \|\cdot\|)$ *consists of a vector space $\mathscr{V}$ and a norm $\|\cdot\|$ on $\mathscr{V}$.*

Axiom 1 requires that the length of a vector is non-negative and that only the zero vector has length zero. Axiom 2 is a version of the triangle inequality for a vector space (cf., axiom 3, Definition 3.2.1). Axiom 3 (homogeneity) requires that the length of the vector λx should be $|\lambda|$ times the length of the vector x, which is in accord with our intuitive idea of multiplication by a scalar 'stretching' a vector.

A normed vector space may be given a natural metric by letting

$$d(x, y) = \|x - y\|.$$

It is not difficult to show (Exercise 3.3,1(a)) that this distance does satisfy the metric axioms. Furthermore

$$d(x, \theta) = \|x - \theta\| = \|x\|$$

so that the norm of any element is equal to its distance from the origin.

Thus, a normed space is a particular type of metric space and we may apply all our previous concepts associated with metric spaces to normed vector spaces. It is important to note that there are metrics (on a vector space) which are not generated by a norm in the above sense (one example is the metric $\tilde{d}$ of Example 3.2.1(a)). The relation of vector spaces, metric spaces and normed vector spaces is shown diagrammatically in Fig. 3.5.

As for metric spaces, a single vector space $\mathscr{V}$ may generate several different, normed vector spaces having different norm functions,

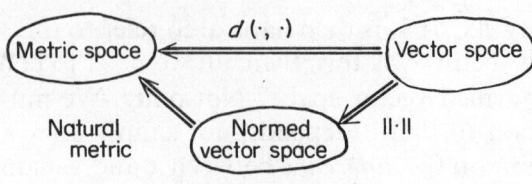

Fig. 3.5

$(\mathscr{V}, \|\cdot\|_1), (\mathscr{V}, \|\cdot\|_2)$, etc. Keeping this proviso in mind, it is usually convenient to denote $(\mathscr{V}, \|\cdot\|)$ simply by $\mathscr{V}$. A (normed) subspace of a normed vector space $\mathscr{V}$ is always taken to mean an algebraic subspace of $\mathscr{V}$ having the same norm function as the parent space.[†]

Since a normed vector space is a metric space, we can consider the questions of convergence of sequences, completeness, completion, separability, denseness and compactness to be settled. For example, in a normed vector space, a sequence of vectors $\{x^{(k)}\}$ converges to x if $\|x^{(k)} - x\| \to 0$ as $k \to \infty$. This is called *convergence with respect to the norm* or *strong convergence*.

A complete (with respect to the norm) normed vector space is called a *Banach space*. A subspace of a Banach space is itself a Banach space if, and only if, it is closed (p. 123).

We shall now consider a few examples of normed vector spaces; we have already, in effect, met these as examples in Section 3.2.

Example 3.3.1

(a) On $\mathscr{R}_n$ (or $\mathscr{C}_n$) we define a family of norms by

$$\|x\|_p = \left\{ \sum_{i=1}^{n} |\xi_i|^p \right\}^{1/p}, \quad 1 \leq p < \infty,$$

and

$$\|x\|_\infty = \max_i (|\xi_i|).$$

Convergence with respect to the norm is equivalent to componentwise convergence. The spaces $(\mathscr{R}_n, \|\cdot\|_p)$, $(\mathscr{C}_n, \|\cdot\|_p)$, $1 \leq p < \infty$, are Banach spaces. They are separable.

(b) On the space of sequences for which $\sum_{i=1}^{\infty} |\xi_i|^p < \infty$ we define the norm

$$\|x\|_p = \left\{ \sum_{i=1}^{\infty} |\xi_i|^p \right\}^{1/p}, \quad 1 \leq p < \infty.$$

This normed vector space is a Banach space usually denoted by l_p. These spaces are separable.

[†] In many texts the term *subspace* of a normed vector space is used to mean *closed algebraic subspace* [26, 28, 35].

(c) On the space of bounded sequences we define the norm
$$\|x\|_\infty = \sup_i \{|\xi_i|\}:$$
this is a Banach space usually denoted by l_∞. This space is *not* separable.

(d) The space of continuous functions on $[0, 1]$ with the uniform metric is a Banach space usually denoted by $\mathscr{C}[0, 1]$. Similarly, the Banach space of n-times continuously differentiable functions on $[0, 1]$ with
$$\|x\| = \max_{t \in [0,1]} \sum_{k=0}^{n} \left|\frac{d^k x}{dt^k}\right|$$
is usually denoted by $\mathscr{C}^{(n)}[0, 1]$. These spaces are separable.

(e) For the functions $x(t)$ defined on $[0, 1]$ for which
$$\int_0^1 |x(t)|^p \, dt < \infty, \quad 1 \leq p < \infty,$$
we define the norm
$$\|x\|_p = \left\{\int_0^1 |x(t)|^p \, dt\right\}^{1/p}$$

As we have already indicated (in Example 3.2.4), completion of this space involves the notion of Lebesgue integration; the reader will find a brief resumé of the theory in Appendix 3.C. With this proviso we can say that the space of functions which are pth power integrable with the indicated norm is a Banach space. In this space we cannot distinguish between functions whose values differ on a set of points in $[0, 1]$ of *measure zero*. By this we roughly mean that the set is so small as to have 'zero length': for example, an isolated point has measure zero and the union of a countable set of isolated points has measure zero. To overcome this difficulty we simply identify functions which differ only on a set of measure zero; we say that they are equal *almost everywhere*. The function $x(t)$ is, in this sense, only a typical representative of an equivalence class. These spaces are denoted by $\mathscr{L}_p[0, 1]$.

Many other normed vector spaces follow in an obvious way from previous examples of metric spaces. A summary of the common Banach spaces is given on p. 158–159.

It is worth noting here the following properties of *finite-dimensional* normed spaces:
1. every finite-dimensional normed vector space is complete (i.e., is a Banach space);
2. every subspace is closed;

3. boundedness implies total boundedness, hence a bounded subspace is compact.

Schauder basis

When discussing the implications of separability of a metric space it was pointed out that a dense countable set could be looked upon as a generalization of basis from finite- to infinite-dimensional spaces. This idea can be formally developed to yield what is called a Schauder basis for a separable Banach space.

Definition 3.3.2 *The countable set of vectors $\{x^n\}$ in a Banach space $\mathscr{V}$ is called a* Schauder basis *if, for every $x \in \mathscr{V}$, there is a unique series $\sum_{n=1}^{\infty} \xi_n x^n$ such that*

$$\lim_{N \to \infty} \left\| x - \sum_{n=1}^{N} \xi_n x^n \right\| = 0.$$

As for a Hamel basis, the vectors x^n must be linearly independent (i.e., no *finite* linear combination is the zero vector) otherwise the representation is not unique. The expansion of x is formally written

$$x = \sum_{n=1}^{\infty} \xi_n x^n.$$

A Banach space having a Schauder basis is separable, but the converse is not true. An example has been given [55] of a separable Banach space which does not have a Schauder basis.

Example 3.3.2

(a) In the family of spaces l_p (Example 3.3.1(b)), the unit vectors
$$e^1 = (1, 0, 0, 0, \ldots) \quad e^2 = (0, 1, 0, \ldots)$$
form a basis. Let $x = (\xi_1, \xi_2, \ldots, \xi_n, \ldots)$ and let
$$x^N = \sum_{n=1}^{N} \xi_n e^n = (\xi_1, \xi_2, \ldots, \xi_N, 0, 0, \ldots),$$
then we have
$$\lim_{N \to \infty} \| x - x^{(N)} \| = \lim_{N \to \infty} \left\{ \sum_{n=N+1}^{\infty} |\xi_n|^p \right\}^{1/p} = 0.$$

(b) The unit vectors are not a basis for the space of bounded sequences, l_∞. If we attempt to use the argument given in (a) we find the difficulty that

$$\|x - x^{(N)}\| = \sup_{n>N} |\xi_n|$$

may not tend to zero as $N \to \infty$. But we know that this space is not separable (Example 3.2.7(d)), so the result is not unexpected.

(c) The space $\mathscr{C}[0,1]$ is separable (Example 3.2.7(b)). A basis for this space can be constructed which consists of the functions t and $(1-t)$ together with functions which are in the form of isosceles triangles of unit height and with base in the intervals $(l/2^k, (l+1)/2^k)$, $k = 1, 2, 3, \ldots$; $l = 0, 1, 2, \ldots 2^{k-1}$ [23]. The sequence of partial sums $x^{(N)}$ clearly consists of polygons whose vertices lie on the curve $x(t)$ at equal intervals of t in $[0, 1]$ (see Fig. 3.6).

Before leaving this topic it would be as well to point out the difference between a Schauder basis and a *spanning set*[†] for an infinite-dimensional vector space. If $\{x^{(k)}\}$ is a set of vectors in $\mathscr{V}$ then the *algebraic span* of $\{x^k\}$ (Definition 2.3.2) consists of all finite linear combinations of the x^k. If $\{x^k\}$ is an infinite set, however, we can also consider the limits of all sequences of vectors which lie in the algebraic span of $\{x^k\}$: note that this extension needs the notion of convergence. Technically, we are now considering the closure of the algebraic span so that we refer to this space as the *closed span* of $\{x^k\}$. It is easy to see that the closed span is a *subspace* of $\mathscr{V}$. We would say that a set S of vectors was a *spanning set for* $\mathscr{V}$ if every vector in $\mathscr{V}$ was contained in the closed span of S.

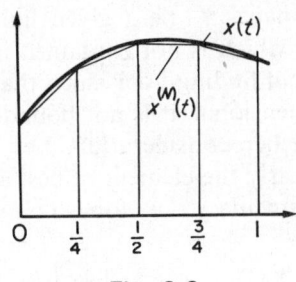

Fig. 3.6

[†] A spanning set $\{x^k\}$ is often referred to as a *complete set* satisfying the condition: for any $\varepsilon > 0$ there exists N and scalars $\{\xi_k\}$ such that

$$\|x - \sum_{k=1}^{N} \xi_k x^k\| < \varepsilon.$$

(See also Exercise 3.3,5). We shall always apply the description 'complete' to those spaces having the Cauchy sequence property.

The difference between spanning set and (Schauder) basis can be illustrated in the following way. Suppose $\{x^k\}$ is a countable set, then every vector in the closed span of $\{x^k\}$ can be represented in the form

$$x = \lim_{l \to \infty} x^{(l)} = \lim_{l \to \infty}(\xi_1^{(l)} x^1 + \xi_2^{(l)} x^2 + \ldots + \xi_n^{(l)} x^n),$$

whereas every vector in the space with basis $\{x^k\}$ takes the form

$$x = \sum_{k=1}^{\infty} \xi_k x^k.$$

Consider the set of functions $1, t, t^2, \ldots$ defined on $[0, 1]$. The closed span is the space of continuous functions on $[0, 1]$ with the uniform metric since we have already seen that (finite) polynomials are dense in this space (p. 124). Considered as a Schauder basis the functions $1, t, t^2, \ldots$, however, yield only the space of functions on $[0, 1]$ which have convergent power series expansions, that is the space of analytic functions on $[0, 1]$.

Example 3.3.3 (Best approximation) We saw in Example 3.2.10 that, if M is a compact subset of a metric space (X, d) and $x \in X$, there exists at least one point in M which is closest to x. As has been pointed out, the requirement of compactness is severe and we cannot often utilize this result. However, every finite-dimensional bounded subspace of a normed space is compact so, in this very important case, we can prove the existence of a solution to the 'best approximation problem' in a normed space.

Let $\mathscr{V}$ be a normed space, $\mathscr{W}$ be a given finite-dimensional subspace and $x \in \mathscr{V}$ a given vector which is not contained in $\mathscr{W}$. The 'best approximation problem' consists of finding $y \in \mathscr{W}$ such that $\|x - y\|$ is a minimum. Although $\mathscr{W}$ is finite-dimensional, it is not bounded, so we cannot exploit compactness without further consideration. Let us consider any $y_0 \in \mathscr{W}$ and let $\delta = \|y_0 - x\|$. Clearly the element of best approximation should be sought amongst those elements $y' \in \mathscr{W}$ for which $\|y' - x\| < \delta$.
Now

$$\|y'\| \leqslant \|y' - x\| + \|x\| \leqslant \|x\| + \delta = \Delta,$$

say, hence we may seek for the element of best approximation in the bounded subset $\|y'\| < \Delta$.

Since $\mathscr{W}$ is finite-dimensional, this subset is compact and it follows from the continuity of the norm function that there exists a $y \in \mathscr{W}$ such that $\inf \|y - x\|$ is actually attained. We have not discussed uniqueness of the element of best approximation, and this must be further investigated in any particular case. To show how nonuniqueness can arise, consider the vector

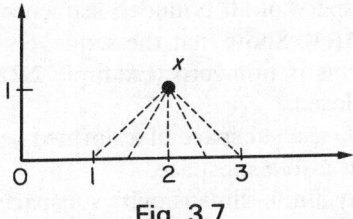

Fig. 3.7

space $\mathscr{R}_2$ with the norm $\|x\| = \max_i |\xi_i|$ with $\mathscr{W}$ the subspace of vectors with basis $(1, 0)$: let x be the point $(2, 1)$ (Fig. 3.7). Every vector $y = (a, 0)$ with $1 \leq a \leq 3$ has $\|y - x\| = 1$, which is clearly the minimum value of $\|x - y\|$; hence this minimum is attained for all vectors $(a, 0)$, $1 \leq a \leq 3$, and in the sense used here they are all elements of best approximation.

Let $\mathscr{W}$ have the basis $\{y^k\}$; then the problem of best approximation is reduced to finding the coefficients η_k for which

$$\left\| \sum_{k=1}^{n} \eta_k y^k - x \right\|$$

attains its least value. Finding the η_k is easy only in those spaces in which the norm is derived from an inner product (*see* Section 3.6).

Exercises 3.3

1. (a) Show that the axioms for a metric are satisfied by $d(x, y) = \|x - y\|$.
   (b) A real-valued function $|x|$ on a vector space $\mathscr{V}$ is a *seminorm* (or *pseudonorm*) if it satisfies axioms 2 and 3 of Definition 3.3.1 but satisfies a weaker form of axiom 1 in which 'if, and only if,' is replaced by 'if'. On the space of continuously differentiable functions on $[0, 1]$ show that

   $$|x| = \int_0^1 |x'(t)| \, dt$$

   is a seminorm and that

   $$\|x\| = \int_0^1 |x(t)| \, dt + |x'(t)| \, dt$$

   is a norm.
   (c) Show that the unit ball $B(0, 1) = \{x \mid \|x\| \leq 1\}$ is convex (Definition 2.3.6). Use this result to show that, in $\mathscr{R}_2$, the function

   $$\|\cdot\|_p = (|x_1|^p + |x_2|^p)^{1/p}, \quad p < 1,$$

   is not a norm.
   (d) Prove that $\|x - y\| \geq |\|x\| - \|y\||$.

2. (a) Consider the space of all bounded sequences with the norm $\|\cdot\|_\infty$ (Example 3.3.1(c)). Show that the sequences in which only a finite number of terms is non-zero (Example 2.2.2(b)) form a subspace which is not closed.
   (b) Show that if $\mathscr{W}$ is a subspace of a normed vector space $\mathscr{V}$ then the closure of $\mathscr{W}$ is also a subspace.
   (c) Show that any finite-dimensional subspace of a normed vector space is closed.

3. Let $(\mathscr{V}, \|\cdot\|_\mathscr{V})$, $(\mathscr{U}, \|\cdot\|_\mathscr{U})$ be normed vector spaces (over the same field). Show that the product space $\mathscr{V} \times \mathscr{U}$ is a normed vector space under any of the following norms:
   (i) $\|\cdot,\cdot\| = \|\cdot\|_\mathscr{V} + \|\cdot\|_\mathscr{U}$;
   (ii) $\|\cdot,\cdot\| = (\|\cdot\|_\mathscr{V}^p + \|\cdot\|_\mathscr{U}^p)^{1/p}$, $1 \leq p < \infty$;
   (iii) $\|\cdot,\cdot\| = \max\{\|\cdot\|_\mathscr{V}, \|\cdot\|_\mathscr{U}\}$.

4. Let $\mathscr{V}$ be a vector space with two norm functions $\|\cdot\|_1, \|\cdot\|_2$. These norms are said to be *equivalent* if there exist positive constants m and M such that
$$m\|x\|_2 \leq \|x\|_1 \leq M\|x\|_2 \quad \forall x \in \mathscr{V}.$$
   (a) Show that all sequences $\{x^{(k)}\}$ in $\mathscr{V}$ which converge to x under one norm also converge to x under the other.
   (b) Show that the norms $\|\cdot\|_p$, $1 \leq p < \infty$, on $\mathscr{C}_n$ are equivalent (to componentwise convergence—see, for example, Exercise 3.2,5(a)).
   (c) Given the inequality
$$(\alpha_1 + \ldots + \alpha_n)^p \geq \alpha_1^p + \alpha_2^p + \ldots + \alpha_n^p,$$
   $\alpha_i > 0$, show that the norm $\|x\|_p$ on $\mathscr{C}_n$ is a decreasing function of p. Compute the norm of the n-tuple $(1, 1, 1, \ldots, 1)$ for $p = 1, 2, \infty$.
   (d) Consider the space of piecewise continuous functions on $[0, \infty)$. By considering functions which are 1 for $t < t_0$ and zero elsewhere show that neither of the norms $\|\cdot\|_1, \|\cdot\|_2$ (cf., Example 3.3.1(e)) is bounded by a constant times the other, and hence that they are not equivalent on an unbounded interval.
   (e) Prove that, on the space of sth–power integrable functions on $[0, 1]$,
$$\left\{\int_0^1 |x(t)|^r dt\right\}^{1/r} \leq \left\{\int_0^1 |x(t)|^s dt\right\}^{1/s}, \quad 1 \leq r < s.$$
   What form does this inequality take on the interval $[\alpha, \beta]$? (Hint: use Hölders inequality, Appendix 3.A, to show
$$\int_0^1 |x(t)| dt \leq \left\{\int_0^1 |x(t)|^p dt\right\}^{1/p}$$
   then take $p = s/r$ and $x(t) = (z(t))^r$.

LINEAR TRANSFORMATIONS ON NORMED VECTOR SPACES 143

Hence show that $\|\cdot\|_p$ is an increasing function of p and that convergence with respect to $\|\cdot\|_s$ implies convergence with respect to $\|\cdot\|_r$. Show by counter-example that the norms $\|\cdot\|_s, \|\cdot\|_r$ are not equivalent. Compute the norm of t^2, $t\in[0,1]$ for $p = 1, 2, \infty$.

5. Contrast the definition of Schauder basis and spanning set, viz.,

   (i) $\quad x = \lim_{n\to\infty} \sum_{i=1}^{n} \xi_i x^i$;

   (ii) for any $\varepsilon > 0$ there exist n and scalars $\{\xi_i\}$ such that

   $$\left\| x - \sum_{i=1}^{n} \xi_i x^i \right\| < \varepsilon.$$

   (a) Show that a Schauder basis for $\mathscr{V}$ is also a spanning set for $\mathscr{V}$.
   (b) If a spanning set is also a Schauder basis show that, for every $x \in \mathscr{V}$, there are linear combinations

   $$x^{(l)} = \sum_{i=1}^{n} \xi_i^{(l)} x^i$$

   approximating x in which the scalars $\xi_i^{(l)}$ are independent of l.
   (c) Consider the countable set of vectors $\{x^i\} \in l_2$ given by

   $$x^i = (1, 0, 0, 0, \ldots, 0, 1, 0, \ldots),$$

   where x^i has a one in the first and the ith places. Show that the vector $(1, 0, 0, 0, \ldots)$ is in the closed span of the $\{x^i\}$. (Hint: try taking $\xi_i = 1/i$.)

6. Let $\sum_{n=1}^{\infty} x^n(t)$ be a series of continuous functions on $[0, 1]$. Assume that there exist positive numbers M_n such that

   $$\max_{t\in[0,1]} |x^n(t)| \leq M_n, \quad n = 1, 2, \ldots,$$

   and such that the series $\sum_{n=1}^{\infty} M_n$ converges. Prove that the series $\sum_{n=1}^{\infty} x^n(t)$ converges uniformly in $[0, 1]$. (This is the Weierstrass M-test for uniform convergence.)

3.4 Linear transformations on normed vector spaces

The naturalness of considering linear transformations on vector spaces was discussed in Section 2.5. Such transformations preserve the two vector space operations of addition and scalar multiplication. The addition of a norm to the vector space structure allows us to develop the theory of linear transformations rather further.

Perhaps it is as well to reiterate that by discussing linear transformations, we appear to be severely restricting the class of operator equations we can tackle. The principle of contraction mapping discussed in Section 3.10 and the development of operator derivatives in Chapter 4 serve as a counterbalance to the apparent concentration on the linearity assumption. Nevertheless, it is true to say that non-linear operator equations pose difficulties which are an order of magnitude greater than those associated with linear equations and the body of techniques available to deal with them is rather limited. The techniques which are available are largely based on the replacement of non-linear operators with a sequence of (locally) linear approximating operators. So perhaps the concentration on linear operators is justified rather more than one might suppose, thereby softening the cynical view that it is merely convenient for the mathematician.

We have already met the idea of continuity of an operator on a metric space (Definition 3.2.10). This idea carries over immediately to a normed space when $d(x, y)$ is replaced by $\|x - y\|$. By saying that $T: \mathscr{V} \to \mathscr{U}$ is continuous at $x_0 \in \mathscr{V}$ we mean that, for every $\varepsilon > 0$, there exists $\delta > 0$ such that if $\|x - x_0\| < \delta$ then $\|Tx - Tx_0\| < \varepsilon$.[†] When the operator is linear we can associate continuity with a property called 'boundedness', and this in turn leads on to the idea of the norm of a linear transformation.

Let $T: \mathscr{V} \to \mathscr{U}$ be a linear transformation which is continuous at the origin: then we can show that T is continuous everywhere in $\mathscr{V}$. We have, by hypothesis, if $\|x\| < \delta$, then $\|Tx\| < \varepsilon$: writing $x = x' - x_0$ gives

$$\|T(x' - x_0)\| = \|Tx' - Tx_0\| < \varepsilon, \quad \text{if } \|x' - x_0\| < \delta$$

which means that T is continuous at x_0, an arbitrary point of $\mathscr{V}$. We may therefore speak of a *continuous linear transformation*. We now make a connection between continuity and boundedness.

Definition 3.4.1 *Let $T: \mathscr{V} \to \mathscr{U}$ be a linear transformation: T is said to be* bounded *if there is a real number $K > 0$ such that*

$$\|Tx\| \leqslant K\|x\| \quad \forall x \in \mathscr{V}.$$

Theorem 3.4.1 *A necessary and sufficient condition that a linear operator $T: \mathscr{V} \to \mathscr{U}$ be continuous is that it is bounded.*

[†] Recall that the norms are those in $\mathscr{V}$ and $\mathscr{U}$ and should strictly be denoted by $\|\cdot\|_\mathscr{V}$ and $\|\cdot\|_\mathscr{U}$.

Proof (a) Assume T is bounded; then
$$\|Tx - Tx_0\| = \|T(x - x_0)\| \leq K\|x - x_0\|;$$
hence $\|x - x_0\| < \delta$ implies $\|Tx - Tx_0\| \leq K\delta = \varepsilon$ and T is continuous.

(b) Assume T is continuous and that it is not bounded. That is to say, we can find, for each positive integer n, a vector $x^{(n)}$ such that
$$\|Tx^{(n)}\| > n\|x^{(n)}\|$$
or, in another form,
$$\|Ty^{(n)}\| > 1,$$
where
$$y^{(n)} = \frac{x^{(n)}}{n\|x^{(n)}\|}.$$
Clearly
$$\|y^{(n)}\| = 1/n \to 0$$
as $n \to \infty$ so that
$$\lim_{n \to \infty} y^{(n)} = \theta.$$
But
$$\|Ty^{(n)}\| > 1$$
so that
$$\lim_{n \to \infty} Ty^{(n)} \neq \theta$$
and T is not continuous at the origin. The contradiction proves the boundedness of T. ■

Example 3.4.1

(a) Let T be the linear operator defined by
$$Tx(t) = \int_0^1 k(t - \tau)x(\tau)d\tau, \quad k \in \mathscr{L}_1[-1, 1], \quad t \in [0, 1].$$

We shall show that T is a bounded (and therefore continuous) transformation of $\mathscr{L}_1[0, 1]$ onto itself. We have

$$\|Tx\|_1 = \int_0^1 dt \left| \int_0^1 k(t-\tau)x(\tau)d\tau \right|$$

$$\leq \int_0^1 dt \int_0^1 |k(t-\tau)| \, |x(\tau)| d\tau$$

$$= \int_0^1 |x(\tau)| d\tau \int_0^1 |k(t-\tau)| dt$$

upon interchanging the order of integration.[†]

Now for every $\tau \in [0,1]$,

$$\int_0^1 |k(t-\tau)| dt = \int_{-\tau}^{1-\tau} |k(\tau')| d\tau':$$

define a function

$$k_1(\tau') = \begin{cases} 0, & -1 \leq \tau' \leq -\tau, \\ k(\tau'), & -\tau \leq \tau' \leq 1-\tau, \\ 0, & 1-\tau \leq \tau' \leq 1, \end{cases}$$

then since $k \in \mathscr{L}_1[-1,1]$ we see that,

$$\int_0^1 |k(t-\tau)| dt \leq K \qquad \forall \tau \in [0,1]$$

and

$$\|Tx\|_1 \leq K \|x\|_1.$$

(b) Let T be the linear operator

$$Tx \equiv \frac{dx}{dt}$$

defined on the space of continuously differentiable functions $x(t)$ on $[0,1]$. If we consider this space as a subspace of the space of continuous functions $\mathscr{C}[0,1]$ with the norm

$$\|x\| = \max_{t \in [0,1]} |x(t)|,$$

then T maps $\mathscr{D}(T) \subset \mathscr{C}[0,1]$ into $\mathscr{C}[0,1]$.

T is not bounded, however: for example, for $x(t) = \sin nt$,

$$Tx = n \cos nt$$

giving

$$\|Tx\| = n\|x\|$$

and the right-hand side can be made arbitrarily large.

[†] This requires that

$$\int_0^1 \int_0^1 k(t-\tau) d\tau \, dt < \infty$$

(Fubini's theorem [37]).

Let us now consider the space of continuously differentiable functions on [0, 1] with the norm

$$\|x\| = \max_{t\in[0,1]} \{|x(t)| + |dx/dt|\}$$

(Example 3.3.1(d)). Now if we define $T \equiv d/dt$ as a mapping $\mathscr{C}^{(1)}[0, 1] \to \mathscr{C}[0, 1]$ then T is clearly bounded (and therefore continuous).

These examples illustrate two rather general points:

1. integral operators are usually bounded, whereas differential operators are very often not;
2. operators are not fully defined unless the domain is specified so that whether an 'operator' is bounded or unbounded is dependent on the domain of definition.

We see that the descriptions 'bounded' and 'continuous' can be applied interchangeably. The set of bounded linear transformations $T: \mathscr{V} \to \mathscr{U}$ is itself a vector space whose elements are linear transformations (Section 2.5). Can we define a norm function on this space which will convert it into a normed vector space?

Definition 3.4.2 *Let $T: \mathscr{V} \to \mathscr{U}$ be a bounded linear transformation, that is,*

$$\|Tx\| \leq K\|x\|.$$

The smallest value of K which satisfies this inequality is denoted $\|T\|$ and called the norm *of T. We then have*

$$\|Tx\| \leq \|T\| \|x\|.$$

It is readily verified that this norm for operators satisfies the axioms for a norm function (Exercise 3.4,1(c)) and that we may therefore talk of the vector space of bounded linear transformations $T: \mathscr{V} \to \mathscr{U}$. Notice that the operator norm is subordinate to the norms of the vector spaces $\mathscr{V}$ and $\mathscr{U}$. The normed vector space of linear transformations $T: \mathscr{V} \to \mathscr{U}$ is usually denoted by $\mathscr{L}(\mathscr{V}, \mathscr{U})$. It can be shown that if $\mathscr{U}$ is a Banach space then $\mathscr{L}(\mathscr{V}, \mathscr{U})$ is a Banach space [25].

From the definition of operator norm we have

$$\|T\| = \sup_{x \neq \theta} \|Tx\|/\|x\|,$$

since K is an upper bound to $\|Tx\|/\|x\|$ and we want the least upper

bound or supremum. Alternatively (Exercise 3.4,1(d)),
$$\|T\| = \sup_{\|x\|=1} \|Tx\|.^\dagger$$

Example 3.4.2

(a) Let $T: \mathscr{R}_n \to \mathscr{R}_m$ have the matrix representation $[t_{ij}]$, $i = 1, \ldots, m$, $j = 1, \ldots, n$ (cf., Example 2.5.1). In the spaces $\mathscr{R}_n$, $\mathscr{R}_m$ we choose the norm
$$\|x\|_\infty = \max_k (|\xi_k|).$$

We shall find the induced operator norm for T which we denote by $\|T\|_\infty$. Let $\|x\|_\infty = 1$, then
$$\max_{1 \leq k \leq n} |\xi_k| = 1$$
and
$$\|Tx\|_\infty = \max_{1 \leq i \leq m} \left(\sum_{j=1}^n |t_{ij} \xi_j| \right) \leq \max_{1 \leq i \leq m} \sum_{j=1}^n |t_{ij}| |\xi_j|$$
$$\leq \max_{1 \leq i \leq m} \left(\sum_{j=1}^n |t_{ij}| \right);$$
that is
$$\|Tx\|_\infty \leq \max_{1 \leq i \leq m} \left(\sum_{j=1}^n |t_{ij}| \right), \quad \|x\|_\infty = 1,$$
so that
$$\|T\|_\infty \leq \max_{1 \leq i \leq m} \left(\sum_{j=1}^n |t_{ij}| \right).$$

Now T is a continuous mapping of the compact set $\|x\|_\infty = 1$, so the image set $\{Tx \mid \|x\|_\infty = 1\}$ is compact: further, the norm function is continuous and will map Tx onto a closed and bounded interval of the real line. Hence $\|T\|_\infty$ will actually attain its supremum on the set $\|x\|_\infty = 1$, so that if we can find a vector x with $\|x\|_\infty = 1$ such that
$$\|T\|_\infty = \max_{1 \leq i \leq m} \left(\sum_{j=1}^n |t_{ij}| \right)$$
we may conclude that this is the value of the required norm. Suppose the maximum row sum occurs for $i = k$ and let the components of

---

† A set of linear transformations $T_k: \mathscr{V} \to \mathscr{U}$ is said to be *uniformly bounded* if $\|T_k\| < M$.

$\tilde{x}$ be

$$\tilde{\xi} = \begin{cases} \operatorname{sgn} t_{kj}, & t_{kj} \neq 0, \\ 0, & t_{kj} = 0. \end{cases}$$

Then $\|\tilde{x}\|_\infty = 1$, and by a direct calculation,

$$\|T\tilde{x}\|_\infty = \sum_{j=1}^n |t_{kj}|.$$

We may conclude that

$$\|T\|_\infty = \max_{1 \leq i \leq m} \left(\sum_{j=1}^n |t_{ij}| \right),$$

the maximum row sum.

(b) By a similar argument we would find that, for the norm

$$\|x\|_1 = \sum_{k=1}^{m,n} |x_k|,$$

the induced norm for $T: \mathscr{R}_n \to \mathscr{R}_m$ is the maximum column sum, viz.,

$$\|T\|_1 = \max_{1 \leq i \leq n} \left(\sum_{i=1}^m |t_{ij}| \right).$$

Both of the above norms apply in complex spaces if the modulus is interpreted appropriately.

Inverse operators

In Section 2.5 we saw that if a linear transformation, T, is one-to-one $\mathscr{D}(T) \to \mathscr{R}(T)$ then it possesses a linear inverse $T^{-1}: \mathscr{R}(T) \to \mathscr{D}(T)$. The requirement that T be one-to-one is equivalent to the requirement that $\mathscr{N}(T) = \mathcal{O}$. These considerations naturally apply to linear transformations on normed spaces, but we can add to them in two important ways. First, we can determine whether T^{-1} is continuous and second we can find a condition on the norm of T which will guarantee the existence of T^{-1} as an alternative to the (algebraic) condition, $\mathscr{N}(T) = \mathcal{O}$.

Example 3.4.3 Consider the results of Examples 2.5.6 and 3.4.1(b). The former shows that the inverse of the operator

$$L(\cdot) \equiv \int_0^t \cdot \, dt :$$

$\mathscr{C}[0,1] \to \mathscr{C}[0,1], x(0) = 0$, is the operator $L^{-1}(\cdot) \equiv d\cdot/dt$. The operator

L is clearly bounded and therefore continuous, but, as we saw in Example 3.4.1(b), L^{-1} is unbounded on the subspace of continuously differentiable functions on $\mathscr{C}[0,1]$.

The following theorem allows us to state a condition on a (possibly unbounded) linear transformation which guarantees that T^{-1} exists and, furthermore, is bounded.

Theorem 3.4.2 *Let $T: \mathscr{V} \to \mathscr{U}$ be a linear transformation over the normed spaces $\mathscr{V}$ and $\mathscr{U}$. If there exists a constant $m > 0$ such that*
$$\|Tx\| \geq m\|x\| \quad \forall x \in \mathscr{V},$$
then T has a continuous inverse $T^{-1}: \mathscr{R}(T) \to \mathscr{V}$ and
$$\|T^{-1}y\| \leq \frac{1}{m}\|y\| \quad \forall y \in \mathscr{R}(T).$$
Furthermore,
$$\|T^{-1}\| \leq 1/m.$$

Proof Suppose m exists. $Tx = 0$ implies $x = 0$, since otherwise the inequality would not be satisfied: hence $\mathscr{N}(T) = \mathcal{O}$ and T^{-1} exists.

Now if $Tx = y$,
$$\|T^{-1}y\| = \|x\| \leq \frac{1}{m}\|Tx\| = \frac{1}{m}\|y\|,$$
T^{-1} is bounded and therefore continuous: clearly
$$\|T^{-1}\| = \sup_{\|y\|=1} \|T^{-1}y\| \leq 1/m.$$
Conversely, if T^{-1} exists and is continuous
$$\|x\| = \|T^{-1}y\| \leq \|T^{-1}\|\|y\| = \|T^{-1}\|\|Tx\|$$
and we can take
$$m = 1/\|T^{-1}\|. \blacksquare$$

Example 3.4.4 Suppose one is faced with the problem of solving the functional equation
$$Tx = y,$$
where it is known that T^{-1} exists and is bounded; a unique solution exists, $y = T^{-1}x$.

While it may be difficult to find the inverse T^{-1}, it may be possible to find the inverse of a neighbouring operator T_0: T and T_0 should be close in the sense that

$$\|T - T_0\| < \frac{1}{\|T_0^{-1}\|}.$$

Given this condition then it can be shown (Exercise 3.4,2(e)) that

$$\|T^{-1} - T_0^{-1}\| < \frac{\|T_0^{-1}\|^2 \|T - T_0\|}{1 - \|T_0^{-1}\| \|T - T_0\|};$$

hence if $x_0 = T_0^{-1} y$ we have

$$\|x - x_0\| = \|T^{-1} y - T_0^{-1} y\| \leq \|T^{-1} - T_0^{-1}\| \|y\|$$

$$< \frac{\|T_0^{-1}\|^2 \|T - T_0\|}{1 - \|T_0^{-1}\| \|T - T_0\|} \|y\|,$$

which gives an estimate for the error $\|x - x_0\|$.

Suppose T is a square matrix and the elements of the matrix are subject to errors which are at most of magnitude ε. If T_0 is the matrix actually used for the computation we can assess the possible error in the solution. If we use the l_1 or l_∞ norm for T (cf., Example 3.4.2) then

$$\|T - T_0\| \leq n\varepsilon$$

and we find

$$\|x - x_0\| < \frac{\|T_0^{-1}\|^2 n\varepsilon}{1 - \|T_0^{-1}\| n\varepsilon} \|y\|.$$

Compact operators

The idea of a compact space (Definitions 3.2.8 and 3.2.9) was introduced by asking whether we could find, for a space, a *finite* subset of points such that every point of the space was 'near' a point of the finite set. In this sense a compact space is 'not much bigger' than a finite-dimensional space. Linear transformations on finite-dimensional spaces (matrices) have relatively simple and well-understood properties: general linear transformations on normed spaces, on the other hand, may exhibit properties which are quite different and of a much more involved nature. Since a compact space is nearly a finite-dimensional space we may hope that transformations whose range is compact might have properties analogous to those of matrix operators. These hopes are largely borne out.

Definition 3.4.3 *Let* $T: \mathscr{V} \to \mathscr{U}$ *be a linear transformation over the Banach spaces* $\mathscr{V}, \mathscr{U}$. *T is said to be* compact *(or* completely continuous) *if it maps every bounded set in $\mathscr{V}$ into a compact set in $\mathscr{U}$. Equivalently, we may say that T maps the closed unit ball in $\mathscr{V}$ into a compact set in $\mathscr{U}$. A compact transformation is continuous.*

As for ordinary compactness, the requirement that a transformation be compact is quite severe; however, a significant class of operators is compact, and consequently their properties are fairly extensively understood. Some examples are given below.

The 'near finite-dimensionality' of a compact operator can be made more precise in the following terms. If $T:\mathscr{V} \to \mathscr{U}$ is compact then, for any $\varepsilon > 0$, there exists a finite-dimensional subspace $\mathscr{W} \subset \mathscr{R}(T)$ such that

$$\inf\{\|Tx - w\| \,|\, w \in \mathscr{W}\} \leqslant \varepsilon \|x\| \quad \forall x \in \mathscr{V};$$

that is to say, every point in $\mathscr{R}(T)$ is 'near' some point of $\mathscr{W}$. This result follows almost immediately from the existence of an ε-net for (the closure of) the range space of T.

Another way of categorizing compact operators is by considering them as having the effect of 'smoothing' a sequence. Let $\{x^{(n)}\}$ be a bounded sequence in $\mathscr{V}$, then $\{Tx^{(n)}\}$ is a bounded sequence in a compact subspace of $\mathscr{U}$ and hence contains a convergent subsequence. Hence we see that a compact operator takes a non-convergent sequence and smoothes it sufficiently to make it converge.

Example 3.4.5

(a) Obviously all transformations on finite-dimensional spaces are compact.

(b) Since every bounded interval of the real line is compact, then every bounded linear functional $l: \mathscr{V} \to \mathscr{R}$ is compact.

(c) The identity $I: \mathscr{V} \to \mathscr{V}$ is continuous but not compact on an infinite-dimensional space (Exercise 3.4,4(a)).

(d) Let $\mathscr{V} = \mathscr{U} = \mathscr{C}[0,1]$: then (Exercise 3.4,4(d))

$$T(\cdot(t)) = \int_0^1 k(t,\tau)\cdot(\tau)\,d\tau$$

is a compact operator if $k(t,\tau)$ is continuous on the square $[0,1] \times [0,1]$.

(e) If, in (d), $\mathscr{V} = \mathscr{U} = \mathscr{L}_2[0,1]$ then T is compact if [23, 29, 32]

$$\int_0^1 dt \int_0^1 k^2(t,\tau)\,d\tau < \infty.$$

Compact integral operators of the above type have properties which are

LINEAR TRANSFORMATIONS ON NORMED VECTOR SPACES

very similar to matrix properties; we shall pursue this point further in Section 3.9.

As a simple example, consider the integral operator with kernel

$$k(t, \tau) = \begin{cases} t - \tau, & 0 \leq \tau \leq t \leq 1, \\ 0, & 0 \leq t \leq \tau \leq 1, \end{cases}$$

operating on the bounded but non-convergent sequence of functions $\sin n\pi t$ in $\mathscr{L}_2[0, 1]$. We have

$$T(\sin nt) = \frac{t}{n\pi} - \frac{\sin n\pi t}{n^2 \pi^2}$$

and the resulting sequence is clearly convergent in $\mathscr{L}_2[0, 1]$. This integral operator is the inverse of the operator

$$Sx = \frac{d^2 x}{dt^2}$$

with $x(0) = dx(0)/dt = 0$.

Exercises 3.4

1. (a) Prove that every linear operator on a finite-dimensional space is continuous.
   (b) Show that $\|Tx\| \leq \|T\| \|x\|$ (Definition 3.4.2).
   (c) Show that the norm of Definition 3.4.2 satisfies the axioms for a norm function.
   (d) Show that
      (i) $\|T\| = \sup_{\|x\| \leq 1} \|Tx\|$,
      (ii) $\|T\| = \sup_{\|x\| = 1} \|Tx\|$,
   (e) If $T \in \mathscr{L}(\mathscr{V}, \mathscr{U})$, prove that the null space $\mathscr{N}(T)$ is a closed subspace of $\mathscr{V}$.
   (f) Let $T: \mathscr{C}[0, 1] \to \mathscr{C}[0, 1]$ be the linear transformation
      $$Tx(t) = \int_0^1 k(t, \tau) x(\tau) d\tau,$$
      where $k(t, \tau)$ is continuous in $[0, 1] \times [0, 1]$. Show that
      $$\|T\| \leq \max_{t \in [0, 1]} \int_0^1 |k(t, \tau)| d\tau.$$
      (It is possible to show that the reverse of this inequality also holds; hence the right-hand side *is* the norm of T—*see*, for example, [23].)

2. Consider the normed vector space $\mathscr{L}(\mathscr{V}, \mathscr{V})$ of continuous transfor-

mations $T: \mathscr{V} \to \mathscr{V}$. Since we may form the composition ST, $\mathscr{L}(\mathscr{V},\mathscr{V})$ is a linear algebra (p. 41) with identity I and $\|I\| = 1$.

(a) Show that the norm in $\mathscr{L}(\mathscr{V},\mathscr{V})$ has the property $\|ST\| \leq \|S\|\|T\|$ and, more generally, that $\|T^n\| \leq \|T\|^n$.

(b) Show that if $T \in \mathscr{L}(\mathscr{V},\mathscr{V})$ is such that $\|T\| < 1$ then $I - T$ has an inverse,

$$(I - T)^{-1} = \sum_{n=0}^{\infty} T^n \quad \text{and} \quad \|(I - T)^{-1}\| \leq \frac{1}{1 - \|T\|}.$$

(Hint: show that $\sum_{n=0}^{\infty} T^n$ converges in $\mathscr{L}(\mathscr{V},\mathscr{V})$ using the Weierstrass M-test—Exercise 3.3,6.)

(c) The series $\sum_{n=0}^{\infty} T^n$ of (b) is called a *Neumann* series. What is a sufficient condition for $(\lambda I - T)^{-1}$ to have a convergent Neumann series? Write down the series.

(d) Verify that the inverse of the transformation $S: \mathscr{C}[0,1] \to \mathscr{C}[0,1]$ defined by

$$Sx = x(t) - \int_0^1 (t - \tau)^2 x(\tau) d\tau$$

has a convergent Neumann expansion (*see* Exercise 3.4.1 (f)). By using the first one or two terms of the series, find an approximation to the solution of the equation $Sx = 1$.

(e) If $T_0 \in \mathscr{L}(\mathscr{V},\mathscr{V})$ has an inverse T_0^{-1} and $T \in \mathscr{L}(\mathscr{V},\mathscr{V})$ is such that

$$\|T - T_0\| < 1/\|T_0^{-1}\|$$

show that T^{-1} exists and furthermore that

$$\|T^{-1} - T_0^{-1}\| \leq \frac{\|T_0^{-1}(T - T_0)\|}{1 - \|T_0^{-1}(T - T_0)\|} \|T_0^{-1}\|$$

$$< \frac{\|T_0^{-1}\|^2 \|T - T_0\|}{1 - \|T_0^{-1}\|\|T - T_0\|}.$$

(f) Suppose we wish to solve $Tx = y$ but, due to error, we are solving $(T + \delta T)(x + \delta x) = y$. If $\|\delta T\| < 1/\|T^{-1}\|$, show that

$$\frac{\|\delta x\|}{\|x\|} \leq \frac{\gamma}{1 - \gamma \|\delta T\|/\|T\|} \frac{\|\delta T\|}{\|T\|},$$

where $\gamma = \|T\|\|T^{-1}\|$ is called the *condition number* of T. (This result, although attractive looking, is not very practical because of the difficulty of estimating γ, $\|\delta T\|$, $\|T\|$.)

3. Let $\mathscr{C}^{(2)}[0,1]$ denote the space of twice continuously differentiable functions on $[0,1]$ with norm

$$\|x\|^{(2)} = \left\{ \int_0^1 \sum_{k=0}^{2} \left|\frac{d^k x}{dt^k}\right|^2 dt \right\}^{1/2}$$

and $\mathscr{C}^{(0)}[0,1]$ the space of continuous functions on $[0,1]$ with norm

$$\|x\|^{(0)} = \left\{\int_0^1 |x|^2 \, dt\right\}^{1/2}.$$

Show that if $x(0) = x'(1) = 0$ there exists a constant K such that

$$\|x\|^{(2)} \leqslant K \|x''\|^{(0)}$$

and hence deduce that the equation $x''(t) = y$, with $x(0) = x'(1) = 0$ has a unique solution $x \in \mathscr{C}^{(2)}[0,1]$ if $y \in \mathscr{C}^{(0)}[0,1]$.

4. (a) Verify that the identity operator is not compact on an infinite-dimensional space.
   (b) Let $S \in \mathscr{L}(\mathscr{V}, \mathscr{V})$: let $T \in \mathscr{L}(\mathscr{V}, \mathscr{V})$ be compact. Show that TS, ST are compact.
   (c) Use (b) to show that the compact operator $T \in \mathscr{L}(\mathscr{V}, \mathscr{V})$ cannot have a bounded inverse except when $\mathscr{V}$ is finite-dimensional.
   (d) Show that the operator $T: \mathscr{C}[0,1] \to \mathscr{C}[0,1]$ defined by

$$T(\cdot(t)) = \int_0^1 k(t, \tau) \cdot (\tau) \, d\tau$$

   is compact if $k(t, \tau)$ is continuous on $[0,1] \times [0,1]$.
   (Hint: show that

$$\int_0^1 |k(t_1, \tau) - k(t_2, \tau)| \, ds < \varepsilon,$$

   and hence that $|Tx(t_1) - Tx(t_2)| \leqslant \varepsilon \|x\|$ for $|t_1 - t_2| < \delta$. Now use the Arzela–Ascoli theorem (p. 129).)

5. Find the integral operator which is the inverse of the operator $Sx = -d^2x/dt^2$ on $[0,1]$ with $x(0) = x(1) = 0$ (cf., Examples 2.5.6, 3.4.5(e)).

3.5 The dual of a normed space

The linear transformations which map a vector space $\mathscr{V}$ into the spaces $\mathscr{R}$ or $\mathscr{C}$ are called linear functionals and were introduced and defined in Section 2.6. The linear functionals $\mathscr{V} \to \mathscr{R}$ (or $\mathscr{C}$) themselves form a vector space which we called the algebraic dual (or conjugate) of $\mathscr{V}$. If $\mathscr{V}$ is a normed space then we can discuss the continuity, boundedness and norm of a linear functional as a special case of a linear transformation.

Thus the linear functional l on $\mathscr{V}$ is *continuous* if $l(x^{(k)})$ converges to $l(x)$ (in $\mathscr{R}$ or $\mathscr{C}$) whenever $x^{(k)} \to x$ in $\mathscr{V}$. If l is continuous then we know that it is bounded on $\mathscr{V}$ (Theorem 3.4.1) and we may define

the norm of l as
$$\|l\| = \sup_{\|x\|=1} |l(x)|.$$

Example 3.5.1

(a) In $\mathscr{R}_n$ every linear functional takes the form
$$y(x) = \sum_{i=1}^{n} \eta^i \xi_i :$$
it is readily verified that this functional is continuous.

(b) On the space $\mathscr{C}[0,1]$ of continuous functions the functional $l(x) = x(\tfrac{1}{2})$ is continuous.

(c) Let $\mathscr{V}$ be the space of sequences in which only a finite number of terms is non-zero with norm $\|\cdot\|_\infty$. Define, for $x = (\xi_1, \xi_2, \ldots, \xi_n, 0, 0, 0, \ldots)$, the functional
$$l(x) = \sum_{k=1}^{n} k\xi_k :$$
this functional is linear but unbounded.

(d) Let us find the general form of a bounded linear functional on the space l_1, the space of convergent sequences with norm
$$\|x\|_1 = \sum_{i=1}^{\infty} |\xi_i|.$$

This space has the Schauder basis (Example 3.3.2) $e^1 = (1,0,0,0,\ldots)$, $e^2 = (0,1,0,0,\ldots), \ldots$.

If l is a linear functional on l_1 then, as for a finite-dimensional space (Section 2.6), l is determined by its values on the unit vectors e^k: let
$$l(e^k) = \eta^k,$$
then
$$l(x) = \sum_{k=1}^{\infty} \eta^k \xi_k.$$

Now
$$|\eta^k| = |l(e^k)| \leq \|l\|\,\|e^k\|_1 = \|l\|$$
for all $k = 1, 2, \ldots$; hence the vector $y = (\eta^1, \eta^2, \ldots, \eta^n, \ldots)$ belongs to the space l_∞ of infinite sequences with
$$\|y\|_\infty = \sup_i \{|\eta^i|\} \leq \|l\|.$$

Conversely, from

$$l(x) = \sum_{k=1}^{\infty} \eta^k \xi_k$$

we have

$$|l(x)| \leq \sup_k \{|\eta^k|\} \sum_{k=1}^{\infty} |\xi_k| = \|y\|_\infty \|x\|_1.$$

Since $\|l\|$ is the smallest constant satisfying the inequality

$$|l(x)| \leq K\|x\|_1,$$

we conclude that

$$\|l\| \leq \|y\|_\infty.$$

The preceding results imply that

$$\|l\| = \|y\|_\infty$$

and that, therefore, the general form of a linear functional on l_1 corresponds to a vector in the space l_∞.

It should not be assumed from this example that a linear functional on l_∞ corresponds to a vector in l_1 (see p. 158).

If we consider the bounded linear functionals l on a normed vector space $\mathscr{V}$ then we can see that they themselves form a normed linear space having the induced operator norm $\|l\|$. This brings us to

Definition 3.5.1 *Let $\mathscr{V}$ be a normed vector space. The space of all bounded functionals on $\mathscr{V}$ is called the* normed dual (*or* conjugate) *of $\mathscr{V}$ and is denoted by $\mathscr{V}^*$. The norm of an element $l \in \mathscr{V}^*$ is given by*

$$\|l\| = \sup_{\|x\|=1} |l(x)|, \quad x \in \mathscr{V}.$$

We use the same symbol, $\mathscr{V}^*$, for the normed dual of $\mathscr{V}$ as we did for the algebraic dual: this should not lead to confusion, and indeed when we mean the normed dual we shall as a general rule omit the adjective 'normed'.

As before, it will often be convenient to denote elements of $\mathscr{V}^*$ by the (vector) symbols x^* or y and to denote the values of $x^*(x)$ or $y(x)$ by $[x^*, x]$ or $[y, x]$, respectively (p. 52).

Surprisingly enough, $\mathscr{V}^*$ is not only a normed space but also a Banach space.

Theorem 3.5.1 *The dual of a normed vector space is a Banach space.*

Table of common Banach spaces and their duals

Designation	Description	Dual space and representation of linear functionals†	Reflexive	Countable dense subsets
$\mathscr{R}_n, \mathscr{C}_n$ ($\mathscr{V}_n$)	Space of real or complex n-tuples with norm $\|x\|_p = \left\{\sum_{i=1}^n |\xi_i|^p\right\}^{1/p}$, $1 \leq p < \infty$. $\|x\|_\infty = \max_i\{|\xi_i|\}$ For every p, convergence with respect to norm is equivalent to componentwise convergence.	Dual is $\mathscr{R}_n, \mathscr{C}_n$ with norm $\|x\|_q$, $1/p + 1/q = 1$. $l_y(x) = \sum_{i=1}^n \eta^i \xi_i$, $y = (\eta^1, \eta^2, \ldots, \eta^n) \in \mathscr{R}_n, \mathscr{C}_n$.	Yes	n-tuples of rational numbers
l_p	All sequences $\{\xi_i\}$ for which the norm $\|x\|_p = \{\sum_i |\xi_i|^p\}^{1/p}$, $1 \leq p < \infty$ is finite.	For $1 < p < \infty$ dual is l_q, $1/p + 1/q = 1$. Dual of l_1 is l_∞. $l_y(x) = \sum_i \eta^i \xi_i$; $y = \{\eta^i\}$, $y \in l_q$.	Yes	Sequences with finite numbers of non-zero entries which are rational numbers
l_∞		$\|x\|_\infty = \sup_i\{|\xi_i|\}$	Dual of l_∞ (see [38]).	No	Not separable
c	All sequences $\{\xi_i\}$ such that $\lim_i \xi_i$ exists, $\|x\| = \sup_i\{|\xi_i|\}$	Dual is l_1	No	
$\mathscr{C}[0, 1]$‡	All continuous functions on $[0, 1]$ with norm $\|x\| = \max_{t \in [0,1]}\{|x(t)|\}$	Dual is $\mathscr{B}_N[0, 1]$, $l_y(x) = \int_0^1 x(t) dy(t)$, $y \in \mathscr{B}_N[0, 1]$	No	Polynomials on $[0, 1]$ with rational coefficients. Polygons (Example 3.3.2(c)).

THE DUAL OF A NORMED SPACE

Space	Description	Dual space†	Separable	Notes
$\mathscr{B}[0,1]$	All functions on $[0,1]$ of bounded variation with norm $\|x\| = |x(0+)| + V[x(t)]$	See [38].	No	Not separable
$\mathscr{B}_N[0,1]$	Those functions in $\mathscr{B}[0,1]$ such that $x(0+) = 0$, x is continuous on right.			
$\mathscr{L}_p[0,1]$‡	All functions on $[0,1]$ such that the norm $\|x\|_p = \left\{\int_0^1 |x(t)|^p\right\}^{1/p}$, $1 \leq p < \infty$ is finite.	For $1 < p < \infty$, dual is $\mathscr{L}_q[0,1]$, $1/p + 1/q = 1$. Dual of $\mathscr{L}_1[0,1]$ is $\mathscr{L}_\infty[0,1]$, $l_y(x) = \int_0^1 x(t)y(t)dt$, $y \in \mathscr{L}_q[0,1]$	Yes No	Functions continuous on $[0,1]$, hence polynomials with rational coefficients. For $\mathscr{L}_p(-\infty,\infty)$ rational step functions which vanish outside finite collection of rational intervals. Note: bounded functions with compact§ support or infinitely differentiable functions with bounded support are dense in $\mathscr{L}_p$ (not countable).
$\mathscr{L}_\infty[0,1]$	All functions on $[0,1]$ such that $|x(\tau)| \leq M$ almost everywhere with norm $\|x\|_\infty = \inf\{M : |x(t)| \leq M \text{ a.e. on } [0,1]\}$	See [38].	No	Not separable

There are generally extensions of the above spaces to vector-valued functions.
References: [25, 27, 29, 30, 35, 38, 41].
† 'Dual space is' means that dual space is isometrically isomorphic to the named space.
‡ The interval $[0,1]$ can be replaced by any interval, finite or infinite, or by any measurable set.
§ A function $x(t)$ has compact support in S if there exists a compact set $M \subset I$ such that $x(t) = 0$ in $S - M$.

Proof We have simply to show that $\mathscr{V}^*$ is complete. Let $\{l^{(k)}\}$ be a Cauchy sequence in $\mathscr{V}^*$, that is,

$$\|l^{(j)} - l^{(k)}\| \to 0 \quad \text{as } j, k \to \infty.$$

For any $x \in \mathscr{V}$, the sequence of scalars $l^{(k)}(x)$ is Cauchy since

$$|l^{(j)}(x) - l^{(k)}(x)| \leq \|l^{(j)} - l^{(k)}\| \, \|x\|$$

so that, for every $x \in \mathscr{V}$, there exists $l \in \mathscr{V}^*$ such that

$$\lim_k l^{(k)}(x) = l(x).$$

It is not difficult to see that the l so defined is a bounded linear functional and that $\|l - l^{(k)}\| \to 0$ as $k \to \infty$; that is, $l^{(k)} \to l$ in $\mathscr{V}^*$. ■

If we consider the various specific Banach spaces listed on p. 158–159, we may legitimately ask what concrete form the functionals of the various dual spaces take. The elucidation of such forms is one of the central achievements of functional analysis: the technical machinery required to establish the results is far beyond the scope of this text. The table summarizes the structure of the duals of the Banach spaces listed there.

Example 3.5.2 We note (p. 158) that the dual of the space l_p is the space l_q where $1/p + 1/q = 1$.

This type of statement is to be understood in the following sense. The linear functional l on l_p is defined in terms of the scalar sequence $(\eta^1, \eta^2, \ldots, \eta^n, \ldots)$ (cf., Example 3.5.1(d)): that is to say, there is a one-to-one correspondence between the linear functionals on l_p and the elements of the vector space l_q. The dual space $(l_p)^*$ is the same as (or isometrically isomorphic to) l_q (cf., the finite-dimensional case on p. 51).

The dual of $\mathscr{L}_p$, the space of pth-power integrable functions defined on $[0, 1]$, is $\mathscr{L}_q$: this means that there is a one-to-one correspondence between the linear functionals on $\mathscr{L}_p$ and the functions y in the space $\mathscr{L}_q$, $l_y \leftrightarrow y$ in the sense that the value of l_y at $x \in \mathscr{L}_p$ is given by

$$l_y(x) = \int_0^1 xy \, dt$$

and furthermore that

$$\|l_y\| = \|y\|_q.$$

Extension of linear functionals

One would often wish to construct a bounded linear functional having certain properties which are convenient for a particular purpose. It may be relatively easy to construct such a functional on a subspace $\mathscr{W} \subset \mathscr{V}$, but the question then arises as to whether the functional can be extended to the whole of $\mathscr{V}$. By an extension we mean the following: let l be a linear functional defined on $\mathscr{W} \subset \mathscr{V}$, then l_E is an *extension* of l to the whole of $\mathscr{V}$ if, for every $w \in \mathscr{W}$,

$$l_E(w) = l(w)$$

and l_E is defined on the whole of $\mathscr{V}$. Now the norm of l is given by

$$\|l\|_{\mathscr{W}} = \sup_{\|w\|=1} |l(w)|, \qquad w \in \mathscr{W} \subset \mathscr{V}.$$

The norm of the extension l_E is

$$\|l_E\|_{\mathscr{V}} = \sup_{\|x\|=1} |l_E(x)|, \qquad x \in \mathscr{V},$$

and it will be clear from the definition of the norm of a transformation that

$$\|l_E\|_{\mathscr{V}} \geq \|l\|_{\mathscr{W}}.$$

In fact, a celebrated result in analysis called the Hahn–Banach theorem tells us that it is always possible to find an extension of l, l_E, for which $\|l_E\|_{\mathscr{V}} = \|l\|_{\mathscr{W}}$.

For example, in a separable space we may define a linear functional on a dense subspace and we then know that we can, as a matter of practice, apply that linear functional, without change of norm, to all elements of the space. But the theorem does not require us to use anything as 'large' as a dense subspace, and applies for all subspaces whatever their form.

The proof of the Hahn–Banach theorem, particularly for non-separable spaces, is long and quite difficult; we shall not give it here, but merely state the result and refer the sceptical reader to suitable texts [23, 24, 27, 30, 36].

Theorem 3.5.2 *Let l be a bounded linear functional defined on a subspace $\mathscr{W}$ of a normed vector space $\mathscr{V}$. Then there exists a bounded linear functional l_E defined on $\mathscr{V}$ which is an extension of l, having the same norm as l on $\mathscr{W}$.*

Corollary 1 *Let $x \in \mathscr{V}$: then there exists a bounded linear functional l, defined on all of $\mathscr{V}$ such that,*

(i) $l(x) = \|x\|$;
(ii) $\|l\| = 1$.

An alternative statement of this result is that there exists l such that $l(x) = \|l\| \|x\|$.

Proof of Corollary 1 Consider the one-dimensional subspace $\mathscr{W}$ spanned by x and define l on $\mathscr{W}$ by $l(\lambda w) = \lambda \|w\|$: clearly l is a bounded linear functional with $\|l\| = 1$ and $l(\lambda w) = \lambda \|w\| = \|\lambda w\|$. This functional can be extended to the whole of the space without change of norm, giving $l_E(x) = \|x\|$ with $\|l_E\| = 1$. ■

Corollary 2 *If all bounded linear functionals vanish on a given vector then the vector must be the zero vector.*

Proof of Corollary 2 Of all linear functionals applied to the vector, one must have the norm of the vector as its value; this implies that the vector is the zero vector. ■

The Hahn–Banach theorem is one of the principal tools used in characterizing the dual space of a given vector space: many of the results given on p. 158–159 are crucially dependent on the theorem.

Example 3.5.3 We note (p. 158) that the dual of the space of continuous functions $\mathscr{C}[0, 1]$ is the space $\mathscr{B}_N[0, 1]$, the space of normalized functions of bounded variation on $[0, 1]$. We have not so far discussed this normed space, and we shall now attempt to make good this omission.

Most of the functions that one meets in practice do not oscillate too rapidly. If one adds up, as it were, their excursions over a finite interval, the resulting number is finite: we say such functions are 'of bounded variation'. Such functions have the nice property that they can be expressed as the sum of monotonic functions. Let us begin with this idea.

A monotonic function $x(t)$ defined on the interval $[0, 1]$ is an increasing (decreasing) function in the sense that $x(t) \leq x(s)$, $(x(t) \geq x(s))$ for all t, $s \in [0, 1]$ whenever $t \leq s$. If $x(t)$ exhibits a jump at $t = c$ then $x(-c) < x(+c)$, $(x(-c) > x(+c))$ if x is increasing (decreasing), where $x(-c)$, $x(+c)$ denote the left-hand and right-hand limits of $x(t)$ at c, respectively. Now a monotonic function can have an infinite number of points of discontinuity, but it can be shown (by generating from each discontinuity c the set of non-intersecting open intervals $(x(-c), x(+c))$) that the set of points of discontinuity is countable. Let $x(t)$ be a bounded monotonic function on $[0, 1]$ and form the countable sum

$$\sum_n |x(+c_n) - x(-c_n)|,$$

where the c_n are the points of discontinuity. Precisely because $x(t)$ is bounded and monotonic it is clear that this sum must be finite.

We wish now to extend this property to a larger class of functions and in this way we arrive at functions of bounded variation. Let $x(t)$ be a real- or complex-valued function defined on the closed interval $[0, 1]$. Consider all *finite* partitionings, P, of $[0, 1]$,

$$P: 0 = t_0 < t_1 \ldots < t_{n-1} < t_n = 1,$$

and with each partition associate the sum

$$\sigma_p[x] = \sum_{i=1}^{n} |x(t_i) - x(t_{i-1})|.$$

We say $x(t)$ is *of bounded variation* on $[0, 1]$ if there exists a finite constant M such that

$$\sigma_p[x] \leq M$$

for all finite partitions of $[0, 1]$. The *total variation*, V, of x on $[0, 1]$ is the smallest number that can be used in this inequality, that is to say,

$$V[x] = \sup_P \sigma_p[x].$$

Every bounded monotonic function is of bounded variation: if $x(t)$ is increasing on $[0, 1]$ then

$$V[x] = x(1) - x(0).$$

A continuous function on $[0, 1]$ with only a finite number of relative maxima and minima is clearly of bounded variation.

As an example of a continuous, bounded function which is not of bounded variation, consider a function x taking the values $x(t) = t$ for $t = 1, \frac{1}{3}, \frac{1}{5}, \ldots$ and $x(t) = 0$ at $\frac{1}{2}, \frac{1}{4}, \ldots$ and at 0; for example, $t|\sin(2\pi/t)|$ or the piecewise linear function graphed in Fig. 3.8. Take P to be the partition

$$0 < \frac{1}{2n} < \frac{1}{2n-1} < \ldots < \frac{1}{2} < 1,$$

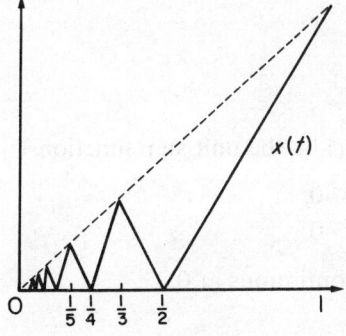

Fig. 3.8

then
$$\sigma_p[x] = \frac{2}{2n-1} + \frac{2}{2n-3} + \ldots + \frac{2}{3} + 1 > 1 + \frac{1}{2} + \ldots + \frac{1}{n},$$

which is not bounded with increasing n.

The relation of functions of bounded variation to monotonic functions is embodied in the statement that every real function of bounded variation can be expressed as the difference of two bounded increasing functions (Exercise 3.5, 3(d)).

The space $\mathscr{B}[0, 1]$ is defined as the vector space of all functions of bounded variation on $[0, 1]$ with the norm given by

$$\|x\| = |x(0)| + V[x].$$

Axioms 1 and 3 of Definition 3.3.1 are clearly fulfilled; the triangle inequality needs to be demonstrated (Exercise 3.5, 3(e)). The space $\mathscr{B}[0, 1]$ is complete and hence is a Banach space [39].

We now need to make some brief remarks on the Riemann–Stieltjes integral. The ordinary Riemann integral

$$\int_0^1 x(t) dt$$

represents the limit of the 'Riemann sums'

$$\sum_{i=1}^n x(\tau_i)(t_i - t_{i-1}), \qquad \tau_i \in [t_{i-1}, t_i],$$

for a sequence of increasingly fine partitions of the interval $[0, 1]$. By an extension of this idea we associate with every partition the sum

$$\sum_{i=1}^n x(\tau_i)(y(t_i) - y(t_{i-1})), \qquad \tau_i \in [t_{i-1}, t_i]:$$

if the limit of these sums for a sequence of increasingly fine partitions exists it is called *the (Riemann–) Stieltjes integral of x with respect to y on $[0, 1]$* and is written

$$\int_0^1 x \, dy = \int_0^1 x(t) dy(t).$$

For example, let $1_+(t)$ be the unit step function

$$1_+(t) = \begin{cases} 0, & t \leq 0, \\ 1, & t > 0; \end{cases}$$

then, provided $x(t)$ is continuous at 0,

$$\int_0^1 x(t) d1_+(t) = x(0).$$

For, in any partition, the quantities $(y(t_i) - y(t_{i-1}))$ vanish except in the extreme left-hand interval and the value there is unity.

We now (without proof) make a statement which (almost) justifies the claim that the dual of $\mathscr{C}[0, 1]$ is the space of functions of bounded variation on $[0, 1]$. Let $l_y \in \mathscr{C}^*[0, 1]$; then there exists a function $y(t) \in \mathscr{B}[0, 1]$ such that, for all $x(t) \in \mathscr{C}[0, 1]$,

$$l_y(x) = \int_0^1 x(t) \, dy(t)$$

and (cf., Example 3.5.2)

$$\|l_y\| = V[y].$$

This is the *Riesz representation theorem*, the proof of which leans heavily on the Hahn–Banach theorem [23, 24, 27, 30].

The foregoing statement indicates a certain correspondence between functionals from $\mathscr{C}^*[0, 1]$ and functions of bounded variation. However, this correspondence is not well-defined, for, if $y(t)$ corresponds to l_y, so does $y(t)$ + constant. Further, a function of bounded variation which differs from $y(t)$ only at its points of discontinuity is indistinguishable from y. For example, the functional $l(x) = x(\frac{1}{2})$ can be represented by a $y(t)$ which is zero on $[0, \frac{1}{2})$, unity on $(\frac{1}{2}, 1]$, and has any value between zero and unity at $t = \frac{1}{2}$.

To overcome this difficulty we delineate a certain subspace of $\mathscr{B}[0, 1]$ in the following way. All we need to do is to assign the value of $y(t)$ at points of discontinuity in some way and we do this by making it equal to the right-hand limit. Thus the function $y(t) \in \mathscr{B}[0, 1]$ is said to be normalized if $y(0) = 0$ and if, for all $t \in [0, 1]$, $y(t) = y(t + 0)$. The collection of normalized functions of bounded variation is a subspace of $\mathscr{B}[0, 1]$ and is denoted by $\mathscr{B}_N[0, 1]$. As a matter of practice, the need for normalization is not often crucial.

An important use of the Hahn–Banach theorem in real vector spaces is in optimization problems, and in this context the theorem is given a somewhat different emphasis. We saw in Section 2.6 (p. 54) that the set of vectors $\{x \mid l(x) = \alpha\}$ is, for some non-zero functional l, a hyperplane in $\mathscr{V}$. Thus we can, if we wish, associate the functionals on $\mathscr{V}$, not with elements of $\mathscr{V}^*$ but with hyperplanes in $\mathscr{V}$ itself. The plane $l(x) = \alpha$ divides $\mathscr{V}$ into two parts; those elements for which $l(x) > \alpha$ and those for which $l(x) \leq \alpha$. The plane $l(x) = \|l\|$ possesses the property that the unit ball $\{x \mid \|x\| \leq 1\}$ is 'to one side of' the plane: for we have

$$l(x) \leq \|l\| \|x\| \leq \|l\|, \qquad x \in B(0, 1).$$

Furthermore the plane 'touches' the closed unit ball (Exercise 3.5,1(c)).

In this context the plane $l(x) = \|l\|$ is called a *supporting plane* for the unit ball and its existence is guaranteed by Corollary 1 of Theorem 3.5.2.

Now a ball is an example of a convex set (Definition 2.3.6) so the question arises as to whether a similar result can be established for any convex set in $\mathscr{V}$. Namely, given a convex set in a (real) vector space, does there exist a hyperplane which 'lies to one side of' the set and, furthermore, does the convex set have a supporting plane?

The answer to both these questions is 'Yes', and the result is often referred to as the geometric form of the Hahn–Banach theorem. The reader is referred to [24, 36] for further details. For a general convex set a supporting hyperplane is characterized by the equation $l(x) = \alpha_l$, where α_l is the infimum of the sequence of constants such that the set is contained in the half-space $l(x) < \alpha$. The suffix l indicates that α is dependent upon l (there is no such dependence in the case of the ball). We can clearly always assume, if we wish, that $\|l\| = 1$. In the following examples $\mathscr{V}$ is real.

Example 3.5.4

(a) The norm of a vector $x \in \mathscr{V}$ can be related to the values of a functional of unit norm, thus

$$\|x\| = \max_{\|l\|=1} |l(x)|.$$

Let S^* denote the unit sphere in $\mathscr{V}^*$, that is, the set of linear functionals on $\mathscr{V}$ having unit norm. Clearly $|l(x)| \leq \|x\|$ for all $l \in S^*$, but we ask whether there is at least one element in S^* for which equality holds. Corollary 1 of Theorem 3.5.2 tells us that there is.

(b) The result in (a) can be generalized so as to give the minimum distance from a point to a subspace. In Example 3.3.3 (Best approximation) we considered the problem of finding the minimum distance from a point $x \in \mathscr{V}$ to a given finite-dimensional subspace $\mathscr{W}$. Because a bounded subset of $\mathscr{W}$ was compact we could assert that the minimum was actually attained for some point in $\mathscr{W}$.

If $\mathscr{W}$ is infinite-dimensional, then although

$$d(x, \mathscr{W}) = \inf_{w \in \mathscr{W}} \|x - w\|$$

exists, there need be no element w_0 for which $d = \|x - w_0\|$. However, it can be shown that

$$d(x, \mathscr{W}) = \inf_{w \in \mathscr{W}} \|x - w\| = \max_{\substack{\|l\|=1 \\ l \in W^0}} l(x),$$

where the maximum is taken over all linear functionals of unit norm in

$\mathscr{W}^0 \subset \mathscr{V}^*$, the annihilator of $\mathscr{W} \subset \mathscr{V}$ (p. 54). That is to say, a linear functional $l \in \mathscr{W}^0$ does exist whose value at x is the required minimum distance.

The necessity for l to be in $\mathscr{W}^0$ can be seen in the following way. Suppose for some $w \in \mathscr{W}$ that $\|x - w\| \leq d + \varepsilon$; if $l \in \mathscr{W}^0$,

$$l(x) = [l, x] = [l, x - w]$$
$$\leq \|l\| \|x - w\|$$
$$\leq d + \varepsilon,$$

and we may conclude, since ε is arbitrary, that $l(x) \leq d$. Then we need only show that there actually exists $\tilde{l} \in \mathscr{W}^0$ such that $\tilde{l}(x) = d$, and this we may infer from the Hahn–Banach theorem [36]. It can be seen that if a minimal solution w_0 does exist, then

$$l(x - w_0) = \|l\| \|x - w_0\|$$

(cf., Corollary 1, Theorem 3.5.2).

Notice that the problem of finding a minimum in $\mathscr{V}$ has been transformed into that of finding a maximum in $\mathscr{V}^*$. This type of equivalence (often referred to as duality) is often of considerable utility in applications (*see also* Example 2.9.3).

A result of this type also exists for the distance from a point to a convex set [36].

Using the idea of a linear functional as a hyperplane we can give a simple geometric interpretation of this result. From Fig. 3.9 we can see that the minimum distance from a point to a convex set is equal to the maximum of the distances from the point to hyperplanes separating the point and the set.

Since the dual of $\mathscr{V}$, namely $\mathscr{V}^*$, is itself a vector space, we are at liberty to consider the elements of $\mathscr{V}^*$ as vectors and to apply to them linear functionals. In the following discussion it will be convenient to adopt the bracket notation first introduced in Section 2.6 (p. 52), whereby we write

$$x^*(x) = [x^*, x], \quad x^* \in \mathscr{V}^*, \quad x \in \mathscr{V}.$$

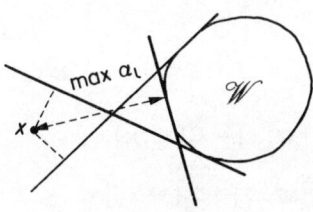

Fig. 3.9

Suppose, instead of generating a series of scalars by varying x while holding x^* fixed, we let x^* vary in $\mathscr{V}^*$ while holding x fixed. In this way, we effectively define a linear functional on a space of linear functionals.[†] Let this linear functional be denoted by x^{**} with value

$$x^{**}(x^*) = [x^{**}, x^*] = [x^*, x].$$

Thus

$$|x^{**}(x^*)| \leqslant \|x\| \|x^*\|$$

or

$$\|x^{**}\| \leqslant \|x\|$$

and we see that the functionals x^{**} are bounded (in the sense of operators on $\mathscr{V}^*$). The space of all bounded functionals on $\mathscr{V}^*$ is called the *second dual* of $\mathscr{V}$ and is denoted $\mathscr{V}^{**}$.

The correspondence

$$[x^{**}, x^*] = [x^*, x]$$

sets up a mapping $\Phi : \mathscr{V} \to \mathscr{V}^{**}$, and furthermore, by Corollary 1 of Theorem 3.5.2, there exists $x^* \in \mathscr{V}^*$ such that

$$[x^*, x] = \|x^*\| \|x\|,$$

so we deduce that

$$\|x^{**}\| = \|x\|.$$

The mapping Φ is thus norm preserving (i.e., an isometry and consequently one-to-one) and in this sense identifies elements of $\mathscr{V}$ and $\mathscr{V}^{**}$: it is commonly called the *natural embedding* of $\mathscr{V}$ into $\mathscr{V}^{**}$, for it is important to note that $\mathscr{V} \subseteq \mathscr{V}^{**}$. With this proviso, we can consider the second dual space to be indistinguishable from the parent space $\mathscr{V}$. In many important cases the mapping Φ is onto $\mathscr{V}^{**}$ and the identity is complete. We have

Definition 3.5.2 *A normed vector space is said to be* reflexive *if* $\mathscr{V} = \mathscr{V}^{**}$.

[†] Since

$$[\lambda_1 x_1^* + \lambda_2 x_2^*, x] = \lambda_1 [x_1^*, x] + \lambda_2 [x_2^*, x]$$

and

$$[x^*, \alpha_1 x_1 + \alpha_2 x_2] = \alpha_1 [x^*, x_1] + \alpha_2 [x^*, x_2]$$

the operation $[\cdot, \cdot]$ is linear in each of its arguments (bilinear).

All results proved for a vector space $\mathscr{V}$ and its dual $\mathscr{V}^*$ can be stated in a symmetrical form if $\mathscr{V}$ is reflexive.

Example 3.5.5

(a) In Section 2.6 (p. 53) we defined algebraically reflexive spaces and noted that the finite-dimensional spaces $\mathscr{V}_n$, $\mathscr{V}_n^*$ were indistinguishable (isomorphic). Finite-dimensional spaces are also clearly reflexive in the above sense and so are *isometrically* isomorphic.

(b) Since $l_p^* = l_q, 1/p + 1/q = 1, 1 < p < \infty$ (p. 158), then
$$l_p^{**} = l_q^* = l_p, \quad 1 < q < \infty,$$
and similarly
$$\mathscr{L}_p^{**} = \mathscr{L}_p.$$

(c) The spaces l_1 and $\mathscr{L}_1$ are not reflexive (*see* p. 158–159).

Weak convergence

In a normed vector space we have defined convergence of a sequence of vectors in terms of convergence of the number sequence of norms. For reasons which will become clear, we refer to convergence with respect to the norm as *strong convergence*.

The introduction of bounded linear functionals on the vector space $\mathscr{V}$ allows us to define another type of convergence which is less stringent than convergence with respect to the norm and, surprisingly perhaps, proves to be of great usefulness in the approximate solution of operator equations: this type of convergence is called *weak convergence*.

Definition 3.5.3 *The sequence $\{x^{(k)}\}$ in a normed vector space $\mathscr{V}$ is said to* converge weakly *to x, written $x^{(k)} \rightharpoonup x$,[†] if for every $l \in \mathscr{V}^*$,*
$$l(x^{(k)}) \to l(x).$$

A weak limit is unique. Every weakly convergent sequence is bounded [26].

If $x^{(k)} \to x$ then $x^{(k)} \rightharpoonup x$, since
$$|l(x^{(k)}) - l(x)| \leq \|l\| \|x^{(k)} - x\| \to 0:$$
the converse is not generally true.

[†] The notation $x^{(k)} \xrightarrow{W} x$ is also used.

Example 3.5.6

(a) In l_2 the sequence of unit vectors $e^1 = (1, 0, 0, \ldots)$, $e^2 = (0, 1, 0, \ldots, 0)$ tends weakly to zero since, for any functional $y = (\eta^1, \eta^2, \eta^3, \ldots) \in l_2^* = l_2$, $y(e^j) = \eta^j \to 0$ as $j \to \infty$ while $\|e^j\| \nrightarrow 0$.

(b) In $\mathscr{L}_2[0, 2\pi]$ we have
$$l_g(x) = \int_0^{2\pi} x(t)g(t)\,dt, \qquad g \in \mathscr{L}_2[0, 2\pi].$$
Consider the sequence,
$$x^{(k)} = (\sin kt)/\pi$$
then
$$l_g(x^{(k)}) = \frac{1}{\pi} \int_0^{2\pi} (\sin kt)g(t)\,dt.$$

Now $l_g(x^{(k)})$ will be recognized as the kth sine coefficient b_k in the Fourier series for $g(t)$ in $(0, 2\pi)$ and (Exercise 3.6, 8—the Riemann–Lebesgue lemma) $\lim_{k \to \infty} b_k = 0$, hence we may deduce that
$$x^{(k)} = (\sin kt)/\pi \rightharpoonup 0$$
in $(0, 2\pi)$. However,
$$\|x^{(k)} - 0\|^2 = \frac{1}{\pi^2} \int_0^{2\pi} \sin^2 kt\, dt = \frac{1}{\pi}$$
and $x^{(k)} \nrightarrow 0$.

(c) If $\mathscr{V}$ is a finite-dimensional normed space then strong convergence and weak convergence are equivalent (Exercise 3.5, 6(b)).

(d) We saw in Section 3.4 (p. 152) that a compact transformation maps a bounded sequence to a convergent sequence. Hence a compact operator maps a weakly convergent sequence to a strongly convergent sequence.

(e) Pointwise convergence of functions and componentwise convergence of number sequences are not covered by convergence with respect to a norm (strong convergence). However, we can associate these cases with weak convergence. Specifically, we have [30]:
  (i) in l_p, $1 \leq p < \infty$; that a sequence $\{x^{(k)}\}$ is weakly convergent implies it is bounded and componentwise convergent;
  (ii) in $\mathscr{C}[0, 1]$; that a sequence $\{x^{(k)}\}$ is weakly convergent implies it is bounded and pointwise convergent.

(f) In Chapter 5 we shall find that we can define the weak derivative of a function which does not have a derivative in the usual sense by using the notion of weak convergence. This leads on to the idea of weak or generalized solutions of differential equations, solutions which formally do not lie in the range of the differential operator.

Weak* convergence

By analogy with weak convergence in $\mathscr{V}$ we could define weak convergence in $\mathscr{V}^*$ by saying that the sequence of linear functionals $l^{(k)}$ tends to l if, for every $h \in \mathscr{V}^{**}$,

$$h(l^{(k)}) \to h(l).$$

However, it turns out to be more useful to adopt the notion of so-called 'weak-star' convergence in $\mathscr{V}^*$.

Definition 3.5.4 *The sequence $\{l^{(k)}\}$ in $\mathscr{V}^*$ is said to converge* 'weak-star' *to l, written $l^{(k)} \xrightarrow{*} l$, if, for every $x \in \mathscr{V}$,*

$$l^{(k)}(x) \to l(x).$$

Every weak* convergent sequence of functionals is bounded [26].

The difference between weak convergence in $\mathscr{V}^*$ and weak* convergence arises from the fact that $\mathscr{V} \subset \mathscr{V}^{**}$, and naturally when $\mathscr{V}$ is reflexive the distinction is unimportant. It also follows that when $\mathscr{V}$ is not reflexive, weak* convergence does not imply weak convergence in $\mathscr{V}^*$.

In the definition of weak* convergence we required that $l^{(k)}(x) \to l(x)$ *for all* $x \in \mathscr{V}$. The need to consider every element of $\mathscr{V}$ can be waived for those spaces for which the closed span of some (infinite) set $\{x^{(k)}\}$ of linearly independent vectors is dense in $\mathscr{V}$ (cf., p. 139). In that case, we can assert that $l^{(k)} \xrightarrow{*} l$ if

1. the sequence $\|l^{(k)}\|$ is bounded, and
2. $l^{(k)}(x) \to l(x)$ for every x in the closed span of $\{x^{(k)}\}$.

That is, we can establish weak-star convergence using only a 'relatively small' subset of the vectors of $\mathscr{V}$.

Example 3.5.7 (Mechanical quadrature) Let $x(t) \in \mathscr{C}[0, 1]$ and consider the definite integral

$$s(x) = \int_0^1 x(t)\,dt:$$

s is clearly a linear functional on $\mathscr{C}[0, 1]$ and we shall be interested in finding a sequence of functionals $s_n(x)$ of the form

$$s_n(x) = \sum_{k=0}^{n} a_k^{(n)} x(t_k^{(n)}), \qquad n = 1, 2, \ldots,$$

such that

$$s_n(x) \xrightarrow{*} s(x).$$

The values of $s_n(x)$ are then (a sequence of) approximations to the value of the integral $s(x)$. The s_n are quadrature formulae, and the process is called mechanical quadrature, or numerical integration. Typical concrete examples are the trapezoidal rule, Simpson's rule and Gaussian quadrature.

There is considerable freedom in choosing the form of the s_n, but the examples mentioned above (and others) all spring from the basic idea of making $s_n(x) = s(x)$ whenever $x(t)$ is a polynomial of degree n or less. Such a condition serves to relate the coefficients $a_k^{(n)}$ and the nodes $t_k^{(n)}$ by the system of equations,

$$\int_0^1 t^i \, dt = \sum_{k=0}^n a_k^{(n)} (t_k^{(n)})^i, \qquad i = 0, 1, \ldots, n.$$

These are $(n+1)$ equations relating $(2n+2)$ parameters, and we can either choose the nodes (trapezoidal, Simpson's rule) or a combination of the nodes and coefficients (Gaussian formulae) [39].

We may now ask what conditions are necessary on the coefficients $a_k^{(n)}$ in order that $s_n \xrightarrow{*} s$. Now

$$|s_n(x)| \leqslant \sum_{k=0}^n |a_k^{(n)}| \, |x(t_k^{(n)})|$$

$$\leqslant \left(\sum_{k=0}^n |a_k^{(n)}| \right) \max |x(t)|$$

$$= \left(\sum_{k=0}^n |a_k^{(n)}| \right) \|x\|$$

so that the s_n are bounded provided

$$\sum_{k=0}^n |a_k^{(n)}| \leqslant M$$

for all n, where M is some constant. Furthermore, by construction, $s_n(x) = s(x)$ for every x in the closed span of the polynomials t^i, $i = 0, \ldots, n$. But the polynomials are dense in $\mathscr{C}[0, 1]$ and hence we may immediately conclude that the quadrature formulae $s_n(x)$ will converge if

$$\sum_{k=0}^n |a_k^{(n)}| \leqslant M$$

for all n.

If the coefficients $a_k^{(n)}$ are all non-negative then no condition whatsoever is required on the coefficients to ensure convergence. For, if $x(t) = 1$ then,

for any n,
$$s_n(x) = s(x)$$
or
$$\sum_{k=0}^{n} a_k^{(n)} = 1.$$

Taking $M = 1$ and using the fact that $a_k^{(n)} > 0$ we see that the above condition for convergence is unconditionally satisfied.

The dual of a linear transformation

The algebraic dual (or conjugate) of a linear transformation was introduced in Section 2.7. Recall (Definition 2.7.1 *et seq.*) that if $T: \mathscr{V} \to \mathscr{U}$ then we define a linear functional $z^* \in \mathscr{V}^*$ by

$$z^*(x) = x^*(Tx), \qquad x^* \in \mathscr{U}^*,$$

and then write $z^* = T'x^*$, $T': \mathscr{U}^* \to \mathscr{V}^*$: we showed that T' is linear.

We shall now show that if T is bounded (and therefore continuous) then z^* is a bounded linear functional and that this implies that T' is bounded.

We have
$$\begin{aligned} |z^*(x)| &= |x^*(Tx)| \\ &\leq \|x^*\| \, \|Tx\| \\ &\leq \|x^*\| \, \|T\| \, \|x\| \end{aligned}$$
so that
$$\|z^*\| \leq \|x^*\| \, \|T\|$$
and z^* is bounded. But this implies that
$$\|T'x^*\| \leq \|T\| \, \|x^*\|$$
or
$$\|T'\| \leq \|T\|,$$
and T' is bounded.

However, we can go further than this and actually assert that equality holds in the last statement. According to Corollary 1 of Theorem 3.5.2 (Hahn–Banach), there exists $x^* \in \mathscr{U}^*$ such that $\|x^*\| = 1$ and $x^*(Tx) = \|Tx\|$. Whence
$$\begin{aligned} \|Tx\| = |[x^*, Tx]| &= |[T'x^*, x]| \\ &\leq \|T'x^*\| \, \|x\| \\ &\leq \|T'\| \, \|x\| \end{aligned}$$

since $\|x^*\| = 1$. We conclude that

$$\|T\| \leq \|T'\|$$

which, coupled with the converse inequality above, implies that

$$\|T'\|_{\mathcal{U}^*,\mathcal{V}^*} = \|T\|_{\mathcal{U},\mathcal{V}}.$$

The subscripts are appended to draw attention to the fact that the norms of T, T' are with respect to 'conjugate' norm pairs.

The following rules for manipulation of the dual are easily verified (for (iv) see Exercise 3.5,8(a)):

(i) for any scalar λ, $(\lambda T)' = \lambda T'$;
(ii) if $T, S \in \mathcal{L}(\mathcal{V}, \mathcal{U})$, $(T + S)' = T' + S'$;
(iii) if $T \in \mathcal{L}(\mathcal{V}, \mathcal{U})$, $S \in \mathcal{L}(\mathcal{U}, \mathcal{W})$ and ST exists, then $(ST)' = T'S'$;
(iv) if $T \in \mathcal{L}(\mathcal{V}, \mathcal{V})$ and T^{-1} exists, then $(T')^{-1} \in \mathcal{L}(\mathcal{V}^*, \mathcal{V}^*)$ exists and $(T')^{-1} = (T^{-1})'$.

A number of very useful relationships exist between the range and null spaces of T and T'. Before continuing, we would advise the reader to refer to the section on annihilators (Section 2.6, p. 54 et seq.). The subset $S^0 \subset \mathcal{V}^*$ consists of those bounded linear functionals which map every element of a subset $S \subset \mathcal{V}$ to zero: the set S^0 is, in fact, a subspace of $\mathcal{V}^*$, and we can further show that it is a *closed subspace* of $\mathcal{V}^*$. That is to say, S^0 contains all its limits (Exercise 3.5,8(b)).

By analogy with annihilators of subsets in $\mathcal{V}$ we can likewise define the annihilator of a subset $S^* \subset \mathcal{V}^*$ as the subspace $^0S^* \subset \mathcal{V}$; this subspace is closed. We further have the result that if $\mathcal{W}$ is a closed subspace of $\mathcal{V}$ then [30]

$$^0(\mathcal{W}^0) = \mathcal{W}.$$

Theorem 3.5.3 *Let $T \in \mathcal{L}(\mathcal{V}, \mathcal{U})$ and let $T' \in \mathcal{L}(\mathcal{U}^*, \mathcal{V}^*)$ be its dual. Then*

1. $\mathcal{R}^0(T) = \mathcal{N}(T')$;
2. $\overline{\mathcal{R}(T)} = {}^0\mathcal{N}(T')$;
3. ${}^0\mathcal{R}(T') = \mathcal{N}(T)$;
4. $\overline{\mathcal{R}(T')} \subset \mathcal{N}^0(T)$.

Proof We shall prove only the first of these results and merely remark on the remainder.

Let $x^* \in \mathcal{N}(T')$ and $y \in \mathcal{R}(T)$, then $y = Tx$ for some $x \in \mathcal{V}$ and
$$x^*(y) = [x^*, Tx] = [T'x^*, x] = 0$$
so that
$$\mathcal{N}(T') \subset \mathcal{R}^0(T).$$
Now let $y^* \in \mathcal{R}^0(T)$, then, for every $x \in \mathcal{V}$,
$$[y^*, Tx] = 0 = [T'y^*, x]$$
which implies that
$$\mathcal{R}^0(T) \subset \mathcal{N}(T').$$

The closure operation in results 2 and 4 is required because, as we have seen, an annihilator is always closed. If T actually has a closed range the operation is clearly superfluous. This is certainly so in the important case when the range is finite-dimensional.

Equality holds in result 4 if $\mathcal{V}$ is (norm) reflexive or if $\mathcal{R}(T')$ is finite-dimensional. ∎

What of the dual of the dual of T, $(T')'$? In the general case we cannot say more than that $(T')' \in \mathcal{L}(\mathcal{V}, \mathcal{U})$ extends $T \in \mathcal{L}(\mathcal{V}, \mathcal{U})$ in the sense that these transformations coincide only for vectors in a subset of $\mathcal{V}$. However, if $\mathcal{V}$ is (norm) reflexive, then $(T')' = T$.

Example 3.5.8

(a) Let T be a transformation $\mathcal{R}_n \to \mathcal{R}_m$ with matrix $[t_{ij}]$: we already know (Section 2.8, p. 69) that the matrix of $T': \mathcal{R}_m^* \to \mathcal{R}_n^*$ has the representation $[t_{ji}]$ (relative to dual basis sets in $\mathcal{R}_m^*, \mathcal{R}_n^*$).

Example 3.4.2 showed that if we choose the norm
$$\|x\|_\infty = \max_k (|\xi_k|)$$
in $\mathcal{R}_n$ and $\mathcal{R}_m$ then $\|T\|_\infty$ is given by the maximum (modulus) *row* sum of $[t_{ij}]$. Now the dual spaces $\mathcal{R}_m^*, \mathcal{R}_n^*$ will have the norm
$$\|x^*\|_1 = \sum_k |\xi^k|$$
and $\|T'\|_1$ is given by the maximum (modulus) *column* sum of $[t_{ji}] = [t_{ij}]^T$. This is an illustration of the result $\|T\|_{\mathcal{V},\mathcal{U}} = \|T'\|_{\mathcal{V}^*,\mathcal{U}^*}$.

The discussion in Section 2.8 (p. 65) pointed out that

(i) $\mathcal{R}(T)$ is spanned by the column $\{t_i\}_j$ of $[t_{ij}]$;
(ii) the vectors $\{t_j\}_i$ span $\mathcal{N}^0(T)$;

similarly,

   (iii) $\mathscr{R}(T')$ is spanned by the columns $\{t_j\}_i$ of $[t_{ji}]$;
   (iv) the vectors $\{t_i\}_j$ span $^0\mathscr{N}(T')$.

Hence we immediately obtain results 2 and 4 of Theorem 3.5.3. Results 1 and 3 follow from the isomorphism of $\mathscr{R}_n, \mathscr{R}_n^*$ and $\mathscr{R}_m, \mathscr{R}_m^*$.

(b) Let $\mathscr{V} = \mathscr{U} = l_2$. For $x = (\xi_1, \xi_2, \ldots) \in \mathscr{V}$ define the 'shift' operator $T: \mathscr{V} \to \mathscr{U}$ by
$$Tx = (0, \xi_1, \xi_2, \ldots).$$
For any $y^* = (\eta^1, \eta^2, \ldots) \in \mathscr{U}^*$,
$$[y^*, Tx] = \sum_{i=1}^{\infty} \xi_i \eta^{i+1} = [T'y^*, x],$$
which implies that
$$T'y^* = (\eta^2, \eta^3, \ldots),$$
a 'reverse shift' operator.

The null space of T', $\mathscr{N}(T')$ consists of the vectors $(\alpha, 0, 0, \ldots)$: the annihilator of $\mathscr{R}(T)$ clearly consists of the same set of vectors. This is an illustration of result 1 of Theorem 3.5.3.

(c) A projection (operator) on a vector space $\mathscr{V}$ is an idempotent linear transformation of $\mathscr{V}$ into itself (Definition 2.5.3). A projection, P, is associated with a direct decomposition of $\mathscr{V}$ such that $\mathscr{V} = \mathscr{U} \oplus \mathscr{W}$, where $\mathscr{U} = \mathscr{R}(P)$ and $\mathscr{W} = \mathscr{N}(P)$. The foregoing ideas are purely algebraic: if $\mathscr{V}$ is a Banach space we define a *projection* to satisfy the additional requirement of being a continuous linear transformation. The most important consequence of imposing continuity is to make $\mathscr{U}$ and $\mathscr{W}$ closed subspaces. The null space of any continuous transformation is closed, so $\mathscr{W} = \mathscr{N}(P)$ is obviously closed (Exercise 3.4,1(e)).

Now $I - P$ is the projection of $\mathscr{V}$ onto $\mathscr{W}$ along $\mathscr{U}$ and, being the difference of two continuous transformations, is continuous; but $\mathscr{U} = \mathscr{N}(I - P)$ and hence $\mathscr{U}$ is closed.

A projection in a Banach space then induces a direct sum decomposition
$$\mathscr{V} = \mathscr{U} \oplus \mathscr{W} = \mathscr{R}(P) \oplus \mathscr{N}(P)$$
in which both $\mathscr{U}$ and $\mathscr{W}$ are closed.

Let P' be the dual of P. Then Theorem 3.5.3 implies that

1. P' is a (continuous) projection;
2. $\mathscr{V}^* = \mathscr{W}^0 \oplus \mathscr{U}^0 = \mathscr{R}(P') \oplus \mathscr{N}(P')$.

This result effectively extends (iii) of Theorem 2.6.2 to the case of infinite-dimensional but closed spaces. That is to say, a (closed) direct sum decomposition

$$\mathscr{V} = \mathscr{U} \oplus \mathscr{W}$$

induces the corresponding (closed) direct sum decomposition

$$\mathscr{V}^* = \mathscr{W}^0 \oplus \mathscr{U}^0$$

in the dual space.

Given a closed subspace $\mathscr{U}$ of $\mathscr{V}$, is there always a *continuous* projection such that $\mathscr{U}$ has a closed direct complement in $\mathscr{V}$? The answer is generally speaking, 'No', unless $\mathscr{U}$ is finite-dimensional (Exercise 3.5,8(c)) or $\mathscr{V}$ is a Hilbert space (Section 3.6).

Exercises 3.5

1. (a) Find the norms of the continuous linear functionals on $\mathscr{C}[0,1]$ given by
    (i) $l(x) = x(0)$,
    (ii) $l(x) = \int_0^1 x(t) dt$,
    (iii) $l(x) = \int_0^1 t x(t) dt$.

    (Hint: change an inequality into an equality by considering constant functions.)
   (b) Show that the set of all non-negative continuous linear functionals on the real normed vector space $\mathscr{V}$ is a convex cone in $\mathscr{V}^*$ (Exercise 2.3, 7).
   (c) Show that the norm of a linear functional l equals $1/d$, where d is the distance from the hyperplane $l(x) = 1$ to the origin (Exercise 2.6, 3(b)). (Hint: $d = \inf_{l(x)=1} \|x\|$; show $d \geq 1/\|l\|$, then use definition of $\|l\|$ to show $d \leq 1/\|l\|$.)

2. Show that a subset S of a normed vector space $\mathscr{V}$ is a spanning set for $\mathscr{V}$ if the only continuous linear functional which vanishes on S is the zero functional. (Hint: if S is a spanning set, show that $l = \theta^*$: if not there exists $x \in \mathscr{V}$ a non-zero distance from $\bar{S}$ and there exists l with $\|l\| \neq 0$—Example 3.5.4(b).)

3. (a) Show that the sum, difference and product of two functions of bounded variation are of bounded variation (Example 3.5.3).
   (b) Show that $V_0^1[x] = V_0^\tau[x] + V_\tau^1[x], 0 < \tau < 1$, where $V_\alpha^\beta[x]$ denotes the total variation of x on $[\alpha, \beta]$.
   (c) Show that the function $v(t) = V_0^t[x]$ is
    (i) continuous at every point at which x is continuous;
    (ii) bounded and increasing.

(d) Prove that a real function $x(t)$ of bounded variation on $[0, 1]$ can be expressed as the difference of two bounded increasing functions. (Hint: write $x(t) = v(t) - [v(t) - x(t)]$, $v(t)$ as in (c), and show that $v - x$ is increasing.)

(e) Verify that $|x(0)| + V[x]$ is a norm on the vector space of functions of bounded variation.

4. (a) Consider the space $\mathscr{C}[0, 1]$ of real continuous functions $x(t)$ on $[0, 1]$ with the uniform norm. The Hahn–Banach theorem guarantees the existence of functions of bounded variation $y(t)$ on $[0, 1]$ (see Example 3.5.3) such that

$$l_y(x) = \int_0^1 x(t) \, dy(t)$$
$$= \|x\| \|l_y\| = \max_{t \in [0,1]} \{|x(t)|\} V[y]$$

Let P be the non-empty (finite or infinite) set of points in $[0, 1]$ at which $|x(t)| = \|x\|$. Show that the functions of bounded variation satisfying the above condition are

(i) constant outside the set P;

(ii) non-decreasing (non-increasing) on P if $x(t) > 0$ (< 0); and hence are step functions if P is finite.

(b) Let $p(t)$ be the polynomial of degree n or less that best approximates $x(t) \in \mathscr{C}[0, 1]$ in the sense of the uniform norm. There does exist such a polynomial since we are minimizing the distance from x to the $(n + 1)$-dimensional space $\mathscr{P}_{n+1}$ of nth degree polynomials (Example 3.3.3). Let $\|x(t) - p(t)\| = d$ and let P be the set of points in $[0, 1]$ at which $|x(t) - p(t)| = d$. Show that P must contain at least $(n + 2)$ points. This is Chebyshev or minimax approximation [18] (Hint: since a polynomial $p_0(t)$ nearest $x(t)$ does exist, then (Example 3.5.4) $l(x - p_0) = \|l\| \|x - p_0\|$; $l \in \mathscr{P}_{n+1}^0 \subset \mathscr{C}^*[0, 1]$. Assume number of points in P is less than $n + 2$ and use part (a) to obtain a contradiction.)

5. (a) Show that every reflexive normed linear space is complete. (Hint: use Theorem 3.5.1.)

(b) Show that if $\mathscr{V}$ is reflexive then, given $x^* \in \mathscr{V}^*$, there exists $x \in \mathscr{V}$ such that $[x^*, x] = \|x^*\| \|x\|$ (cf., Corollary 1 of Theorem 3.5.2).

6. (a) Show that
(i) a weak limit is unique;
(ii) if $x^{(k)} \rightharpoonup x$, $y^{(k)} \rightharpoonup y$ then $x^{(k)} + y^{(k)} \rightharpoonup x + y$ and $\lambda x^{(k)} \rightharpoonup \lambda x$ for any scalar λ.

(b) Show that on a finite-dimensional normed space strong convergence is equivalent to weak convergence. (Hint: show that weak convergence is componentwise convergence.)

(c) Show that the sequence of unit vectors e^i does not converge weakly in l_1.

7. Consider the sequence $\{l^{(k)}\}$ of continuous linear functionals on a Banach space $\mathscr{V}$. The norm of each $l^{(k)}$ is finite but there is no guarantee that the norms might not form an increasing sequence. The *uniform boundedness principle* states that if the sequence $\{|l^{(k)}(x)|\}$ is bounded for all $x \in \mathscr{V}$ then the sequence $\{\|l^{(k)}\|\}$ is bounded. Here is an example of a normed space in which this is not true.

Let $\mathscr{P}$ be the space of polynomials $x(t) = \sum_{k=0}^{\infty} \alpha_k t^k$, where only the first $m(x)$ coefficients are non-zero, and take $\|x\| = \max\{|\alpha_k|\,|\,k = 1, 2, \ldots\}$. Let $l^{(k)}(x) = \sum_{j=0}^{k-1} \alpha_j$; these are continuous linear functionals on $\mathscr{P}$. Show that $|l^{(k)}(x)| \leq (m(x) + 1)\|x\|$, $k = 1, 2, \ldots$, and hence that $l^{(k)}(x)$ is bounded, but by taking $x(t) = \sum_{j=0}^{k-1} t^j$ show that $\|l^{(k)}\| \geq k$ and hence that $\{\|l^{(k)}\|\}$ is unbounded.

8. (a) Prove that if $T \in \mathscr{L}(\mathscr{V}, \mathscr{V})$ has a bounded inverse then $(T')^{-1} = (T^{-1})'$. (Hint: show that T' is one-to-one and onto.)

(b) Prove that $S^0 \subset \mathscr{V}^*$ is a closed subspace. (Hint: consider the limit of a sequence of functionals $l^{(k)}(x)$, $x \in S$.)

(c) Prove that if $\mathscr{U}$ is a finite-dimensional subspace of $\mathscr{V}$ then there is a *closed* subspace $\mathscr{W} \subset \mathscr{V}$ such that $\mathscr{V} = \mathscr{U} \oplus \mathscr{W}$. (Hint: combine the Corollary of Theorem 2.6.2 with a result similar to (b).)

(d) Let $\mathscr{V}$ be the space of sequences $\{\xi_i\}$ in which only a finite number of terms is non-zero (Example 2.2.2(b)) with the norm $\|x\| = \sum_i |\xi_i|$. Let $\mathscr{U}$ be the subspace consisting of the vectors with $\xi_i = 0$, $i = 1, 3, 5, 7, \ldots$, and let $\mathscr{W}$ be the subspace consisting of the vectors with $\xi_2 = \xi_1, \xi_4 = 2\xi_3, \xi_6 = 3\xi_5, \xi_8 = 4\xi_7, \ldots$. Show that
   (i) $\mathscr{V} = \mathscr{U} \oplus \mathscr{W}$;
   (ii) $\mathscr{U}$ is closed;
   (iii) $\mathscr{W}$ is not closed.

9. (a) Let $\mathscr{V}$ be a complex vector space. A functional $f : \mathscr{V} \times \mathscr{V} \to \mathscr{C}$ is called a *sesquilinear functional* if
   (i) $f(\alpha x + \beta y, z) = \alpha f(x, z) + \beta f(y, z)$;
   (ii) $f(x, \gamma y + \delta z) = \bar{\gamma} f(x, y) + \bar{\delta} f(y, z)$.

If $\mathscr{V}$ is real f is called a *bilinear* functional. The functional is said to be Hermitean (symmetric if real) if $f(x, y) = \overline{f(y, x)}$ and positive if $f(x, x) \geq 0$ for all x. The functional $\tilde{f}(x) = f(x, x) : \mathscr{V} \to \mathscr{C}$ is called *the quadratic form* associated with f.

Show that
   (i) $f(x, y) = \tilde{f}(\tfrac{1}{2}(x + y)) - \tilde{f}(\tfrac{1}{2}(x - y)) + i\tilde{f}(\tfrac{1}{2}(x + iy)) - i\tilde{f}(\tfrac{1}{2}(x - iy))$ (i.e., a sesquilinear functional can be inferred from its quadratic form);
   (ii) f is symmetric if and only if $\tilde{f}$ is real-valued;
   (iii) $|f(x, y)|^2 \leq \tilde{f}(x)\tilde{f}(y)$ if f is positive.

(b) Show that if $f(x, y)$ is continuous on $\mathscr{V} \times \mathscr{V}$ then there exists $K \geq 0$ such that $|f(x, y)| \leq K \|x\| \|y\|$. (Hint: f is continuous at $(0, 0)$.)

(c) Verify that every continuous sesquilinear functional $f(x, y) : \mathscr{V} \times \mathscr{V} \to \mathscr{C}$ is associated with a unique linear transformation $T_f \in \mathscr{L}(\mathscr{V}, \mathscr{V}^*)$ such that

$$f(x, y) = [T_f x, y] \quad \forall\, x, y \in \mathscr{V}.$$

Show that the sesquilinear functional $f(x, y) : \mathscr{V}_n \times \mathscr{V}_n \to \mathscr{C}$ determines, relative to reciprocal basis sets $\{x^i\}, \{x_i\}$ in $\mathscr{V}_n, \mathscr{V}_n^*$, a matrix $[t_{ij}]^T$ with $t_{ij} = f(x^i, x^j)$.

3.6 Inner product spaces

The spaces l_2, $\mathscr{L}_2$ have the property of being self-reflexive; that is to say, their duals are isometrically isomorphic to the spaces themselves and thus indistinguishable from them. The dual of a transformation $T : l_2 \to l_2$, for example, is therefore another transformation over the same space. Linear functionals over l_2 (or $\mathscr{L}_2$) are, in the above sense, indistinguishable from elements of l_2 (or $\mathscr{L}_2$) and therefore the value of a linear functional on such a space could be considered as being generated by an internal operation involving two elements of the space. The annihilator of a subspace in l_2, for example, is itself a subspace of l_2.

The real self-reflexive spaces l_2 (and $\mathscr{R}_n$ with norm $\|\cdot\|_2$) express the norm (or length) of a vector x in terms of the sum of squares of its components, viz.,

$$\|x\|_2 = \left(\sum_i \xi_i^2 \right)^{1/2},$$

which is analogous to the usual definition for the length of a vector in ordinary three-dimensional Euclidean space. Is there, then, also an analogy for the scalar or dot product of two ordinary vectors $\mathbf{x} = (\xi_1, \xi_2, \xi_3)$ and $\mathbf{y} = (\eta_1, \eta_2, \eta_3)$ defined by

$$\mathbf{x} \cdot \mathbf{y} = (\xi_1 \eta_1 + \xi_2 \eta_2 + \xi_3 \eta_3)?$$

In accordance with the remark made above concerning linear functionals, we could interpret the right-hand side of this expression, in $\mathscr{R}_3$ for example, as the value of a linear functional (η_1, η_2, η_3) on the vector (ξ_1, ξ_2, ξ_3) *when the basis set in the space (and its dual) is self-reciprocal*, that is to say, when

$$x^i(x^j) = [x^i, x^j] = \delta_i^j$$

(cf., p. 51 *et seq.*). With this interpretation we have, in effect, defined a scalar-valued internal operation on pairs of elements in $\mathscr{R}_3$. Furthermore

$$(\mathbf{x} \cdot \mathbf{x})^{1/2} = (\xi_1^2 + \xi_2^2 + \xi_3^2)^{1/2} = \|x\|,$$

the Euclidean 'norm' of x.

All these considerations suggest that there may be profit in considering a class of normed spaces in which we may define an inner operation between pairs of elements analogous to the dot product, and which yields the norm of an element as a special case. The operation we seek is called the *inner product*; the resulting spaces *inner product spaces*. These spaces could loosely be described as the subset of normed spaces whose geometry is Euclidean.

Given the foregoing motivation for introducing an inner product, we now proceed on an axiomatic basis.

Definition 3.6.1 *An inner product on a vector space $\mathscr{V}$ is a scalar-valued function $\langle x, y \rangle$, defined for all ordered pairs of vectors $x, y \in \mathscr{V}$ and which satisfies the following axioms:*

1. $\langle x, y \rangle = \overline{\langle y, x \rangle} \quad \forall x, y \in \mathscr{V}$;
2. $\langle \alpha x + \beta y, z \rangle = \alpha \langle x, z \rangle + \beta \langle y, z \rangle \quad \forall x, y, z \in \mathscr{V}$ *and scalars* α, β;
3. $\langle x, x \rangle > 0; \langle x, x \rangle = 0$ *if, and only if,* $x = \theta$.

The following property follows from axioms 1 and 2:

4. $\langle x, \gamma y + \delta z \rangle = \bar{\gamma} \langle x, y \rangle + \bar{\delta} \langle x, z \rangle$.

The bar denotes complex conjugate.

An inner product space, *denoted by* $(\mathscr{V}, \langle \cdot, \cdot \rangle)$ *consists of a vector space $\mathscr{V}$ and an inner product $\langle \cdot, \cdot \rangle$ on $\mathscr{V}$.*

Axiom 1 imposes the property of *conjugate symmetry* on the inner product; this property is essential if $\langle x, x \rangle$ is to be real, as is implied by axiom 3. Axiom 2 makes the inner product linear in its first argument. Note from 4 that the inner product is conjugate linear in its second argument (*see* Exercise 3.5.9(a)). A real inner product space is often called a *Euclidean space*, while a complex inner product space is often called a *unitary space*.

There is clearly a similarity between the notations used for the inner product $\langle x, y \rangle$ and the bracket form for linear functionals $[x, x^*]$. Indeed, for *real spaces* their properties are formally identical, provided we make the very important distinction that the former is an operation involving only $\mathscr{V}$, while the latter involves $\mathscr{V}$ and its dual $\mathscr{V}*$.[†] We shall return to this point when we discuss so-called orthonormal basis sets.

We have indicated that the real quantity $(\langle x, x \rangle)^{1/2}$ is a natural norm for an inner product space. Axioms 1 and 3 of Definition 3.3.1 are clearly satisfied, and we need only verify that the triangle inequality holds. To do this, we need an important inequality which we shall often find useful.

Theorem 3.6.1 (The Schwarz inequality) *In any inner product space $\mathscr{V}$,*

$$|\langle x, y \rangle|^2 \leq \langle x, x \rangle \langle y, y \rangle \qquad \forall x, y \in \mathscr{V}.$$

Proof If x or $y = \theta$ the inequality holds trivially. Otherwise, for all scalars α,

$$0 \leq \langle x - \alpha y, x - \alpha y \rangle$$
$$= \langle x, x \rangle - \alpha \langle y, x \rangle - \bar{\alpha} \langle x, y \rangle + |\alpha|^2 \langle y, y \rangle.$$

In particular, if

$$\alpha = \langle x, y \rangle / \langle y, y \rangle$$

we have

$$0 \leq \langle x, x \rangle - |\langle x, y \rangle|^2 / \langle y, y \rangle$$

or

$$|\langle x, y \rangle|^2 \leq \langle x, x \rangle \langle y, y \rangle. \quad \blacksquare$$

We can now show that $\langle x, x \rangle^{1/2}$ satisfies the triangle inequality and can therefore be taken as the norm, $\|x\|$, of x. We have,

$$0 \leq \|x + y\|^2 = \langle x + y, x + y \rangle$$
$$= \|x\|^2 + 2 \operatorname{Re} \langle x, y \rangle + \|y\|^2$$
$$\leq \|x\|^2 + 2|\langle x, y \rangle| + \|y\|^2$$
$$\leq \|x\|^2 + 2\|x\| \|y\| + \|y\|^2$$
$$= (\|x\| + \|y\|)^2$$

upon using the Schwarz inequality.

A familiar result of elementary geometry is that the sum of the

---

[†] In this book, the order of factors in the function bracket $[\cdot, \cdot]$ can be reversed without change in value. In some treatments, the functional bracket is taken to be 'conjugate bilinear' in its entries, as for the inner product.

squares of the sides of a parallelogram equals the sum of the squares of the diagonals: this so-called *'parallelogram law'* also holds in an inner product space, viz.,

$$\|x+y\|^2 + \|x-y\|^2 = 2\|x\|^2 + 2\|y\|^2$$

(Exercise 3.6,1(b)).

Example 3.6.1 (A selection of inner product spaces)

(a) The space $\mathscr{C}_n$ with the inner product of two vectors $x = (\xi_1, \xi_2, \ldots, \xi_n)$ and $y = (\eta_1, \eta_2, \ldots, \eta_n)$ defined by

$$\langle x, y \rangle = \sum_{i=1}^{n} \xi_i \bar{\eta}_i;$$

$$\|x\|^2 = \sum_{i=1}^{n} |\xi_i|^2.$$

This space is called a unitary (complex) or Euclidean (real) n-dimensional space. We shall denote it by the symbol $\mathscr{E}_n$.

(b) The space l_2 with the inner product

$$\langle x, y \rangle = \sum_{i=1}^{\infty} \xi_i \bar{\eta}_i;$$

$$\|x\|^2 = \sum_{i=1}^{\infty} |\xi_i|^2.$$

The Hölder inequality (Appendix 3.A) guarantees that the inner product is finite when $x, y \in l_2$.

(c) The space $\mathscr{L}_2[0, 1]$ of square integrable functions on $[0, 1]$ with

$$\langle x, y \rangle = \int_0^1 x(t) \bar{y}(t) dt;$$

$$\|x\|^2 = \int_0^1 |x(t)|^2 dt.$$

The Hölder inequality again guarantees that $\langle x, y \rangle < \infty$.

(d) Let Γ be a compact region in $\mathscr{R}_n$ and let $\mathscr{C}(\Gamma)$ be the space of complex-valued continuous functions on Γ. Take

$$\langle x, y \rangle = \int_\Gamma x \bar{y} \, d\Gamma;$$

$$\|x\|^2 = \int_\Gamma |x|^2 \, d\Gamma.$$

(e) Let Γ be an open region in $\mathscr{R}_2$ and let $\mathscr{C}^{(1)}(\Gamma)$ be the space of real-valued functions having continuous first partial derivatives in Γ. Take

$$\langle x, y \rangle = \int_\Gamma \left(xy + \frac{\partial x}{\partial t_1} \frac{\partial y}{\partial t_1} + \frac{\partial x}{\partial t_2} \frac{\partial y}{\partial t_2} \right) d\Gamma;$$

$$\|x\|^2 = \int_\Gamma \left(x^2 + \left(\frac{\partial x}{\partial t_1}\right)^2 + \left(\frac{\partial x}{\partial t_2}\right)^2 \right) d\Gamma.$$

(f) Let Γ be an open region in $\mathscr{R}_n$. Let $\mathscr{C}^{(m)}(\Gamma)$ denote the space of functions which have continuous partial derivatives (in all variables) up to and including order m. Let $k = (k_1, k_2, \ldots, k_n)$ be a vector with integral entries, and let

$$|k| = \sum_{i=1}^n k_i.$$

Define the differential operator D^k by

$$D^k x = \frac{\partial^{|k|} x}{\partial t_1^{k_1} \partial t_2^{k_2} \cdots \partial t_n^{k_n}}.$$

Take

$$\langle x, y \rangle^{(m)} = \int_\Gamma \sum_{|k| \leq m} D^k x \, \overline{D^k y} \, d\Gamma;$$

$$\|x\|_2^{(m)} = \left\{ \int_\Gamma \sum_{|k| \leq m} |D^k x|^2 \, d\Gamma \right\}^{1/2}.$$

Example (e) illustrates the case $m = 1$. We shall discuss this type of inner product space in some detail in Section 5.4.

Since an inner product space is a special type of normed space, we can ask whether the space is complete or not. An inner product space which is complete with respect to the norm induced by the inner product is called a *Hilbert space*. Every inner product space has a completion. In Example 3.6.1, parts (a), (b) and (c) show examples of Hilbert spaces.

It will be convenient in the sequel to denote a Hilbert space by $\mathscr{H}$.

Orthogonality

We can give a rather significant geometric interpretation of the Schwarz inequality in a real inner product space. The quantity

$$\langle x, y \rangle / \|x\| \, \|y\|$$

always has modulus less than 1: by analogy with ordinary three-dimensional vectors we could define the cosine of the 'angle between x and y' by

$$\cos \widehat{xy} = \langle x, y \rangle / \|x\| \|y\|.$$

Again, by analogy with ordinary three-dimensional space we would refer to the vectors x and y as being orthogonal if $\cos \widehat{xy} = 0$; that is, if their dot product vanishes. While the idea of angle is of very limited usefulness in real inner product spaces (and not at all in complex spaces), the concept of orthogonality is of extreme importance.

Definition 3.6.2 *In an inner product space $(\mathscr{V}, \langle \cdot, \cdot \rangle)$ the vectors x, y are said to be* orthogonal, *written $x \perp y$, if $\langle x, y \rangle = 0$. The vector x is said to be orthogonal to the subset $S \subset \mathscr{V}$, written $x \perp S$, if x is orthogonal to every $y \in S$. The subsets $S \subset \mathscr{V}$ and $R \subset \mathscr{V}$ are orthogonal, $S \perp R$, if $\langle y, z \rangle = 0$ for all $y \in S, z \in R$. The set of vectors $S^{\perp} = \{y \in \mathscr{V} \mid y \perp x \; \forall x \in S\}$ is called the* orthogonal complement *of S.*

The following properties consequent upon Definition 3.6.2 may be noted:

1. The zero vector θ is orthogonal to every x.
2. $x \perp x$ if, and only if, $x = \theta$.
3. If $x \perp y^{(k)}$ and $y^{(k)} \to y$ (in the sense of the induced norm), then $x \perp y$ (Exercise 3.6.1(a)).
4. (The Pythagorean theorem) If $x \perp y$ then
   $$\|x + y\|^2 = \|x\|^2 + \|y\|^2.$$
5. If $x \perp y^1, y^2, \ldots, y^n$, then x is orthogonal to every element in the span of the $\{y^i\}$.

In Examples 3.3.3 and 3.5.4(b) we considered the problem of finding the minimum distance $d(x, \mathscr{W})$ from a point $x \in \mathscr{V}$ to the subspace $\mathscr{W} \subset \mathscr{V}$. If $\mathscr{W}$ is compact there actually exists $y_0 \in \mathscr{W}$ such that $d(x, \mathscr{W}) = d(x, y_0)$: if $\mathscr{W}$ is not compact then y_0 need not exist, but we saw that we could relate $d(x, \mathscr{W})$ to the maximum norm of functionals in $\mathscr{W}^0 \subset \mathscr{V}^*$. In a Hilbert space $\mathscr{H}$ the situation is simpler. Since a Hilbert space is self-reflexive, one would expect the role of $\mathscr{W}^0$, the annihilator of $\mathscr{W} \subset \mathscr{H}$, to be assumed by a subspace of $\mathscr{H}$ which is orthogonal to $\mathscr{W}$. Part (c) of Example 3.5.8 would then suggest that there might exist a projection, P, on $\mathscr{H}$ inducing the

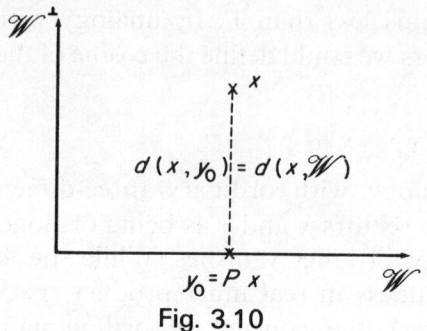

Fig. 3.10

direct sum decomposition

$$\mathscr{H} = \mathscr{W} \oplus \mathscr{W}^\perp.$$

Such a projection would be called orthogonal and the problem of finding $d(x, \mathscr{W})$ would be resolved into that of finding the orthogonal projection of x onto $\mathscr{W}$ (Fig. 3.10).

We shall now establish these conjectures.

Theorem 3.6.2 (The projection theorem) *Let $\mathscr{W}$ be a closed subspace of a Hilbert space $\mathscr{H}$ and let $x \in \mathscr{H}$ be a vector not in $\mathscr{W}$. Then there exists a unique vector $y_0 \in \mathscr{W}$ such that $d(x, \mathscr{W}) = \|x - y_0\|$.*

A necessary and sufficient condition that y_0 is the unique minimizing vector is that $(x - y_0) \perp \mathscr{W}$.

Proof Let $d = \inf_{y \in \mathscr{W}} \|x - y\|$ and choose a sequence of vectors $y^{(k)} \in \mathscr{W}$ such that

$$\lim_{k \to \infty} \|x - y^{(k)}\| = d.$$

By the parallelogram law

$$\begin{aligned}
\|y^{(k)} - y^{(j)}\|^2 &= 2\|y^{(k)} - x\|^2 + 2\|y^{(j)} - x\|^2 \\
&\quad - \|(y^{(k)} - x) + (y^{(j)} - x)\|^2 \\
&= 2\|y^{(k)} - x\|^2 + 2\|y^{(j)} - x\|^2 \\
&\quad - 4\left\|\frac{y^{(k)} + y^{(j)}}{2} - x\right\|^2.
\end{aligned}$$

Since $\mathscr{W}$ is a linear subspace, $\frac{1}{2}(y^{(k)} + y^{(j)}) \in \mathscr{W}$ for all k, j and, from the meaning of d, it follows that

$$\left\|\frac{y^{(k)} + y^{(j)}}{2} - x\right\| \geq d,$$

giving
$$\|y^{(k)} - y^{(j)}\|^2 \leqslant 2\|y^{(k)} - x\|^2 + 2\|y^{(j)} - x\|^2 - 4d^2.$$
Since
$$\|y^{(k)} - x\|^2 \to d^2$$
as $k \to \infty$, we may conclude that $\{y^{(k)}\}$ is a Cauchy sequence in $\mathscr{W}$. But $\mathscr{W}$ is a closed subspace of the complete space $\mathscr{H}$, hence the sequence has a limit $y_0 \in \mathscr{W}$. By continuity of the norm, $d = \|y_0 - x\|$.

Suppose there is another vector $z_0 \in \mathscr{W}$ such that $d = \|z_0 - x\|$. Then
$$\|y_0 - z_0\|^2 = 2\|y_0 - x\|^2 + 2\|z_0 - x\|^2 - \|y_0 + z_0 - 2x\|^2$$
$$= 4d^2 - 4\left\|\frac{y_0 + z_0}{2} - x\right\| \leqslant 0,$$
which implies that y_0 is unique.

For the last part of the theorem we first show that if y_0 is the minimizing vector then $x - y_0 \perp \mathscr{W}$. Let $y \in \mathscr{W}$ and λ any scalar, then $y_0 + \lambda y \in \mathscr{W}$, so that
$$\|x - (y_0 + \lambda y)\|^2 \geqslant \|x - y_0\|^2$$
and, interpreting the norm as an inner product,
$$-\bar{\lambda}\langle x - y_0, y\rangle - \lambda\langle y, x - y_0\rangle + |\lambda|^2\|y\|^2 \geqslant 0.$$
With $\lambda = \varepsilon, \varepsilon > 0$ this becomes
$$2\,\mathrm{Re}\,\langle x - y_0, y\rangle \leqslant \varepsilon\|y\|^2$$
and with $\lambda = i\varepsilon, \varepsilon > 0$ it becomes
$$2\,\mathrm{Im}\,\langle x - y_0, y\rangle \leqslant \varepsilon\|y\|^2.$$
Since $\varepsilon > 0$ is arbitrary, it follows that
$$\langle x - y_0, y\rangle = 0 \quad \text{or} \quad x - y_0 \perp \mathscr{W}.$$

This proves necessity: to prove sufficiency suppose $x - y_0 \perp \mathscr{W}$, then, for any $y \in \mathscr{W}$,
$$\|x - y\|^2 = \|(x - y_0) + (y_0 - y)\|^2$$
$$= \|x - y_0\|^2 + \|y_0 - y\|^2$$
so that $\|x - y\| \geqslant \|x - y_0\|$ for $y \neq y_0$.

Notice that we have used the completeness of $\mathscr{H}$ and that the inner product structure of $\mathscr{H}$ is embodied in the parallelogram law. ∎

Corollary (Exercise 3.6,2(e)) *If $\mathscr{W}$ is a closed subspace of a Hilbert space $\mathscr{H}$, then every $x \in \mathscr{H}$ has a unique decomposition*

$$x = y + z, \qquad y \in \mathscr{W}, \quad z \in \mathscr{W}^\perp.$$

The transformation $P : Px = y$ is a projection such that $\mathscr{W} = \mathscr{R}(P)$ and $\mathscr{W}^\perp = \mathscr{N}(P)$ are mutually orthogonal closed subspaces with

$$\mathscr{H} = \mathscr{W} \oplus \mathscr{W}^\perp.$$

For each $x \in \mathscr{H}$, Px is the unique element of $\mathscr{W}$ which minimizes the distance from x to $\mathscr{W}$. The vector $y = Px$ is called the orthogonal projection of x onto $\mathscr{W}$ (Fig. 3.10).

Example 3.6.2

(a) Let us develop the remark made at the end of Example 3.3.3 relating to the determination of a vector $x \in \mathscr{H}$ which is nearest to a *finite-dimensional* subspace $\mathscr{W} \subset \mathscr{H}$.

If $\mathscr{W}$ has the basis $\{y^k\}$ then the condition that the element

$$y_0 = \sum_{i=1}^{n} \eta_i y^i$$

is the element minimizing the distance from x to $\mathscr{W}$ is, from Theorem 3.6.2,

$$\langle y_0 - x, y^j \rangle = 0, \qquad j = 1, \ldots, n, \quad \text{or}$$

$$\sum_{i=1}^{n} \eta_i \langle y^i, y^j \rangle = \langle x, y^j \rangle, \qquad j = 1, \ldots, n,$$

which is a set of n simultaneous equations for the coordinates $\eta_i, i = 1, \ldots, n$. These equations are often called the *normal equations* for the problem of best approximation.

According to Theorem 3.6.2, the vector y_0 is unique, which implies that the normal equations have a unique solution. This, in turn, implies that the so-called *Gram determinant*

$$\det[\langle y^i, y^j \rangle]$$

does not vanish. It is easy to show that the Gram determinant does not vanish if the y^k are linearly independent.

Another type of minimum norm problem which leads to a finite set of linear equations is the following. Find $x_0 \in \mathscr{H}$ having minimum norm and satisfying the n constraints

$$\langle x_0, y^i \rangle = \lambda_i, \qquad i = 1, \ldots, n.$$

If $\mathscr{W}$ is the (closed) subspace spanned by the $\{y^i\}$, then its orthogonal

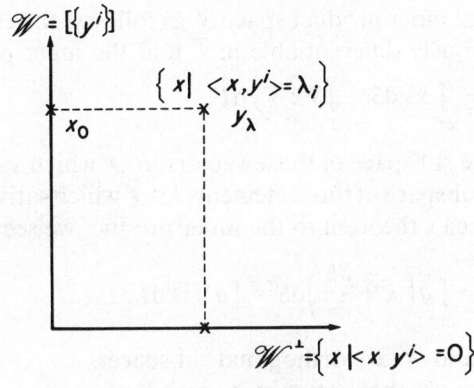

Fig. 3.11

complement $\mathscr{W}^\perp$, of codimension n, consists of those vectors x for which $\langle x, y^i \rangle = 0$. The set $\langle x, y^i \rangle = \lambda_i$ is the coset $y_\lambda + \mathscr{W}^\perp$, a translation of $\mathscr{W}^\perp$, where $y_\lambda \in \mathscr{H}$ is such that $\langle y_\lambda, y^i \rangle = \lambda_i$. The vector x_0 is that element of $y_\lambda + \mathscr{W}^\perp$ having smallest norm; that is to say, x_0 is the orthogonal projection of y_λ onto $\mathscr{W}$ (Fig. 3.11). Since $x_0 \in \mathscr{W}$, we may write

$$x_0 = \sum_{i=1}^n \xi_i y^i,$$

and the n constraint equations become

$$\sum_{i=1}^n \xi_i \langle y^i, y^j \rangle = \lambda_j, \qquad j = 1, \ldots, n,$$

n simultaneous equations for the coordinates ξ_i.

(b) (Dirichlet's principle) This example illustrates the importance of completeness for the existence of an element of minimum norm. A lack of awareness of the need for completeness led to several abortive (or mistaken) attempts to prove the existence of a solution to the so-called Dirichlet problem using calculus of variations techniques, so this is a rather celebrated problem in the history of mathematics.

For simplicity, we shall restrict the discussion to two (physical) dimensions and to Laplace's equation, rather than a more general elliptic operator. Let Γ be an open region in the plane, and $\partial \Gamma$ its smooth boundary. We seek a function, u, in the space of twice continuously differentiable functions in Γ such that

$$\nabla^2 u = 0 \quad \text{in} \quad \Gamma$$

and

$$u = f \quad \text{on} \quad \partial \Gamma.$$

This is (a restricted form of) Dirichlet's problem.

Define a real inner product space $\mathscr{I}$ as follows: the elements of $\mathscr{I}$ are twice continuously differentiable in $\bar{\Gamma}$ and the inner product is

$$\langle x, y \rangle = \int_{\partial \Gamma} xy \, \mathrm{d}S + \int_{\Gamma} \nabla x \cdot \nabla y \, \mathrm{d}\Gamma.$$

Let $\mathscr{G}$ be the subspace of those vectors $g \in \mathscr{I}$ which vanish on $\partial \Gamma$; and let $\mathscr{K}$ be the subspace of those elements $k \in \mathscr{I}$ which satisfy $\nabla^2 k = 0$ in Γ. Applying Green's theorem to the inner product we see that

$$\langle g, k \rangle = \int_{\partial \Gamma} g \left(k + \frac{\partial k}{\partial \nu} \right) \mathrm{d}S - \int_{\Gamma} g \nabla^2 k \, \mathrm{d}\Gamma,$$

and hence $\mathscr{G}$ and $\mathscr{K}$ are orthogonal subspaces.

The solution u is that vector in $\mathscr{K}$ such that (*see* Fig. 3.12)

$$u - f \in \mathscr{G};$$

that is to say, u is the orthogonal projection of f onto $\mathscr{K}$. Put another way, we can see that amongst all elements of $\mathscr{I}$ which equal f on $\partial \Gamma$ the solution u is that having the smallest norm. (u is the 'perpendicular distance' from the origin to the coset $f + \mathscr{G}$.)

This characterization of u by a minimum property is known as *Dirichlet's principle*. What happens if we try to infer the existence of a solution to Dirichlet's problem by using Dirichlet's principle? We may retrace the steps of the proof of Theorem 3.6.2 using the fact that $f + \mathscr{G}$ is a convex set (Exercise 2.3,5(d)) to show that the distance d from the origin to $f + \mathscr{G}$ is positive, and that there exists a sequence $\{u^{(k)}\}$ in $\mathscr{I}$ such that $\|u^{(k)}\| \to d$. However, since $\mathscr{I}$ is not complete we cannot assert that the sequence actually has a limit in $\mathscr{I}$, and so an existence proof fails.

However, if we know a solution to the Dirichlet problem exists, then we can certainly characterize it by Dirichlet's principle.

An alternative way to proceed is to admit weak or generalized solutions to the Dirichlet problem as the limit of the sequences $\{u^{(k)}\}$ in Dirichlet's principle. Since such limits need not be in the domain of the

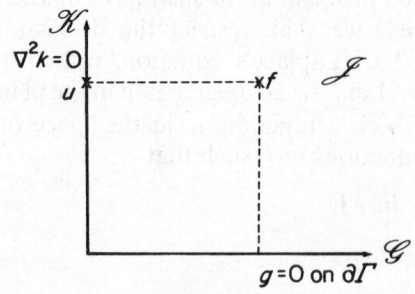

Fig. 3.12

operator ∇^2, they do not satisfy the differential equation in the classical sense: we return to this important point later, in Chapter 5.

Consider the very simple example of Dirichlet's problem in one variable. In the space of twice continuously differentiable functions $u(t)$ in $[0, 1]$ with $u(0) = 0$, $u(1) = \beta \neq 0$,

$$\|u\|^2 = \langle u, u \rangle = \int_0^1 \left(\frac{du}{dt}\right)^2 dt$$

obviously has infimum equal to zero. However, no function in the space actually attains the value zero for $\|u\|$ (cf., Exercise 3.2, 10(c)).

(c) We have several times remarked that the dual of a Hilbert space is the space itself. The projection theorem gives us the means to justify this remark.

If $y_0 \in \mathscr{H}$, then $\langle x, y_0 \rangle$ is a linear functional on $\mathscr{H}$. The Schwarz inequality gives

$$\frac{|\langle x, y_0 \rangle|}{\|x\|} \leq \|y_0\|,$$

so that, for each $y_0 \in \mathscr{H}$, $\langle x, y_0 \rangle$ is a bounded linear functional on $\mathscr{H}$ with norm $\|y_0\|$. (Take $x = y_0$!)

This establishes a natural norm-preserving mapping $\Phi : \mathscr{H} \to \mathscr{H}'$: it is not linear, but satisfies

$$\Phi(\alpha x_1 + \beta x_2) = \bar{\alpha}\Phi(x_1) + \bar{\beta}\Phi(x_2).$$

If we can show that Φ is *onto* then we can effectively identify the dual of a Hilbert space with itself. That is to say, if $x^* \in \mathscr{H}'$ is a bounded linear functional on $\mathscr{H}$, there exists a unique $y_0 \in \mathscr{H}$ such that

$$x^*(x) = \langle x, y_0 \rangle \qquad \forall x \in \mathscr{H}.$$

Now the null space of the functional x^*, $\mathscr{N}(x^*)$, is a closed subspace of $\mathscr{H}$. Let x_0 be a vector in the orthogonal complement $\mathscr{N}^\perp(x^*)$. Take $y_0 = \alpha x_0$, where α is chosen so that

$$x^*(x_0) = \langle x_0, y_0 \rangle,$$

that is,

$$x^*(x_0) = \bar{\alpha}\|x_0\|^2,$$

giving

$$\alpha = \overline{x^*(x_0)}/\|x_0\|^2.$$

Now let us see if this y_0 fulfils the requirements we seek. Let x be any vector in $\mathscr{H}$ and take

$$\beta = x^*(x)/x^*(x_0);$$

then
$$x - \beta x_0 \in \mathcal{N}(x^*)$$
and, since $y_0 \in \mathcal{N}^\perp(x^*)$,
$$\begin{aligned}\langle x, y_0 \rangle &= \langle x - \beta x_0, y_0 \rangle + \langle \beta x_0, y_0 \rangle \\ &= \beta \langle x_0, y_0 \rangle \\ &= \beta x^*(x_0) \\ &= x^*(x),\end{aligned}$$
and hence there exists a one-to-one correspondence $x^* \leftrightarrow y_0$ between linear functionals on $\mathcal{H}$ and vectors in $\mathcal{H}$.[†] The uniqueness of y_0 follows from the fact that if
$$\langle x, y_0 \rangle = \langle x, y_1 \rangle \quad \forall x \in \mathcal{H}$$
then, for the special choice $x = y_0 - y_1$,
$$0 = \langle x, y_0 - y_1 \rangle = \langle y_0 - y_1, y_0 - y_1 \rangle,$$
implying that $y_1 = y_0$.

Thus Hilbert spaces and their duals are indistinguishable metrically and nearly indistinguishable algebraically. The 'nearly' arises from the fact that the mapping is conjugate linear instead of linear, and if we restrict ourselves to real spaces this qualification can be removed.

The correspondence between each linear functional x^* on $\mathcal{H}$ and an element $y_0 \in \mathcal{H}$ means that, in Hilbert space, we can interpret weak convergence (Definition 3.5.3) in terms of the inner product. Specifically, the sequence $\{x^{(k)}\}$ in a Hilbert space $\mathcal{H}$ will *converge weakly* to $x \in \mathcal{H}$, $x^{(k)} \to x$ if
$$\langle x^{(k)}, y \rangle \to \langle x, y \rangle \quad \forall y \in \mathcal{H}.$$

Because $\mathcal{H}$ and $\mathcal{H}^*$ are (nearly) indistinguishable weak* convergence (Definition 3.5.4) is the same as weak convergence.

The orthogonal decomposition of a Hilbert space can be extended to more than one pair of subspaces. Let $\mathcal{W}_n$ be a finite or countable set of subspaces of $\mathcal{H}$ which are mutually orthogonal. Then every $x \in \mathcal{H}$ has a unique representation
$$x = \sum_n P_n(x),$$
where P_n is the (orthogonal) projection of x onto $\mathcal{W}_n$. This result is fairly obvious when n is finite: we shall not prove it when n is infinite

---

[†] This result is often referred to as the *Riesz representation theorem* for a Hilbert space (cf., Example 3.5.3).

(*see*, for example, [30]) since we shall cover essentially the same point in dealing with orthonormal basis sets. We now turn to finite or infinite sets of vectors which are pairwise orthogonal.

Definition 3.6.3 *A subset* $S = \{x^i\}$ *of vectors in the inner product space $\mathscr{V}$ is said to be* an orthogonal set of vectors *if $x^i \perp x^j$ for each pair $x^i, x^j \in S$, $i \neq j$. If, in addition, for each $x^i \in S$, $\|x^i\| = 1$, the set is said to be* orthonormal.

An orthogonal set of vectors $S = \{x^i\}$ is linearly independent. To see this, choose any finite collection of vectors $\{x^\rho\}$ from S and suppose that

$$\sum_\rho \lambda_\rho x^\rho = \theta.$$

Taking the inner product of both sides of this equation with $x^\mu, \mu \in \{\rho\}$, gives

$$\sum_\rho \lambda_\rho \langle x^\rho, x^\mu \rangle = \langle \theta, x^\mu \rangle = 0$$

or

$$\lambda_\mu \langle x^\mu, x^\mu \rangle = 0.$$

Since S is a set of non-zero vectors, we conclude that $\lambda_\rho = 0$ for all ρ, which, from Definition 2.4.1(c), implies linear independence.

Example 3.6.3 (cf., Example 3.6.1)

(a) In the finite-dimensional space $\mathscr{E}_n$ the set of 'unit' vectors $e^1 = (1, 0, 0, \ldots, 0)$, $e^2 = (0, 1, 0, \ldots, 0)$, $e^3 = (0, 0, 1, \ldots, 0), \ldots, e^n = (0, 0, \ldots, 0, 1)$ is orthonormal. Further, since this set is linearly independent, it is an *orthonormal basis* for $\mathscr{E}_n$.

Notice that the definition of inner product on $\mathscr{E}_n$ (Example 3.6.1(a)), namely

$$\langle x, y \rangle = \sum_{i=1}^n \xi_i \bar{\eta}_i,$$

implies that the coordinates ξ_i, η_i are relative to an orthonormal basis. For if

$$x = \sum_{i=1}^n \xi_i e^i \quad \text{and} \quad y = \sum_{i=1}^n \eta_i e^i,$$

$$\langle x, y \rangle = \left\langle \sum_{i=1}^n \xi_i e^i, \sum_{j=1}^n \eta_j e^j \right\rangle$$

$$= \sum_{i=1}^{n} \xi_i \sum_{j=1}^{n} \bar{\eta}_j \langle e^i, e^j \rangle$$

$$= \sum_{i=1}^{n} \xi_i \bar{\eta}_i.$$

(b) In the space l_2, the set of vectors $(1, 0, 0, \ldots), (0, 1, 0, \ldots), (0, 0, 1, 0, \ldots) \ldots$ is orthonormal.

(c) In $\mathscr{L}_2[-\pi, \pi]$ the set of functions

$$x^k = \frac{1}{\sqrt{2\pi}} e^{ikt}, \quad k = 0, \pm 1, \pm 2, \ldots,$$

is orthonormal.

Given a set of linearly independent vectors, there exists a systematic procedure for converting the set into an orthonormal set called the *Gram–Schmidt orthogonalization process*. Given a countable set of linearly independent vectors $\{y^i\}$ we shall construct an orthonormal set $\{x^i\}$ such that, for every n, the subspace spanned by the $\{x^i\}, i = 1, \ldots, n$, is the same as the subspace spanned by the $\{y^i\}, i = 1, \ldots, n$. For the first vector take

$$x^1 = y^1 / \|y^1\|.$$

To continue, let

$$z^2 = y^2 - \langle y^2, x^1 \rangle x^1$$

and then take

$$x^2 = z^2 / \|z^2\|.$$

It is readily verified that $z^2 \perp x^1$ and hence that $x^2 \perp x^1$. Since x^1, x^2 are linear combinations of y^1, y^2, they span the same subspace. Suppose we have in this way generated the orthonormal set $\{x^1, x^2, \ldots, x^{n-1}\}$: we then define

$$z^n = y^n - \sum_{i=1}^{n-1} \langle y^n, x^i \rangle x^i$$

and take

$$x^n = z^n / \|z^n\|.$$

For any $k \leq n - 1$,

$$\langle z^n, x^k \rangle = \langle y^n, x^k \rangle - \sum_{i=1}^{n-1} \langle y^n, x^i \rangle \langle x^i, x^k \rangle$$

$$= \langle y^n, x^k \rangle - \sum_{i=1}^{n-1} \langle y^n, x^i \rangle \delta_k^i$$

$$= 0$$

so that z^n, and hence x^n, is orthogonal to each $x^k, k = 1, \ldots, (n-1)$. It is clear from the recursive process that the set $\{x^i\}$, $i = 1, \ldots, n$, spans the same space as the set $\{y^i\}$, $i = 1, \ldots, n$.

Example 3.6.4

(a) In the space $\mathscr{L}_2[-1, 1]$ consider the countable linearly independent set $\{1, t, t^2, \ldots, t^n, \ldots\}$. We have

$$x^0 = 1/\sqrt{2};$$

$$z^1 = t - \left\{ \int_{-1}^{1} \frac{t\,dt}{\sqrt{2}} \right\} \frac{1}{\sqrt{2}} = t,$$

$$x^1 = \sqrt{\frac{3}{2}} t;$$

$$z^2 = t^2 - \left\{ \int_{-1}^{1} \frac{t^2\,dt}{\sqrt{2}} \right\} \frac{1}{\sqrt{2}} - \left\{ \int_{-1}^{1} \sqrt{\frac{3}{2}} t^3 dt \right\} \sqrt{\frac{3}{2}} t = t^2 - \frac{1}{3},$$

$$x^2 = \frac{1}{2}\sqrt{\frac{5}{2}}(3t^2 - 1);$$

and so on. The process soon becomes unwieldy, but in this case we can recognize the emergence of the normalized Legendre polynomials which are given by the general formula [39]

$$x^k = \sqrt{\frac{2k+1}{2}} \frac{1}{2^k k!} \frac{d^k}{dt^k}[(t^2 - 1)^k].$$

(b) In the space $\mathscr{L}_2[0, \infty)$ the linearly independent set $\{e^{-t/2}, te^{-t/2}, \ldots, t^n e^{-t/2}, \ldots\}$ leads to the orthonormal set

$$x^0 = e^{-t/2};$$

$$z^1 = te^{-t/2} - \left\{ \int_0^\infty te^{-t}\,dt \right\} e^{-t/2} = e^{-t/2}(t-1) = x^1;$$

$$z^2 = t^2 e^{-t/2} - \left\{ \int_0^\infty t^2 e^{-t}\,dt \right\} e^{-t/2} - \left\{ \int_0^\infty t^2(t-1)e^{-t}\,dt \right\} e^{-t/2}(t-1)$$

$$= e^{-t/2}(t^2 - 4t + 2),$$

$$x^2 = \tfrac{1}{2} e^{-t/2}(t^2 - 4t + 2);$$

and so on. Here we are generating the orthonormal set

$$x^k = \frac{1}{k!} e^{-t/2} L_k(t),$$

where $L_k(t)$ is the Laguerre polynomial

$$L_k(t) = e^t(-1)^k \frac{d^k}{dt^k}(t^k e^{-t}).$$

Since the sets $\{e^{-t/2}, te^{-t/2}, \ldots, t^{n-1} e^{-t/2}\}$ and $\{x^0, x^1, \ldots, x^{n-1}\}$ span the same space, we can show directly from the formula for $L_k(t)$ that the x^k are orthogonal. We need only show, for any n, that $x^n \perp t^j e^{-t/2}$ for $j = 0, 1, \ldots, n-1$, that is

$$\int_0^\infty e^{-t} t^j L_n(t) dt = \int_0^\infty t^j \frac{d^n}{dt^n}(t^n e^{-t}) dt = 0$$

for $j = 0, 1, 2, \ldots, n-1$. This is readily verified by repeated integration by parts.

Orthonormal bases and Fourier series

In Example 3.6.2(a) we considered the problem of finding the element of a *finite-dimensional* subspace $\mathscr{W}$ which was closest to a given element of a Hilbert space, $\mathscr{H}$. If, therein, the basis $\{y^k\}$ for $\mathscr{W}$ is orthonormal, then the normal equations are trivially soluble, giving

$$\eta_i = \langle x, y^i \rangle$$

and

$$y_0 = \sum_{i=1}^n \langle x, y^i \rangle y^i.$$

We now seek to extend a result of this type to the closed subspaces generated by infinite sets of orthonormal vectors, and indeed to the whole of $\mathscr{H}$.

We shall go only part of the way to begin with, and consider countable sets of orthonormal vectors. Here the reader is asked to look again at Definition 3.3.2 describing a Schauder basis, for we are essentially going to apply this idea when the set of vectors $\{x^n\}$ is orthogonal. Whereas we cannot extend the idea of Schauder basis in a Banach space to uncountable sets, we shall find that this is possible in Hilbert spaces. Definition 3.3.2 tells us what meaning we are to assign to the statement

$$x = \sum_{n=1}^\infty \xi_n x^n.$$

First, we shall see what conditions are required on the scalars ξ_n in order that the right-hand side is meaningful (i.e., converges to some element of $\mathcal{H}$) when the $\{x^n\}$ are orthonormal. Second, we shall see if we can represent every element in the form of an infinite series using an orthonormal set: when we can do this we naturally refer to the set as an *orthonormal basis for $\mathcal{H}$*.

Theorem 3.6.3a *Let $\{e^k\}$ be a countable orthonormal set in a Hilbert space $\mathcal{H}$. The series*

$$\sum_{i=1}^{\infty} \xi_i e^i$$

converges to a vector $x \in \mathcal{H}$ if, and only if,

$$\sum_{i=1}^{\infty} |\xi_i|^2 < \infty$$

and, if this be the case, $\xi_i = \langle x, e^i \rangle$.

Proof Suppose

$$\sum_{i=1}^{\infty} |\xi_i|^2 < \infty$$

and define

$$x^{(n)} = \sum_{i=1}^{n} \xi_i e^i.$$

Then, for $n < m$,

$$\| x^{(n)} - x^{(m)} \|^2 = \left\| \sum_{i=n+1}^{m} \xi_i e^i \right\|^2 = \sum_{i=n+1}^{m} |\xi_i|^2,$$

which tends to zero as $m, n \to \infty$. Hence $\{x^{(n)}\}$ is a Cauchy sequence and the completeness of $\mathcal{H}$ means that there is a vector $x \in \mathcal{H}$ such that $x^{(n)} \to x$.

Conversely, if the sequence $\{x^{(n)}\}$ converges it is a Cauchy sequence so $\sum_{i=n+1}^{m} |\xi_i|^2 \to 0$ as $m, n \to \infty$. Letting $m \to \infty$ we have $\sum_{i=n+1}^{\infty} |\xi_i|^2 \to 0$ as $n \to \infty$, which implies that

$$\sum_{i=1}^{\infty} |\xi_i|^2 < \infty.$$

In the partial sum

$$x^{(n)} = \sum_{i=1}^{n} \xi_i e^i,$$

take the inner product of both sides with e^j, then

$$\langle x^{(n)}, e^j \rangle = \xi_j, \quad 1 \leq j \leq n.$$

But $x^{(n)} \to x$, and hence, by the continuity of the inner product, $\langle x^{(n)}, e^j \rangle \to \langle x, e^j \rangle$. ∎

The coefficients $\xi_i = \langle x, e^i \rangle$ in the series

$$x = \sum_{i=1}^{\infty} \langle x, e^i \rangle e^i$$

are commonly called the *Fourier coefficients of x*, and the series itself is called the *Fourier expansion of x*.

An important inequality exists between the Fourier coefficients, $\langle x, e^i \rangle$, of x and the norm of x called *Bessel's inequality*.

For any finite n, we have

$$0 \leq \left\| x - \sum_{i=1}^{n} \langle x, e^i \rangle e^i \right\|^2$$

$$= \|x\|^2 - \sum_{i=1}^{n} \langle x, e^i \rangle \langle e^i, x \rangle$$

$$\quad - \sum_{j=1}^{n} \overline{\langle x, e^j \rangle} \langle x, e^j \rangle + \sum_{i=1}^{n} |\langle x, e^i \rangle|^2$$

$$= \|x\|^2 - \sum_{i=1}^{n} |\langle x, e^i \rangle|^2.$$

But this means that no partial sum of the convergent series

$$\sum_{i=1}^{\infty} |\xi_i|^2 = \sum_{i=1}^{\infty} |\langle x, e^i \rangle|^2$$

can exceed $\|x\|^2$; hence we obtain

$$\sum_{i=1}^{\infty} |\langle x, e^i \rangle|^2 \leq \|x\|^2,$$

which is *Bessel's inequality*.

Bessel's inequality guarantees that, for $x \in \mathscr{H}$,

$$\sum_{i=1}^{\infty} |\langle x, e^i \rangle|^2 < \infty$$

and hence (Theorem 3.6.3a) the series

$$\sum_{i=1}^{\infty} \xi_i e^i = \sum_{i=1}^{\infty} \langle x, e^i \rangle e^i$$

converges to some vector in $\mathscr{H}$. However, there is no reason to suppose that the series actually converges to x. The next theorem throws some light on the situation, but before we pass to its consideration, let us recall what we mean by 'the closed linear subspace generated by a set S'. This means that we form $[S]$ (Definition 2.3.2) and then take the closure of this space $[\bar{S}]$ (*see* discussion following Definition 3.3.2, p. 139).

Theorem 3.6.3b *Let $\{e^k\}$ be a countable orthonormal set in $\mathscr{H}$. For any $x \in \mathscr{H}$, the series*

$$\sum_{i=1}^{\infty} \langle x, e^i \rangle e^i$$

converges to an element $\hat{x}$ in the closed linear subspace $\mathscr{S} \subset \mathscr{H}$ generated by the $\{e^k\}$. Furthermore, $x - \hat{x} \perp \mathscr{S}$.

Proof As we have remarked, Bessel's inequality guarantees that the series converges and we see that $\hat{x} \in \mathscr{S}$. For any finite n and for every $j < n$,

$$\langle x - x^{(n)}, e^j \rangle = \langle x - \sum_{i=1}^{n} \langle x, e^i \rangle e^i, e^j \rangle = 0.$$

But $x^{(n)} \to \hat{x}$, so that, by the continuity of the inner product, $(x - \hat{x}) \perp \mathscr{S}$. ∎

We may give a specific characterization of $\hat{x}$ by saying that $\hat{x}$ is the orthogonal projection of $x \in \mathscr{H}$ onto $\mathscr{S}$: that is, we have an operator P defined by

$$Px = \sum_{i=1}^{\infty} \langle x, e^i \rangle e^i$$

which is idempotent and such that $\mathscr{H} = \mathscr{S} \oplus \mathscr{S}^{\perp}$ (Exercise 3.6,6(b)).

We have not been able, as yet, to show that every vector in $\mathscr{H}$ has a

Fourier expansion. However, if the closed span $\mathscr{S}$ of the orthonormal set $\{e^k\}$ was the whole of $\mathscr{H}$, then $(x - \hat{x})$ must be the zero vector, or in other words, $\hat{x} = x$. So we need a further condition on an orthonormal set $\{e^k\}$ before we can express every vector in terms of the $\{e^k\}$.

Definition 3.6.4 *An orthonormal set $\{e^k\}$ in a Hilbert space $\mathscr{H}$ is maximal if, and only if, the only vector orthogonal to each of the e^k is the zero vector.*[†]

Let $\{e^k\}$ be a maximal orthonormal set in $\mathscr{H}$ and let $\mathscr{S}$ be the closed linear subspace generated by the $\{e^k\}$. If $x \in \mathscr{S}^\perp$ then $x \perp e^k$ for all k. But $\{e^k\}$ is maximal so $x = 0$ and $\mathscr{S}^\perp = \mathcal{O}$; hence (Theorem 3.6.2, Corollary) $\mathscr{S} = \mathscr{H}$. The orthogonal projection of x onto $\mathscr{S}$ is then simply the identity transformation and $\hat{x} = x$.

Since every x in $\mathscr{H}$ can then be represented as a Fourier series (cf., Schauder basis) we call a maximal orthonormal set in a Hilbert space an *orthonormal basis* (Exercise 3.6,6(c)).

When $\{e^k\}$ is an orthonormal basis, Bessel's inequality becomes an equality, namely

$$\|x\|^2 = \sum_{i=1}^{\infty} |\langle x, e^i \rangle|^2$$

(Exercise 3.6,6(d)) usually called Parseval's formula. An alternative form is

$$\langle x, y \rangle = \sum_{i=1}^{\infty} \langle x, e^i \rangle \overline{\langle y, e^i \rangle} \qquad \forall x, y \in \mathscr{H}$$

(Exercise 3.6,6(e))

Conversely, if equality holds in Bessel's inequality then this implies that the set $\{e^k\}$ is maximal.

It is possible to show that every inner product space contains a maximal orthonormal set: that is to say, given any orthonormal set in $\mathscr{V}$ there is a maximal set containing the given set. If the space $\mathscr{V}$ is complete, then this means to say that every Hilbert space has an orthonormal basis [25, 30].

We have so far restricted the discussion to finite or countable orthonormal sets and we will remedy this omission presently. But we

---

[†] The term 'complete orthonormal set' is very commonly used instead of maximal (*see also* footnote on p. 139).

should first verify the almost obvious fact that if the inner product space $\mathscr{V}$ is separable (Definition 3.2.7) then every orthonormal set in $\mathscr{V}$ is countable. Geometrically speaking this means that the number of orthogonal 'coordinate vectors' in a separable space cannot be so numerous as to be uncountable.

For any two vectors e^i, e^j,

$$\|e^i - e^j\|^2 = \|e^i\|^2 + \|e^j\|^2 = 2,$$

implying that each pair of vectors is a distance $\sqrt{2}$ apart. Let $\{x^k\}$ be a countable set, everywhere dense in $\mathscr{V}$: this means that, for any $e^i \in \mathscr{V}$, there is an n such that

$$\|e^i - x^n\| < \sqrt{2}/3,$$

say (Definition 3.2.6).

Now

$$\sqrt{2} = \|e^i - e^j\| \leq \|e^i - x^n\| + \|x^n - x^m\| + \|x^m - e^j\|$$
$$< 2\sqrt{2}/3 + \|x^n - x^m\|$$

or

$$\|x^n - x^m\| > \sqrt{2}/3.$$

So what we have shown is that, with any two distinct elements of the set $\{e^i\}$ we may associate two distinct elements of the countable set $\{x^k\}$. This establishes a one-to-one correspondence between the set $\{e^i\}$ and a subset of the countable set $\{x^k\}$. Hence $\{e^i\}$ is countable.

If the orthonormal set $\{x^\rho\}$ is not countable, we have essentially to give a meaning to a sum of the form

$$\sum_\rho \langle x, x^\rho \rangle x^\rho$$

over some indexed set $\{\rho\}$. The surprising fact is that, for an uncountable set $\{x^\rho\}$, 'most' of the scalars $\langle x, x^\rho \rangle$ are zero. In fact, it can be shown that the number of vectors x^μ such that $\langle x, x^\mu \rangle \neq 0$ is countable or finite [25, 29]. This means, in effect, that we may carry over all previous results established only for separable Hilbert spaces to non-separable spaces.

In practice, the majority of Hilbert spaces which prove to be useful in applications are separable.

Example 3.6.5 (Orthonormal bases)

(a) (Cf., Example 3.6.3(b)) In the space l_2, the set of vectors $e^1 = (1, 0, 0, \ldots)$,

$e^2 = (0, 1, 0, \ldots)$, $e^3 = (0, 0, 1, 0, 0, \ldots), \ldots$ is an orthonormal basis.

Every vector $x \in l_2$ is represented by the infinite sequence $(\xi_1, \xi_2, \ldots, \xi_n, \ldots)$ with

$$\sum_{i=1}^{\infty} |\xi_i|^2 < \infty.$$

Consider the vectors

$$x^{(n)} = \sum_{k=1}^{n} \xi_k e^k$$

which are in the subspace generated by $e^1, e^2, \ldots, e^n$. Using the concrete form for the inner product in l_2 (Example 3.6.1(b)), then

$$\|x - x^{(n)}\|^2 = \sum_{i=n+1}^{\infty} |\xi_i|^2,$$

which tends to zero as $n \to \infty$. Hence every vector in l_2 is in the closed span of the $\{e^i\}$.

(b) (Cf., Example 3.6.3(c)) In $\mathscr{L}_2[-\pi, \pi]$, the set of functions

$$x^k = \frac{1}{\sqrt{2\pi}} e^{ikt}, \quad k = 0, \pm 1, \pm 2, \ldots,$$

is maximal (and hence an orthonormal basis).

Let $x(t) \in \mathscr{L}_2[-\pi, \pi]$ and suppose that, for all k, $x \perp x^k$, that is

$$\int_{-\pi}^{\pi} x(t) e^{ikt} \, dt = 0, \quad k = 0, \pm 1, \pm 2, \ldots.$$

We shall show that this implies that $x(t) = 0$. Let

$$z(t) = \int_{-\pi}^{t} x(s) \, ds - \alpha,$$

then $z(t)$ is continuous† and $z'(t) = x(t)$: the scalar α will be determined later.

Integrating by parts, we have, for $k \neq 0$,

$$\int_{-\pi}^{\pi} x(t) e^{ikt} \, dt = 0 = \left[z(t) e^{ikt} \right]_{-\pi}^{\pi} - ik \int_{-\pi}^{\pi} z(t) e^{ikt} \, dt$$

$$= (-1)^k (z(\pi) - z(-\pi)) - ik \int_{-\pi}^{\pi} z(t) e^{ikt} \, dt.$$

Now since, by hypothesis, $x(t)$ is orthogonal to the function e^{ikt}

---

† Actually $z(t)$ is *absolutely continuous* [26] and this allows us to conclude that $z'(t) = x(t)$ almost everywhere.

when $k = 0$,
$$\int_{-\pi}^{\pi} x(s)\,ds = 0$$
or
$$z(\pi) - z(-\pi) = 0$$
and hence, for $k \neq 0$,
$$\int_{-\pi}^{\pi} z(t)e^{ikt}\,dt = 0.$$

By choosing α we can extend this result to $k = 0$: indeed, we simply take α such that
$$2\pi\alpha = \int_{-\pi}^{\pi} dt \int_{-\pi}^{t} x(s)\,ds.$$

Thus far we have shown that our hypothesis leads to
$$\int_{-\pi}^{\pi} z(t)e^{ikt}\,dt = 0, \qquad k = 0, \pm 1, \pm 2, \ldots,$$
where $z(t)$ is *continuous*. We now use the fact that a continuous function can be arbitrarily closely approximated by polynomials (Weierstrass theorem—see p. 123) to show that $z(t) = 0$. Here, since $z(t)$ is periodic there is a *trigonometric polynomial*
$$p(t) = \sum_{k=-n}^{n} \beta_k e^{ikt}$$
such that, for any $\varepsilon > 0$,
$$|z(t) - p(t)| < \varepsilon \qquad \forall t \in [-\pi, \pi].$$
Now
$$\int_{-\pi}^{\pi} z(t)\overline{p(t)}\,dt = \sum_{k=-n}^{n} \bar{\beta}_k \int_{-\pi}^{\pi} z(t)e^{-ikt}\,dt = 0,$$
whence
$$\int_{-\pi}^{\pi} |z(t)|^2\,dt = \int_{-\pi}^{\pi} z(t)\overline{z(t)}\,dt = \int_{-\pi}^{\pi} \overline{z(t)}(z(t) - p(t))\,dt$$
$$\leq \varepsilon \int_{-\pi}^{\pi} |z(t)|.1\,dt \leq \varepsilon \left[\int_{-\pi}^{\pi} |z(t)|^2\,dt \right]^{1/2} \left[\int_{-\pi}^{\pi} dt \right]^{1/2}$$
by the Schwarz inequality.

We have thus shown that
$$\int_{-\pi}^{\pi} |z(t)|^2 \leq 2\pi\varepsilon^2,$$

which implies that $z(t) = 0$. This in turn implies that

$$\int_{-\pi}^{t} x(s)\,ds = \alpha$$

or that $x(t) = 0$ almost everywhere. The qualification 'almost everywhere' here is necessary because in the space $\mathscr{L}_2$ we identify functions which differ only on a set of measure zero (see Example 3.2.4(b)).

For any function $x(t) \in \mathscr{L}_2[-\pi, \pi]$ the Fourier coefficients $\langle x, e^i \rangle$ are in this case the classical Fourier coefficients

$$\xi_k = \langle x, e^k \rangle = \frac{1}{\sqrt{2\pi}} \int_{-\pi}^{\pi} x(t) e^{ikt}\,dt$$

and Parseval's formula is

$$\sum_{k=-\infty}^{\infty} |\xi_k|^2 = \int_{-\pi}^{\pi} |x(t)|^2\,dt.$$

Every $x(t) \in \mathscr{L}_2[-\pi, \pi]$ has the classical expansion

$$x(t) = \frac{1}{\sqrt{2\pi}} \sum_{k=-\infty}^{\infty} \xi_k e^{ikt},$$

but it should be noted (as for all Fourier expansions) that there is no implication of pointwise convergence to the function $x(t)$. The Fourier expansion of $x(t)$ is a 'best approximation' to $x(t) \in \mathscr{L}_2$ in the sense of least squares.

The question of pointwise convergence of Fourier series is another and difficult matter which we do not pursue [44].

The corresponding orthonormal basis for the *real* space $\mathscr{L}_2[-\pi, \pi]$ is, of course, the set

$$\left(\frac{1}{\sqrt{2\pi}}, \frac{1}{\sqrt{\pi}} \cos kt, \frac{1}{\sqrt{\pi}} \sin kt \right), \qquad k = 1, 2, 3, \ldots.$$

(c) By a similar argument it can be shown that the Legendre polynomials (Example 3.6.4(a)) are an orthonormal basis for $\mathscr{L}_2[-1, 1]$.

(d) A somewhat different type of orthonormal basis is represented by the Haar system on the interval $[0, 1]$. The Haar functions are defined by

$$h_0(t) = 1,$$

$$h_n^{(k)}(t) = \begin{cases} \sqrt{2^n}, & t \in ((k-1)/2^n, (k-\tfrac{1}{2})/2^n), \\ -\sqrt{2^n}, & t \in ((k-\tfrac{1}{2})/2^n, k/2^n), \\ 0 & \text{for all other } t \in [0, 1], \end{cases}$$

where $n = 0, 1, 2, \ldots$ and $k = 1, 2, 2^2, \ldots, 2^n$. The first four functions are shown in Fig. 3.13 [32].

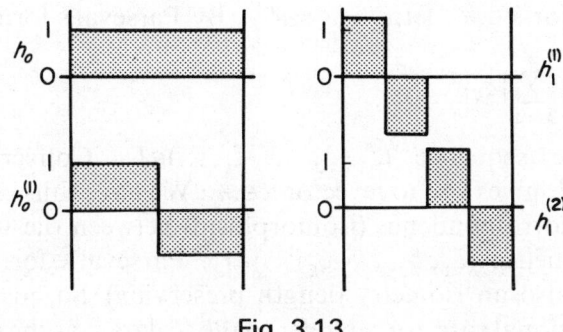

Fig. 3.13

A development of the Haar system leads to another orthonormal basis called the Walsh system which has proved to be of considerable application in the field of communication theory [43].

(e) Let $w(t)$ be a positive *weight* function which is integrable on the interval I. Then one can show that the set of real functions $x(t)$ such that

$$\int_I w(t)(x(t))^2 dt < \infty$$

is a Hilbert space $\mathscr{H}^w$ with inner product

$$\langle x, y \rangle_w = \int w(t)x(t)y(t)dt.$$

A set of functions $\{x^k\}$ such that

$$\langle x^k, x^j \rangle_w = 0, \qquad j \neq k,$$

is said to be orthogonal with respect to the weight function $w(t)$ on I.

From the linearly independent set of polynomials $\{1, t, t^2, t^3, \ldots\}$ the Gram–Schmidt process can be used to generate various families of orthogonal polynomials, for example [39],

   (i) the Legendre polynomials on $[-1, 1]$ with $w(t) = 1$;
   (ii) the Chebyshev polynomials on $[-1, 1]$ with $w(t) = (1 - t^2)^{-1/2}$;
   (iii) the Laguerre polynomials on $[0, \infty)$ with $w(t) = e^{-t}$;
   (iv) the Hermite polynomials on $(-\infty, \infty)$ with $w(t) = e^{-t^2}$.

One of the central results of Chapter 2 was embodied in Theorem 2.4.2, namely that every vector space of (finite) dimension n is isomorphic to $\mathscr{R}_n$ (or $\mathscr{C}_n$). One cannot in general make a meaningful statement of this kind for infinite-dimensional spaces. However, in the case of *separable* Hilbert spaces one can show that every such space is isometrically isomorphic to the space l_2.

The separable Hilbert space $\mathscr{H}$ contains a countable orthonormal basis $\{e^k\}$. For $x \in \mathcal{H}$, let $\xi_k = \langle x, e^k \rangle$. By Parseval's formula,

$$\|x\|^2 = \sum_{k=1}^{\infty} |\xi_k|^2,$$

so that the sequence $(\xi_1, \xi_2, \ldots, \xi_k, \ldots) \in l_2$. Conversely, every sequence in l_2 gives rise to a vector $x \in \mathcal{H}$. We have thus established a one-to-one correspondence (isomorphism) between the vector $x \in \mathcal{H}$ and the sequence $(\xi_1, \xi_2, \ldots, \xi_k, \ldots) \in l_2$. Parseval's formula shows that this is also an isometry (length preserving). So, just as for the finite-dimensional case, we can refer to the scalars ξ_i as the coordinates of x (relative to the orthonormal basis) and follow through our manipulations using the coordinates. For example, by the extended form of Parseval's formula, the inner product of $x, y \in \mathcal{H}$ is given by

$$\langle x, y \rangle = \sum_{i=1}^{\infty} \xi_k \bar{\eta}_k.$$

It should be noted that although the question of verifying that an orthonormal set is a basis resolves itself into showing that the set is maximal, there is no standard method for doing this [45]. However, many maximal orthonormal sets are known to be generated as solutions of certain differential and integral equations; we shall touch on this at a later point.

Exercises 3.6

1. (a) Show that the inner product $\langle x, y \rangle$ is jointly continuous, i.e. continuous in x for fixed y and vice versa. (Hint: use the Schwarz inequality. Note that the inner product is a positive definite Hermitean sesquilinear functional—Exercise 3.5,9.)
    (b) Prove the parallelogram law (p. 183).
    (c) Show that equality holds in the Schwarz inequality if and only if $x = \lambda y$ (i.e., x and y are collinear).
    (d) Verify that if $\mathscr{V}$ is an inner product space and $T: \mathscr{V} \to \mathscr{V}$, then $\langle Tx, y \rangle$ is a sesquilinear functional.
    (e) Show that if $\mathscr{V}^{(1)}, \mathscr{V}^{(2)}$ are inner product spaces, then

    $$\langle x, y \rangle = \langle x_1, y_1 \rangle_{\mathscr{V}^{(1)}} + \langle x_2, y_2 \rangle_{\mathscr{V}^{(2)}}$$

    is an inner product on $\mathscr{V}^{(1)} \times \mathscr{V}^{(2)}$, where $x = (x_1, x_2), y = (y_1, y_2)$ (cf., Exercise 3.2,3).

2. (a) Show that $S^{\perp}$ is a closed subspace of $\mathscr{V}$ (Definition 3.6.2) (cf., Exercise 3.5,8(b)).
    (b) Verify the following statements when S, R are subsets of an inner product space:

(i) $S \cap S^\perp = \{\theta\}$;
(ii) $S \subset (S^\perp)^\perp$;
(iii) if $S \subset R$ then $S^\perp \supset R^\perp$.

(c) Suppose the set S is dense in the inner product space $\mathscr{V}$: show that $S^\perp = \{\theta\}$. (Hint: use the Pythagorean theorem to deduce that if $x \in S^\perp$ then $\|x\| = 0$.) The result implies that if $\langle x, y \rangle = 0$ for all $x \in S$ then $y = \theta$.

(d) Verify that if $\mathscr{U}$ is a closed subspace of the Hilbert space $\mathscr{H}$ then $(\mathscr{U}^\perp)^\perp = \mathscr{U}$ (Hint: apply Theorem 3.6.2 together with (i) and (ii) of (b).)

(e) Let $\mathscr{U}, \mathscr{W}$ be closed subspaces of a Hilbert space $\mathscr{H}$ and suppose $\mathscr{U} \perp \mathscr{W}$: show that $\mathscr{U} + \mathscr{W}$ is closed. Now give a proof for the corollary to Theorem 3.6.2.

(f) Show that the first part of Theorem 3.6.2 is true if 'closed subspace' is replaced by 'closed convex set'.

(g) Verify that an orthogonal projection is continuous, that $\|P\| = 1$, and that $\langle Px, y \rangle = \langle x, Py \rangle$.

(h) The second part of Example 3.6.2(a) implies that there exists a vector of minimum norm in a coset $y + \mathscr{U}$, where $\mathscr{U} \subset \mathscr{H}$ is closed. Prove this from Theorem 3.6.2.

(i) Justify the statement in Example 3.6.2(c):
$$\Phi(\alpha x_1 + \beta x_2) = \bar{\alpha}\Phi(x_1) + \bar{\beta}\Phi(x_2).$$

(j) Consider the space $\mathscr{V}$ of all real continuous functions on $[-1, 1]$ as a subspace of $\mathscr{L}_2[-1, 1]$. What is the orthogonal complement (in $\mathscr{V}$) of the subspace of all odd functions in $\mathscr{V}$?

3. (a) The Gram–Schmidt process (p. 194) can lead to computational difficulties. Let the vectors $\{y^i\}$ by real.
  (i) Show that
  $$\|z^2\| = \|y^2\| (1 - (\widehat{\cos y^1 y^2})^2)^{1/2}$$
  (p. 185) and hence that $x^2 = \gamma z^2$, where $\gamma > 0$ could be very large.

  (ii) The effect of rounding errors in the computation leads to a deterioration in the orthogonality relations. Suppose $\tilde{x}^n$ is a computed approximation to x^n based on the (exact) orthonormal set $\{x^i\}$, $i = 1, \ldots, n-1$: let $x^n - \tilde{x}^n = \delta$. Show that
  $$\delta = \langle \delta, x^n \rangle x^n - \sum_{i=1}^{n-1} \langle \tilde{x}^n, x^i \rangle x^i$$
  and, assuming $\|\tilde{x}^n\| = 1$, deduce that
  $$\langle \delta, x^n \rangle = O(\|\delta\|^2),$$
  $$\langle \tilde{x}^n, x^i \rangle = O(\|\delta\|)$$
  (Appendix 4.A). Neglecting the term $O(\|\delta\|^2)$ show that an

improved approximation to x^n would be the vector

$$\tilde{x} = \tilde{z}^n / \|\tilde{z}^n\|,$$

where

$$\tilde{z}^n = \tilde{x}^n - \sum_{i=1}^{n-1} \langle \tilde{x}^n, x^i \rangle x^i.$$

(This process is known as reorthogonalization and should be applied at each stage of the process.)

(b) Let the orthonormal set $\{e^i\}$ be generated by the Gram–Schmidt process from the linearly independent vectors $\{y^i\}$, $i = 1, \ldots, n$. For any $x \in \mathcal{V}_n$ we have

$$x = \sum_{i=1}^{n} \langle x, e^i \rangle e^i = \sum_{i=1}^{n} \eta_i y^i.$$

Show that the coordinates η_i can be calculated recursively from the (Fourier) coordinates $\langle x, e^i \rangle$. (Hint: y^i is a linear combination of the e^j for $j \leq i$ only.)

4. (a) Show that the functions of a complex variable which are analytic on the unit disc $|t| < 1$ in the (complex) t-plane and for which

$$\int_{|t|<1} |x(t)|^2 \, dt_1 \, dt_2$$

is finite form an inner product space when

$$\langle x, y \rangle = \int_{|t|<1} x \bar{y} \, dt_1 \, dt_2.$$

Show that an orthonormal set in this space is given by

$$x^m(t) = \sqrt{\frac{m}{\pi}} t^{m-1}, \quad m = 1, 2, 3, \ldots,$$

and compare the Fourier series for x with its power series expansion. (Hint: you will need to use Green's theorem in two variables).

(b) Given an orthonormal set $\{x^i\}$ for, say, the interval $[0, 1]$, an orthonormal set for the n-dimensional interval $[0, 1] \times [0, 1] \times [0, 1] \times \ldots$ is given by

$$x^i(t) = x^{i_1}(t_1) x^{i_2}(t_2) \ldots x^{i_n}(t_n),$$

where $i = (i_1, i_2, i_3, \ldots, i_n)$. Verify this.

5. (a) Consider the real transformation $T: \mathcal{E}_n \to \mathcal{E}_m$ and let $y \in \mathcal{E}_m$ be a vector which is not in $\mathcal{R}(T)$: then the equation $Tx = y$ has no solution. However we may seek a 'least squares solution' $\tilde{x}$ in the sense that $\tilde{x}$ is the vector which minimizes $\|T\tilde{x} - y\|$ in $\mathcal{E}_m$. Show (Example

3.6.2(a)) that $\tilde{x}$ is a solution of the equation

$$\langle T\tilde{x} - y, z \rangle = 0 \quad \forall z \in \mathcal{R}(T) \subset \mathscr{E}_m.$$

Recall that a basis for $\mathcal{R}(T)$ is any set of linearly independent columns $\{t_i\}_j$ of the matrix representation $[t_{ij}]$ for T relative to orthonormal basis sets in $\mathscr{E}_n$, $\mathscr{E}_m$. Hence show that the above equation takes the concrete form

$$\sum_{j=1}^{n} \lfloor t_i \rfloor_k \{t_i\}_j \tilde{\xi}_j = \lfloor \eta_i \rfloor \{t_i\}_k, \quad k = 1, \ldots, n,$$

where $\tilde{\xi}_i$, η_i are the coordinates of $\tilde{x} \in \mathscr{E}_n$ and $y \in \mathscr{E}_m$, respectively. This equation may not have a unique solution (why?): if so can you reconcile this with the existence of the unique closest element of Theorem 3.6.2?

(b) (Least squares estimation of data) Suppose it is hypothesized that a physical variable η is linearly dependent on n other physical variables ξ_i, viz.,

$$\eta = \sum_{i=1}^{n} \alpha_i \xi_i.$$

A large number $M > n$ of measurements of η and the ξ_i are made: $(\eta^{(k)}; \xi_1^{(k)}, \xi_2^{(k)}, \ldots, \xi_n^{(k)})$, $k = 1, 2, \ldots, M$. The numbers α_i may be estimated by demanding that the mean square error

$$\frac{1}{M} \sum_{k=1}^{M} \left(\eta^{(k)} - \sum_{i=1}^{n} \alpha_i \xi_i^{(k)} \right)^2$$

is a minimum. By identifying the $(M \times n)$ array $\xi_i^{(k)}$ with the matrix $[t_{ki}]$ in (a), deduce the form of the equations for the α_i (the *normal equations*).

6. (a) Show that if $\mathcal{U}_n$ is a finite-dimensional subspace of the Hilbert space $\mathcal{H}$ then the linear transformation $P: \mathcal{H} \to \mathcal{H}$ defined by

$$Px = \sum_{i=1}^{n} \langle x, e^i \rangle e^i,$$

where $\{e^i\}$ is an orthornormal basis for $\mathcal{U}_n$, is an orthogonal projection of $\mathcal{H}$ onto $\mathcal{U}_n$. (Hint: show that P is idempotent, that $\mathcal{R}(P) = \mathcal{U}_n$, and that $\mathcal{R}^\perp(P) = \mathcal{N}(P)$.)

(b) Extend (a) to the case when $\mathcal{U}$ is the closure of the countable orthonormal set $\{e^i\}$. (Hint: we now need to consider the continuity of P.)

(c) Show that a (countable) orthonormal basis is a Schauder basis for $\mathcal{H}$.

(d) Show that if Parseval's formula holds then $\{e^i\}$ is an orthonormal

basis. (Hint: assume $\{e^i\}$ not maximal and obtain a contradiction.)
   (e) Derive the alternative form of Parseval's formula (p. 200).

7. Consider the set of complex-valued functions $x(t)$ defined on $(-\infty, \infty)$ and satisfying
$$\lim_{T \to \infty} \frac{1}{2T} \int_{-T}^{T} |x^2(t)| \, dt < \infty.$$
Show that these functions form a vector space $\mathscr{A}$ and verify that
$$\langle x, y \rangle = \lim_{T \to \infty} \frac{1}{2T} \int_{-T}^{T} x(t) \bar{y}(t) \, dt$$
is an inner product for $\mathscr{A}$.

Show that this inner product space contains an uncountable orthonormal set $x^\alpha(t) = e^{i\alpha t}$ and hence is not separable.

Show that if $x_p(t)$ is periodic with period τ_x then $x_p \in \mathscr{A}$ and
$$\|x_p\|^2 = \frac{1}{\tau_x} \int_0^{\tau_x} |x_p(t)|^2 \, dt.$$

A function $x_a(t)$ is called *almost periodic* [42, 51] if
  (i) $x_a(t)$ is continuous in $(-\infty, \infty)$;
  (ii) for each $\varepsilon > 0$ there exists $L(\varepsilon)$ so that in each interval of length greater than L a point τ can be found such that $|x_a(t + \tau) - x_a(t)| < \varepsilon$ $\forall t \in (-\infty, \infty)$.

It can be shown [51] that $x_a(t) \in \mathscr{A}$ (the space of almost periodic functions) and that every $x_a \in \mathscr{A}$ can be represented by a (generalized) Fourier series $\sum_\alpha \gamma_\alpha e^{i\alpha t}$ in terms of the uncountable set $\{e^{i\alpha t}\}$.

8. We refer to Example 3.6.5(b). Show that in the real space $\mathscr{L}_2[-\pi, \pi]$ Bessel's inequality takes the form
$$a_0^2 \frac{\pi}{2} + \pi \left(\sum_{k=1}^{\infty} (a_k^2 + b_k^2) \right) \leq \int_{-\pi}^{\pi} x^2(t) \, dt,$$
where
$$a_0 = \frac{1}{\pi} \int_{-\pi}^{\pi} x(t) \, dt, \qquad a_k = \frac{1}{\pi} \int_{-\pi}^{\pi} x(t) \cos kt \, dt, \qquad b_k = \frac{1}{\pi} \int_{-\pi}^{\pi} x(t) \sin kt \, dt.$$
Hence deduce that
$$\lim_{k \to \infty} \int_{-\pi}^{\pi} x(t) \cos kt \, dt = \lim_{k \to \infty} \int_{-\pi}^{\pi} x(t) \sin kt \, dt = 0.$$
This is the Riemann–Lebesgue lemma [9].

9. Verify that the Haar system (Example 3.6.5(d)) is orthonormal.

3.7 Linear transformations on inner product spaces

Since an inner product space is a normed space with norm induced by the inner product, all the properties discussed in Sections 3.4 and 3.5 relating to linear transformations on normed spaces carry over immediately to linear transformations on inner product spaces. However, the crucial fact that the dual of a Hilbert space coincides with the space itself means that the dual of an operator on a Hilbert space is itself an operator on the space. This result has profound implications concerning the structure of linear transformations on Hilbert spaces. In particular, there exists the possibility that the dual of a transformation is the transformation itself, so generalizing the property of a real symmetric (square) matrix that it coincides with its transpose.

Suppose $\mathscr{V}_1, \mathscr{V}_2$ are two real Hilbert spaces with inner products $\langle \cdot, \cdot \rangle_1, \langle \cdot, \cdot \rangle_2$, respectively, and let $T: \mathscr{V}_1 \to \mathscr{V}_2$ be a bounded linear transformation. The dual of T is the bounded linear transformation $T': \mathscr{V}_2^* \to \mathscr{V}_1^*$ defined by (Definition 2.7.1 and p. 173)

$$[T'x^*, x]_{\mathscr{V}_1} = [x^*, Tx]_{\mathscr{V}_2} \qquad \forall \begin{cases} x \in \mathscr{V}_1, \\ x^* \in \mathscr{V}_2^*. \end{cases}$$

Identifying $\mathscr{V}_1^*, \mathscr{V}_2^*$ with $\mathscr{V}_1, \mathscr{V}_2$, respectively, and elements of $\mathscr{V}_1^*, \mathscr{V}_2^*$ with vectors in $\mathscr{V}_1, \mathscr{V}_2$, we can rewrite this definition as

$$\langle T'y, x \rangle_{\mathscr{V}_1} = \langle y, Tx \rangle_{\mathscr{V}_2} \qquad \forall \begin{cases} x \in \mathscr{V}_1, \\ y \in \mathscr{V}_2, \end{cases}$$

where (since $\mathscr{V}_1, \mathscr{V}_2$ are real spaces) we have replaced the bracket notation $[\cdot, \cdot]$ by the inner product $\langle \cdot, \cdot \rangle$ (see Fig. 3.14).

In the special case when T is $\mathscr{V} \to \mathscr{V}$ then T' is also $\mathscr{V} \to \mathscr{V}$ and we could formally identify T' with T if the condition

$$\langle Ty, x \rangle_{\mathscr{V}} = \langle y, Tx \rangle_{\mathscr{V}} \qquad \forall x, y \in \mathscr{V}$$

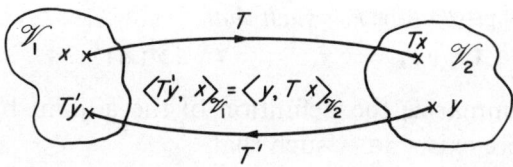

Fig. 3.14

were satisfied. In $\mathscr{R}_n$ with inner product

$$\langle x, y \rangle = \sum_{i=1}^{n} \xi_i \eta_i$$

relative to an orthonormal basis, the transformation $T: \mathscr{R}_n \to \mathscr{R}_n$ is the square matrix $[t_{ij}]$ and the above condition reads

$$\sum_{i=1}^{n} \xi_i \sum_{j=1}^{n} t_{ij} \eta_j = \sum_{i=1}^{n} \eta_i \sum_{j=1}^{n} t_{ij} \xi_j = \sum_{i=1}^{n} \xi_i \sum_{j=1}^{n} t_{ji} \eta_j.$$

This implies that the matrix $[t_{ij}]$ coincides with its transpose, that is to say, is symmetric; so the condition

$$\langle Ty, x \rangle = \langle y, Tx \rangle$$

suggests how we might characterize a 'symmetric' transformation on an infinite-dimensional Hilbert space.

While these observations indicate the path we might follow, the approach through the dual is not as convenient or as general as we can construct within the context of inner product spaces. So we shall eschew this approach and start afresh.

The adjoint of a linear transformation

We shall begin with a definition and then discuss how and why it differs from the simple dual interpretation given above.

Definition 3.7.1 *Let $\mathscr{V}, \mathscr{U}$ be inner product spaces. Let $A: \mathscr{D}(A) \subset \mathscr{V} \to \mathscr{U}$ be an arbitrary mapping, where $\mathscr{D}(A)$, the domain of A, is dense in $\mathscr{V}$. The adjoint of A, written A^*, is the mapping*

$$A^*: \mathscr{D}(A^*) \subset \mathscr{U} \to \mathscr{V}$$

given by

$$\langle Ax, y \rangle_{\mathscr{U}} = \langle x, A^*y \rangle_{\mathscr{V}}$$

for all $x \in \mathscr{D}(A)$, $y \in \mathscr{D}(A^*)$. The set $\mathscr{D}(A^*)$ is defined by

$$\mathscr{D}(A^*) = \{ y \in \mathscr{U} \mid \exists z \in \mathscr{V} \text{ such that}$$
$$\langle Ax, y \rangle_{\mathscr{U}} = \langle x, z \rangle_{\mathscr{V}} \quad \forall x \in \mathscr{D}(A) \}.$$

We may summarize the definition of the adjoint by saying that, if for $y \in \mathscr{U}$, there exists $z \in \mathscr{V}$ such that

$$\langle Ax, y \rangle_{\mathscr{U}} = \langle x, z \rangle_{\mathscr{V}}$$

for all $x \in \mathcal{D}(A)$, then we have defined a mapping $y \to z, A^*y = z$ with $y \in \mathcal{D}(A^*)$.

For a given mapping A there exists at most one z with the above property. For suppose there exists also $w \in \mathscr{V}$ such that

$$\langle Ax, y \rangle_{\mathscr{U}} = \langle x, w \rangle_{\mathscr{V}},$$

then we have

$$\langle x, z \rangle_{\mathscr{V}} = \langle x, w \rangle_{\mathscr{V}}$$

or

$$\langle x, z - w \rangle_{\mathscr{V}} = 0$$

for all $x \in \mathcal{D}(A)$. But $\mathcal{D}(A)$ is, by assumption, dense in $\mathscr{V}$, so that we conclude $z - w = \theta$ or $z = w$ (cf., Exercise 3.6,2(c)). Hence A^* is a well-defined mapping (or function). It is precisely for this reason that the stipulation that the domain of A is dense in $\mathscr{V}$ is made.

Although A is an arbitrary mapping, the adjoint A^* is, perhaps surprisingly, a linear transformation and $\mathcal{D}(A^*)$ a subspace of $\mathscr{U}$. Suppose $y_1, y_2 \in \mathcal{D}(A^*)$ and

$$A^*y_1 = z_1, \quad A^*y_2 = z_2,$$

then, for scalars λ, μ,

$$\begin{aligned}\langle Ax, \lambda y_1 + \mu y_2 \rangle_{\mathscr{U}} &= \langle Ax, \lambda y_1 \rangle_{\mathscr{U}} + \langle Ax, \mu y_2 \rangle_{\mathscr{U}} \\ &= \bar{\lambda} \langle Ax, y_1 \rangle_{\mathscr{U}} + \bar{\mu} \langle Ax, y_2 \rangle_{\mathscr{U}} \\ &= \bar{\lambda} \langle x, A^*y_1 \rangle_{\mathscr{V}} + \bar{\mu} \langle x, A^*y_2 \rangle_{\mathscr{V}} \\ &= \langle x, \lambda z_1 \rangle_{\mathscr{V}} + \langle x, \mu z_2 \rangle_{\mathscr{V}} \\ &= \langle x, \lambda z_1 + \mu z_2 \rangle_{\mathscr{V}} \quad \forall x \in \mathcal{D}(A),\end{aligned}$$

so that

$$(\lambda y_1 + \lambda y_2) \in \mathcal{D}(A^*).$$

It is also clear that

$$A^*(\lambda y_1 + \mu y_2) = \lambda z_1 + \mu z_2 = \lambda A^*y_1 + \mu A^*y_2,$$

so that A^* is a *linear transformation*.

If T is a bounded linear transformation defined on the whole of $\mathscr{V}$ and if $\mathscr{V}$ and $\mathscr{U}$ are real Hilbert spaces[†] then the adjoint of T coincides with the (formal) dual of T as discussed in the introductory paragraph. If $\mathscr{V}, \mathscr{U}$ are complex spaces, the situation is similar in

---

[†] In the interest of continuity we have temporarily abandoned our convention of denoting Hilbert spaces by $\mathscr{H}$.

character, but of course complex conjugates appear where they would not for the dual: for example (p. 174),

$$(\lambda T)^* = \bar{\lambda} T^*$$
$$(T + S)^* = T^* + S^*$$
$$(TS)^* = S^* T^*.$$

The relations between the range and null spaces of T, T^* are (cf., Theorem 3.5.3) [27, 30]

(i) $\mathscr{R}^\perp(T) = \mathscr{N}(T^*)$;
(ii) $\overline{\mathscr{R}(T)} = \mathscr{N}^\perp(T^*)$;
(iii) $\mathscr{R}^\perp(T^*) = \mathscr{N}(T)$;
(iv) $\overline{\mathscr{R}(T^*)} = \mathscr{N}^\perp(T)$.

Furthermore, as for the dual, we can show that

$$\|T^*\|_{\mathscr{U},\mathscr{V}} = \|T\|_{\mathscr{V},\mathscr{U}}$$

(Exercise 3.7,2) and

$$(T^*)^* = T.$$

As a matter of fact, if T is a bounded linear transformation defined only on a dense subset of the Hilbert space $\mathscr{V}$ the situation is exactly as if the domain of definition were the whole of $\mathscr{V}$ (the transformation can be 'extended') [30].

So much for bounded linear transformations. Let us now turn to unbounded linear transformations, for it is in accommodating these that the definition of the adjoint is significantly different from that for the dual. The reader should look again at Examples 3.4.1(a) and (b). There we saw that typically an integral operator is bounded, whereas a differential operator is not: this is in fact a general rule, and we would exclude from consideration the large class of differential operators (on Hilbert spaces) were we to be restricted to bounded operators.

Example 3.7.1 Let us pursue Example 3.4.1(b) further, but now in the context of a Hilbert space setting. We shall consider the linear transformation

$$Tx \equiv \frac{dx}{dt}$$

to be $\mathscr{L}_2[0, 1] \to \mathscr{L}_2[0, 1]$. The domain of definition of T will be that subspace of $\mathscr{L}_2[0, 1]$ for which $dx/dt \in \mathscr{L}_2[0, 1]$ (so that the image vectors are in $\mathscr{L}_2[0, 1]$) and we shall, in addition, impose the boundary value

$x(0) = 0$. Hence

$$\mathscr{D}(T) = \left\{ x(t) \,\middle|\, x \text{ continuously differentiable}; \int_0^1 \left(\frac{dx}{dt}\right)^2 dt < \infty\,; x(0) = 0 \right\}.$$

The subspace $\mathscr{D}(T)$ is dense in $\mathscr{L}_2[0, 1]$ (see p. 159).
Now

$$\langle Tx, y \rangle = \int_0^1 \frac{dx}{dt} y \, dt$$
$$= [xy]_0^1 - \int_0^1 x \frac{dy}{dt} dt$$

so we may conclude that

$$T^*x \equiv -\frac{dx}{dt}$$

with

$$\mathscr{D}(T^*) = \left\{ x(t) \,\middle|\, x \text{ continuously differentiable}; \int_0^1 \left(\frac{dx}{dt}\right)^2 dt < \infty\,; x(1) = 0 \right\}.$$

Notice that the boundary value $x(1) = 0$ for $\mathscr{D}(T^*)$ is chosen so as to make the integrated term $[xy]_0^1$ vanish.

In this example it turned out that $\mathscr{D}(T^*)$ was also dense in $\mathscr{L}_2[0, 1]$. Is this a chance occurrence, or a more general result? To pursue this point we need to introduce the idea of a *closed operator*.

As we have already noted, the definition of the adjoint embraces unbounded, and therefore discontinuous, operators. The vast majority of discontinuous linear operators which occur in practice have a property which, in a sense, compensates for their lack of continuity; this is the property of being closed (or at the least, closeable).

Definition 3.7.2 (Cf., Definition 3.4.1 and preceding remarks) *Let $\mathscr{V}, \mathscr{U}$ be normed vector spaces. The linear transformation*

$$T : \mathscr{D}(T) \subset \mathscr{V} \to \mathscr{U}$$

is said to be closed *if, for every convergent sequence $x^{(k)}$ in $\mathscr{D}(T)$ with $x^{(k)} \to x \in \mathscr{V}$ such that $Tx^{(k)}$ is a convergent sequence in $\mathscr{U}$ with $Tx^{(k)} \to y \in \mathscr{U}$, the following two conditions are satisfied:*

(i) $x \in \mathscr{D}(T)$;
(ii) $y = Tx$.

This definition is obscure on first sight and needs to be enlarged on considerably before its implications are understood. Recall first (Definition 3.2.10) that an operator T is *continuous at* x_0 if $x^{(k)} \to x_0$ implies $Tx^{(k)} \to Tx_0$; second, that continuity of a linear operator at one point guarantees continuity everywhere. We can see that a continuous (i.e., bounded) linear operator is closed provided $\mathscr{D}(T)$ is closed in $\mathscr{V}$.

If $\mathscr{V}$ and $\mathscr{U}$ are Banach spaces, a celebrated result, usually referred to as the *closed graph theorem*[†] [27] tells us that if the domain of T is all of $\mathscr{V}$ then T must be continuous. It follows that the domain of an unbounded closed linear operator cannot be a Banach space, that is to say, $\mathscr{D}(T)$ cannot be a *closed* subspace of $\mathscr{V}$.

Now this result is revealing, for it emphasizes that when we talk of 'every convergent sequence $x^{(k)}$ in $\mathscr{D}(T)$' in Definition 3.7.2 the subsequent conditions are essential. If $\mathscr{D}(T)$ were closed, condition (i) would be unnecessary. We are looking for convergent sequences in $\mathscr{D}(T)$ whose limits are in $\mathscr{D}(T)$ *and* for which $Tx^{(k)} \to y = Tx$. This cannot apply to *every* convergent sequence in $\mathscr{D}(T)$, otherwise $\mathscr{D}(T)$ would be closed and T continuous.

What we are left with, then, is a type of continuity which is restricted to certain subspaces of a normed space; in other words, the difference between a continuous and a closed transformation resides in the specification of the domain of definition.

The following example may help to clarify the situation.

Example 3.7.2 Let $\mathscr{V} = \mathscr{U} = \mathscr{C}[0,1]$ and let $T: \mathscr{D}(T) \subset \mathscr{V} \to \mathscr{U}$ be the differentiation operator

$$Tx \equiv \frac{dx}{dt}$$

with

$$\mathscr{D}(T) = \left\{ x \in \mathscr{V} \,\bigg|\, \frac{dx}{dt} \in \mathscr{C}[0,1] \right\}.$$

We saw in Example 3.4.1(b) that T is unbounded.
Let $x^{(k)} \in \mathscr{D}(T)$, $x^{(k)} \to x$ and

$$Tx^{(k)} = \frac{dx^{(k)}}{dt} \to y \in \mathscr{C}[0,1].$$

This means that $dx^{(k)}/dt$ converges uniformly to y and y is continuous.

[†] We do not introduce the idea of the *graph of an operator* in this book (but *see* p. 11). The graph of $T: \mathscr{V} \to \mathscr{U}$ is the set of points (x, Tx) in the product space $\mathscr{V} \times \mathscr{U}$ and the description 'closed' in Definition 3.7.2 refers to the fact that this set is closed in $\mathscr{V} \times \mathscr{U}$.

Now if the sequence $dx^{(k)}/dt$ converges uniformly then the limit function x is continuously differentiable and $dx/dt = y$.[†]

Therefore $x \in \mathcal{D}(T)$ and $Tx = y$, so that T is closed.

A similar argument shows that $T \equiv d/dt : \mathscr{L}_2[0, 1] \to \mathscr{L}_2[0, 1]$ is unbounded but closed. However, the details of the proof require the concept of absolute continuity which properly belongs within the framework of Lebesgue integration (every continuously differentiable function is absolutely continuous and so are functions satisfying a Lipschitz condition on $[0, 1]$, [26]).

If a linear transformation is not closed it is almost always closeable by suitably modifying or extending the domain of definition. The reader is pretty safe in assuming that he is always dealing with closed linear transformations.

Indeed, the situation may almost be viewed the other way round. We would like to have a property for unbounded operators which is like continuity, but which can hold only in some subspace: very well then, so choose the domain of the operator that the property is assured.

We may now return to the matter of whether the domain of the adjoint transformation T^* is, or is not, dense in $\mathscr{U}$.

We state without proof that

(a) the adjoint of an arbitrary operator A is a closed linear transformation;
(b) if $A : \mathscr{V} \to \mathscr{U}$ is a closed linear transformation, (i) $\mathscr{D}(A^*)$ is dense in $\mathscr{U}$; (ii) $A^{**} = A$, where A^{**} is the adjoint of A^*. (If A is not closed, A^{**} is an *extension* of A [30].)

One further result pertaining to the adjoint is worth setting down (again without proof) [30]. If the inverse A^{-1} of $A : \mathscr{V} \to \mathscr{U}$ exists and $\mathscr{D}(A^{-1})$ is dense in $\mathscr{U}$, then

$$(A^{-1})^* = (A^*)^{-1}.$$

Example 3.7.3 (Examples of adjoints)

(a) Let $T : \mathscr{E}_n \to \mathscr{E}_m$ (Example 3.6.1(a)) be represented by the $(m \times n)$ matrix array $[t_{ij}]$ relative to a pair of orthonormal bases in $\mathscr{E}_n, \mathscr{E}_m$. With the scalar products

$$\langle x, y \rangle_{\mathscr{E}_n} = \sum_{i=1}^{n} \xi_i \bar{\eta}_i, \quad \langle x, y \rangle_{\mathscr{E}_m} = \sum_{i=1}^{m} \xi_i \bar{\eta}_i$$

[†] This is a standard result in analysis—see, for example, [9].

we have

$$\langle Tx, y\rangle_{\mathscr{E}_m} = \sum_{i=1}^{m} \bar{\eta}_i \sum_{j=1}^{n} t_{ij}\xi_j$$

$$= \sum_{j=1}^{n} \xi_j \sum_{i=1}^{m} \overline{(\bar{t}_{ij}\eta_i)}$$

$$= \langle x, T^*y\rangle_{\mathscr{E}_n};$$

hence the matrix representation of T^* (relative to the same pair of bases) is $[\bar{t}_{ji}]$, the *conjugate transpose* of $[t_{ij}]$. Compare this, in the real case, with the dual (p. 68) and notice that the dual matrix representation is relative to a pair of dual basis sets.

(b) In the real space $\mathscr{L}_2(-\infty, \infty)$ let T be the 'right shift' operator

$$Tx(t) = x(t + \lambda),$$

where λ is a fixed number. We have

$$\langle Tx, y\rangle = \int_{-\infty}^{\infty} x(t + \lambda)y(t)\,dt$$

$$= \int_{-\infty}^{\infty} x(t)y(t - \lambda)\,dt$$

$$= \langle x, T^*y\rangle,$$

whence

$$T^*x(t) = x(t - \lambda).$$

In this case we find that the adjoint coincides with the inverse of T, namely a shift to the left. We shall later discuss, in some detail, operators whose adjoints coincide with their inverses.

(c) Let $T: \mathscr{L}_2[0, 1] \to \mathscr{L}_2[0, 1]$ be the integral operator

$$Tx = \int_0^1 k(t, \tau)x(\tau)\,d\tau,$$

then

$$\langle Tx, y\rangle = \int_0^1 \bar{y}(t)\,dt \int_0^1 k(t, \tau)x(\tau)\,d\tau$$

$$= \int_0^1 x(t)\,dt \int_0^1 \overline{(k(\tau, t)y(\tau))}\,d\tau$$

$$= \langle x, T^*y\rangle;$$

hence
$$T^*x = \int_0^1 \overline{k(\tau, t)} x(\tau) d\tau.$$

Note the interchange of variables in the kernel function.

(d) As a development of (c) above, let $T: \mathscr{L}_2[0, 1] \to \mathscr{L}_2[0, 1]$ be a Volterra integral operator,
$$Tx = \int_0^t k(t, \tau) x(\tau) d\tau, \quad t \in [0, 1].$$

Setting $k(t, \tau) = 0$ for $\tau > t$ we recover case (c), thus the kernel of the adjoint vanishes for $\tau < t$, giving
$$T^*x = \int_t^1 \overline{k(\tau, t)} x(\tau) d\tau.$$

A Volterra integral operator can be interpreted physically as giving the 'output' Tx of a linear system to an 'input' x and the fact that, in effect, $k(t, \tau)$ vanishes for $\tau > t$ is a consequence of the causality of the system: that is to say, the output can depend on the present and past inputs to the system but not on future inputs. The adjoint operator, on the other hand, is anti-causal, depending only on the present and future. One could imagine the time to be reversed and the system to be initially disturbed at $t = 1$.

(e) Let T be the differential operator
$$Tx = t \frac{dx}{dt} - 2x$$
with
$$\mathscr{D}(T) = \{x(t) | \frac{dx}{dt} \in \mathscr{L}_2[1, 2]; x(1) = 0\}.$$

We have
$$\langle Tx, y \rangle = \int_1^2 \left(t \frac{dx}{dt} - 2x \right) y \, dt$$
$$= [tyx]_1^2 - \int_1^2 \left(\frac{d}{dt}(ty) + 2y \right) x \, dt,$$

implying that
$$T^*x = -t \frac{dx}{dt} - 3x$$
with
$$\mathscr{D}(T^*) = \left\{ x(t) \left| \frac{dx}{dt} \in \mathscr{L}_2[1, 2]; x(2) = 0 \right. \right\}.$$

This example illustrates a general procedure for deducing the adjoint of a differential operator. Specifically, by an integration by parts, we obtain a relation of the form

$$\langle Tx, y \rangle - \langle x, T^*y \rangle = \text{boundary term},$$

where the right-hand side depends only on the values of $x(t)$, $y(t)$ at the ends of the interval. The boundary values of T being prescribed, the boundary values of T^* are obtained by making the right-hand side vanish.

It is important to recall that the boundary conditions must be homogeneous. If a differential operator be given with inhomogeneous boundary conditions, then the derivation of the adjoint is carried through with those conditions replaced by homogeneous forms.

It will be clear from the above example that the general nth-order differential operator

$$Tx = \sum_{k=0}^{n} p_k(t) \frac{d^k x}{dt^k}$$

has the adjoint

$$T^*x = \sum_{k=0}^{n} (-1)^k \frac{d^k}{dt^k}(p_k(t)x):$$

clearly a variety of boundary values can be associated with the two operators.[†]

(f) The analogue of integration by parts for functions of more than one variable is the divergence theorem. Let Γ be a region of ordinary three-dimensional physical space with smooth boundary $\partial \Gamma$: the divergence theorem states

$$\sum_{i=1}^{3} \int_{\Gamma} \frac{\partial u_i}{\partial t_i} d\Gamma = \sum_{i=1}^{3} \int_{\partial \Gamma} u_i v_i dS,$$

where $u_i(t_j)$ are the components of a vector-valued function $\mathbf{u}$ and v_i are the components of the unit outward normal to $\partial \Gamma$.

Let $\mathcal{H}_1$ be a Hilbert space whose elements ϕ, ψ are scalar functions of position with inner product

$$\langle \phi, \psi \rangle_{\mathcal{H}_1} = \int_{\Gamma} \phi \psi \, d\Gamma.$$

Let $\mathcal{H}_2$ be a Hilbert space whose elements are vector-valued func-

---

[†] Differential operators are sometimes said to be *formally adjoint* if their domains are not chosen so as to make the boundary term in the adjoint condition above vanish; we shall not use this phraseology.

LINEAR TRANSFORMATIONS ON INNER PRODUCT SPACES

tions **u**, **v** of position with inner product

$$\langle \mathbf{u}, \mathbf{v}\rangle_{\mathscr{H}_2} = \int_\Gamma \mathbf{u}\cdot\mathbf{v}\,d\Gamma = \sum_{i=1}^{3}\int_\Gamma u_i v_i\,d\Gamma.$$

Consider the differential (gradient) operator $T: \mathscr{H}_1 \to \mathscr{H}_2$ given by

$$T\phi = \frac{\partial \phi}{\partial t_i} = \text{grad } \phi.$$

The domain of T will consist of that subspace of $\mathscr{H}_1$ whose elements are continuously differentiable in Γ and we shall further impose the boundary value $\phi = 0$ on $\partial\Gamma$.

We have

$$\begin{aligned}
\langle T\phi, \mathbf{v}\rangle_{\mathscr{H}_2} &= \sum_{i=1}^{3}\int_\Gamma \frac{\partial \phi}{\partial t_i} v_i\,d\Gamma \\
&= \sum_{i=1}^{3}\int_\Gamma \frac{\partial}{\partial t_i}(\phi v_i)\,d\Gamma - \sum_{i=1}^{3}\int_\Gamma \phi\frac{\partial v_i}{\partial t_i}\,d\Gamma \\
&= \sum_{i=1}^{3}\int_{\partial\Gamma} \phi v_i \nu_i\,dS - \sum_{i=1}^{3}\int_\Gamma \phi\frac{\partial v_i}{\partial t_i}\,d\Gamma \\
&= \langle \phi, T^*\mathbf{v}\rangle_{\mathscr{H}_1},
\end{aligned}$$

the boundary term vanishing identically.

Hence we see that $T^*: \mathscr{H}_2 \to \mathscr{H}_1$ is the (divergence) operator

$$T^*\mathbf{v} = -\sum_{i=1}^{3}\frac{\partial v_i}{\partial t_i} = -\text{div }\mathbf{v},$$

upon which no boundary conditions are imposed.

If ϕ vanishes only over a part $\partial\Gamma_1$ of $\partial\Gamma$ then it is easy to see that the adjoint operator will be associated with the boundary condition

$$\sum_{i=1}^{3} v_i \nu_i = 0$$

on the remainder $\partial\Gamma_2$ of $\partial\Gamma$.

Similar results apply for (underlying) spaces of higher dimension.

Symmetric and self-adjoint transformations

In the introductory remarks to this section it was pointed out that a symmetric matrix could be characterized by the condition

$$\langle Ty, x\rangle = \langle y, Tx\rangle,$$

implying that it coincides with its adjoint or, put another way, is self-adjoint. We shall now formalize this idea for general linear transformations. (Since the adjoint is a linear transformation there is no point in attempting to define self-adjoint non-linear operators.)

Self-adjoint transformations have particularly pleasant properties and there exists a large body of results describing them. Many of the important operators which arise in physical problems are self-adjoint: roughly speaking, self-adjoint (or symmetric) operators are associated with physical problems which can be described by the behaviour of an 'action potential'. Indeed, we shall see later that with every self-adjoint operator equation we can associate a functional whose stationary points characterize the solution(s) of the equation.

Definition 3.7.3 1. *The unbounded linear transformation* $T: \mathscr{D}(T) \to \mathscr{H}, \mathscr{D}(T)$ *dense in* $\mathscr{H}$ *is symmetric if*

$$\langle Tx, y \rangle = \langle x, Ty \rangle \qquad \forall x, y \in \mathscr{D}(T).^\dagger$$

2. *The bounded linear transformation* $T: \mathscr{H} \to \mathscr{H}$ *is self-adjoint if*

$$\langle Tx, y \rangle = \langle x, Ty \rangle \qquad \forall x, y \in \mathscr{H}.$$

Example 3.7.4

(a) Let $T: \mathscr{E}_n \to \mathscr{E}_n$ be represented by the $(n \times n)$ matrix $[t_{ij}]$ *relative to an orthonormal basis in* $\mathscr{E}_n$. Example 3.7.3(a) then shows that if T is self-adjoint its matrix is *Hermitian*, that is

$$[\overline{t_{ji}}] = [t_{ij}].$$

In the real case the matrix is said to be *symmetric*. Notice that the result only holds for an orthonormal basis.

(b) The bounded real linear transformation

$$Tx = \int_0^1 k(t, \tau) x(\tau) d\tau$$

(cf., Example 3.7.3(c)) is self-adjoint if the kernel $k(t, \tau)$ is symmetric in its variables, viz.,

$$k(t, \tau) = k(\tau, t).$$

For example, the kernel

$$k(t, \tau) = \begin{cases} t(1 - \tau), & 0 \leq t \leq \tau \leq 1, \\ \tau(1 - t), & 0 \leq \tau \leq t \leq 1, \end{cases}$$

is clearly symmetric (*see* Exercise 3.4,5).

---

†In some texts the designation 'symmetric' is used to denote that T^* extends T in the sense that $\mathscr{D}(T^*) \supseteq \mathscr{D}(T)$: we do not make this distinction.

In the complex case we require the kernel to be Hermitian, i.e.
$$k(t, \tau) = \overline{k(\tau, t)}.$$

(c) Let T be the differential operator
$$Tx = -\frac{d^2x}{dt^2}$$
with
$$\mathscr{D}(T) = \left\{ x(t) \left| \frac{d^2x}{dt^2} \in \mathscr{L}_2[0,1]; x(0) = x(1) = 0 \right. \right\}.$$
Then
$$\langle Tx, y \rangle = -\int_0^1 \frac{d^2x}{dt^2} y \, dt$$
$$= -\left[\frac{dx}{dt} y - x \frac{dy}{dt} \right]_0^1 - \int_0^1 x \frac{d^2y}{dt^2} \, dt$$
$$= \langle x, T^*y \rangle$$
so that
$$T^* = T$$
and
$$\mathscr{D}(T^*) = \mathscr{D}(T).$$
T is therefore symmetric.

If, instead of the above boundary values, we stipulate (the initial values)
$$x(0) = \frac{dx(0)}{dt} = 0,$$
T is no longer symmetric.

(d) Let $\mathscr{H}$ be a Hilbert space whose elements ϕ, ψ are scalar functions of position defined in a region Γ with inner product (cf., Example 3.7.3(f))
$$\langle \phi, \psi \rangle = \int_\Gamma \phi \psi \, d\Gamma,$$
and let T be the Laplacian operator
$$T\phi = \nabla^2 \phi = \sum_{i=1}^3 \frac{\partial^2 \phi}{\partial t_i^2}$$

224 METRIC, NORMED AND INNER PRODUCT SPACES

with $\phi = 0$ on $\partial \Gamma$. Then

$$\langle T\phi, \psi \rangle = \int_\Gamma \sum_{i=1}^{3} \frac{\partial}{\partial t_i}\left(\frac{\partial \phi}{\partial t_i}\right) \psi \, \mathrm{d}\Gamma$$

$$= \int_\Gamma \sum_{i=1}^{3} \frac{\partial}{\partial t_i}\left(\psi \frac{\partial \phi}{\partial t_i}\right) \mathrm{d}\Gamma - \int_\Gamma \sum_{i=1}^{3} \frac{\partial \phi}{\partial t_i} \frac{\partial \psi}{\partial t_i} \mathrm{d}\Gamma$$

and, applying the divergence theorem to the first integral,

$$\langle T\phi, \psi \rangle = \int_{\partial \Gamma} \psi \frac{\partial \phi}{\partial \nu} \mathrm{d}S - \int_\Gamma \sum_{i=1}^{3} \frac{\partial}{\partial t_i}\left(\phi \frac{\partial \psi}{\partial t_i}\right) \mathrm{d}\Gamma + \int_\Gamma \sum_{i=1}^{3} \phi \frac{\partial}{\partial t_i}\left(\frac{\partial \psi}{\partial t_i}\right) \mathrm{d}\Gamma$$

where $\partial/\partial \nu$ denotes the outward normal derivative.

Another application of the divergence theorem gives

$$\langle T\phi, \psi \rangle = \int_{\partial \Gamma}\left(\psi \frac{\partial \phi}{\partial \nu} - \phi \frac{\partial \psi}{\partial \nu}\right) \mathrm{d}S + \int_\Gamma \sum_{i=1}^{3} \phi \frac{\partial}{\partial t_i}\left(\frac{\partial \psi}{\partial t_i}\right) \mathrm{d}\Gamma$$

$$= \langle \phi, T^*\psi \rangle,$$

showing that $T \equiv \nabla^2$ in Γ with $\phi = 0$ on $\partial \Gamma$ is a symmetric operator.

Notice that we can write

$$-\nabla^2 \phi \equiv -\mathrm{div}(\mathrm{grad}\, \phi),$$

showing that $-\nabla^2$ is the composition of the operator 'grad' and its adjoint '$-$ div' (Example 3.7.3(f)). This is a general result, as can readily be seen: let T be a linear transformation and T^* its adjoint, then the composition T^*T is clearly symmetric since (see Fig. 3.15)

$$\langle T^*Tx, y \rangle_{\mathscr{H}_1} = \langle Tx, Ty \rangle_{\mathscr{H}_2}$$
$$= \langle x, T^*Ty \rangle_{\mathscr{H}_1}$$

Consider the inner product $\langle Tx, x \rangle$ when T is symmetric. We have, using axiom 1 of Definition 3.6.1,

$$\langle Tx, x \rangle = \overline{\langle x, Tx \rangle}.$$

But since T is symmetric we also have

$$\langle Tx, x \rangle = \langle x, Tx \rangle$$

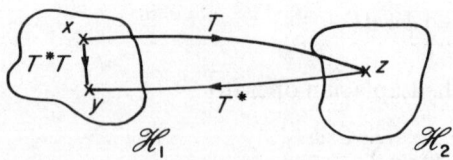

Fig. 3.15

LINEAR TRANSFORMATIONS ON INNER PRODUCT SPACES

and we may conclude that the scalar $\langle Tx, x \rangle$ is real. This leads to two important concepts, those of *positive operator* and *energy product*.

Definition 3.7.4 *The symmetric transformation T is said to be* positive *if*

$$\langle Tx, x \rangle \geq 0 \quad \forall x \in \mathscr{D}(T),$$

positive definite *if* $\langle Tx, x \rangle > 0, x \neq \theta$.

Assuming T to be positive definite, consider the functional $\langle Tx, y \rangle$ with $x, y \in \mathscr{D}(T)$. This scalar we call the *energy product* of x and y: it possesses all the usual properties of the inner product (Definition 3.6.1), as can readily be verified. In like vein we call $\langle Tx, x \rangle^{1/2}$ the *energy norm* of x. These concepts will prove to be of central importance in the approximate solution of symmetric operator equations discussed in Chapter 4.

Example 3.7.5

(a) The differential operator of Example 3.7.4(c), namely,

$$Tx = -\frac{d^2 x}{dt^2}, \quad x(0) = x(1) = 0,$$

is positive definite since

$$\langle Tx, x \rangle = -\int_0^1 \frac{d^2 x}{dt^2} x \, dt$$

$$= -\left[\frac{dx}{dt} x\right]_0^1 + \int_0^1 \left(\frac{dx}{dt}\right)^2 dt,$$

which is positive and vanishes only for $x(t) = 0$.

If the boundary values were changed to $x'(0) = x'(1) = 0$, T would still be positive but no longer positive definite since $\langle Tx, x \rangle$ would vanish for the function $x(t) = $ constant.

(b) An important example of a self-adjoint transformation is an orthogonal projection (*see* Theorem 3.6.2 et seq.).

If $\mathscr{W}$ is a closed subspace of a Hilbert space $\mathscr{H}$ then every $x \in \mathscr{H}$ has a unique decomposition

$$x = x_1 + x_2, \quad x_1 \in \mathscr{W}, \quad x_2 \in \mathscr{W}^\perp.$$

Let $P: \mathscr{H} \to \mathscr{H}$ be the orthogonal projection of x onto $\mathscr{W}$, then

$$\langle Px, y \rangle = \langle x_1, y \rangle$$
$$= \langle x_1, y_1 + y_2 \rangle$$
$$= \langle x_1, y_1 \rangle$$

and similarly
$$\langle x, Py \rangle = \langle x_1, y_1 \rangle$$
so that P is self-adjoint. P is also positive since
$$\langle Px, x \rangle = \langle x_1, x \rangle = \langle x_1, x_1 \rangle \geq 0.$$

Unitary transformations

A mapping or transformation which preserves distances is called an isometry. We can characterize isometric transformations on Hilbert spaces very conveniently using the concept of adjoint transformation.

In terms of the natural norm
$$\|x\| = \langle x, x \rangle^{1/2}$$
the transformation $T : \mathcal{H} \to \mathcal{H}$ will be isometric if
$$\|Tx\| = \|x\| ;$$
that is to say, if
$$\langle Tx, Tx \rangle = \langle x, x \rangle \qquad \forall x \in \mathcal{H}.$$
From this last statement we deduce that
$$\langle x, T^*Tx \rangle = \langle x, x \rangle,$$
$$\langle x, (T^*T - I)x \rangle = 0 \qquad \forall x \in \mathcal{H},$$
and therefore that
$$T^*T = I = TT^*.$$
It also follows that, for all $x, y \in \mathcal{H}$,
$$\langle x, y \rangle = \langle x, T^*Ty \rangle$$
$$= \langle Tx, Ty \rangle.$$

Since an isometric mapping is one-to-one the inverse T^{-1} exists, and it follows from $T^*T = I$ that
$$T^{-1} = T^*.$$
All the above conditions are equivalent and we summarize them in

Definition 3.7.5 *Let U be a bounded linear transformation of the Hilbert space $\mathcal{H}$ onto itself. Then U is* unitary *if*

1. $U^*U = I$, i.e., $U^{-1} = U^*$; or
2. $\langle Ux, Uy \rangle = \langle x, y \rangle \quad \forall x, y \in \mathcal{H}$; or
3. $\|Ux\| = \|x\| \quad \forall x \in \mathcal{H}$.

In the early part of the discussion on orthogonality (p. 185), it was pointed out that the quantity $\langle x, y \rangle / \|x\| \|y\|$ had, in a real inner product space, the geometric connotation of 'the cosine of the angle between the vectors x and y'.

For the unitary transformation U we have

$$\cos \widehat{UxUy} = \frac{\langle Ux, Uy \rangle}{\|Ux\| \|Uy\|} = \frac{\langle x, y \rangle}{\|x\| \|y\|} = \cos \widehat{xy},$$

so that this class of transformation preserves angles in real spaces. Since length is also preserved we may interpret a unitary transformation geometrically as being equivalent to a *rotation*. There is, of course, no such interpretation in the complex case: therein we can instead draw a parallel with the unit circle in the complex plane which is characterized by the identities $|\lambda| = 1$ or $\lambda \bar{\lambda} = \bar{\lambda}\lambda = 1$ (Exercise 3.8,6).

A unitary transformation clearly does not disturb orthogonality so that the image of a set of orthogonal vectors is another orthogonal set. Looked at in this way, we see that the class of unitary transformations carries orthonormal basis sets into orthonormal basis sets and therefore represents the rotation of Cartesian coordinate systems. As we shall see in the following section, the natural changes of basis to consider in Hilbert spaces are those governed by unitary transformations.

Example 3.7.6

(a) The 'right-shift' operator of Example 3.7.3(b) is unitary. We have

$$\langle Tx, Ty \rangle = \int_{-\infty}^{\infty} x(t + \lambda) y(t + \lambda) dt$$

$$= \int_{-\infty}^{\infty} x(t) y(t) dt = \langle x, y \rangle.$$

(b) Let $U: \mathscr{E}_n \to \mathscr{E}_n$ be a unitary operator with matrix $[u_{ij}]$ *relative to an orthonormal basis* for $\mathscr{E}_n$; then it is easy to see that

$$[u_{ij}]^{-1} = [\bar{u}_{ji}] = [\bar{u}_{ij}]^{\mathrm{T}}.$$

The matrix is said to be *unitary*; in the real case *orthogonal*.

(c) An important example of a unitary transformation is seen in the Fourier transform. The Fourier transform of a function $x(t)$ is usually defined by a formula of the type

$$\tilde{x}(\omega) = F(x) = \int_{-\infty}^{\infty} e^{-2\pi i \omega t} x(t) dt$$

with the inversion formula

$$x(t) = F^{-1}(\tilde{x}) = \int_{-\infty}^{\infty} e^{2\pi i t \omega} \tilde{x}(\omega) d\omega.$$

Considering F to be a mapping of $\mathscr{L}_2(-\infty, \infty)$ onto itself, a formal manipulation would give

$$\begin{aligned}
\langle Fx, y \rangle &= \int_{-\infty}^{\infty} \tilde{x}(\omega) y(\omega) d\omega \\
&= \int_{-\infty}^{\infty} y(\omega) d\omega \int_{-\infty}^{\infty} e^{-2\pi i \omega t} x(t) dt \\
&= \int_{-\infty}^{\infty} x(t) dt \int_{-\infty}^{\infty} e^{-2\pi i \omega t} y(\omega) d\omega \\
&= \int_{-\infty}^{\infty} x(t) \bar{\tilde{y}}(t) dt = \langle x, F^* y \rangle,
\end{aligned}$$

so that it appears that $F^* = F^{-1}$ and that F is unitary. The foregoing argument is, however, not at all rigorous since we have

1. not shown that F is a mapping of $\mathscr{L}_2(-\infty, \infty)$ onto itself;
2. not justified the interchange in the order of integration.

The reader will find a rigorous account in [47] (*see also* Section 5.3).

Exercises 3.7

1. If $\mathscr{V}$ is an inner product space deduce that its dual $\mathscr{V}^*$ is a Hilbert space. (Hint: use Theorem 3.5.1.) Hence deduce that we can 'identify' $\mathscr{V}$ with $\mathscr{V}^*$ only if $\mathscr{V}$ is complete. By 'identify' we mean that there exists an isometry $J: \mathscr{V}^* \to \mathscr{V}$.

2. Let T and S be bounded linear transformations on the Hilbert spaces $\mathscr{U}, \mathscr{V}, \mathscr{W}$. Show that
   (i) $(\lambda T)^* = \bar{\lambda} T^*$;
   (ii) $(T + S)^* = T^* + S^*, T, S : \mathscr{V} \to \mathscr{U}$;
   (iii) $(TS)^* = S^* T^*, S : \mathscr{W} \to \mathscr{V}, T : \mathscr{V} \to \mathscr{U}$;
   (iv) $(T^*)^* = T$;

(v) $\|T^*\| \leq \|T\|$ and hence, from (iv), $\|T^*\| = \|T\|$;
(vi) $\|T^*T\| = \|T\|^2$;
(vii) $I^* = I$ and hence, if T is invertible, $(T^{-1})^* = (T^*)^{-1}$;
(viii) if λ is an eigenvalue of T then $\bar{\lambda}$ is an eigenvalue of T^*.

3. The sesquilinear functional $f(x,y): \mathscr{H} \times \mathscr{H} \to \mathscr{C}$ (Exercise 3.5, 9) is said to be *coercive* if there exists $\gamma > 0$ such that $|f(x,x)| \geq \gamma \|x\|^2$ for all $x \in \mathscr{H}$. If f is continuous and l is any bounded linear functional on $\mathscr{H}$, show that there exist $y_1, y_2 \in \mathscr{H}$ such that

$$l(x) = f(x,y_1) = \overline{f(y_2,x)}.$$

This is the Lax–Milgram theorem—it is a generalization of the Riesz representation theorem for Hilbert spaces (p. 192).

(Hint: show first that there exists a unique continuous linear transformation T with a bounded inverse T^{-1} such that

$$\langle x, Ty \rangle = f(x,y), \quad \|T^{-1}\| < 1/\gamma).$$

4. (a) If, in $\mathscr{L}_2[0,1]$, $Tx = x'' + \alpha(t)x' + \beta(t)x$, find T^* if
    (i) $x(0) = x(1) = 0$;
    (ii) $x(0) = x'(0) = 0$;
    (iii) $x(0) = x(1)$, $x'(0) = x'(1)$.
    Under what conditions is $T^* = T$?

   (b) Find T^* if $T: l_2 \to l_2$ is defined by

   $$T(\xi_1, \xi_2, \ldots) = (\eta_1, \eta_2, \ldots),$$

   where

   $$\eta_k = \frac{1}{k} \sum_{j=1}^{k} \xi_j.$$

   (c) Show that the adjoint of an operator $T: \mathscr{H}_1 \to \mathscr{H}_2$ is closed (Hint: consider a sequence of vectors from $\mathscr{D}(T^*)$ and use continuity of the inner product.)

   (d) Let S be an *extension* of T in the sense that $\mathscr{D}(S) \supset \mathscr{D}(T)$ and $S = T$ on $\mathscr{D}(T)$; let $\mathscr{D}(T)$ be dense in $\mathscr{H}$. Show that this implies that T^* is an extension of S^* (i.e., $\mathscr{D}(T^*) \supset \mathscr{D}(S^*)$).
   If T_1, T_2 have dense domains in $\mathscr{H}$, what can you say about $\mathscr{D}(T_1^* + T_2^*)$, $\mathscr{D}(T_2^*T_1^*)$? Compare with rules (ii), (iii) for continuous transformations in Exercise 3.7, 2.

   (e) Let $\mathscr{H}$ be the space of vector-valued functions $\mathbf{u}, \mathbf{v}$ with the inner product (cf., Example 3.7.3 (f))

   $$\langle \mathbf{u}, \mathbf{v} \rangle = \int_\Gamma \mathbf{u} \cdot \mathbf{v} d\Gamma.$$

   Given that curl $\mathbf{u} = \varepsilon_{ijk} \partial u_k / \partial t_j$, where $\varepsilon_{ijk} = 0$ if any of i,j,k

are equal, $+1$ if i, j, k are unequal and in cyclic order, and -1 otherwise, show that the adjoint of the operator $T \equiv \text{curl}$ is $T^* \equiv \text{curl}$ and examine the relation between their domains.

5. Let $\mathscr{H}$ have the orthonormal basis $\{e^i\}$ and let T be a compact linear transformation on $\mathscr{H}$ (Definition 3.4.3). Define the sequence of transformation $T^{(n)}$ by

$$T^{(n)}x = \sum_{i=1}^{n} \sum_{j=1}^{n} \langle x, e^i \rangle \langle Te^i, e^j \rangle e^j.$$

Verify that $T^{(n)}$ is compact and that $\lim_n T^{(n)} = T$ (in the sense that $\|T^{(n)} - T\| \to 0$).

A transformation of the type $T^{(n)}$ is said to be *degenerate* or *finite* (meaning that $\mathscr{R}(T^{(n)})$ is finite).

The above result states that the compact transformation T can be approximated arbitrarily closely by a degenerate transformation. We noted in Example 3.4.5(e) that the integral operator $T \cdot = \int_0^1 k(t, \tau) \cdot d\tau$ on $\mathscr{L}_2[0, 1]$ is compact. Hence there exists a degenerate integral operator $T^{(n)}$ which approximates T. Show that the kernel of this integral operator is the nth partial sum of the (double) Fourier series for $k(t, \tau)$ in $\mathscr{L}_2[0, 1] \times \mathscr{L}_2[0, 1]$ (see [34, 46]). Find the Fourier sine series for the kernel of Example 3.7.4(b) and compute an approximation to $T(-1)$. Compare with the exact solution.

6. (a) Show that if T, S are self-adjoint then ST is self-adjoint only if $TS = ST$.
   (b) If T is positive definite show that the solution of $Tx = y$, if it exists, is unique.
   (c) Show that if $T \in \mathscr{L}(\mathscr{H}, \mathscr{H})$ then
      (i) $T^*T = TT^*$ is positive;
      (ii) $(I + T)$ is non-singular if T is positive.
   (d) Prove that if T is self-adjoint then

   $$\|T\| = \sup_{\|x\|=1} \{|\langle Tx, x \rangle|\}$$

   in the following way: let the right-hand side equal γ_T, say, then
      (i) show that $\gamma_T \leq \|T\|$;
      (ii) show that for any real $\lambda > 0$

   $$4\|Tx\|^2 = \langle T(\lambda x + \lambda^{-1}Tx), (\lambda x + \lambda^{-1}Tx) \rangle$$
   $$- \langle T(\lambda x - \lambda^{-1}Tx), (\lambda x - \lambda^{-1}Tx) \rangle$$
   $$\leq \gamma_T \|\lambda x + \lambda^{-1}Tx\|^2 + \gamma_T \|\lambda x - \lambda^{-1}Tx\|^2$$

   and apply the parallelogram law (p. 183) to obtain

   $$4\|Tx\|^2 \leq 2\gamma_T(\lambda^2 \|x\|^2 + \lambda^{-2}\|Tx\|^2);$$

(iii) now take $\|x\| = 1$ and $\lambda^2 = \|Tx\|$ to show $\|T\| \leq \gamma_T$.
(e) If $f(x, y)$ is a continuous, Hermitean sesquilinear functional, show that the transformation defined by $\langle x, Ty \rangle = f(x, y)$ is self-adjoint (cf., Exercise 3.7,3).
(f) Show that the functional $\langle Tx, y \rangle$ with $x, y \in \mathscr{D}(T)$, T positive definite is an inner product on $\mathscr{D}(T)$ and that

$$|\langle Tx, y \rangle|^2 \leq \langle Tx, x \rangle \langle Ty, y \rangle.$$

Hence verify that $\langle Tx, x \rangle^{1/2}$ is a norm on $\mathscr{D}(T)$ (see Section 4.4).

(g) (i) Show that if P is the orthogonal projection of x onto $\mathscr{W} \subset \mathscr{H}$ (Example 3.7.5(b)) then $\|P\| = 1$. (Hint: show $\|P\| \leq 1$, then choose $x \in \mathscr{W}$.)
(ii) Let P_1, P_2 be the orthogonal projections on the closed subspaces $\mathscr{W}_1, \mathscr{W}_2 \subset \mathscr{H}$, respectively. Show that $P_1 P_2$ is a projection if and only if $P_1 P_2 = P_2 P_1$ (cf., Exercise 3.7,6(a)) and that $\mathscr{R}(P_1 P_2) = \mathscr{W}_1 \cap \mathscr{W}_2$.
(iii) A pair of orthogonal projections P_1, P_2 are said to be *orthogonal to each other* if $\mathscr{W}_1 \perp \mathscr{W}_2$. Show that this condition is expressed by $P_1 P_2 = P_2 P_1 = 0$ and hence verify that in this case $P_1 + P_2$ is a projection with $\mathscr{R}(P_1 + P_2) = \mathscr{W}_1 \oplus \mathscr{W}_2$. Extend this result to n mutually orthogonal projections.

(h) Verify that the least squares solution of Exercise 3.6,5(a) satisfies $\langle T^*T\tilde{x} - T^*y, x \rangle = 0 \ \forall x \in \mathscr{E}_n$ which has the matrix representation

$$[t_{ij}]^T [t_{ij}] \{\tilde{\xi}_j\} = [t_{ij}]^T \{n_j\}.$$

3.8 Linear transformations on finite-dimensional inner product spaces

In this section we illustrate some of the foregoing material in the simpler context of finite-dimensional unitary or Euclidean spaces. We also take the opportunity to introduce, as special cases, some of the results on spectral theory to be dealt with in Section 3.9.

In a way, this section is a development of Section 2.8, and the reader is advised to compare and contrast the results presented here with those of that section. We may pursue the discussion entirely in terms of vector spaces of real (complex) n-tuples: every linear transformation is, of course, represented by a matrix array.

There is little profit in discussing the general case of transformations over spaces of different dimension, since the inner product yields no significant advantage over the dual formulation of Section 2.8. So we shall restrict the discussion to linear transformations $T: \mathscr{E}_n \to \mathscr{E}_n$ having square ($n \times n$) matrix representations.

Change of basis

In Section 2.8 we saw how two matrix representations of a linear transformation $T: \mathscr{V}_n \to \mathscr{V}_n$ were related when referred to different basis sets: this led to the idea of similarity (Definition 2.8.2). The element added to this situation by the inner product structure is that of orthogonality; hence if we are to make progress it is clearly expedient to consider now only changes of orthonormal basis sets, that is to say, rotation of (Cartesian) axes.

Let $\{x^i\}, \{\hat{x}^i\}$ be two orthonormal basis sets for $\mathscr{E}_n$: as in Section 2.8 we write (p. 70)

$$x^j = \sum_{i=1}^{n} l_{ij} \hat{x}^i, \quad j = 1, \ldots, n,$$

and the coordinates of the vector

$$x = \sum_{i=1}^{n} \xi_i x^i = \sum_{i=1}^{n} \hat{\xi}_i \hat{x}^i$$

are related by (p. 70)

$$\hat{\xi}_i = \sum_{j=1}^{n} l_{ij} \xi_j$$

or, in matrix terms,

$$\{\hat{\xi}_i\} = [l_{ij}]\{\xi_j\}.$$

But, by hypothesis, this is an isometric transformation, so that the matrix $[l_{ij}]$ is a representation of a unitary transformation for which (Example 3.7.6(b))

$$[l_{ij}]^{-1} = [\bar{l}_{ji}].$$

Hence the coordinates of x relative to the basis sets $\{x^i\}$ and $\{\hat{x}^i\}$ are related by

$$\{\hat{\xi}_i\} = [l_{ij}]\{\xi_j\}, \quad \{\xi_i\} = [\bar{l}_{ij}]^T\{\hat{\xi}_j\}.$$

The columns of the matrix $[\bar{l}_{ij}]^T$ are the coordinate sets of the 'new' basis vectors relative to the 'old'.

Example 3.8.1

(a) In $\mathscr{E}_2$ an anticlockwise rotation of Cartesian axes through the angle θ

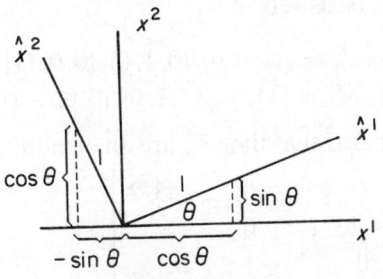

Fig. 3.16

is represented by (*see* Fig. 3.16),

$$[l_{ij}]^T = \begin{bmatrix} \cos\theta & -\sin\theta \\ \sin\theta & \cos\theta \end{bmatrix}$$

so that

$$\hat{\xi}_1 = \cos\theta\, \xi_1 + \sin\theta\, \xi_2,$$
$$\hat{\xi}_2 = \sin\theta\, \xi_1 + \cos\theta\, \xi_2.$$

Notice that

$$|\det [l_{ij}]^T| = |\det [l_{ij}]| = 1:$$

this is a general result, for if any matrix $[l_{ij}]$ is unitary (or orthogonal)

$$\det [l_{ij}] \det [\bar{l}_{ij}]^T = \det([l_{ij}][\bar{l}_{ij}]^T)$$
$$= \det I$$
$$= 1$$

and since

$$\det [\bar{l}_{ij}]^T = \overline{(\det [l_{ij}])}$$

we obtain

$$|\det [l_{ij}]| = 1.$$

In the case of an *orthogonal* matrix $[l_{ij}]$ we say that the transformation is a rotation if $\det [l_{ij}] = +1$ and a reflection if $\det [l_{ij}] = -1$. In $\mathscr{E}_2$, for example, for a rotation through $\pi/2$,

$$[l_{ij}]^T = \begin{bmatrix} 0 & -1 \\ 1 & 0 \end{bmatrix}$$

with $\det [l_{ij}]^T = 1$, while reversing the direction of the x^1 axis only is represented by

$$[l_{ij}]^T = \begin{bmatrix} -1 & 0 \\ 0 & 1 \end{bmatrix}$$

with $\det [l_{ij}]^T = -1$.

(b) Consider the two basis sets in $\mathscr{E}_3$:

'old' basis, $\{x^i\} = \{(1,0,0); (0,1,0); (0,0,1)\}$
'new' basis, $\{\hat{x}^i\} = \{(\tfrac{1}{2}, -\sqrt{3}/2, 0); (0, 0, -1); (\sqrt{3}/2, \tfrac{1}{2}, 0)\}$.

It is readily verified that the $\{\hat{x}^i\}$ are an orthonormal set and that

$$[l_{ij}]^T = \begin{bmatrix} \tfrac{1}{2} & 0 & \sqrt{3}/2 \\ -\sqrt{3}/2 & 0 & \tfrac{1}{2} \\ 0 & -1 & 0 \end{bmatrix}$$

is an orthogonal matrix with $\det [l_{ij}]^T = +1$.

The coordinates of the vector (ξ_1, ξ_2, ξ_3) relative to the new basis are given by

$$\begin{bmatrix} \hat{\xi}_1 \\ \hat{\xi}_2 \\ \hat{\xi}_3 \end{bmatrix} = \begin{bmatrix} \tfrac{1}{2} & -\sqrt{3}/2 & 0 \\ 0 & 0 & -1 \\ \sqrt{3}/2 & \tfrac{1}{2} & 0 \end{bmatrix} \begin{bmatrix} \xi_1 \\ \xi_2 \\ \xi_3 \end{bmatrix}.$$

At this point it is convenient to discuss the evaluation of the inner product $\langle x, y \rangle$ when the coordinates of x and y are referred to a non-orthonormal basis. Thus, suppose x, y are referred to the (general) basis set $\{x^i\}$ so that

$$x = \sum_{i=1}^{n} \xi_i x^i, \qquad y = \sum_{i=1}^{n} \eta_i x^i.$$

Let $\{e^i\}$ be an orthonormal basis with

$$x = \sum_{i=1}^{n} \hat{\xi}_i e^i, \qquad y = \sum_{i=1}^{n} \hat{\eta}_i e^i$$

and let the matrix of the transformation relating $\{x^i\}, \{e^i\}$ be $[l_{ij}]$, so that

$$\{\hat{\xi}_i\} = [l_{ij}]\{\xi_j\}, \qquad \{\hat{\eta}_i\} = [l_{ij}]\{\eta_j\}.$$

Now

$$\langle x, y \rangle = \sum_{i=1}^{n} \hat{\xi}_i \bar{\hat{\eta}}_i = \lfloor \hat{\xi}_i \rfloor \{\bar{\hat{\eta}}_i\} = \lfloor \xi_j \rfloor [l_{ij}]^T [\bar{l}_{ij}] \{\bar{\eta}_j\}$$
$$= \lfloor \xi_i \rfloor [h_{ij}] \{\bar{\eta}_j\},$$

where

$$[h_{ij}] = [l_{ij}]^T [\bar{l}_{ij}].$$

Of course, if $[l_{ij}]$ is unitary, $[h_{ij}]$ is the identity matrix. The above form for the inner product is similar to that for the evaluation of linear functionals on $\mathscr{V}_n$, where $[h_{ij}]$ plays the rôle of the fundamental matrix (*see* Example 2.8.2). Again we see the naturalness and convenience of always using orthonormal basis sets in inner product spaces.

Unitary similarity

Recall that square matrices which are different representations of a single linear transformation $T: \mathscr{V}_n \to \mathscr{V}_n$ are said to be similar (Definition 2.8.2): two representations are connected by

$$[\hat{t}_{ij}] = [l_{ij}][t_{ij}][l_{ij}]^{-1}.$$

If we now restrict ourselves to changes of orthonormal bases we have:

Definition 3.8.1 *Square matrices of order n which are different representations of a single linear transformation T on the inner product space $\mathscr{E}_n$ when only changes in orthonormal basis sets are considered are said to be* **unitarily similar**.[†] *If $[t_{ij}]$ is a square matrix of order n and $[u_{ij}]$ is a unitary matrix (of order n) then $[\hat{t}_{ij}]$ is unitarily similar to $[t_{ij}]$ if*

$$[\hat{t}_{ij}] = [u_{ij}][t_{ij}][\bar{u}_{ij}]^T.$$

The first point to note is that if $[t_{ij}]$ is Hermitian (T is self-adjoint) then $[\hat{t}_{ij}]$ is Hermitian since

$$[\hat{t}_{ij}]^T = \overline{([u_{ij}][t_{ij}][\bar{u}_{ij}]^T)^T} = [u_{ij}][t_{ij}][\bar{u}_{ij}]^T = [\hat{t}_{ij}],$$

so that changes in orthonormal basis do not destroy the Hermitian character of the matrix: once again we see the convenience of orthonormal basis sets.

In section 2.8 we developed the theory of canonical forms through similarity using properties associated with the eigenvalues and eigenvectors of $[t_{ij}]$: we shall now do the same for unitary similarity, but it turns out that the situation is very much simpler.

Eigenvalues and eigenvectors of Hermitian matrices

Recall (p. 44) that the eigensolutions (λ, x) of the linear transformation $T: \mathscr{E}_n \to \mathscr{E}_n$ are the solutions of

$$Tx = \lambda x:$$

[†] Orthogonally similar if real. Compare with congruence—Exercise 3.8,1(a).

relative to a basis for $\mathscr{E}_n$ this equation takes the form

$$[t_{ij}]\{\xi_j\} = \lambda\{\xi_j\}.$$

As in Section 2.8, we denote the subspace spanned by the eigenvectors $x^{(i)}$ belonging to the eigenvalue λ_i by $\mathscr{E}^{(i)} \subset \mathscr{E}_n$ (cf., Section 2.5, p. 44).

We show first of all that the eigenvalues of a self-adjoint transformation $T : \mathscr{E}_n \to \mathscr{E}_n$ (and therefore of the associated Hermitian matrix) are real. Take the complex conjugate of

$$\langle Tx, x \rangle = \langle \lambda x, x \rangle = \lambda \langle x, x \rangle$$

to give

$$\overline{\langle Tx, x \rangle} = \bar{\lambda} \langle x, x \rangle.$$

But

$$\overline{\langle Tx, x \rangle} = \langle x, Tx \rangle = \langle Tx, x \rangle$$

since T is self-adjoint; hence we see, by subtraction, that

$$(\lambda - \bar{\lambda})\langle x, x \rangle = 0$$

and since

$$\langle x, x \rangle = \|x\|^2 \neq 0$$

we conclude that $\lambda = \bar{\lambda}$ and that λ is real.

Second, we show that eigenspaces belonging to *distinct* eigenvalues λ_i, λ_j are orthogonal.

Consider the scalar equations

$$\langle Tx^{(i)}, x^{(j)} \rangle = \lambda_i \langle x^{(i)}, x^{(j)} \rangle$$

and

$$\langle x^{(i)}, Tx^{(j)} \rangle = \bar{\lambda}_j \langle x^{(i)}, x^{(j)} \rangle = \lambda_j \langle x^{(i)}, x^{(j)} \rangle.$$

Since T is self-adjoint, the left-hand sides are equal and, by subtraction, we obtain

$$(\lambda_i - \lambda_j)\langle x^{(i)}, x^{(j)} \rangle = 0,$$

from which we conclude that if $\lambda_i \neq \lambda_j$ then

$$\langle x^{(i)}, x^{(j)} \rangle = 0$$

and

$$\mathscr{E}^{(i)} \perp \mathscr{E}^{(j)}, \quad i \neq j.$$

The reader should compare this result with that relating the subspace $\mathscr{V}^{(i)}$ and its annihilator in the case of non-self-adjoint transformations (p. 89).

Let $\mathscr{U} \subset \mathscr{E}_n$ be the direct sum of the eigenspaces $\mathscr{E}^{(i)}$ of T; that is

$$\mathscr{U} = \sum_{k=1}^{r} {}_{\oplus} \mathscr{E}^{(k)},$$

where $r \leq n$ is the number of distinct eigenvalues of T. A diagonizable transformation $T: \mathscr{V}_n \to \mathscr{V}_n$ (Definition 2.8.3) is one for which the eigenspaces of T actually span the whole of $\mathscr{V}_n$. The major simplicity of self-adjoint transformations is that they are all diagonizable: furthermore, their eigenvectors (suitably scaled) form an orthonormal basis for $\mathscr{E}_n$ and hence every matrix of a self-adjoint transformation T is unitarily similar to the diagonal matrix of the eigenvalues of T (repeated where necessary).

Let us now justify these assertions. Since

$$\mathscr{U} = \sum_{k=1}^{r} {}_{\oplus} \mathscr{E}^{(k)},$$

every vector $u \in \mathscr{U}$ can be expressed in the form

$$u = \sum_{k=1}^{r} x^{(k)}.$$

Then

$$Tu = \sum_{k=1}^{r} Tx^{(k)} = \sum_{k=1}^{r} \lambda_k x^{(k)} \in \mathscr{U}.$$

Let $w \in \mathscr{U}^\perp$, then $\langle w, u \rangle = 0$ and

$$\langle Tw, u \rangle = \langle w, Tu \rangle = 0 \qquad \forall u \in \mathscr{U};$$

thus $Tw \in \mathscr{U}^\perp$ whenever $w \in \mathscr{U}^\perp$. That is to say, $\mathscr{U}^\perp$ is *invariant* with respect to T(p. 83): hence if T_r is the *restriction* (Exercise 2.5,13) of T to $\mathscr{U}^\perp$ we can consider the eigenvalue equation

$$Tw = \mu w$$

in $\mathscr{U}^\perp$.

If $\dim \mathscr{U}^\perp > 0$ then T_r (p. 82) must have at least one eigenvalue: but in that case, since T_r coincides with T on $\mathscr{U}^\perp$, that eigenvalue must coincide with one of the $\lambda^{(j)}$. But all eigenvectors for which $Tw = \lambda^{(j)} w$ are contained in $\mathscr{U}$, so we conclude that $\dim \mathscr{U}^\perp \ngtr 0$ and that $\mathscr{U} = \mathscr{E}_n$. This means that the algebraic and geometric multiplicities of every eigenvalue $\lambda^{(i)}$ are equal (p. 82).

So we have shown that the eigenspaces $\mathscr{E}^{(i)}$ span $\mathscr{E}_n$. If, within each eigenspace, we choose an orthonormal basis, then we have, overall, an orthonormal basis for $\mathscr{E}_n$ relative to which T is represented by the diagonal matrix of eigenvalues.

In matrix terms if $[t_{ij}]$ is a representation for T and the columns of the matrix $[\xi_{ij}]$ are the (normalized) eigenvectors (relative to that basis) then

$$[\lambda_i \delta_i^j] = [\bar{\xi}_{ij}]^T [t_{ij}] [\xi_{ij}].$$

Since the eigenvalues are real, they may be ordered, so we can see that we have deduced a *unique canonical form with respect to unitary similarity* for a self-adjoint transformation.

The foregoing result is often called the *spectral theorem* for the following reason. Let P_j be the orthogonal projection of $\mathscr{E}_n$ onto $\mathscr{E}^{(j)}$. The projections P_i are mutually orthogonal and since the $\mathscr{E}^{(i)}$ span $\mathscr{E}_n$ we have (Exercise 3.7,6(g)(iii))

$$I = \sum_{k=1}^{r} P_k.$$

Furthermore, the restriction of T to $\mathscr{E}^{(j)}$ is the dilatation $\lambda_j I$, so T can be represented by the sum

$$T = \sum_{k=1}^{r} \lambda_k P_k$$

(cf., p. 83). This sum is referred to as the *spectral resolution of T*.

It is of importance to note that if T is real then all the above holds in a real space; that is to say, the canonical representation is available without the necessity of using complex arithmetic. Since most physical systems which throw up self-adjoint models are couched in terms of real numbers, such systems can be resolved spectrally entirely in terms of real numbers (contrast Examples 2.9.4 and 3.10.4).

Although the foregoing discussion has been pursued for self-adjoint transformations, the main results actually apply to a somewhat wider class of transformations which includes self-adjoint transformations as a special (although extremely important) case; these are *normal* transformations. The transformation $T : \mathscr{V} \to \mathscr{V}$ is *normal* if

$$TT^* = T^*T.$$

The eigenvalues of a normal transformation are not necessarily real, but the spectral theorem is still true (Exercise 3.8,2).

Example 3.8.2 We shall find the eigensolutions of the symmetric matrix
$$[t_{ij}] = \begin{bmatrix} 2 & 0 & 0 \\ 0 & 3 & 1 \\ 0 & 1 & 3 \end{bmatrix}.$$
The characteristic polynomial is
$$\det([t_{ij}] - \lambda[\delta_i^j]) = (2 - \lambda)^2(4 - \lambda)$$
with roots $\lambda_1, \lambda_2 = 2, \lambda_3 = 4$. For $\lambda_3 = 4$ the eigenvector is the solution of
$$\begin{bmatrix} -2 & 0 & 0 \\ 0 & -1 & 1 \\ 0 & 1 & -1 \end{bmatrix} \begin{bmatrix} \xi_1^{(3)} \\ \xi_2^{(3)} \\ \xi_3^{(3)} \end{bmatrix} = \begin{bmatrix} 0 \\ 0 \\ 0 \end{bmatrix},$$
which is the vector $(0, 1/\sqrt{2}, 1/\sqrt{2})$ upon normalization.

For $\lambda = 2$ we obtain the matrix
$$\begin{bmatrix} 0 & 0 & 0 \\ 0 & 1 & 1 \\ 0 & 1 & 1 \end{bmatrix}$$
of rank 1 so that (as we should expect) the nullity is 2. Two independent solutions are $(1, 0, 0)$ and $(0, 1, -1)$. These are already orthogonal: if they were not we would retain the vector $(1, 0, 0)$ and seek another vector of the form
$$y = \beta(1, 0, 0) + (0, 1, -1)$$
which was orthogonal to $(1, 0, 0)$. Upon normalization we obtain the basis $(1, 0, 0), (0, 1/\sqrt{2}, -1/\sqrt{2})$. The matrix $[\xi_{ij}]$ of eigenvectors is
$$\begin{bmatrix} 1 & 0 & 0 \\ 0 & 1/\sqrt{2} & 1/\sqrt{2} \\ 0 & -1/\sqrt{2} & 1/\sqrt{2} \end{bmatrix}$$
and it is readily verified that
$$[\xi_{ij}]^T [t_{ij}] [\xi_{ij}]$$
is a diagonal matrix with entries $2, 2, 4$.

Example 3.8.3

(a) In Example 3.4.2 we found the induced operator norm for matrix transformations $T: \mathscr{V}_n \to \mathscr{V}_m$ when we used either the $p = 1$ or $p = \infty$ norms. Now let us find the norm induced by the Euclidean norm $p = 2$ for matrix transformations $T: \mathscr{E}_n \to \mathscr{E}_n$.

We begin with the basic definition
$$\|T\|_s = \max_{\|x\|_2 = 1} \|Tx\|_2.$$

But
$$\|Tx\|_2^2 = \langle Tx, Tx \rangle$$
$$= \langle x, T^*Tx \rangle,$$

where the transformation T^*T is positive (Exercise 3.7,6(c)(i)). Writing $S = T^*T$, we will show that if μ is an eigenvalue of S then

$$0 \leqslant \mu_{\min} \leqslant \mu \leqslant \mu_{\max},^\dagger$$

where the bounds

$$\mu_{\min} = \min_{\|x\|_2 = 1} \langle Sx, x \rangle, \qquad \mu_{\max} = \max_{\|x\|_2 = 1} \langle Sx, x \rangle$$

are themselves eigenvalues of S.

Suppose $Sx = \mu x$, $\|x\|_2 = 1$, then

$$\langle Sx, x \rangle = \langle \mu x, x \rangle = \mu$$

and as x ranges over $\mathscr{E}_n$ we have

$$\min_{\|x\|_2 = 1} \langle Sx, x \rangle \leqslant \mu \leqslant \max_{\|x\|_2 = 1} \langle Sx, x \rangle.$$

Now let

$$\mu_{\max} = \max_{\|x\|_2 = 1} \langle Sx, x \rangle,$$

then if we are to show that $\mu_{\max}$ is actually an eigenvalue of S we need to show that, for some x,

$$(\mu_{\max} I - S)x = 0.$$

Let $\tilde{x}$ be a vector such that

$$\mu_{\max} = \langle S\tilde{x}, \tilde{x} \rangle$$

and let $S_m = \mu_{\max} I - S$. Now, for $\|x\|_2 = 1$,

$$\langle S_m x, x \rangle = \mu_{\max} - \langle Sx, x \rangle \geqslant 0$$

so that S_m is positive and, for an arbitrary vector x and scalar α,

$$0 \leqslant \langle S_m(\tilde{x} + \alpha x), (\tilde{x} + \alpha x) \rangle$$
$$= \alpha \langle S_m x, \tilde{x} \rangle + \bar{\alpha} \langle S_m \tilde{x}, x \rangle + |\alpha|^2 \langle S_m x, x \rangle,$$

since

$$\langle S_m \tilde{x}, \tilde{x} \rangle = 0.$$

Now put $\alpha = \beta \langle S_m \tilde{x}, x \rangle$, where β is real, then

$$\bar{\alpha} = \beta \langle x, S_m \tilde{x} \rangle = \beta \langle S_m x, \tilde{x} \rangle$$

---

† This result applies to self-adjoint operators on general Hilbert spaces if max, min are replaced by sup, inf (see p. 251).

and we obtain
$$0 \leq \beta |\langle S_m \tilde{x}, x \rangle|^2 (2 + \beta \langle S_m x, x \rangle).$$

For β negative and numerically small the right-hand side of this inequality could be negative: hence we must conclude that
$$\langle S_m \tilde{x}, x \rangle = 0$$
and, since x is arbitrary, that
$$S_m \tilde{x} = 0$$
or
$$S\tilde{x} = \mu_{max} \tilde{x}$$
and μ_{max} is actually an eigenvalue of S. A similar argument applies to μ_{min}.

Now we can return to the question of the induced matrix norm. Let μ_{max} be the largest eigenvalue of T^*T, then, for $\|x\|_2 = 1$,
$$\|Tx\|_2^2 = \langle x, T^*Tx \rangle \leq \mu_{max}$$
and we see that
$$\|T\|_s = \max_{\|x\|_2 = 1} \|Tx\|_2 = \sqrt{\mu_{max}}.$$

The quantity $\sqrt{\mu_{max}} = \{$largest eigenvalue of $T^*T\}^{1/2}$ is generally called the *spectral radius* of T and the above norm is called the *spectral norm* of T (hence the subscript s). If T is itself Hermitian, then its spectral radius is equal to the largest eigenvalue.

(b) The spectral norm, albeit the induced operator norm for l_2, is not easily calculated and a more useful norm is obtained by treating the square matrices $[t_{ij}] : \mathscr{E}_n \to \mathscr{E}_n$ as vectors having n^2 elements and using the Euclidean norm. This leads to the definition
$$\|T\|_2 = \|[t_{ij}]\|_2 = \left[\sum_{i,j=1}^n |t_{ij}|^2 \right]^{1/2}.$$

It is not difficult to show that, for two square matrices T, S,
$$\|TS\|_2 \leq \|T\|_2 \|S\|_2$$
and that
$$\|Tx\|_2 \leq \|T\|_2 \|x\|_2.$$

This last condition shows that
$$\|T\|_s \leq \|T\|_2.$$

The norm $\|T\|_2$ is said to be a matrix norm consistent with the vector norm $\|x\|_2$, but it cannot be an induced operator norm for l_2, since it can be shown that the induced norm of the identity matrix must be unity, whereas we have

$$\|I\|_2 = \sqrt{n}.$$

Exercises 3.8

1. (a) Two square matrices $[t_{ij}], [\hat{t}_{ij}]$ are said to be *conjunctive* (*congruent* if real) if $[t_{ij}] = [l_{ij}]^T [\hat{t}_{ij}][\bar{l}_{ij}]$ for some non-singular matrix $[l_{ij}]$.
    Relative to the non-orthonormal basis $\{\hat{x}^i\}$ in $\mathscr{E}_n$ the inner product $\langle x, y \rangle$ is given by (p. 234)

    $$\langle x, y \rangle = \sum_{i,j=1}^{n} \hat{\xi}_i \hat{h}_{ij} \bar{\hat{\eta}}_j.$$

    Verify that the elements of the matrix $[\hat{h}_{ij}]$ are given by $\hat{h}_{ij} = \langle \hat{x}^i, \hat{x}^j \rangle$. Now let the basis be changed to the (non-orthonormal) basis $\{x^i\}$ through $x^j = \sum_{i=1}^{n} l_{ij} \hat{x}^i$: show that the matrix $[h_{ij}]$ is conjunctive to $[\hat{h}_{ij}]$, i.e.,

    $$[h_{ij}] = [l_{ij}]^T [\hat{h}_{ij}][\bar{l}_{ij}].$$

   (b) Show that the product of (any number of) unitary matrices is unitary. Verify that the matrix

   $$[l_{ij}]_1^T = \begin{bmatrix} 1 & 0 & 0 \\ 0 & \cos\theta_1 & -\sin\theta_1 \\ 0 & \sin\theta_1 & \cos\theta_1 \end{bmatrix}$$

   represents a rotation in $\mathscr{E}_3$ of θ_1 about the e^1 axis (Example 3.8.1(a)).
       Write down matrices which represent (positive) rotations θ_2, θ_3 about the e^2 and e^3 axes.
       Hence find the orthogonal matrix which represents a rotation of θ_3 about e^3 followed by a rotation of θ_2 and then a rotation of θ_1 about the *carried positions* of the e^2 and e^1 axes. (Many interesting and useful results of this type pertaining to mechanics can be found in [17].)

   (c) Show that the sesquilinear functional $f(x, y)$ on $\mathscr{E}_n$ determines uniquely (relative to the orthonormal basis $\{x^i\}$ in $\mathscr{E}_n$) a matrix $[t_{ij}]$ with $t_{ij} = f(x^i, x^j)$ and such that

   $$f(x, y) = \lfloor \xi_i \rfloor [t_{ij}] \{\bar{\eta}_j\}:$$

   compare with the result of Exercise 3.5, 9(c).

   (d) Answer the question in Exercise 2.5,11 if T is a self-adjoint transformation on $\mathscr{E}_n$.

(e) Find the representation of the matrix

$$\begin{bmatrix} 2 & 1 & 3 \\ 1 & 4 & 3 \\ 3 & 3 & 6 \end{bmatrix}$$

relative to the orthogonal basis $(4, -1, 5), (-1, 1, 1), (2, 3, -1)$.

2. Let $T: \mathscr{E}_n \to \mathscr{E}_n$ be a normal transformation: Prove that $\|Tx\| = \|T^*x\|$ for all $x \in \mathscr{E}_n$.
   Show that
   (i) if (λ, x) is an eigensolution of T then $(\bar{\lambda}, x)$ is an eigensolution of T^*;
   (ii) the eigenspaces $\mathscr{E}^{(i)}$ are mutually orthogonal;
   (iii) the eigenspaces span $\mathscr{V}_n$.

3. Consider the following when $[t_{ij}]: \mathscr{E}_n \to \mathscr{E}_n$ is non-singular:
   (i) $[t_{ij}][\bar{t}_{ij}]^T$ is Hermitian, positive definite and hence unitarily similar to a diagonal matrix $[d_i^2 \delta_i^j], d_i > 0$, i.e.,

   $$[u_{ij}][t_{ij}][\bar{t}_{ij}]^T [\bar{u}_{ij}]^T = [d_i^2 \delta_i^j];$$

   (ii) $[\bar{u}_{ij}]^T [d_i \delta_i^j][u_{ij}] = [h_{ij}]$ is Hermitian, positive definite;
   (iii) $[v_{ij}] = [h_{ij}]([\bar{t}_{ij}]^T)^{-1}$ is unitary;
   (iv) $[t_{ij}] = [h_{ij}][v_{ij}]$.

   Thus $[t_{ij}]$ is expressed as the product of a positive definite Hermitian matrix $[h_{ij}]$ and a unitary matrix $[v_{ij}]$: this is called the *polar decomposition* of $[t_{ij}]$. The polar decomposition exists for any bounded linear transformation on a Hilbert space [29].

4. (a) Find the eigenvalues and eigenvectors of the matrices

   (i) $\begin{bmatrix} 2 & 1 & 3 \\ 1 & 2 & 3 \\ 3 & 3 & 20 \end{bmatrix}$, (ii) $\begin{bmatrix} 2 & 2 & 1 \\ 2 & 5 & 2 \\ 1 & 2 & 2 \end{bmatrix}$

   and verify that the canonical form is the diagonal matrix of eigenvalues. Write down the matrices representing the projection operators P_k (p. 238) and verify that their sum is the identity operator.
   (b) Verify that if $T: \mathscr{E}_n \to \mathscr{E}_n$ is positive definite then every eigenvalue $\lambda_k > 0$.
   (c) Let $[t_{ij}], [s_{ij}]$ be two $(n \times n)$ symmetric matrices and let $[t_{ij}]$ be positive definite.
      (i) Show that there exists a non-singular matrix $[p_{ij}]$ such that

      $$[p_{ij}][t_{ij}][p_{ij}]^T = [\delta_i^j].$$

(Hint: first transform $[t_{ij}]$ to canonical form and then apply a 'scaling' transformation.)

(ii) Show that $[p_{ij}][s_{ij}][p_{ij}]^T$ is symmetric and hence there exists an orthogonal transformation $[u_{ij}]$ such that $[u_{ij}][p_{ij}][s_{ij}][p_{ij}]^T[u_{ij}]^T$ is canonical.

(iii) Hence show that the two matrices can be transformed simultaneously to the identity matrix and a diagonal matrix, respectively.

5. (a) (i) We refer to the power method for the computation of the dominant eigenvalue of a matrix (Exercise 2.8,21), viz.,

$$T^r x = \lambda_1^r \sum_{k=1}^{n} \xi_k \left(\frac{\lambda_k}{\lambda_1}\right)^r x^{(k)}; \qquad T^{r+1}x \to \lambda_1 T^r x.$$

The component ratio $[(T^{r+1}x)_i/(T^r x)_i]$ differs from λ_1 by a quantity $O(|\lambda_2/\lambda_1|^r)$. If $T: \mathscr{E}_n \to \mathscr{E}_n$ is symmetric, more rapid convergence to λ_1 can be obtained by using the Rayleigh quotient. Show that

$$\left(\frac{\langle T^r x, T^{r+1}x \rangle}{\langle T^r x, T^r x \rangle} - \lambda_1\right) = O\left(\left|\frac{\lambda_2}{\lambda_1}\right|^{2r}\right).$$

Use the ordinary power method and the Rayleigh quotient to find approximations to the dominant eigenvalue of the matrix

$$\begin{bmatrix} -0.030 & -0.242 & -0.603 & 0.178 \\ -0.242 & 0.860 & -0.343 & 0.393 \\ -0.603 & -0.343 & 1.350 & 0.251 \\ 0.178 & 0.393 & 0.251 & 2.630 \end{bmatrix}.$$

(ii) The power method can be developed so that the roots λ_2, λ_3 can be computed in succession. The eigenvalue λ_2 is found by constructing a matrix whose eigensolutions are $(0; e^r)$ and $(\lambda_i; x^{(i)})$, $i = 2, 3, \ldots, n$.

Suppose $\lambda_1, x^{(1)}$ have been found (by iteration) and let $x^{(1)} = (\xi_1^{(1)}, \xi_2^{(1)}, \ldots, \xi_n^{(1)})$ be normalized so that the element of largest magnitude is unity: let this be the rth element. Form a matrix $E^{(1)}$ whose rth row is $\lfloor \xi_i^{(1)} \rfloor$ but is otherwise null and form $S^{(1)} = I - E^{(1)}$.

Now show that $S^{(1)}$ is the projection onto the subspace spanned by $\{x^{(2)}, \ldots, x^{(i)}, \ldots, x^{(n)}\}$, along the subspace spanned by $\{e^r\}$. Hence show that the matrix $T^{(1)} = TS^{(1)}$ has eigensolutions $(0; e^r), (\lambda_i; x^{(i)})$, $i = 2, 3, \ldots, n$. Notice that the rth column of $T^{(1)}$ is null and hence we may find λ_2 by iterating on the $(n-1) \times (n-1)$ submatrix of $T^{(1)}$ obtained by removing the rth row and rth column. We may continue in this way to construct $T^{(2)} = T^{(1)}S^{(2)}$; however, numerical accuracy deteriorates rapidly. Can this procedure fail?

Find the second largest eigenvalue of the matrix in (i), again employing the Rayleigh quotient.

(b) (i) Let $[s_{ij}]$ be a real symmetric (2×2) matrix and let $[r_{ij}]$ be the rotation matrix (Example 3.8.1(a))

$$[r_{ij}] = \begin{bmatrix} \cos\theta & \sin\theta \\ -\sin\theta & \cos\theta \end{bmatrix}$$

with $\tan 2\theta = 2s_{12}/(s_{11} - s_{22})$, $-\pi/4 \leqslant \theta \leqslant \pi/4$. Verify that $[r_{ij}][s_{ij}][r_{ij}]^T$ is a diagonal matrix.

(ii) Let $[t_{ij}]$ be a real symmetric $(n \times n)$ matrix and let $[u_{ij}]$ be the *two-dimensional rotation* matrix with elements $u_{rr} = u_{ss} = \cos\theta$, $u_{rs} = -u_{sr} = \sin\theta$, $u_{ii} = 1$, $i \neq r, s$, and $u_{ij} = 0$ otherwise. Verify that with $\tan 2\theta = 2t_{rs}/(t_{rr} - t_{ss})$ the matrix $[\hat{t}_{ij}] = [u_{ij}][t_{ij}][u_{ij}]^T$ has zeros in the (r, s) and (s, r) places, and further that

$$\hat{t}_{ij} = t_{ij}, \quad i \neq r, s, \; j \neq r, s,$$
$$\left.\begin{array}{l}\hat{t}_{ri} = \hat{t}_{ir} = t_{ir}\cos\theta + t_{is}\sin\theta, \\ \hat{t}_{si} = \hat{t}_{is} = -t_{ir}\sin\theta + t_{is}\cos\theta,\end{array}\right\} \quad i \neq r, s,$$
$$\hat{t}_{rr} = t_{rr}\cos^2\theta + 2t_{rs}\cos\theta\sin\theta + t_{ss}\sin^2\theta,$$
$$\hat{t}_{ss} = t_{rr}\sin^2\theta - 2t_{rs}\cos\theta\sin\theta + t_{ss}\cos^2\theta.$$

The essence of *Jacobi's method* for the calculation of eigenvalues of symmetric matrices is to apply a sequence of two-dimensional rotations selecting at each step r, s so as to reduce to zero the numerically largest element t_{rs}. In this way an approximation is obtained to the diagonal matrix of eigenvalues. (Of course application of subsequent rotations renders non-zero elements which at a previous stage have been made zero.) Apply this method to the matrix in Exercise 3.8,4(a)(i).

(iii) Show that the above sequence of rotations will reduce off-diagonal elements by the following argument. Let

$$\Phi([t_{ij}]) = \sum_{\substack{i=1 \\ j \neq i}}^{n}\sum_{j=1}^{n} t_{ij}^2 = \left\|[t_{ij}]\right\|_2^2 - \sum_{i=1}^{n} t_{ii}^2$$

(Example 3.8.3(b)): Φ is the Euclidean norm of $[t_{ij}]$ less the sum of the squares of the diagonal elements.

Show that (for a rotation on r, s)

$$\Phi([\hat{t}_{ij}]) = \Phi([t_{ij}]) - 2t_{rs}^2.$$

Now t_{rs} is numerically the largest of the $n(n-1)$ off-diagonal elements, so that

$$t_{rs}^2 \geq \Phi([t_{ij}])/n(n-1);$$

hence show that

$$\Phi([\hat{t}_{ij}]) \leq \Phi([t_{ij}])(1 - 2/n(n-1))$$
$$< \Phi([t_{ij}]) \exp(-2/n(n-1)).$$

A direct application of the Jacobi method generally produces very slow convergence of $[t_{ij}]$ to diagonal form: modifications of the method which are more successful computationally have been developed by Givens and Householder [18, 19, 21].

6. (a) Show that the eigenvalues of an orthogonal transformation $T: \mathscr{E}_n \to \mathscr{E}_n$ have modulus 1 and hence take the form $+1, -1$ or $e^{i\theta}$. Furthermore, show that if $(e^{i\theta}; x^{(j)})$ is an eigensolution then so is $(e^{-i\theta}; \bar{x}^{(j)})$, and that the eigenspaces belonging to $e^{i\theta}$ and $e^{-i\theta}$ are orthogonal.

   (b) If $e^{i\theta} \neq \pm 1$, write the eigenvector $x = y + iz$: show that

   $$Ty = \cos\theta\, y - \sin\theta\, z,$$
   $$Tz = \sin\theta\, y + \cos\theta\, z.$$

   (c) Show that a real orthonormal basis for $\mathscr{E}_n$ can be constructed from the direct sum of
   (i) orthonormal bases for the eigenspaces belonging to $+1, -1$;
   (ii) orthornormal bases $\{\sqrt{2}y_1^{(j)}, \sqrt{2}z_1^{(j)}, \sqrt{2}y_2^{(j)}, \sqrt{2}z_2^{(j)}, \ldots\}$ for the eigenspaces $\mathscr{V}^{(j)} \oplus \mathscr{V}^{(-j)}$ belonging to $e^{i\theta_j}$ and $e^{-i\theta_j}$ and hence show that, relative to this basis, U takes the form

$$\begin{bmatrix} 1 & & & & & & & & & & \\ & 1 & & & & & & & & & \\ & & \ddots & & & & & & & & \\ & & & 1 & & & & & & & \\ & & & & 1 & & & & & & \\ & & & & & -1 & & & & & \\ & & & & & & -1 & & & & \\ & & & & & & & \ddots & & & \\ & & & & & & & & -1 & & \\ & & & & & & & & & \cos\theta_1 & -\sin\theta_1 \\ & & & & & & & & & \sin\theta_1 & \cos\theta_1 \\ & & & & & & & & & & \ddots \\ & & & & & & & & & & & \cos\theta_k & -\sin\theta_k \\ & & & & & & & & & & & \sin\theta_k & \cos\theta_k \end{bmatrix}$$

This is a canonical form for orthogonal matrices when we

restrict ourselves to *real* vector spaces (see the remark on p. 90).

(Hint: in (c) first show that the (real and complex) eigenvectors of T span $\mathscr{E}_n$ using an argument similar to that on p. 237; that is to say, a unitary transformation is diagonalizable.)

3.9 An outline of spectral theory

The eigenvalues and eigenvectors of a linear transformation were originally characterized in Section 2.5 by the geometric analogy of the dilatation operator. In each eigenspace $\mathscr{V}^{(\rho)}$ of the transformation $T: \mathscr{V} \to \mathscr{V}$ the effect of T is simply to 'stretch' each vector by the factor λ_ρ. The spectral resolution of T given in the last section lays bare the structure of a self-adjoint transformation by expressing it as a direct sum of simple dilatation operators acting on pairwise orthogonal subspaces. For non-self-adjoint transformations on finite-dimensional spaces a similar result holds if we use the reciprocal basis sets consisting of the eigenvectors of T and T'.

In this section we shall restrict discussion to self-adjoint or symmetric transformations on a Hilbert space. It is easy to show that if $T: \mathscr{H} \to \mathscr{H}$ is symmetric then

1. the eigenvalues of T are real;
2. eigenspaces of T belonging to distinct eigenvalues are orthogonal.

Is it possible, then, to express T in spectral form as in the finite-dimensional case, but allowing perhaps for an infinite (convergent) sum rather than a finite one? Unfortunately, the answer is a firm 'No': only for certain types of (symmetric) transformation does there exist a direct analogue of the spectral resolution, and it will be partly the aim of this section to delineate these. In general, the situation is very much more complicated than in the finite-dimensional case, and in order to aid discussion it is best to alter our point of view, moving away from the geometric idea of dilatation and towards the algebraic idea of the inverse operator.

So let us consider when the linear transformation $(\lambda I - T)$ possesses an inverse. In the finite-dimensional case the situation is clear — if λ is an eigenvalue of T then $\mathscr{N}(\lambda I - T)$, the null space of $(\lambda I - T)$, is not empty and the inverse does not exist. There are at most n such eigenvalues if $\mathscr{V}$ is n-dimensional, so there are at most n isolated values of λ for which an inverse of $(\lambda I - T)$ does not exist.

248 METRIC, NORMED AND INNER PRODUCT SPACES

In the infinite-dimensional case, it comes as no surprise that there may be an infinity of such values of λ, and we must ask whether they are a countable set. Even given that $(\lambda I - T)^{-1}$ exists, we need to qualify that statement and ask whether $(\lambda I - T)^{-1}$ is a bounded (and therefore continuous) transformation. Notice that none of these questions arises in the finite-dimensional case.

In order to deal with this more extensive situation we introduce the *spectrum of T*, denoted by $s(T)$, and the *resolvent set of T*, denoted by $r(T)$.

Definition 3.9.1 1. *If $\mathscr{R}(\lambda I - T)$ is dense in $\mathscr{H}$ and $(\lambda I - T)^{-1}$ exists and is bounded (i.e. continuous), then λ belongs to the resolvent set, $r(T)$, of T: values of λ in the resolvent set are often called regular. The transformation $(\lambda I - T)^{-1}$ is called the resolvent of T.*

2. *The complement of the resolvent set (in the scalar field) is the spectrum, $s(T)$, of T.*

It is obvious that the eigenvalues of T belong to $s(T)$; the set of eigenvalues is called the *point* (or *discrete*) spectrum $Ps(T)$ of T.

The remainder of the spectrum is conventionally subdivided into two further disjoint sets—the continuous and residual spectra of T [29]. We shall not define these here, since the residual spectrum in particular will not concern us in dealing with self-adjoint transformations. The description *continuous spectrum* refers to the fact that the points of this set may form a continuum, for example, a whole interval of the real line [82].

Example 3.9.1
(a) Consider the transformation $T: l_2 \to l_2$ defined by

$$T(\xi_1, \xi_2, \ldots) = (\xi_1, \xi_2/2, \xi_2/3, \ldots).$$

This transformation clearly has the eigensolutions

$$(1/k; e^k), \quad k = 1, 2, \ldots,$$

where e^k is the kth unit vector. Consider the equation

$$(\lambda I - T)x = y:$$

since the e^k are a basis for l_2 we may write

$$y = \sum \langle y, e^k \rangle e^k = \sum \eta_k e^k$$

and the solution x is then clearly given by

$$x = \sum \xi_k e^k$$

where

$$\xi_k = \frac{\eta_k}{(\lambda - 1/k)}.$$

Hence the inverse of $(\lambda I - T)$ exists for $\lambda \neq 1/k$. However, we should note that the eigenvalues form a sequence whose limit is zero, and we might suspect that, although the inverse may exist for $\lambda = 0$, it is not bounded. That this is indeed the case may be seen by exhibiting T^{-1}, viz.,

$$T^{-1}(\xi_1, \xi_2, \ldots) = (\xi_1, 2\xi_2, 3\xi_3, \ldots).$$

The resolvent set of T consists of the real line excluding the points $\lambda = 1/k$ and the point 0. The spectrum of T consists of the discrete spectrum $\lambda = 1/k$ together with the continuous spectrum consisting of the point 0 (*see* remark on p. 252 on parasitic spectrum).

(b) Let $T: \mathscr{L}_2(-\infty, \infty) \to \mathscr{L}_2(-\infty, \infty)$ be defined by

$$Tx = \int_{-\infty}^{t} e^{-(t-\tau)} x(\tau) d\tau.$$

Taking $x(t) = e^{i\omega t}$ we find

$$T(e^{i\omega t}) = \int_{-\infty}^{t} e^{-(t-\tau)} e^{i\omega \tau} d\tau$$

$$= e^{i\omega t}/(1 + i\omega),$$

so is $e^{i\omega t}$ an eigenvector of T with eigenvalue $(1 + i\omega)^{-1}$? The answer is 'No', because the function $e^{i\omega t} \notin \mathscr{L}_2(-\infty, \infty)$.

(c) Let $T: \mathscr{L}_2[0, 1] \to \mathscr{L}_2[0, 1]$ be defined by

$$Tx = tx(t).$$

Now

$$(\lambda I - T)x = (\lambda - t)x(t) = 0$$

implies that $x(t) = 0$ except possibly at $t = \lambda$. But functions in $\mathscr{L}_2$ which differ on a set of measure zero are deemed to be equal, so that we conclude that $x(t) = 0$ and hence $\lambda \in [0, 1]$ is not an eigenvalue of T. Hence the inverse of $\lambda I - T$ exists and from

$$(\lambda I - T)x = y, \quad y \in \mathscr{L}_2$$

or

$$(\lambda - t)x(t) = y(t):$$

we have

$$x(t) = \frac{y(t)}{\lambda - t}$$

However, this inverse is not bounded since for $y \in \mathscr{L}_2[0,1]$ the function
$$\frac{y(t)}{\lambda - t} \notin \mathscr{L}_2[0,1]$$
for any $\lambda \in [0,1]$. Hence in this case the spectrum of T consists of the closed interval $[0,1]$.

In passing, we may note that it can be shown quite generally that the spectrum of any linear transformation is a closed set [27].

(d) Let $T: \mathscr{L}_2(-\infty, \infty) \to \mathscr{L}_2(-\infty, \infty)$ be defined by
$$Tx = \frac{dx}{dt}.$$
Can we find eigensolutions of T? The only solutions of
$$\lambda x - \frac{dx}{dt} = 0$$
are the functions $\alpha e^{\lambda t}$ and these functions are not in $\mathscr{L}_2(-\infty, \infty)$ except when $\alpha = 0$. So T has no eigensolutions, $(\lambda I - T)^{-1}$ exists, and it remains to be seen for which values of λ, $(\lambda I - T)^{-1}$ is continuous. As in the previous example, we consider the solution of
$$\lambda x - \frac{dx}{dt} = y, \qquad y \in \mathscr{L}_2(-\infty, \infty).$$
It is not difficult to verify, by differentiation, that
$$x(t) = -\int_{-\infty}^{t} e^{\lambda(t-\tau)} y(\tau) d\tau, \qquad \text{Re } \lambda < 0,$$
$$x(t) = \int_{t}^{\infty} e^{\lambda(t-\tau)} y(\tau) d\tau, \qquad \text{Re } \lambda > 0.$$
We can show that $x \in \mathscr{L}_2(-\infty, \infty)$ if $y \in \mathscr{L}_2(-\infty, \infty)$ and, furthermore, that the indicated transformations are continuous.

However, if Re $\lambda = 0$ we have a different situation since the lack of an exponential 'damping' function at infinity means that $x(t)$ may not be in $\mathscr{L}_2(-\infty, \infty)$. For example, with $\lambda = 0$,
$$x(t) = -\int_{-\infty}^{t} y(\tau) d\tau:$$
now suppose that $y(t) \sim 1/t^{3/2}$ for large (positive) t, then
$$x(t) \sim \frac{1}{t^{1/2}},$$
a behaviour at infinity which excludes $x(t)$ from $\mathscr{L}_2(-\infty, \infty)$. Hence the inverse is not bounded for Re $\lambda = 0$ and we surmise that in the complex

plane the resolvent set consists of all λ with Re $\lambda \neq 0$ while the spectrum is the whole of the imaginary axis, Re $\lambda = 0$.

(e) Let us move temporarily away from a Hilbert space context, and consider the transformation of (c) over the space of bounded real functions in $[0, 1]$. In this case we do not identify functions which differ at an isolated point, so we could say that every point in $[0, 1]$ is an eigenvalue of T with eigenfunction $x(t) = 0$ for $t \neq \lambda$, and $x(\lambda) \neq 0$. The spectrum of T is the same set as in (c), but the viewpoint is somewhat different. The term continuous spectrum is sometimes used in this sense. For example, for the transformation

$$Tx = -\frac{d^2 x}{dt^2}$$

on the space of bounded, twice differentiable functions on $(-\infty, \infty)$ every non-negative number λ is an eigenvalue with eigenvectors $\cos\sqrt{\lambda}t$, $\sin\sqrt{\lambda}t$. Compare this with the same transformation on the bounded interval $[0, \pi]$ (Example 2.5.8(ii)); there we had the countable set of eigenvalues $n^2, n = 1, 2, 3\ldots$, with eigenvectors $\sin nt$. Extending the interval to infinity has had the effect of increasing the density of eigenvalues to form a continuum. Whereas in a bounded interval a function can be expanded in terms of a Fourier series in $\sin nt$, in the infinite interval the Fourier series becomes a Fourier integral representation.

We now pass on to a brief survey of results available for self-adjoint operators (many of these results apply to normal operators). We know that the scalar $\langle Tx, x \rangle$ is real and we define

$$m(T) = \inf_{\|x\|=1} \langle Tx, x \rangle, \qquad M(T) = \sup_{\|x\|=1} \langle Tx, x \rangle.$$

The first result we note is that the spectrum of T lies in the closed interval $[m(T), M(T)]$ of the real axis: the endpoints actually belong to $s(T)$. (This is consistent with the fact that the spectrum is a closed set.)[†]

For a self-adjoint transformation T there exists a direct sum decomposition of the space $\mathcal{H}$ into two subspaces $\mathcal{H}_1$ and $\mathcal{H}_2$ such that on $\mathcal{H}_1$ the transformation has only a point spectrum and on $\mathcal{H}_2$ only a continuous spectrum (here we mean the *restriction* of T to $\mathcal{H}_1$ and $\mathcal{H}_2$) [28, 61].

This last result suggests that, at least on a subspace of $\mathcal{H}$, T could be represented in spectral form. For a certain class of transformations,

---

[†] The proof is almost identical to that given in Example 3.8.3(a).

the *compact* self-adjoint transformations, $\mathcal{H}_1$ is the whole of $\mathcal{H}$ and T has a pure point spectrum: these transformations are very similar to transformations on finite-dimensional spaces (*see* Definition 3.4.3 — a large class of integral operators are compact).

We quote, without proof, some results pertaining to a compact transformation T [27–29].

A perhaps surprising fact is that for every (non-zero) eigenvalue λ_ρ, $\mathcal{N}(\lambda_\rho I - T)$ is finite-dimensional. Furthermore, the eigenvalues of T are countable and $\lambda = 0$ is the only possible limit point. This means to say that any sequence of eigenvalues cannot have a limit other than 0. The importance of this latter fact is that, when T^{-1} exists, it may be ill-conditioned because of the fact that it has eigenvalues arbitrarily close to zero. This type of situation is sometimes referred to as T having a *parasitic spectrum* [48].

For compact self-adjoint transformations then, there exists a countable set of orthogonal projections $\{P_i\}$ such that

$$I = \sum_r P_r$$

and

$$T = \sum_r \lambda_r P_r.$$

We have not yet given a meaning to a (convergent) sum of operators: the partial sums

$$T_N = \sum_{r=1}^{N} \lambda_r P_r$$

converge with respect to the operator norm in the sense that

$$\lim_{N \to \infty} \| T - T_N \| = 0.$$

If, as in the finite-dimensional case, we choose orthonormal basis sets for each of the eigenspaces $\mathcal{H}^{(r)}$ and if the Fourier expansion of $x \in \mathcal{H}$ relative to this basis, $\{x^{(i)}\}$, for $\mathcal{H}$ is

$$x = \sum_r \langle x, x^{(r)} \rangle x^{(r)},$$

then

$$Tx = \sum_r \lambda_r \langle x, x^{(r)} \rangle x^{(r)}.$$

The inverse transformation $(\lambda I - T)^{-1}, \lambda \neq \lambda_r$, can readily be

expressed in terms of the eigenvalues and eigenvectors. Suppose
$$(\lambda I - T)x = y,$$
then
$$\lambda x - y = Tx = \sum_r \lambda_r \langle x, x^{(r)} \rangle x^{(r)}:$$
taking the inner product with $x^{(s)}$ gives
$$\lambda \langle x, x^{(s)} \rangle - \langle y, x^{(s)} \rangle = \lambda_s \langle x, x^{(s)} \rangle$$
and
$$\langle x, x^{(s)} \rangle = \frac{\langle y, x^{(s)} \rangle}{\lambda - \lambda_s}$$
so that
$$x = \frac{1}{\lambda} y + \frac{1}{\lambda} \sum_r \lambda_r \frac{\langle y, x^{(r)} \rangle}{\lambda - \lambda_r} x^{(r)}$$
provided $\lambda \neq 0$.

We can reformulate this result in terms of the projections P_r and thus obtain a spectral representation for $(\lambda I - T)^{-1}$, $\lambda \neq \lambda_r$. We have
$$x = (\lambda I - T)^{-1} y = \frac{1}{\lambda} y + \frac{1}{\lambda} \sum_r \frac{\lambda_r P_r y}{\lambda - \lambda_r}$$
and using the fact that $\sum_r P_r = I$ we may write
$$(\lambda I - T)^{-1} y = \frac{1}{\lambda} \sum_r \left(P_r + \frac{\lambda_r P_r}{\lambda - \lambda_r} \right) y$$
and hence
$$(\lambda I - T)^{-1} = \sum_r \frac{P_r}{\lambda - \lambda_r}.$$
In particular, if $\lambda = 0$ is not an eigenvalue of T (T is invertible)
$$T^{-1} = \sum_r \frac{1}{\lambda_r} P_r.$$

We have naturally specified that $\lambda \neq \lambda_r$ in the above expression. However, we notice that if $\lambda = \lambda_s$ say, we could avoid the singular term(s) in the sum by stipulating that $\langle y, x^{(s)} \rangle = 0$, that is, that y is in the orthogonal complement of $\mathcal{H}^{(s)}$. However, in this case the solution x is not unique, for to any x we can add a vector in $\mathcal{H}^{(s)}$; the solution is thus the coset $x + \mathcal{H}^{(s)}$. So we are not in contradiction with the existence or otherwise of $(\lambda I - T)^{-1}$ for *all* $y \in \mathcal{R}(\lambda I - T)$.

This last result is a version of the so-called *Fredholm alternative* for the solution of the operator equation

$$(\lambda I - T)x = y.$$

Fredholm originally developed his results in connection with integral equations which satisfy the conditions for compact operators [30, 46, 49].

Although our main concern is with self-adjoint and symmetric transformations, we mention in passing that *compact* non-self-adjoint transformations have eigenvalue–eigenfunction expansions which are analogous to those discussed above. As in the finite-dimensional case (p. 89 and Example 2.9.4) the expansion involves the dual transformation which is also compact. The null spaces $\mathcal{N}(\lambda_\rho I - T)$ are again finite-dimensional, but as we have seen in connection with the Jordan form we may also need to consider the spaces $\mathcal{N}((\lambda_\rho I - T)^n)$. We even have the results (as above) that the spectrum is a pure point spectrum and the number of eigenvalues is countable. The Fredholm alternative takes a similar form except that now the right-hand side must be annihilated by an eigenspace of T' for a solution to exist if λ is an eigenvalue [29, 50].

Now let us move on from compact self-adjoint transformations to a restricted class of symmetric transformations. These are transformations T such that there exists a scalar $\tilde{\lambda}$ in the resolvent set of T for which $(\tilde{\lambda}I - T)^{-1}$ is compact and self-adjoint. The point here is that essentially we wish to discuss (symmetric) differential operators: now a large class of these has inverses which are integral operators which are compact. In particular, the theory of this class of operator is based on the construction of Green's functions for the inverse integral transformation. Symmetric operators with compact resolvent have a spectral representation exactly analogous to that for compact self-adjoint operators. Once again, the number of points in its spectrum is countable but now, of course, since the transformation is unbounded, the magnitude of the eigenvalues increases without bound. The most celebrated type of transformation of the above type is the so-called Sturm–Liouville second-order differential operator (Example 3.10.5).

The restriction to compact transformations or transformations with compact resolvent ensures that the spectrum consists entirely of eigenvalues. Let us now discuss in a very heuristic way what happens when we have a continuous spectrum—again restricting ourselves to self-adjoint operators. First, let us write the spectral resolution of the transformation T in a different form. Assume that the (real) eigenvalues of a (compact) transformation T are arranged in ascending order $\lambda_1 < \lambda_2 < \lambda_3 \ldots$ and define a new set of projection operators by

$$E_{\lambda_0} = 0, \quad E_{\lambda_1} = P_1, \quad E_{\lambda_2} = P_1 + P_2,$$
$$E_{\lambda_i} = P_1 + P_2 + \ldots + P_i.$$

Then

$$T = \sum_r \lambda_r P_r$$
$$= \sum_r \lambda_r (E_{\lambda_r} - E_{\lambda_{r-1}})$$
$$= \sum_r \lambda_r \Delta E_{\lambda_r}$$

if we write

$$\Delta E_{\lambda_r} = E_{\lambda_r} - E_{\lambda_{r-1}}$$

Now this suggests what might happen if we have a continuous spectrum: the summation could become an integral of the form

$$\int \lambda \, \mathrm{d} E_\lambda,$$

in much the same way that a Fourier series becomes a Fourier integral as the frequency discrimination is indefinitely refined.

We have already pointed out (p. 251) that, for a self-adjoint transformation, there is a direct sum decomposition of $\mathscr{H}$ into two subspaces on one of which, $\mathscr{H}_1$, T has only a point spectrum and on the other, $\mathscr{H}_2$, a continuous spectrum. The restriction of T to $\mathscr{H}_1$ will have a resolution like that for a compact transformation while on $\mathscr{H}_2$ an integral representation will hold. As a matter of fact, the spectral representation can be entirely expressed in integral form if the integral is interpreted in the Riemann–Stieltjes sense (Example 3.5.3): where the spectrum is not continuous the function E_λ exhibits a jump at each isolated eigenvalue. Hence, for a bounded self-adjoint transformation we can always write

$$T = \int_{m(T)}^{M(T)} \lambda \, \mathrm{d} E_\lambda$$

while for a symmetric transformation the only obvious difference is that the limits will be $\pm \infty$ [41, 50, 51].

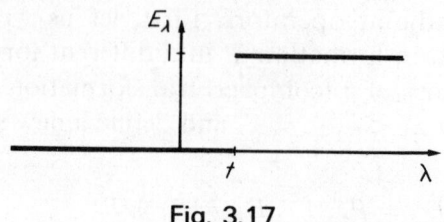

Fig. 3.17

Example 3.9.2

(a) Let the symmetric (matrix) transformation $T: \mathscr{E}_n \to \mathscr{E}_n$ have eigenvalues $\lambda_1 < \lambda_2 < \ldots < \lambda_n$ with associated eigenvectors $x^{(i)}$. For $\lambda_r \leqslant \lambda < \lambda_{r+1}$, E_λ is the transformation which projects onto the r-dimensional subspace spanned by the $\{x^{(i)}\}, i = 1, \ldots, r : E_\lambda = 0$ if $\lambda < \lambda_1$ and $E_\lambda = I$ if $\lambda \geqslant \lambda_n$. E_λ is a step function and is defined for all (real) λ.

(b) Let T be the operator of Example 3.9.1(c). Take $E_\lambda x(t) = z(t)$, where

$$z(t) = \begin{cases} 0, & t > \lambda, \\ x(t), & t \leqslant \lambda, \end{cases}$$

while $E_\lambda x = 0$ for $\lambda < 0$ and $E_\lambda x = x$ for $\lambda \geqslant 1$.

Now we have

$$T = \int_0^1 \lambda \, dE_\lambda,$$

where for *any fixed* t E_λ is the step function sketched in Fig. 3.17. Evaluated as a Stieltjes integral this gives

$$T = t.$$

3.10 Application examples

Example 3.10.1 (The virtual work identity) Example 2.9.1 dealt with the relationships between the internal forces and displacements and the external forces and displacements in a pin-jointed structure. Let us extend this example to the case of a continuous elastic material suffering small strain and small displacements.

Let the material occupy the region Γ in ordinary three-dimensional space and let the (smooth) boundary be denoted by $\partial \Gamma$. If we refer to a Cartesian axis system in physical space then the state of internal stress at each point of the material is described by a symmetric stress tensor which we may represent by the 3×3 matrix $[\tau^{ij}]$, while the state of strain is described by the symmetric tensor $[\varepsilon_{ij}]$.[†] Let the material be subject to a surface traction f^i (per unit of surface area): the equation of equilibrium is

$$\sum_{j=1}^{3} \frac{\partial \tau^{ij}}{\partial t_j} = 0 \quad \text{in} \quad \Gamma$$

and

$$\sum_{j=1}^{3} \tau^{ij} v_j = f^i \quad \text{on} \quad \partial\Gamma,$$

where v_j is the outward normal to $\partial\Gamma$. If e_i is the displacement at a general point then the relationship between (small) displacement and strain is given by

$$\varepsilon_{ij} = \frac{1}{2}\left(\frac{\partial e_i}{\partial t_j} + \frac{\partial e_j}{\partial t_i}\right).$$

Let $\mathscr{E}$ denote the space of symmetric strain tensors $[\varepsilon_{ij}]$ and $\mathscr{D}^*$ the space of surface displacement vectors ${}_s e_i$. Let $\mathscr{E}^*$ be identified with the space of symmetric stress tensors such that

$$\int_\Gamma \sum_{i,j=1}^{3} \varepsilon_{ij} \tau^{ij} d\Gamma < \infty$$

and let $\mathscr{D}$ be identified with the space of surface traction vectors f^i such that

$$\int_{\partial\Gamma} \sum_{i=1}^{3} {}_s e_i f^i dS < \infty.$$

We begin with the analogue of the linear functional $t(e)$ of Example 2.9.1, viz.,

$$[\varepsilon, \tau] = \int_\Gamma \sum_{i,j=1}^{3} \varepsilon_{ij} \tau^{ij} d\Gamma$$

$$= \frac{1}{2} \int_\Gamma \sum_{i,j=1}^{3} \left(\frac{\partial e_i}{\partial t_j} + \frac{\partial e_j}{\partial t_i}\right) \tau^{ij} d\Gamma$$

$$= \int_\Gamma \sum_{i,j=1}^{3} \frac{\partial e_i}{\partial t_j} \tau^{ij} d\Gamma$$

[†] We are restricting our discussion to small displacement: when displacement is not small we need to use a curvilinear coordinate system and the notation we use reflects this fact. At each point in space the tangent vectors to the (local) curvilinear coordinate curves represent a basis for three-dimensional physical space: at each point there also exists a reciprocal basis set (p. 52 et seq.) In this sense the stress tensor is doubly contravariant, the strain tensor doubly covariant. For small displacement the distinction between covariant and contravariant components disappears [52].

upon using the fact that $\tau^{ij} = \tau^{ji}$. We now write

$$\frac{\partial e_i}{\partial t_j}\tau^{ij} = \frac{\partial}{\partial t_j}(e_i \tau^{ij}) - e_i \frac{\partial \tau^{ij}}{\partial t_j}$$

and apply the divergence theorem (Example 3.7.3(f)) to give

$$[\varepsilon, \tau] = \sum_{i=1}^{3} \int_{\Gamma} {}_s e_i f^i \, dS = [{}_s e, f],$$

which is a statement of the virtual work identity for a continuous material.

As a matter of fact, the analysis of this example and that of Example 2.9.1 could be carried through equally well in a Hilbert space context, since we are using only real spaces and we can identify $\mathscr{E}$ and $\mathscr{E}^*$ with each other and with the Hilbert space $\mathscr{L}_2$ on Γ; that is, the space of symmetrical tensors whose scalar product is square integrable over Γ. In a similar way we can identify $\mathscr{D}$ and $\mathscr{D}^*$ with $\mathscr{L}_2$ on $\partial \Gamma$.[†]

The reader may care to note that the analogue of the dual (adjoint) operators L, L' of Example 2.9.1 are the gradient and divergence operators

$$L(e_i) = \frac{\partial e_i}{\partial t_j} \quad \text{and} \quad L^*(\tau^{ij}) = -\sum_{j=1}^{3} \frac{\partial \tau^{ij}}{\partial t_j},$$

respectively (cf., Example 3.7.4(d)).

As in Example 2.9.1 a solution of the elastic boundary-value problem is not possible until a constitutive equation relating stress and strain in the material is given (the analogue of T in Example 2.9.1).

Example 3.10.2 (Principal axes) Let us fix our attention on the symmetrical tensors introduced in the last example—the stress and strain tensors. If we consider, say, the strain tensor to be represented by a symmetric 3×3 matrix $[\varepsilon_{ij}]$ then we may expect that, by a rotation of axes, we can reduce the matrix to a diagonal form (*see* p. 238 and Example 3.8.2).

Suppose then that the eigenvalues of $[\varepsilon_{ij}]$ are the scalars $\varepsilon_\text{I}, \varepsilon_\text{II}, \varepsilon_\text{III}$ and the eigenvectors are the vectors $\{l_i\}^\text{I}, \{l_i\}^\text{II}, \{l_i\}^\text{III}$; what do these physically represent? Recall that the strain tensor is a function of position and that we are considering conditions at *one point in space*: the diagonal elements of the strain tensor represent changes in length of an element aligned along a coordinate axis (the direct strains) while the off-diagonal elements represent the change in the angle between two such line elements (the shear strains). What we have achieved then is an orientation of the (Cartesian) axis system *at the point* such that we have only direct strains; the eigenvectors $\{l_i\}^\text{I}$, etc., are the direction cosines of these new axes relative to the original. The new axes are called the *principal axes of strain* at the point, and the three direct strains, the *principal strains*. A similar analysis holds for stress.

[†] Strictly speaking, we should refer to Sobolev spaces–Section 5.4.

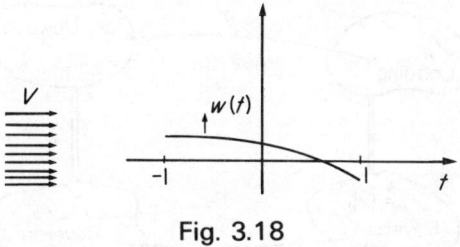

Fig. 3.18

Example 3.10.3 (Reverse flow theorem) In Example 3.10.1 we could use either a dual space or an inner product space formulation. We look now at an example in which a dual space formulation is the more illuminating.

We consider a thin aerofoil placed in a uniform fluid flow and whose projected chordline occupies the interval $[-1, 1]$. If the shape of the camber line is specified by the function $\zeta(t)$ then the tangential flow condition leads to the specification that, in $[-1, 1]$, the vertical component of the fluid velocity (upwash) $w(t)$ is given by (Fig. 3.18)

$$w(t) = \frac{d\zeta}{dt}$$

in suitable non-dimensional variables.

The momentum change in the fluid as it passes over the aerofoil leads to a pressure loading $p(t)$ sustained by the surface: the loading is related to the upwash by the singular integral equation

$$w(t) = Tp(t) = -\frac{1}{\pi} \int_{-1}^{1} \frac{p(\tau)d\tau}{t - \tau},$$

where the Cauchy principal value is understood.

The null space of T is non-empty and in order to make T invertible it is necessary to impose the so-called Kutta condition, which stipulates that the loading vanishes at the trailing edge, $t = 1$, of the aerofoil. By considering the behaviour of the integral in the neighbourhood of $t = -1$ it can be shown that the loading $p(t)$ exhibits an inverse half-power singularity at $t = -1$. Hence p is not square integrable and we consider T to be a linear transformation from $\mathscr{L}_p$ to $\mathscr{L}_q$, where $p < 2$ and $q > 2$ [53, 54].

Introducing the dual spaces $\mathscr{L}_p^*$, $\mathscr{L}_q^*$ (see Fig. 3.19) we may immediately deduce that the dual transformation $T': \mathscr{L}_q^* \to \mathscr{L}_p^*$ is given by

$$w^*(t) = T'p^*(t) = -\frac{1}{\pi} \int_{-1}^{1} \frac{p^*(\tau)d\tau}{\tau - t}.$$

Physically we can interpret this as a reversal of the uniform fluid flow and we make T' one-to-one by insisting that $p^*(t)$ vanishes at $t = -1$: it will, of course, be singular at $t = 1$.

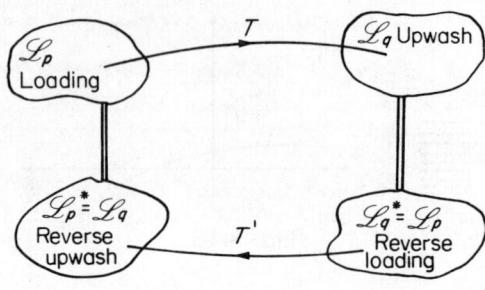

Fig. 3.19

Now let us write down formally the definition of the dual transformation, viz.

$$[Tp, p^*] = [p, T'p^*] \quad \text{or} \quad [w, p^*] = [p, w^*],$$

i.e.,

$$\int_{-1}^{1} w(t)p^*(t)dt = \int_{-1}^{1} p(t)w^*(t)dt.$$

If we interpret this statement in physical terms it says that the integral of the forward flow upwash times the reverse loading is equal to the integral of the forward loading times the reverse flow upwash: this is called the *reverse flow theorem* and can be generalized to finite wings and to unsteady flows. In a sense it is an identity like the virtual work identity, and has the same status [56].

Say we wish to find the total lift

$$L = \int_{-1}^{1} \tilde{p}(t)dt$$

on the aerofoil due to the (given) upwash $\tilde{w}(t)$. Choose $w^*(t) = 1$, then

$$L = \int_{-1}^{1} \tilde{p}(t)dt = \int_{-1}^{1} \tilde{w}(t)p_1^*(t)dt,$$

where $p_1^*(t)$ is the loading on the aerofoil in the reverse flow when the upwash is constant and of magnitude unity: this may be an easier problem to solve than the original.

Example 3.10.4 Conservative systems in physics are characterized by the fact that there exists an energy functional for the system whose value is conserved. We shall take this matter up in some detail in Chapter 4; meanwhile, we consider the case of a mechanical system having a finite number of degrees of freedom (i.e. Lagrangian coordinates) and having only mass and stiffness elements.

The equations of motion of such a system take the form (cf., Example 2.9.4)

$$[m_{ij}]\left\{\frac{d^2\xi_j}{dt^2}\right\} + [k_{ij}]\{\xi_j\} = \{\zeta_i(t)\},$$

where the 'mass' matrix $[m_{ij}]$ is positive definite and the 'stiffness' matrix $[k_{ij}]$ is positive: the $\zeta_i(t)$ are given functions of time defined for $t > 0$ and the system is subject to initial conditions

$$\xi_i(0) = \xi_i^0, \quad \frac{d\xi_i(0)}{dt} = \dot{\xi}_i^0.$$

It is, of course, possible to rewrite the system equation as

$$\frac{d^2\xi_i}{dt^2} + [m_{ij}]^{-1}[k_{ij}]\{\xi_j\} = [m_{ij}]^{-1}\{\zeta_i(t)\}$$

and to use the analysis of Example 2.9.4, but this destroys the symmetry of the original formulation. It is better to deal with the analysis in the following way.

Consider the generalized eigenvalue problem

$$(\lambda[m_{ij}] + [k_{ij}])\{\xi_j\} = \theta$$

or in operator terms

$$(\lambda M + K)x = \theta$$

or

$$\lambda Mx = -Kx.$$

Now $\langle \lambda Mx, x \rangle = \lambda \langle Mx, x \rangle = -\langle Kx, x \rangle$, so that

$$\lambda = -\frac{\langle Kx, x \rangle}{\langle Mx, x \rangle}$$

is the quotient of two real scalars and is hence real. Furthermore, since $\langle Mx, x \rangle > 0$ and $\langle Kx, x \rangle \geq 0$ we see that $\lambda \leq 0$ and we may write $\lambda = -\omega^2$.

Again, using the fact that M is positive definite we may introduce the weighted inner product $\langle Mx, y \rangle$ into $\mathcal{R}_n$. The vectors x and y are orthogonal, $x \perp y$, with respect to this inner product if $\langle Mx, y \rangle = 0$.

Let $(\omega_k^2; x^{(k)})$ be an eigensolution, that is to say,

$$(\omega_k^2 M - K)x^{(k)} = \theta.$$

We have, for any two eigensolutions,

$$\omega_k^2 \langle Mx^{(k)}, x^{(j)} \rangle = \langle Kx^{(k)}, x^{(j)} \rangle$$

and

$$\omega_j^2 \langle Mx^{(j)}, x^{(k)} \rangle = \langle Kx^{(j)}, x^{(k)} \rangle,$$

which, upon subtraction, give

$$(\omega_k^2 - \omega_j^2)\langle Mx^{(j)}, x^{(k)}\rangle = 0.$$

Hence we see that the eigenvectors $x^{(k)}$ are orthogonal and can be made orthonormal by a suitable change of scale. Suppose this to be done, then we see that

$$\langle Kx^{(j)}, x^{(k)}\rangle = \delta_k^j \omega_j^2$$

and in terms of the basis $\{x^{(k)}\}$ the system equation reduces to (compare Exercise 3.8, 4(c))

$$\frac{d^2\mu_k}{dt^2} + \omega_k^2 \mu_k = \lfloor \xi_i \rfloor^{(k)} \{\zeta_i(t)\}, \qquad k = 1, \ldots, n,$$

with

$$\mu_i^0 = \lfloor \xi_i \rfloor^{(k)} \{\xi_i^0\}, \qquad \dot{\mu}_i^0 = \lfloor \xi_i \rfloor^{(k)} \{\dot{\xi}_i^0\} \qquad \text{where } \mu_k^{(t)} \text{ is the } k\text{th coordinate of } x.$$

In this context the eigenvectors $\{\xi_i\}^{(k)}$ are called the *normal modes* of the system and the numbers ω_k the corresponding *natural frequencies*. The system reduces to the superposition of a set of simple oscillators. Taking $\xi_i^0 = 0$, $\dot{\xi}_i^0 = 0$, $i = 1, \ldots, n$, the solution for each μ_k is given by

$$\mu_k(t) = \frac{1}{\omega_k}\int_0^t \sin\omega_k(t - \tau)\lfloor \xi_i \rfloor^{(k)}\{\zeta_i(\tau)\}d\tau.$$

The solution does not exist if the vector $\{\zeta_i(t)\}$ is harmonic with frequency ω_k, say $\zeta_i(t) = \tilde{\zeta}_i \sin\omega_k t$, unless

$$\lfloor \xi_i \rfloor^{(k)} \{\tilde{\zeta}_i\} = 0$$

(cf. the Fredholm alternative p. 254).

As in Example 2.9.4, the procedure used is equivalent to a similarity transformation of M and K: in this case, the eigenvectors of the adjoint coincide with those of the operator. For a large system we do not need to compute all eigensolutions: normally in practice we are interested only in the lower values of natural frequency.

Computational techniques exist for dealing directly with the generalized eigenvalue problem $(\lambda M + K)x = 0$. Such a facility is of great importance when the matrices M and K are sparse (i.e. contain many zeros) as is often the case in finite element analysis (Section 4.6) since computer storage can be minimized.

Example 3.10.5 (Sturm–Liouville operators) We remarked on p. 254 that an important class of symmetric differential operators have inverses which are compact integral operators. It turns out that operators of this type are an important source of orthonormal basis sets whose elements are the (denumerable) eigenvectors of the operator. In this example we shall give a somewhat limited discussion of the second-order Sturm–Liouville operator defined on an interval of the real line.

Consider the second-order differential operator

$$Tx \equiv -\frac{d}{dt}\left(p(t)\frac{dx}{dt}\right) + q(t)x(t),$$

defined on $\mathscr{L}_2[0,1]$, where the functions $p(t)$, $q(t)$ are continuous on $[0,1]$, $p(t)$ has a continuous derivative and $p(t) > 0$. We shall subject the functions $x(t)$ to the boundary conditions

$$x(0) = x(1) = 0.^\dagger$$

We shall consider the domain $\mathscr{D}(T)$ of T to consist of functions satisfying the boundary conditions and having continuous first and second derivatives on $[0,1]$.

We show first of all that T is symmetric. Let $x, y \in \mathscr{D}(T)$, then

$$\langle Tx, y \rangle = \int_0^1 \left[-\frac{d}{dt}\left(p(t)\frac{dx}{dt}\right) + q(t)x(t)\right]y(t)dt$$

$$= \int_0^1 \left[p(t)\frac{dx}{dt}\frac{dy}{dt} + q(t)x(t)y(t)\right]dt$$

$$= \int_0^1 \left[-\frac{d}{dt}\left(p(t)\frac{dy}{dt}\right) + q(t)y(t)\right]x(t)dt$$

$$= \langle x, Ty \rangle.$$

Assuming, in addition to $p(t) > 0$ that $q(t) \geq 0$, then we have

$$\langle Tx, x \rangle = \int_0^1 \left[p(t)\left(\frac{dx}{dt}\right)^2 + q(t)x^2\right]dt > 0$$

so that we conclude that T is positive definite and that therefore its (real) eigenvalues are all positive. Since zero is not an eigenvalue the inverse, T^{-1}, exists and we now proceed to exhibit the form of this operator.

Denote by $u(t)$, $v(t)$ the solutions of the equation

$$Tx = \theta$$

which respectively satisfy the conditions $u(0) = 0$, $v(1) = 0$. These solutions must be linearly independent, for otherwise we should have $u(t) = \alpha v(t)$, which would yield $u(1) = 0$, implying that $Tu = \theta$ for $u(t) \in \mathscr{D}(T)$: but this is impossible, since T is invertible.

---

† It is here that our treatment is, for simplicity, limited: more generally we should consider homogeneous boundary conditions of the form $(a_i x + b_i(dx/dt)) = 0$, $i = 1, 2$, at $x = 0$ and $x = 1$.

Since $u(t)$, $v(t)$ are independent, their Wronskian

$$w(t) = u\frac{dv}{dt} - v\frac{du}{dt}$$

does not vanish. By direct differentiation one can readily verify that

$$-\frac{dw}{dt} = \left(\frac{dp/dt}{p}\right)w$$

so that

$$w(t) = \frac{w_0}{p(t)},$$

where w_0 is an arbitrary constant.

Now we introduce on the square $[0, 1] \times [0, 1]$ the function $g(t, \tau)$ of two variables:

$$g(t, \tau) = \begin{cases} -\dfrac{1}{w_0} u(t)v(\tau), & t \leqslant \tau, \\ -\dfrac{1}{w_0} u(\tau)v(t), & t \geqslant \tau. \end{cases}$$

The function $g(t, \tau)$ is continuous on the whole square and furthermore is symmetrical in its arguments, $g(t, \tau) = g(\tau, t)$.

Now consider the integral operator

$$x(t) = \int_0^1 g(t, \tau) y(\tau) d\tau,$$

where $y(t)$ is a continuous function. It is straightforward but rather tedious to show by direct differentiation that

$$Tx = y$$

and it is easy to see that

$$x(0) = x(1) = 0.$$

(In verifying this computation the reader should use the result for the Wronskian already derived.) Hence the integral operator

$$S(\cdot) \equiv \int_0^1 g(t, \tau) \cdot d\tau$$

is the inverse T^{-1} of T: the function $g(t, \tau)$ is called the Green's function for T. The operator S is compact (Example 3.4.5(d)) and self-adjoint and therefore possesses a denumerable set of eigenvalues and eigenvectors (Section 3.9) which are mutually orthogonal. If the eigensolutions of S are $(\lambda_n; x^{(n)})$ then the eigensolutions of the operator T are $(1/\lambda_n; x^{(n)})$ (p. 45).

Since the operator S is compact and self-adjoint it has a spectral representation

$$S = \sum_n \lambda_n P_n$$

and consequently, as we saw on p. 253, S^{-1} has the spectral representation

$$S^{-1} = T = \sum_n \frac{1}{\lambda_n} P_n.$$

The operator S is bounded and positive and thus every eigenvalue λ_n lies in some finite interval $(0, M]$ of the real line, while the eigenvalues of T, $\mu_n = 1/\lambda_n$, are contained in the semi-infinite interval $[1/M, \infty)$.

Any function $\tilde{x} \in \mathscr{D}(T)$ may be expanded in the Fourier series

$$\tilde{x} = \sum_n \langle \tilde{x}, x^{(n)} \rangle x^{(n)},$$

and this series is absolutely and uniformly convergent [44]. But the space of twice-differentiable functions which vanish at $0, 1$ contains the space of infinitely differentiable functions with bounded support (p. 159): this space of functions is itself dense in $\mathscr{L}_2[0, 1]$ and hence we may conclude that *every* function $x \in \mathscr{L}_2[0, 1]$ is representable as the Fourier series

$$x = \sum_n \langle x, x^{(n)} \rangle x^{(n)},$$

where convergence is now with respect to the norm in $\mathscr{L}_2$. That is, the eigenfunctions of a Sturm–Liouville operator are maximal in $\mathscr{L}_2$.

As a simple example of a Sturm–Liouville operator consider, on $[0, 1]$,

$$Tx \equiv -\frac{d^2 x}{dt^2}$$

with $x(0) = x(1) = 0$. The eigensolutions of T are given from

$$\frac{d^2 x}{dt^2} + \lambda x = 0, \qquad x(0) = x(1) = 0$$

and are clearly given by $(n^2 \pi^2, \sqrt{2} \sin n\pi t)$. Hence we can expand any $x \in \mathscr{L}_2[0, 1]$ in the (usual) Fourier series

$$x = 2 \sum_n \left(\int_0^1 x(t) \sin n\pi t \, dt \right) \sin n\pi t.$$

For example, the series

$$2 \sum_n \left(\int_0^1 t \sin n\pi t \, dt \right) \sin n\pi t$$

is convergent in mean-square to the function $x(t) = t$, but not uniformly, since every eigenfunction $\sin n\pi t$ vanishes at $t = 1$.

If $\tilde{x} \in \mathcal{D}(T)$,

$$T\tilde{x} = -\frac{d^2\tilde{x}}{dt^2} = \sum_n \lambda_n \langle \tilde{x}, x^{(n)} \rangle x^{(n)}$$

$$= \sum n^2\pi^2 \langle \tilde{x}, \sqrt{2}\sin n\pi t \rangle \sqrt{2}\sin n\pi t$$

$$= 2\pi^2 \sum_n \left(\int_0^1 \tilde{x}(t) \sin n\pi t \, dt \right) n^2 \sin n\pi t,$$

which is the result we would, of course, obtain by term-by-term differentiation.

The Green's function for T is readily verified to be (Exercise 3.4,5).

$$g(t,\tau) = \begin{cases} (1-\tau)t, & 0 \leqslant t \leqslant \tau \leqslant 1, \\ (1-t)\tau, & 0 \leqslant \tau \leqslant t \leqslant 1, \end{cases}$$

by direct differentiation. It can also be verified that $\sqrt{2}\sin n\pi t$ is an eigenvector of T^{-1} with eigenvalue $1/n^2\pi^2$. For any $x \in \mathscr{L}_2[0,1]$ (p. 253)

$$T^{-1}x = 2 \sum_n \left(\int_0^1 x(t) \sin n\pi t \, dt \right) \frac{1}{n^2\pi^2} \sin n\pi t.$$

If the function $p(t)$ in the Sturm–Liouville operator vanishes at one of the end points of $[0,1]$ or the interval on which T is defined is infinite, then the operator is said to be *singular*. For this type of problem we take the inner product to be

$$\langle x, y \rangle = \int_I w(t) x(t) y(t) dt,$$

where $w(t) > 0$ is an appropriate weight function.

For example, the Chebyshev polynomials (Example 3.6.5(e)) are the eigenvectors of the differential operator

$$Tx = (1-t^2)\frac{d^2x}{dt^2} - t\frac{dx}{dt} + n^2 x$$

defined on $[-1,1]$ with $w(t) = (1-t^2)^{-1/2}$ [57].

Example 3.10.6 (Contraction mappings) A common method for solving a large class of operator equations is the iteration method. Suppose, in a general metric space, we are required to find a solution, $\tilde{x}$, of the operator equation

$$x = Tx.$$

We begin with an arbitrary element $x^{(0)}$ and take, recursively,

$$x^{(1)} = Tx^{(0)},$$
$$x^{(2)} = Tx^{(1)},$$
$$x^{(3)} = Tx^{(2)},$$
$$\vdots$$

We may ask: 'Does the sequence $\{x^{(k)}\}$ converge (with respect to the

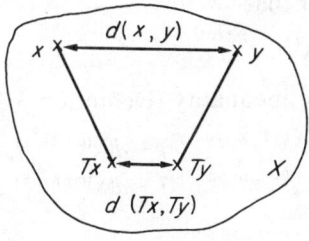

Fig. 3.20

metric) and if so, is the limit the solution point $\tilde{x}$ of $x = Tx$?' We can answer these questions by introducing the concept of a *contraction mapping* in a metric space.

Let $T: X \to X$ be a mapping of a metric space (X, d) into itself. T is a *contraction* (or a *contraction mapping*) if there is a real number $\alpha < 1$ such that

$$d(Tx, Ty) \leqslant \alpha d(x, y) \quad \forall x, y \in X.^{\dagger}$$

If T is a contraction, then it means (*see* Fig. 3.20) that the distance between a pair of image points is less than the distance between the points. We should note (Definition 3.2.10) that every contraction operator is continuous.

In the context of the iterative scheme outlined above, we would have

$$d(x^{(2)}, x^{(1)}) < d(x^{(0)}, x^{(1)}),$$
$$d(x^{(3)}, x^{(2)}) < d(x^{(2)}, x^{(1)}) < d(x^{(0)}, x^{(1)}),$$

so that we would suspect that $\{x^{(n)}\}$ is a Cauchy sequence and therefore, if the space (X, d) is complete, convergent. This proves to be the case, and indeed the limit is the solution point $\tilde{x}$. In this context the point $\tilde{x}$ is called a *fixed point* of T for obvious reasons. We should also note that the solution of an operator equation of the form $Sx = 0$ can always be converted into a fixed point by taking $Tx = x + Sx$: the fact that this conversion can be accomplished in different ways is of considerable practical importance because the resulting T may or may not be a contraction and, if it is, one conversion may produce a smaller value of α than another. We shall now show that if T is a contraction and X is complete then the sequence of iterates $\{x^{(n)}\}$ converges to $\tilde{x}$ and, furthermore, $\tilde{x}$ is unique: we shall also find an estimate for the *rate of convergence*.

For the general iterative scheme

$$x^{(n+1)} = Tx^{(n)},$$

we have

$$d(x^{(n+1)}, x^{(n)}) = d(Tx^{(n)}, Tx^{(n-1)})$$
$$\leqslant \alpha d(x^{(n)}, x^{(n-1)})$$

---

† If an operator satisfies a condition of this type but we do not impose the condition that $\alpha < 1$, then T is said to satisfy a *Lipschitz condition with constant* α.

and by induction we infer that
$$d(x^{(n+1)}, x^{(n)}) \leq \alpha^n d(x^{(1)}, x^{(0)}).$$

Now using the triangle inequality (Definition 3.2.1, axiom 3) repeatedly,
$$\begin{aligned}
d(x^{(n+k)}, x^{(n)}) &\leq d(x^{(n+k)}, x^{(n+k-1)}) + d(x^{(n+k-1)}, x^{(n)}), \\
&\leq d(x^{(n+k)}, x^{(n+k-1)}) + d(x^{(n+k-2)}, x^{(n)}) + d(x^{(n+k-2)}, x^{(n)}) \\
&\vdots \\
&\leq d(x^{(n+k)}, x^{(n+k-1)}) + \ldots + d(x^{(n+1)}, x^{(n)}) \\
&\leq (\alpha^{n+k-1} + \ldots + \alpha^n) d(x^{(1)}, x^{(0)}) \\
&= (1 + \alpha + \ldots + \alpha^{k-1}) \alpha^n d(x^{(1)}, x^{(0)}) \\
&\leq \alpha^n \sum_{i=0}^{\infty} \alpha^i d(x^{(1)}, x^{(0)})
\end{aligned}$$

since, by hypothesis, $\alpha < 1$. Hence
$$d(x^{(n+k)}, x^{(n)}) \leq \frac{\alpha^n}{1-\alpha} d(x^{(1)}, x^{(0)}).$$

This shows that $d(x^{(m)}, x^{(n)}) \to 0$ as $n, m = n + k \to \infty$ and hence that $\{x^{(n)}\}$, is a Cauchy sequence: since by hypothesis (X, d) is complete, $\lim \{x^{(n)}\} = \tilde{x}$ exists in X.

We now show that the limit $\tilde{x}$ is a fixed point of T. We have
$$\begin{aligned}
d(x^{(n+1)}, T\tilde{x}) &= d(Tx^{(n)}, T\tilde{x}) \\
&\leq \alpha d(x^{(n)}, \tilde{x});
\end{aligned}$$
hence $d(x^{(n+1)}, T\tilde{x}) \to 0$ as $n \to \infty$, or
$$\tilde{x} = \lim_{n \to \infty} x^{(n+1)} = T\tilde{x}$$

and $\tilde{x}$ is a fixed point of T.

Now assume that there are two elements $\tilde{x}, \tilde{y} \in X$ which are fixed points of T:
$$d(\tilde{x}, \tilde{y}) = d(T\tilde{x}, T\tilde{y}) \leq \alpha d(\tilde{x}, \tilde{y}).$$

Equality cannot hold, since $\alpha < 1$, hence we conclude that $d(\tilde{x}, \tilde{y}) = 0$, $\tilde{y} = \tilde{x}$ and the fixed point is unique.

If, in the previously derived inequality,
$$d(x^{(n+k)}, x^{(n)}) \leq \frac{\alpha^n}{1-\alpha} d(x^{(1)}, x^{(0)}),$$

we allow $k \to \infty$ we obtain an estimate for the rate of convergence, namely
$$d(x^{(n)}, \tilde{x}) \leq \frac{\alpha^n}{1-\alpha} d(x^{(1)}, x^{(0)}).$$

This throws light on our earlier remark (p. 267) that of the rearrangements of $Sx = 0$ to give $Tx = x$ the one with the smallest value of α is the best. Furthermore, the error is reduced by a good initial estimate of $\tilde{x}$.

In the midst of showing that $\tilde{x}$ is the fixed point we made the statement

$$d(x^{(n+1)}, \tilde{x}) \leqslant \alpha d(x^{(n)}, \tilde{x}),$$

which means to say that the error at the $(n+1)$th iteration is less than a fixed multiple, $\alpha < 1$, of the error at the nth iteration: an iterative method in which the error behaves in this fashion is said to be a *first-order method* and the convergence is said to be *linear*. We shall discuss iterative methods with faster rates of convergence in Chapter 4.

Since a closed subspace of a complete metric space is itself a complete metric space (p. 123), the whole of the foregoing discussion applies to this case provided that the image points also lie in the subspace.

Naturally, if we are working in a Banach space the distance functional is the norm. Now

$$\begin{aligned} \|Tx - Ty\| &= \|T(x - y)\| \\ &\leqslant \|T\| \|x - y\| \end{aligned}$$

so that if $\|T\| < 1$, T is certainly a contraction.

The type of situation we have outlined above is generically referred to as a *fixed-point principle*. There exist other (technically much more involved) fixed-point principles, notably those associated with the names of Brouwer and Schauder [32, 35]. For example, in Brouwer's principle, T need only be continuous, but its domain and range must be contained in a convex compact subset of $\mathscr{R}_n$.

Let us look at one or two simple applications of the fixed point principle for contraction operators.

(a) Let $x(t)$ be a real differentiable function defined on the interval $[\alpha, \beta]$

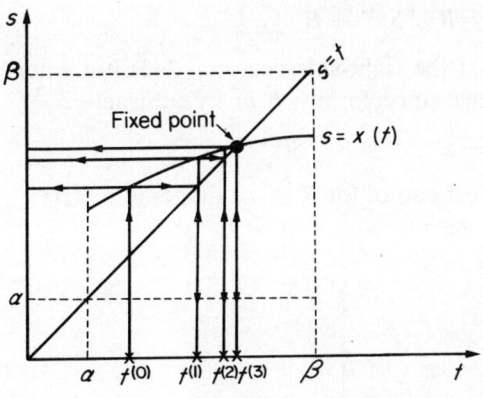

Fig. 3.21

of the real line and let the values of $x(t)$ lie in $[\alpha, \beta]$. The fixed point $x(\tilde{t}) = \tilde{t}$ has the geometric interpretation shown in Fig. 3.21.

With $d(x, y) = |x - y|$ we have, by the mean value theorem,

$$|x(t_1) - x(t_2)| = |x'(\tau)(t_1 - t_2)|$$
$$= |x'(\tau)||t_1 - t_2|, \qquad \alpha < \tau < \beta,$$

so that $x(t)$ is a contraction provided $|x'(t)| \leq k, k < 1$, for all $t \in [\alpha, \beta]$. Figure 3.21 shows the convergence of $t^{(n)}$ to $\tilde{t}$. In practice, the nearer one can estimate initially the fixed point $\tilde{t}$, the shorter the interval $[\alpha, \beta]$ can be.

(b) Let us now consider iterative methods for solution of the set of linear algebraic equations

$$[t_{ij}]\{\xi_j\} = \{\eta_i\}; \qquad i, j = 1, \ldots, n,$$

which is a matrix representation of the operator equation

$$Tx = y; \qquad T : \mathscr{R}_n \to \mathscr{R}_n.$$

Most iterative methods are based on a *splitting* of T, thus

$$T = R - S,$$

so that the equation becomes

$$Rx = Sx + y$$

and if R is chosen to be invertible

$$x = R^{-1}Sx + R^{-1}y.$$

We naturally choose R so that its inverse is easily determined.

Iteration then proceeds according to the scheme

$$x^{(n+1)} = R^{-1}Sx^{(n)} + R^{-1}y.$$

Notice that the right-side vector y does not enter into the question of convergence since, for any n, m, by subtraction

$$x^{(n+1)} - x^{(m+1)} = R^{-1}S(x^{(n)} - x^{(m)}).$$

The simplest choice for R is the diagonal matrix

$$\begin{bmatrix} t_{11} & & & \\ & t_{22} & & \\ & & \ddots & \\ & & & t_{nn} \end{bmatrix}$$

which leads to the iterative scheme

$$\xi_i^{(n+1)} = \frac{1}{t_{ii}}\left(\eta_i - \sum_{\substack{j=1\\j\neq i}}^{n} t_{ij}\xi_j^{(n)}\right), \quad i = 1, \ldots, n.$$

This is usually called *Jacobi iteration*. Convergence is assured if

$$\|R^{-1}S\| = \|R^{-1}(R-T)\|$$
$$= \|I - R^{-1}T\|$$
$$< 1$$

for any operator norm in $\mathscr{R}_n$. For example, if we choose the ∞-norm (Example 3.3.1(a))

$$\|x\| = \max_i(|\xi_i|)$$

the induced operator norm is the 'maximum row sum' (Example 3.4.2(a)) so that

$$\|I - R^{-1}T\| = \max_i \sum_{\substack{j=1\\j\neq 1}}^{n}\left(\left|\frac{t_{ij}}{t_{ii}}\right|\right).$$

A matrix for which

$$|t_{ii}| > \sum_{\substack{j=1\\j\neq 1}}^{n} |t_{ij}|$$

for each i is said to be *strictly diagonally dominant*. We then see that the Jacobi scheme converges if $[t_{ij}]$ is strictly diagonally dominant.

Diagonal dominance is also a sufficient condition for convergence of the Gauss–Seidel scheme in which T is split into lower and upper triangular matrices [18, 32]:

$$\sum_{j=1}^{i} t_{ij}\xi_j^{(n+1)} = -\sum_{j=i+1}^{n} t_{ij}\xi_j^{(n)} + \eta_i, \quad i = 1, \ldots, n.$$

We have emphasized in these examples the numerical aspects of contraction operators. However, the various fixed-point principles play an equally, if not more important role in establishing the existence and uniqueness of solutions of general non-linear operator equations.

Appendix 3.A. The Hölder and Minkowski inequalities

Lemma *Let α, β be non-negative real numbers. Then*

$$\alpha\beta \leq \frac{\alpha^p}{p} + \frac{\beta^q}{q},$$

where $1 < p < \infty$ and $1/p + 1/q = 1$.

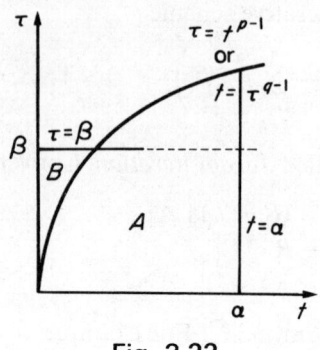

Fig. 3.22

Proof Consider the curve

$$\tau = t^{p-1}$$

(*see* Fig. 3.22).

The area of region A, bounded by the t-axis, the curve and the line $t = \alpha$, is given by

$$\int_0^\alpha t^{p-1}\,dt = \frac{\alpha^p}{p}$$

The area of the region B, bounded by the τ-axis, the curve and the line $\tau = \beta$, is given by

$$\int_0^\beta \tau^{q-1}\,d\tau = \frac{\beta^q}{q}.$$

Clearly the area of $A \cup B$ is greater than the area of the rectangle $\alpha\beta$: this yields the inequality. ∎

The Hölder inequality

Let

$$\|x\|_p = \left(\sum_{i=1}^n |\xi_i|^p\right)^{1/p}:$$

then for any $x, y \in \mathscr{V}_n$

$$\frac{|\xi_i|}{\|x\|_p}\frac{|\eta_i|}{\|y\|_p} \leq \frac{|\xi_i|^p}{p\|x\|_p^p} + \frac{|\eta_i|^q}{q\|y\|_q^q}.$$

Summing from 1 to n we have

$$\sum_{i=1}^{n} \frac{|\xi_i|}{\|x\|_p} \frac{|\eta_i|}{\|y\|_p} \leq \frac{1}{p\|x\|_p^p} \sum_{i=1}^{n} |\xi_i|^p + \frac{1}{q\|y\|_q^q} \sum_{i=1}^{n} |\eta_i|^q$$
$$= \frac{1}{p} + \frac{1}{q} = 1,$$

whence

$$\sum_{i=1}^{n} |\xi_i \eta_i| = \sum_{i=1}^{n} |\xi_i||\eta_i| \leq \|x\|_p \|y\|_q,$$

which is a finite form of the *Hölder inequality*.

A straightforward extension to infinite sums (which of course converge) is given by allowing n to tend to infinity.

Now let $x(t)$, $y(t)$ be functions defined on an interval I and let

$$\|x\|_p = \left\{ \int_I |x(t)|^p dt \right\}^{1/p}.$$

Beginning with

$$\frac{|x(t)|}{\|x\|_p} \frac{|y(t)|}{\|x\|_q} \leq \frac{|x(t)|^p}{\|x\|_p^p} + \frac{|y(t)|^p}{\|y\|_q^q}$$

we obtain, by integration,

$$\int_I |x(t) y(t)| dt \leq \|x\|_p \|y\|_q.$$

The special case $p = q = 2$ is called the *Schwarz inequality*.

These results also hold for $p = 1$, $q = \infty$.

The Minkowski inequality

We note that

$$(|\alpha| + |\beta|)^p = (|\alpha| + |\beta|)^{p-1}|\alpha| + (|\alpha| + |\beta|)^{p-1}|\beta|:$$

putting $\alpha = \xi_i$, $\beta = \eta_i$ and summing over i from 1 to n we have

$$\sum_{i=1}^{n} |\xi_i \pm \eta_i|^p \leq \sum_{i=1}^{n} (|\xi_i| + |\eta_i|)^p = \sum_{i=1}^{n} (|\xi_i| + |\eta_i|)^{p-1}|\xi_i|$$
$$+ \sum_{i=1}^{n} (|\xi_i| + |\eta_i|)^{p-1}|\eta_i|.$$

We now apply the Hölder inequality to each sum on the right to give

$$\sum_{i=1}^{n} (|\xi_i| + |\eta_i|)^p \leq \left(\sum_{i=1}^{n} (|\xi_i| + |\eta_i|)^{(p-1)q}\right)^{1/q} \left(\sum_{i=1}^{n} |\xi_i|^p\right)^{1/p}$$

$$+ \left(\sum_{i=1}^{n} (|\xi_i| + |\eta_i|)^{(p-1)q}\right)^{1/q} \left(\sum_{i=1}^{n} |\eta_i|^p\right)^{1/p}$$

$$= \left(\sum_{i=1}^{n} (|\xi_i| + |\eta_i|)^p\right)^{1/q} \left(\sum_{i=1}^{n} |\xi_i|^p\right)^{1/p}$$

$$+ \left(\sum_{i=1}^{n} (|\xi_i| + |\eta_i|)^p\right)^{1/q} \left(\sum_{i=1}^{n} |\eta_i|^p\right)^{1/p}$$

since $(p-1)q = p$. Dividing both sides by

$$\left(\sum_{i=1}^{n} (|\xi_i| + |\eta_i|)^p\right)^{1/q}$$

we obtain

$$\left(\sum_{i=1}^{n} |\xi_i \pm \eta_i|^p\right)^{1/p} \leq \left(\sum_{i=1}^{n} (|\xi_i| + |\eta_i|)^p\right)^{1/p}$$

$$\leq \left(\sum_{i=1}^{n} |\xi_i|^p\right)^{1/p} + \left(\sum_{i=1}^{n} |\eta_i|^p\right)^{1/p},$$

which is the *Minkowski inequality*.

The result holds for infinite sums and for integrals as in the Hölder inequality. We have proved Minkowski's inequality for $1 < p < \infty$, but it holds in an elementary way also for $p = 1$, hence we have

$$\|x \pm y\|_p \leq \|x\|_p + \|y\|_p, \qquad 1 \leq p < \infty.$$

Appendix 3.B. Topological and metric spaces

In this book the notion of *closed set* is used only in the sense of a subset of a metric space which contains the limit of every convergent sequence in the set. However, closed and open sets are much more basic entities than this definition would imply and this short appendix is designed to give the reader an informal account of topological spaces whose building bricks are open sets. We shall motivate the discussion by stating some properties enjoyed by open sets in a metric space and then indicate that these 'properties' could themselves be used as axioms to free us from the metric and so define a topological space.

This approach is the wrong way round, since we are arguing from the particular to the general, but at least we begin from a (more or less) familiar point.

In a metric space we may characterize an *open set G* by using *open balls* (Definition 3.2.3): G is open if, given any $x \in G$, there exists a positive real number r such that

$$B(x, r) \subseteq G;$$

that is to say, if each point of G is the centre of some open ball contained in G.

Alternatively, using the idea of closed set as stated above, we may say that G is open if its complement G' is closed. Both of these definitions rely on distance. Distance is also used when we define a *continuous mapping* on a metric space (Definition 3.2.10).

However, a theorem true for metric spaces is the following [25, 30, 59]. Let X, Y be metric spaces and f a mapping $X \to Y$: f is continuous (in the 'metric space' sense) if, and only if, $f^{-1}(G)$ is open in X whenever G is open in Y. So continuity of a mapping can be described solely in terms of open sets without direct reference to a metric. Can we free open sets from the metric as well? The answer is that we can, and we look to the following properties of *open sets in a metric space*:

1. Any union of open sets is open.
2. Any *finite* intersection of open sets is open.

Now we abandon the metric and define a topology for a space in terms of entities (i.e. open sets) which satisfy these two properties, henceforth taken to be axioms. A *class of subsets* of a set S satisfying the axioms is said to be a *topology*, and the subsets are called *open sets*. Continuity of a mapping is then described on a topological space by using the open mapping property given above. In the foregoing sense a metric space is an example of a topological space wherein the open sets are characterized by using the metric.

A *neighbourhood of a point or set* in a topological space is an open set which contains the point or set. A point $x \in S$ is a *limit point*[†] of a subset A if each of its neighbourhoods contains a point of A different from x. The *closure*, $\bar{A}$, of A is the union of A and the set of all its *limit points*. Throughout this book the expression *limit* is frequently found, but nowhere the expression *limit point*: the two are not the

---

[†] Sometimes *cluster point* or *accumulation point*.

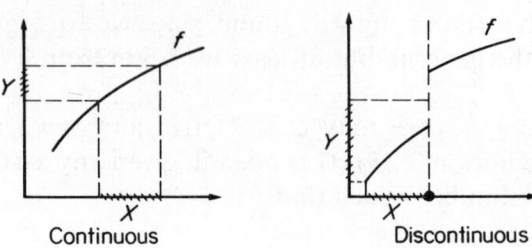

Fig. 3.23

same, as a simple example will demonstrate. On the real line the sequence $\{1, 1, 1, \ldots\}$ is convergent with *limit* 1, but the set contains only the point 1 and this is *not* a *limit point* of the set. The important point to grasp is this: a sequence of points in a set is not a subset of the set, it is a function defined on the positive integers *with values in the set* and is listed $\{x^{(1)}, x^{(2)}, \ldots, x^{(n)}, \ldots\}$ (Example 1.2.4(g)). Of course, the elements of the sequence generate a set in a metric (and hence topological) space and this set may have limit points. In fact, if a convergent sequence in a metric space has infinitely many *distinct* points, then its *limit* is a *limit point* of the set of points of the sequence. For example, $\{1, \frac{1}{2}, \frac{1}{3}, \ldots\}$ has 0 as both limit and limit point.

We end this brief outline by giving a physical or geometric interpretation of topology. Topology is concerned with those qualitative properties of spatial configurations that are independent of size, location and shape—a 'rubber sheet' geometry. We think of deformations like stretching and bending, but *without tearing*. This last condition means that points which are 'neighbours' in one configuration are 'neighbours' in another and we should recognize this effectively as a description of *continuity* (of a mapping). Consider the very simple example illustrated in Fig. 3.23, where $f: X \to Y$ is an ordinary function of one variable. If f is continuous and Y is open then $X = f^{-1}(Y)$ is also open. However, when f is discontinuous the inverse image of the open set Y is not open but includes the end point marked by a heavy dot.

Appendix 3.C. The Lebesgue integral

The Riemann approach to integration is the one adopted in almost every elementary book on calculus; there is no doubt that, from a practical point of view, this approach is all that is needed. However, the Riemann integral has certain drawbacks in a theoretical sense and it is subsumed in the more general Lebesgue theory of integration. We give here a short and informal account of the Lebesgue integral [26, 27, 29, 31, 37, 60].

Recall first of all that the Riemann integral of a function $x(t)$ defined on the interval $[0, 1]$ is defined with the aid of upper and lower Riemann sums on a partition of $[0, 1]$ (*see* Fig. 3.24(a)).

The function $x(t)$ lies between the two step functions $s_u^{(n)}(t), s_l^{(n)}(t)$ such that, for all $t \in [0, 1]$,

$$s_u^{(n)}(t) \geqslant x(t) \geqslant s_l^{(n)}(t),$$

where n refers to the fineness of the partition

$$0 = t_0 < t_1 < t_2 \ldots < t_n = 1.$$

If the upper and lower sums

$$S_u = \lim_{n \to \infty} \sum_{i=1}^{n} s_u(t_i) |t_i - t_{i-1}|, \qquad S_l = \lim_{n \to \infty} \sum_{i=1}^{n} s_l(t_i) |t_i - t_{i-1}|$$

exist and tend to the same limit then we say f is (Riemann) integrable and the value of the integral is the limit. This definition clearly leads to difficulty for a function such as

$$x(t) = \begin{cases} 1, & t \text{ rational,} \\ 0, & t \text{ irrational,} \end{cases} \qquad t \in [0, 1].$$

Another area in which difficulty arises is when we have to consider infinite sequences of integrals when we might wish to 'pass the limit under the integral sign'.

Another way of performing the integration is to partition the range of x rather than the domain; that is, the vertical axis rather than the

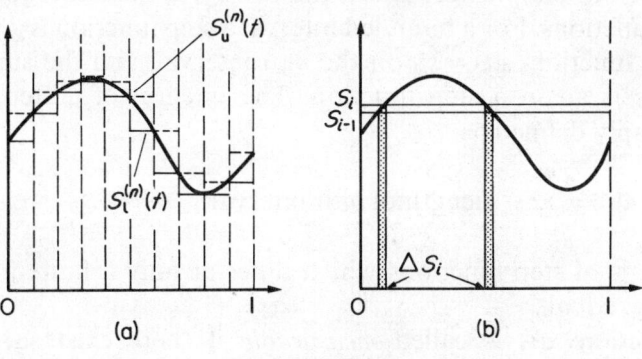

Fig. 3.24

horizontal as in Fig. 3.24(b). The area under the curve can then be seen to be a summation of the form

$$\sum_{i=1}^{n} \eta_i \Delta(s_i),$$

where $s_{i-1} < \eta_i < s_i$ and $\Delta(s_i)$ is the *total length of the intervals* containing points which are the preimages of the values of x lying in the interval $s_{i-1} < x(t) < s_i$. This approach to the integral is in the spirit of the Lebesgue theory and it may be intuitively evident to the reader that the function exhibited above which was not Riemann integrable might be Lebesgue integrable with the value

$1 \times$ (length of intervals in which t is rational)
$+ 0 \times$ (length of intervals in which t is irrational).

Since the rationals are isolated points in a sea of irrationals, the 'length of the rationals' is zero and the value of the integral is zero. The original development of the Lebesgue integral depended heavily on being able to measure the 'total length of intervals' and the resulting structure is called measure theory.

All we shall need in what follows is the primitive idea of a null set (or set of measure zero). A set, N, of real numbers is a *null set* if for every $\varepsilon > 0$ there exists a countable family F_ε of open intervals such that the intervals jointly cover N and their total length does not exceed ε. Every countable set $\{x_n\}$ is null since we may enclose each point x_n within an open interval of length $\varepsilon/2^n, n = 1, 2, 3, \ldots$, and the total of lengths is clearly less than ε: in particular, the rationals are a null set. We say a property holds *almost everywhere* (a.e.) if there is a null set $N \subset S$ such that the property holds on $S - N$.

Instead of following the traditional route based on measure, we shall base our simple account of the theory of the Lebesgue integral on step functions. For a bounded interval a step function is a piecewise constant function, $s(t) = s_m$ on the mth interval, and the sum of step functions is again a step function. The integral of a step function $s(t)$ is simply defined as

$$\int_0^1 s(t) dt = \sum_{j=1}^{m} s_m \text{ (length of } m\text{th interval)}.$$

Integrals of step functions which differ at only a finite number of points are equal.

A function $x(t)$ is called *measurable* if there exists a sequence $\{s^{(n)}\}$ of step functions such that $s^{(n)}(t) \to x(t)$ pointwise almost everywhere in $[0, 1]$. For example, continuous and piecewise continuous functions are certainly measurable. The limit functions of pointwise almost everywhere convergent sequences of measurable functions are measurable. Indeed, one is unlikely to meet a non-measurable function and their construction requires considerable mathematical ingenuity.

A function $x(t)$ is said to be *Lebesgue integrable* over $[0, 1]$ if there exists a sequence $\{s^{(n)}\}$ of step functions such that

$$\lim_{n \to \infty} s^{(n)}(t) = x(t) \quad \text{a.e. on} \quad [0, 1],$$

$$\lim_{m,n \to \infty} \int |s^{(n)}(t) - s^{(m)}(t)| \, dt = 0;$$

in that case the integral of $x(t)$ is

$$\int_0^1 x(t) dt = \lim_{n \to \infty} \int_0^1 s^{(n)}(t) dt.$$

Functions which are integrable are taken to be equal if they differ only on a null set.

The step functions form a vector space and the completion of this space gives the space $\mathscr{L}_1[0, 1]$ of Lebesgue integrable functions on $[0, 1]$ with norm

$$\|x\|_1 = \int_0^1 |x(t)| \, dt.$$

The members of $\mathscr{L}_1[0, 1]$ can be characterized by the statement that every integrable function on $[0, 1]$ is the limit (in the mean) of a sequence of step functions. Step functions can be approximated in the mean by continuously differentiable functions and even by infinitely differentiable functions, so that each of these sets is dense in $\mathscr{L}_1[0, 1]$ (p. 159).

Lastly, we will make a remark concerning the concept of measure, to which we have already alluded. Whereas Lebesgue originally based his integral on measure, the path followed here allows measure to be defined in terms of integral. Let S be a subset of the real line: the *characteristic function* of S is the function $\mathscr{X}(t)$ which has the value 1 on S and 0 elsewhere. A set is said to be *measurable* if its characteristic function is measurable and the measure $m(S)$ is defined by

$$m(S) = \int_{-\infty}^{\infty} \mathscr{X}(t) dt.$$

The measure of an interval is its length and the measure of the union of disjoint intervals is the sum of the lengths. A set has *measure zero* if, and only if, it is a null set; for example, if S is the set of rational numbers in $[0, 1]$ then

$$\mathscr{X}(t) = \begin{cases} 1, & t \text{ rational,} \\ 0, & \text{otherwise,} \end{cases}$$

and

$$m(S) = \int_{-\infty}^{\infty} \mathscr{X}(t)dt = 0.$$

4 Calculus of operators and operator equations

4.1 Introduction

The preceding two chapters give an elementary account of the most basic ideas of functional analysis. There are very many ways in which these basic concepts can be developed. In this chapter we follow a path which is essentially directed towards the approximate solution of operator equations. An operator equation is usually the product of a more or less idealized mathematical model of a real, physical problem and it is clearly of great importance that we should have systematic methods for approximating a solution. Of course, before attempting to find solutions it is prudent to ascertain whether solutions exist and how numerous they are. Functional analysis has made, and is making possible notable advances in the task of proving existence and uniqueness of solutions of operator equations. However, any description of this aspect of functional analysis would be extensive and more often than not difficult, and this is not a question we shall pursue in any depth.

In a sense, since much of this chapter is devoted to approximation, the material is a part of numerical analysis. Here again, functional analysis is a very natural tool and many treatises on numerical analysis are, from the outset, couched in functional analytic terms. The types of approximation we shall discuss, while very important, form only a small part of a huge subject.

There has been a heavy emphasis on linear operators in our treatment so far, and this gives the impression perhaps that the methods of functional analysis are of little use where non-linear operators are involved. This is certainly not the case. Linear analysis has severe limitations, but it must be conceded that it is surprisingly successful in analysing physical problems. Such success is in no small measure due to the fact that it is very often possible to approximate a non-linear operator *locally* by a linear operator.

For an ordinary function of one variable the relevant result is Taylor's theorem which requires, as we know, the existence of at least some of the derivatives of the function at a point. Can we do the same for operators on normed vector spaces? The answer is that, in large part, we can as soon as we develop a suitable concept of differentiation. In other words, we need to develop a calculus of operators.

There are often two distinct ways in which the physical laws describing a particular field model may be expressed mathematically. One is to apply the laws locally and so derive pointwise partial differential equations; the other is to apply equivalent forms of the laws in a global sense over finite control volumes of the field. In steady, incompressible fluid flow, for example, the conservation of mass can be expressed locally as the vanishing of the divergence of the velocity or globally as zero net mass flux across the boundary of some chosen control volume.

There is, of course, a close connection between global and local mathematical statements of a given physical problem and, historically speaking, the *variational calculus* is the source of the early theory. The variational calculus provides a correspondence between finding the minimum (or maximum) value of a real functional defined on a vector space and an associated differential equation. While the two formulations are theoretically equivalent, their impact on the question of finding approximate solutions is quite different. As we shall see, the global rather than the local formulation lends itself more naturally to approximation methods in which the error is measured in an integral or mean sense rather than pointwise. In addition, for a certain class of operator equations, we are able to obtain upper and lower bounds on an integral property of the solution. It must be obvious that this is an extremely valuable feature of an approximation method.

For those operator equations which do not correspond to an extremum principle in the sense of the variational calculus it is still possible to obtain a global formulation by the use of so-called *projection methods*. Here the operator equation is satisfied only in a projectional or weak sense; again, the measure of error is expressed in an integral or mean sense.

The body of known results on the conditions for the convergence of projection methods is much more sparse than the results available for the classical extremum principles associated with the variational calculus: the difficulties are particularly severe for non-linear operator equations. We give in this chapter some of the simpler results concerning convergence and the linked question of bounds on error. However, for the analyst faced with a particular problem thrown up from a physical situation, the primary need is to apply approximation methods to obtain solutions. He often has neither the technical skill nor the interest to verify at every stage the legitimacy of his techniques. Many convergence theorems require the satisfaction of conditions which are hideously impractical. It is probably true to say that many analysts have applied, and will continue to apply, the type of approximation methods described in this chapter very successfully without being too concerned about finer points. The development of the finite element method, an automated technique for the computer analysis of a whole range of field problems, was almost entirely in the hands of engineers until fairly recently; now there is intense activity toward providing mathematical justification, in retrospect, for the manifold applications of the method.

Throughout this chapter we shall deal almost entirely with operators on real spaces; the extension to complex spaces, where necessary, is straightforward.

4.2 Calculus of operators

We begin by reviewing the basic features of differentiation of ordinary functions of one and several real variables.

For the function $f(x)$ of one variable, the derivative $f'(\tilde{x})$ at $\tilde{x}$ is defined to be

$$f'(\tilde{x}) = \lim_{\lambda \to 0} \frac{f(\tilde{x} + \lambda) - f(\tilde{x})}{\lambda}$$

when this limit exists. If the derivative exists for all x in some open interval of $\mathscr{R}$ (or union of open intervals) then we have defined a new function of x, $f'(x)$, the derived function. For the function $f(x)$, $x = (\xi_1, \xi_2, \ldots, \xi_n)$, we define n partial derivatives by, in turn, treating every variable to be fixed bar one, so reducing the problem to n applications of the definition for one variable. The partial derivative of $f(x)$ at $\tilde{x} = (\tilde{\xi}_1, \tilde{\xi}_2, \ldots, \tilde{\xi}_n)$ with respect to ξ_j is defined to be

$$D_j f(\tilde{x}) = \lim_{\lambda \to 0} \frac{f(\tilde{\xi}_1, \tilde{\xi}_2, \ldots, \tilde{\xi}_j + \lambda, \ldots, \tilde{\xi}_n) - f(\tilde{\xi}_1, \tilde{\xi}_2, \ldots, \tilde{\xi}_j, \ldots, \tilde{\xi}_n)}{\lambda}$$

when this limit exists. Again, if the partial derivatives exist for all x in some open region of $\mathscr{R}_n$ then we have defined n new derived functions, $D_j f(x)$. Alternative notations for $D_j f(x)$, which we shall sometimes use, are

$$\frac{\partial f}{\partial \xi_j} \quad \text{and} \quad f_{\xi_j}.$$

For the case of a single real variable the existence of the derivative $f'(\tilde{x})$ implies continuity of the function $f(x)$ at $\tilde{x}$. For several variables, on the other hand, the existence of all the partial derivatives at $\tilde{x}$ does not imply continuity of $f(x)$ at $\tilde{x}$. For example, let

$$f(\xi_1, \xi_2) = \begin{cases} \xi_1 + \xi_2, & \xi_1 = 0 \text{ or } \xi_2 = 0, \\ 1, & \text{otherwise:} \end{cases}$$

now

$$D_1 f(0,0) = \lim_{\lambda \to 0} \frac{f(\lambda, 0) - f(0,0)}{\lambda} = 1$$

and similarly $D_2 f(0,0) = 1$, but it is clear that the function is not continuous at $(0, 0)$. It may be objected that we have chosen our coordinate directions badly and that the existence, or otherwise, of partial derivatives in other directions would give an indication of continuity. In fact, this is not so and the existence of the partial derivatives of a function in every direction at a point still does not imply continuity of the function.

In order to overcome this difficulty, classical analysis introduces the differential (or differential function) at a point. Let us begin with the case of a single variable: for every point $\tilde{x}$ at which f' exists and for every $s \in \mathscr{R}$ we define the function

$$df(\tilde{x}; s) = f'(\tilde{x})s$$

called the differential of f (at $\tilde{x}$): notice that this function is linear in the variable s. The graph of the differential can be given a geometric interpretation as the tangent line to $f(x)$ at $\tilde{x}$ referred to a local axis system with origin at $(\tilde{x}, f(\tilde{x}))$ (Fig. 4.1). Now by the definition of $f'(\tilde{x})$ we have

$$\left| \frac{f(\tilde{x} + s) - f(\tilde{x})}{s} - f'(\tilde{x}) \right| < \varepsilon$$

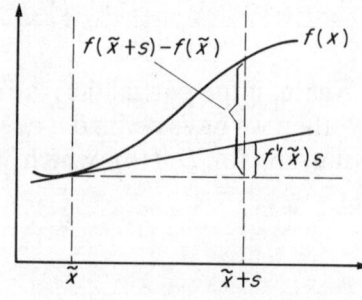

Fig. 4.1

for $|s|$ sufficiently small; hence it follows that, for every $\varepsilon > 0$, there exists an interval I containing $\tilde{x}$ such that, for $\tilde{x} + s \in I$,

$$|(f(\tilde{x}+s) - f(\tilde{x})) - \mathrm{d}f(\tilde{x};s)| < \varepsilon |s|.$$

We can characterize this inequality by saying that $\mathrm{d}f(\tilde{x};s)$ is *a uniform linear approximation to f in the neighbourhood of $\tilde{x}$*.

The extension of the differential to n real variables yields the definition

$$\mathrm{d}f(\tilde{x};s) = \sum_{j=1}^{n} D_j f(\tilde{x})\sigma_j, \qquad s = (\sigma_1, \sigma_2, \ldots, \sigma_n),^{\dagger}$$

and we insist on the satisfaction of

$$|(f(\tilde{x}+s) - f(\tilde{x})) - \mathrm{d}f(\tilde{x};s))| < \varepsilon \|s\|_2$$

which characterizes $\mathrm{d}f(\tilde{x};s)$ as a uniform linear approximation to f in the neighbourhood of $\tilde{x}$.

We are now able to make the required connection between differentiability and continuity (for several variables) if we define differentiability of $f(x)$ at $\tilde{x}$ to mean that $\mathrm{d}f(\tilde{x};s)$ exists; that is to say, $f(\tilde{x})$ is continuous at $\tilde{x}$ if f is differentiable at $\tilde{x}$. For the previous example,

$$f(\xi_1, \xi_2) = \begin{cases} \xi_1 + \xi_2, & \xi_1 = 0 \text{ or } \xi_2 = 0, \\ 1, & \text{otherwise,} \end{cases}$$

it is clear that there is no linear approximation at $(0,0)$ in the sense of the differential so that, for this function, $\mathrm{d}f$ simply does not exist.

The question of the existence of the differential of f at a point $\tilde{x}$ is not settled by the existence of the partial derivatives, as we have seen: however, if the partial derivatives are *continuous* at $\tilde{x}$ then this serves as a sufficient condition for the differentiability of f and this test is the one we would tend to use in practice [9].

The Gateaux and Fréchet derivatives

So much for classical analysis. The discussion has pointed out the shortcomings of the partial derivative and suggested a solution in the shape of the differential. From our point of view, faced with the need

---

†It is common to write the differential in the 'mnemonic' form

$$\mathrm{d}f = \frac{\partial f}{\partial \xi_1}\mathrm{d}\xi_1 + \frac{\partial f}{\partial \xi_2}\mathrm{d}\xi_2 + \ldots + \frac{\partial f}{\partial \xi_n}\mathrm{d}\xi_n,$$

but this notation is confusing and inappropriate, as the sequel should show.

to extend the idea of differentiation to operators, the form of the differential strongly suggests the direction our path should take. Let us view the differential

$$df(\tilde{x};s) = \sum_{j=1}^{n} D_j f(\tilde{x})\sigma_j$$

as the value of a linear functional defined on $\mathscr{R}_n$ with basis $\{e^i\}$. Every linear functional on $\mathscr{R}_n$ is represented, as we know (p. 158) by a vector in $\mathscr{R}_n$, so that this functional is represented by the vector

$$(D_1 f(\tilde{x}), D_2 f(\tilde{x}), \ldots, D_n f(\tilde{x})),$$

which we would recognize as the gradient vector of f at $\tilde{x}$, $\nabla f(\tilde{x})$. The above specifically refers to Cartesian axes, but we can clearly extend the development to cover any basis in $\mathscr{R}_n$; of course, in that case the gradient vector will be referred to the corresponding dual basis in $\mathscr{R}_n$.[†] The value of the linear functional is, as we would expect, invariant to change of basis.

Now if we consider $f(x)$ to be a mapping $f : \mathscr{R}_n \to \mathscr{R}$, then through that mapping we have generated a linear mapping which, being a functional on $\mathscr{R}_n$, is also $\mathscr{R}_n \to \mathscr{R}$: let us call this linear mapping $df(x)$ and represent its value on the vector s by $df(x;s)$.

But how is all this related to the customary limit definition of the derivative? Let us go back to one variable, wherein $f : \mathscr{R} \to \mathscr{R}$. In this space vectors and scalars are formally indistinguishable, but within our usual convention of using italic letters for vectors and Greek letters for scalars we can make a distinction. Using the limit definition for $f'(\tilde{x})$ (p. 283) we have

$$df(\tilde{x};s) = \lim_{\lambda \to 0} \frac{f(\tilde{x}+\lambda) - f(\tilde{x})}{\lambda} s$$

and replacing λ by λs gives

$$df(\tilde{x};s) = \lim_{\lambda \to 0} \frac{f(\tilde{x}+\lambda s) - f(\tilde{x})}{\lambda s} s$$

$$= \lim_{\lambda \to 0} \frac{f(\tilde{x}+\lambda s) - f(\tilde{x})}{\lambda}.$$

[†] The gradient vector refers specifically to the point $\tilde{x}$ and if we consider s referred to a 'local' axis system then we are free to use dual basis sets defined locally at every point in the domain of f. This is the situation we meet in tensor analysis when $\mathscr{R}_n$ is the physical space $\mathscr{R}_3$ and leads to the concepts of covariant and contravariant derivatives [52].

Now we are adding the 'vector' λs to the 'vector' $\tilde{x}$ so a limit of this form will make sense not only for $f : \mathscr{R} \to \mathscr{R}$ but for any mapping from one vector space to another provided the latter space is normed. This brings us to

Definition 4.2.1 *Consider an operator $T : \mathscr{V} \to \mathscr{U}$ where $\mathscr{V}$ is a vector space and $\mathscr{U}$ is a normed vector space, Let $x \in \mathscr{D}(T) \subset \mathscr{V}$ and let $s \in \mathscr{V}$: if the limit*

$$dT(x;s) = \lim_{\lambda \to 0} \frac{T(x + \lambda s) - T(x)}{\lambda}$$

exists, it is called the Gateaux *or* weak differential *of T at x in the direction s. The limit is to be understood in the sense of convergence with respect to the norm in $\mathscr{U}$. The differential may exist for some s and fail to exist for others: if the differential exists at x for all s we say that T is* Gateaux differentiable *at x.*

Note that the definition implies that there is some open subset of $\mathscr{D}(T)$ containing x since it is necessary that $x + \lambda s \in \mathscr{D}(T)$ for some λ sufficiently small.

The Gateaux differential is homogeneous in s in the sense that

$$dT(x; \alpha s) = \alpha dT(x; s)$$

but is not, in general, linear in s. As a simple example, consider the operator $T : \mathscr{R}_2 \to \mathscr{R}$ defined by

$$T(\xi_1, \xi_2) = \begin{cases} \xi_1^2(1 + 1/\xi_2), & \xi_2 \neq 0, \\ 0, & \xi_2 = 0, \end{cases}$$

then

$$dT(\theta; s) = \lim_{\lambda \to 0} \frac{(\lambda \sigma_1)^2 (1 + 1/\lambda \sigma_2)}{\lambda} = \frac{\sigma_1^2}{\sigma_2},$$

which is not linear: nor, it should be noted, is $dT(\theta; s)$ a continuous operator (in s).[†] It is important to note that $dT(x; s) \in \mathscr{U}$.

Example 4.2.1

(a) Let $\mathscr{V} = \mathscr{R}_n$ and consider the functional $f(x) : \mathscr{R}_n \to \mathscr{R}$. If e^j is the jth standard basis vector in $\mathscr{R}_n$ then the Gateaux differential of f at x in

---

[†] The name Gateaux differential is sometimes reserved for the case when $dT(x;s)$ is linear and continuous; otherwise it is called the Gateaux variation. The Gateaux differential generalizes the concept of directional derivative familiar in classical analysis [9].

the direction e^j is

$$df(x;e^j) = \lim_{\lambda \to 0} \frac{f(\xi_1, \xi_2, \ldots, \xi_j + \lambda, \ldots, \xi_n) - f(\xi_1, \xi_2, \ldots, \xi_j, \ldots, \xi_n)}{\lambda}$$

$$= D_j f(x),$$

the jth partial derivative of f at x, provided this derivative exists.

(b) Let $\mathscr{V} = \mathscr{C}[0,1]$ and let $f: \mathscr{C}[0,1] \to \mathscr{R}$ be the functional

$$f(x) = \int_0^1 g(x(t), t)\,dt$$

wherein it is assumed that $g_x(x, t)$, the partial derivative of g with respect to x, exists and is continuous at (x, t). Then

$$df(x;s) = \lim_{\lambda \to 0} \frac{1}{\lambda}\left(\int_0^1 g(x + \lambda s, t)\,dt - \int_0^1 g(x, t)\,dt\right)$$

$$= \lim_{\lambda \to 0} \int_0^1 \frac{g(x + \lambda s, t) - g(x, t)}{\lambda}\,dt.$$

We may interchange the integration and limit operations, giving

$$df(x;s) = \int_0^1 \lim_{\lambda \to 0}\left(\frac{g(x + \lambda s, t) - g(x, t)}{\lambda}\right)dt.$$

Now, for fixed t,

$$\lim_{\lambda \to 0} \frac{g(x + \lambda s, t) - g(x, t)}{\lambda s} = g_x(x, t)$$

so that finally we have

$$df(x;s) = \int_0^1 g_x(x, t)\,s(t)\,dt.$$

In this case $df(x;s)$ is linear in s.

In Exercise 4.2,1(b) you are asked to show that

$$dT(x;s) = \frac{d}{d\lambda} T(x + \lambda s)\Big|_{\lambda=0}:$$

this formula is often more convenient for the calculation of dT than the basic definition. The reader should obtain the results of the above example using this formula.

We have seen that the Gateaux differential at x need be neither linear nor continuous in s. Nor does the existence of the Gateaux differential at $\tilde{x}$ ensure continuity of T at $\tilde{x}$ (Exercise 4.2,1(c)). In fact, the existence of the Gateaux differential is a rather weak requirement, basically because we have not demanded that $\mathscr{V}$ be normed: there being no means of defining continuity of T(cf., Definition 3.2.10) we cannot relate differentiability of T to continuity of T.

So now let us go forward on the basis that $\mathscr{V}$ is also a normed vector space. Suppose $dT(x;s)$ is linear and continuous in s for some $x \in \mathscr{V}$, then we may write

$$dT(x;s) = \lim_{\lambda \to 0} \frac{T(x + \lambda s) - T(x)}{\lambda} = T'^w(x)s.$$

The operator $T'^w(x)$ is, by definition, a mapping $\mathscr{V} \to \mathscr{U}$ and being linear and continuous (and therefore bounded) we may conclude that,

$$T'^w(x) \in \mathscr{L}(\mathscr{V}, \mathscr{U}).$$

This operator is called the *Gateaux* or *weak derivative of T at x*.

When $T'^w(x)$ exists it is certainly true that

$$T(x + \lambda s) - T(x) = T'^w(x)\lambda s + \varepsilon(x, s, \lambda),$$

where $\varepsilon/\lambda \to 0$ as $\lambda \to 0$ with x, s fixed. However, the convergence may not be uniform with respect to s and in that case T cannot be approximated by a linear operator with uniform accuracy in the neighbourhood of x. If we further demand uniform convergence then we arrive at the strong derivative.

Definition 4.2.2 *Let $\mathscr{V}, \mathscr{U}$ be normed vector spaces. An operator $T: \mathscr{V} \to \mathscr{U}$ is Fréchet differentiable at $x \in \mathscr{D}(T) \subset \mathscr{V}$ if there exists a continuous linear operator $T'(x) \in \mathscr{L}(\mathscr{V}, \mathscr{U})$ such that, for all $s \in \mathscr{V}$,*

$$T(x + s) - T(x) = T'(x)s + \varepsilon(x;s)$$

with

$$\lim_{\|s\|_{\mathscr{V}} \to 0} \frac{\|\varepsilon(x;s)\|_{\mathscr{U}}}{\|s\|_{\mathscr{V}}} = 0.$$

The operator $T'(x)$ is called the Fréchet *or* strong derivative of T at x. *The Fréchet derivative at x is unique.*

If we write the definition of the Fréchet derivative in a slightly different form it follows almost at once that existence of the Fréchet derivative of T at x implies continuity of T at x. We have

(Definition 3.2.10)

$$\|T(x+s) - T(x)\|_\mathscr{U} = \|T'(x)s + \varepsilon(x;s)\|_\mathscr{U}$$
$$\leqslant \|T'(x)s\|_\mathscr{U} + \|\varepsilon(x;s)\|_\mathscr{U}$$
$$\leqslant \|s\|_\mathscr{V}(\|T'(x)\|_{\mathscr{V},\mathscr{U}} + \|\varepsilon(x;s)\|_\mathscr{U}/\|s\|_\mathscr{V}),$$

and for $\|s\|_\mathscr{V}$ sufficiently small, the right-hand side of the inequality can be made arbitrarily small.

If the strong (Fréchet) derivative of T exists at x then so does the weak (Gateaux) derivative and they are equal. Suppose $T'(x)$ to exist, then

$$\|T(x+s) - T(x) - T'(x)s\| = \|\varepsilon(x;s)\|$$

and

$$\lim_{\|s\|_\mathscr{V} \to 0} \frac{\|T(x+s) - T(x) - T'(x)s\|_\mathscr{U}}{\|s\|_\mathscr{V}} = 0$$

or, putting $s = \lambda t, t \in \mathscr{V}$,

$$\lim_{\lambda \to 0} \frac{\|T(x+\lambda t) - T(x) - T'(x)\lambda t\|}{\lambda} = 0$$

so that

$$\lim_{\lambda \to 0} \frac{T(x+\lambda t) - T(x)}{\lambda} = T'^w(x)t = T'(x)t.$$

While the existence of $T'^w(x)$ does not imply the existence of $T'(x)$, we do have the following result. It is the analogue of the result we quoted earlier (for classical analysis) that continuity of the partial derivatives is a sufficient condition for differentiability of a function of several variables.

Theorem 4.2.1 *If $T'^w(x)$ exists in a neighbourhood of x and is continuous (with respect to the norm in $\mathscr{L}(\mathscr{V}, \mathscr{U})$) at x, then $T'(x)$ exists.*

Proof Let $T: \mathscr{V} \to \mathscr{U}$ have a Gateaux derivative at every point of the line segment $L(x, x+s)$.[†]

Let $l \in \mathscr{U}^*$ and let

$$g(\alpha) = l(T(x + \alpha s)), \quad 0 < \alpha < 1,$$

then

$$g'(\alpha) = l(T'^w(x + \alpha s)s)$$

[†] By the line segment $L(x, y)$ we mean the set of vectors,
$$L(x, y) = \{z \mid z = \alpha x + (1-\alpha)y, 0 < \alpha < 1\}.$$

and, by the ordinary mean value theorem, $g(1) - g(0) = g'(\alpha)$, which leads to

$$l(T(x+s) - T(x)) = l(T'^w(x + \alpha s)s).$$

Adding $-l(T'^w(x)s)$ to both sides gives

$$l(T(x+s) - T(x) - T'^w(x)s) = l((T'^w(x + \alpha s) - T'^w(x))s).$$

By the Hahn–Banach theorem there exists $l \in \mathscr{U}^*$ such that $\|l\| = 1$ and $l(y) = \|y\|$, $y \in \mathscr{U}$ (see Corollary 1, Theorem 3.5.2). Choosing this functional gives

$$\|T(x+s) - T(x) - T'^w(x)s\| \leq \|T'^w(x + \alpha s) - T'^w(x)\| \|s\|.$$

Since the operator T'^w is continuous at x (with respect to the operator norm) there exists $\delta > 0$ such that

$$\frac{\|T(x+s) - T(x) - T'^w(x)s\|}{\|s\|} < \varepsilon$$

whenever $\|s\| < \delta$. This completes the proof. ∎

Example 4.2.2

(a) Let $f : \mathscr{R}_n \to \mathscr{R}_m$. Then f is a vector-valued function $(f_1(x), f_2(x), \ldots, f_m(x))$ of $x = (\xi_1, \xi_2, \ldots, \xi_n)$. If $f'^w(x)$ exists then $f'^w(x) \in \mathscr{L}(\mathscr{R}_n, \mathscr{R}_m)$ and is therefore represented relative to a pair of dual basis sets by the $(m \times n)$ matrix $[t_{ij}(x)]$. Hence, for $1 \leq i \leq m$,

$$\lim_{\lambda \to 0} \frac{f_i(x + \lambda s) - f_i(x)}{\lambda} = \sum_{k=1}^{n} t_{ik}(x) \sigma_k.$$

Since this is true for all $s \in \mathscr{R}_n$ we may take $\sigma_j = 1$, $\sigma_k = 0$, $k \neq j$, to give

$$D_j f_i(x) = t_{ij}.$$

Therefore $[t_{ij}]$ is the Jacobian matrix $[D_j f_i(x)]$ evaluated at x.

In the special case $f : \mathscr{R}_n \to \mathscr{R}$ ($m = 1$), $f'^w(x) : \mathscr{R}_n \to \mathscr{R}$ is the gradient of f,

$$\nabla f(x) = (D_1 f(x), D_2 f(x), \ldots, D_n f(x)).$$

If the partial derivatives are continuous at x we may conclude that $f'(x) = f'^w(x)$ and

$$|f(x+s) - f(x) - [\nabla f(x), s]| < \varepsilon \|s\|$$

for sufficiently small $\|s\|$.

(b) Let us show that the differential in Example 4.2.1(b) can be expressed as a Fréchet derivative. We found

$$df(x;s) = \int_0^1 g_x(x(t), t)s(t)dt$$

so that the Gateaux derivative $f'^w(x): \mathscr{C}[0,1] \to \mathscr{R}$ of f at x is

$$f'^w(x)\cdot = \int_0^1 g_x(x,t)\cdot dt.$$

Now

$$|f(x+s) - f(x) - f'^w(x)s| = \left|\int_0^1 \{g(x+s,t) - g(x,t) - g_x(x,t)s(t)\}dt\right|$$

while for fixed t we have, from the mean value theorem,

$$g(x+s, t) - g(x, t) = g_x(\hat{x}, t)s,$$

where $\hat{x}$ lies on the line segment $L(x, x+s)$; hence

$$|f(x+s) - f(x) - f'^w(x)s| = \left|\int_0^1 (g_x(\hat{x}, t) - g_x(x, t))s(t)dt\right|$$

$$\leqslant \int_0^1 |g_x(\hat{x}, t) - g_x(x, t)||s(t)|dt.$$

The assumption of (uniform) continuity of g_x in x and t means that there exists $\delta > 0$ such that for $\|s(t)\| = \max_{t \in [0,1]}|s(t)| < \delta$, $|g_x(x+s, t) - g_x(x, t)| < \varepsilon$. Thus for $\|s\| < \delta$,

$$|f(x+s) - f(x) - f'^w(x)s| < \varepsilon\|s\|,$$

which shows that $f'^w(x) = f'(x)$.

(c) To some extent the existence or otherwise of the Fréchet derivative is dependent on the norms adopted in $\mathscr{V}$ and $\mathscr{U}$. As an example, consider the function $f: \mathscr{V} \to \mathscr{R}$ defined by

$$f(x) = \int_0^1 g(t, x(t), x'(t))dt,$$

where g is continuous and has continuous second partial derivatives. The differential of f at x in the direction s is

$$df(x; s) = \int_0^1 \{g_x(t, x, x')s + g_{x'}(t, x, x')s'\}dt,$$

which is linear in s.

We cannot assert that f has a Gateaux derivative $f'^w(x)$ since we need to show that $df(x; s)$ is continuous in s. We can consider $\mathscr{V}$ to be a subspace of the space $\mathscr{C}[0, 1]$ with $\|s\| = \max_{t \in [0,1]}|s(t)|$ or the space

$\mathscr{C}^{(1)}[0,1]$ with
$$\|s\| = \max_{t\in[0,1]} (|s(t)| + |s'(t)|).$$

If $\mathscr{V} \subset \mathscr{C}[0,1]$ then $df(x;s)$ is not continuous (cf., Example 3.4.1(b)): if $\mathscr{V} \subset \mathscr{C}^{(1)}[0,1]$ then $df(x;s)$ is continuous and, furthermore, with the hypotheses on the function g, the operator $f'^w(x)$ is continuous at x [63]. Hence (Theorem 4.2.1) the function $f: \mathscr{C}^{(1)}[0,1] \to \mathscr{R}$ is Fréchet differentiable for every $x \in \mathscr{C}^{(1)}[0,1]$.

An integration by parts gives the alternative form
$$df(x;s) = \int_0^1 \{g_x(t,x,x') - \frac{d}{dt}(g_{x'}(t,x,x'))\} s(t) dt$$
$$+ [g_{x'}(t,x,x')s(t)]_{t=a}^{t=b}.$$

In ordinary calculus the mean value theorem asserts that $f(x+s) - f(x) = f'(x+\alpha s)s$, $0 < \alpha < 1$, whenever f' exists in $(x, x+s)$; hence
$$|f(x+s) - f(x)| \leq \sup_{0 < \alpha < 1} |f'(x+\alpha s)| |s|.$$

An analogous result holds for operators having a Gateaux derivative on the line segment $L(x, x+s)$ and indeed we almost developed the required result in the course of the proof of Theorem 4.2.1; we find (Exercise 4.2, 3(a)) that
$$\|T(x+s) - T(x)\| \leq \sup_{0 < \alpha < 1} \|T'^w(x+\alpha s)\| \|s\|.$$

We conclude this section by stating one or two manipulative results for derivatives which are familiar in ordinary calculus. It is clear from the definition of the Fréchet derivative that if T_1 and T_2 are both Fréchet differentiable at x then
$$(\alpha T_1 + \beta T_2)'(x) = \alpha T'_1(x) + \beta T'_2(x).$$

For the composition $g \circ f$ of ordinary functions of one variable we have the chain rule
$$(g \circ f)'(x) = g'(f(x))f'(x).$$

The analogous rule for operators is the following: let $\mathscr{W}, \mathscr{V}, \mathscr{U}$ be normed vector spaces, let $S: \mathscr{W} \to \mathscr{V}$ be Fréchet differentiable at $x \in \mathscr{W}$ and let $R: \mathscr{V} \to \mathscr{U}$ be Fréchet differentiable at $y = Sx \in \mathscr{V}$. Then $T = RS$ is Fréchet differentiable at x and
$$T'(x) = R'(Sx)S'(x),^\dagger \qquad T'(x) \in \mathscr{L}(\mathscr{W}, \mathscr{U}).$$

[†] There is a weaker form of the chain rule in which S is only Gateaux differentiable in $\mathscr{W}$ [64].

294 CALCULUS OF OPERATORS AND OPERATOR EQUATIONS

In normed vector spaces the operation corresponding to ordinary integration for functions of one variable is called direct integration.

Definition 4.2.3 *Suppose* $T : \mathscr{V} \to \mathscr{U}$, *where* $\mathscr{V}, \mathscr{U}$ *are normed vector spaces. Given* $x_0, x_1 \in \mathscr{V}$ *if*

$$\int_0^1 T(x_0 + \lambda(x_1 - x_0)) \, d\lambda$$

exists, then it is called the direct (Riemann) integral *of T on the line segment* $L(x_0, x_1)$ *and is written*[†]

$$\int_{x_0}^{x_1} T(x) \, dx.$$

How does this definition relate to the one with which we are familiar? Take $\mathscr{V} = \mathscr{U} = \mathscr{R}$, then if we write

$$t = x_0 + \lambda(x_1 - x_0)$$

we obtain for the ordinary function f

$$\int_{x_0}^{x_1} f(x) \, dx = \frac{1}{x_1 - x_0} \int_{x_0}^{x_1} f(t) \, dt$$

so that this represents a normalized or mean-value integral.

The commonest situation in which we will require integration is when we know that T is continuously Fréchet differentiable; in that case we have the analogue of the *fundamental theorem of the calculus*, namely [63, 64]

$$Tx_1 - Tx_0 = \int_{x_0}^{x_1} T'(x)(x_1 - x_0) \, dx$$

$$= \int_0^1 T'(x_0 + \lambda(x_1 - x_0))(x_1 - x_0) \, d\lambda.$$

Higher derivatives

As for ordinary functions, so with operators we may consider the 'derivative of a derivative'. But before we can do this we need to clear up a point which, in ordinary calculus, is invariably obscured. The (Fréchet) derivative $f'(\tilde{x})$ of the ordinary function $f(x) = x^3$,

---

[†] More generally the integral may be taken along a polygonal path.

say, at the point $\tilde{x}$ we write as $3\tilde{x}^2$ so that the differential of x^3 at $\tilde{x}$ is given by $(3\tilde{x}^2)\sigma$ (p. 284). However, we very often omit reference to the point $\tilde{x}$ and write $f'(x) = 3x^2$, meaning x to be any number in some assigned interval. In this way we are really concealing the fact that we are setting up a correspondence or mapping $x \to f'(x)$ whereby, for a given function $f(x)$, we assign to the point x the linear operator $3x^2$. This mapping we call simply the derivative of f: it is this mapping which we then use to define the 'higher derivatives' of f.

For a given operator $T: \mathscr{V} \to \mathscr{U}$, the Fréchet derivative at x is unique (Definition 4.2.2) and the 'operation' of forming the Fréchet derivative $T'(x)$ at x may be considered to be a mapping from $\mathscr{V}$ to the space of linear operators $\mathscr{L}(\mathscr{V}, \mathscr{U})$. Accordingly, the operator

$$T': \mathscr{V} \to \mathscr{L}(\mathscr{V}, \mathscr{U})$$

which assigns to $x \in \mathscr{V}$ the (linear) operator $T'(x)$ is called the *Fréchet derivative of T*.

Definition 4.2.4 *Let $T: \mathscr{V} \to \mathscr{U}$ be Fréchet differentiable in a neighbourhood of $x \in \mathscr{V}$. If the Fréchet derivative of $T': \mathscr{V} \to \mathscr{L}(\mathscr{V}, \mathscr{U})$ at x exists it is called the* Fréchet (*or* strong) second derivative of T at x *and is written $T''(x)$.*

So much for a formal definition, but what is the nature of the operator $T''(x)$? First, we should note that since $T': \mathscr{V} \to \mathscr{L}(\mathscr{V}, \mathscr{U})$ then $T'': \mathscr{V} \to \mathscr{L}(\mathscr{V}, \mathscr{L}(\mathscr{V}, \mathscr{U}))$. In order to interpret the space $\mathscr{L}(\mathscr{V}, \mathscr{L}(\mathscr{V}, \mathscr{U}))$ let us proceed formally using the limit definition for the derivative; for $T: \mathscr{V} \to \mathscr{U}$, $T'(x)$ is given by

$$T'(x)s_1 = \lim_{\lambda \to 0} \frac{T(x + \lambda s_1) - T(x)}{\lambda}, \qquad x, s_1 \in \mathscr{V}.$$

Hence for $x, s_1, s_2 \in \mathscr{V}$ we should expect

$$T''(x)s_1 s_2 = \lim_{\lambda \to 0} \frac{T'(x + \lambda s_1)s_2 - T'(x)s_2}{\lambda},$$

which suggests that $T''(x)$ is an operator which is linear in s_1 and s_2. But this in turn suggests that we are perhaps dealing here with a mapping $\mathscr{V} \times \mathscr{V} \to \mathscr{U}$, where $\mathscr{V} \times \mathscr{V}$ is the Cartesian product of $\mathscr{V}$ with itself, having the usual induced norm (Exercise 3.3,3). It is easy to see that the collection of mappings $\mathscr{V} \times \mathscr{V} \to \mathscr{U}$ which are linear and continuous in each of their arguments (bilinear) forms a vector space and this space can be appropriately normed: let us denote this space by $\mathscr{L}(\mathscr{V} \times \mathscr{V}, \mathscr{U})$. Then it is true (although we shall not prove it here) that the spaces $\mathscr{L}(\mathscr{V} \times \mathscr{V}, \mathscr{U})$ and $\mathscr{L}(\mathscr{V}, \mathscr{L}(\mathscr{V}, \mathscr{U}))$ are isometrically isomorphic (see, for example, [63]).

Having introduced the second derivative we shall now, without proof, state the following 'quadratic' form of the mean value theorem [64]:

$$\| T(x+s) - T(x) - T'(x)s \| \leq \frac{1}{2} \sup_{0 < \alpha < 1} \| T''(x + \alpha s) \| \, \| s \|^2.$$

By far the commonest application of the foregoing development is to functionals, and we shall develop this in the following section. However, the next example treats a case which is more or less familiar in classical analysis in the context of Taylor's theorem for a vector function of several variables. It extends Example 4.2.2(a).

Example 4.2.3

(a) Let $f: \mathscr{R}_n \to \mathscr{R}_m$: we saw that $f'(x)$ was represented by the matrix $[D_j f_i(x)]$. We know that $f''(x)$ must be represented by a bilinear operator $B: \mathscr{R}_n \times \mathscr{R}_n \to \mathscr{R}_m$, so first let us try to deduce the form of B. Just as the effect of a linear transformation $\mathscr{R}_n \to \mathscr{R}_m$ is characterized by its operation on the basis vectors in $\mathscr{R}_n$, so we should expect to characterize B through the vectors $Be^i e^j \in \mathscr{R}_m$. Let us write

$$Be^i e^j = (b_{ij}^{(1)}, b_{ij}^{(2)}, \ldots, b_{ij}^{(m)}), \quad 1 \leq i, j \leq n,$$

then since B is, by hypothesis, linear in each of its arguments, we have, for any $s, t \in \mathscr{R}_n$,

$$Bst = B \sum_{i=1}^{n} \sigma_i e^i \sum_{j=1}^{n} \tau_j e^j = \sum_{i,j=1}^{n} \sigma_i \tau_j Be^i e^j,$$

whence if $Bst = r$, where $r = (\rho_1, \rho_2, \rho_3, \ldots, \rho_m)$, then

$$\rho_k = \sum_{i,j=1}^{n} b_{ij}^{(k)} \sigma_i \tau_j, \quad 1 \leq k \leq m.$$

Hence

$$Bst = (\lfloor \sigma_i \rfloor [b_{ij}]^{(1)} \{\tau_j\}, \lfloor \sigma_i \rfloor [b_{ij}]^{(2)} \{\tau_j\}, \ldots, \lfloor \sigma_i \rfloor [b_{ij}]^{(m)} \{\tau_j\})$$

and we can say that a bilinear mapping $\mathscr{R}_n \times \mathscr{R}_n \to \mathscr{R}_m$ is represented by an m-tuple whose elements are $(n \times n)$ matrices and whose value is as given above.

The kth element of the m-vector $f''(x)st$ is accordingly given by

$$(f''(x)st)_k = \sum_{i,j=1}^n b_{ij}^{(k)} \sigma_i \tau_j$$

$$= \lim_{\lambda \to 0} \frac{(f'(x+\lambda s)t - f'(x)t)_k}{\lambda}$$

$$= \lim_{\lambda \to 0} \frac{\sum_{j=1}^n D_j f_k(x+\lambda s)\tau_j - \sum_{j=1}^n D_j f_k(x)\tau_j}{\lambda}$$

$$= \sum_{j=1}^n \tau_j \lim_{\lambda \to 0} \frac{D_j f_k(x+\lambda s) - D_j f_k(x)}{\lambda}.$$

Since the above is true for all $s, t \in \mathscr{V}$ we may obtain $b_{ij}^{(k)}$ by taking $t = e^j$, $s = e^i$, so giving

$$b_{ij}^{(k)} = \lim_{\lambda \to 0} \frac{D_j f_k(x + \lambda e^i) - D_j f_k(x)}{\lambda}$$

$$= D_{ji} f_k(x), \quad i,j = 1,\ldots,n, \quad k = 1,\ldots,m.$$

The matrix $[b_{ij}]^{(k)}$ is thus the $(n \times n)$ matrix of second partial derivatives of f_k, the so-called *Hessian of f_k*.

(b) In Example 4.2.2(c) we saw that the functional $f: \mathscr{C}^{(1)}[0,1] \to \mathscr{R}$ defined by

$$f(x) = \int_0^1 g(t, x(t), x'(t)) dt$$

has the Fréchet derivative

$$f'(x)s = \int_0^1 \{g_x(t, x, x')s + g_{x'}(t, x, x')s'\} dt$$

provided g is continuous and possesses continuous second partial derivatives.

From the basic definition of second derivative (p. 295) we have

$$f''(x)s_1 s_2 = \lim_{\lambda \to 0} \frac{f'(x + \lambda s_1)s_2 - f'(x)s_2}{\lambda}$$

$$= \int_0^1 \{g_{xx}(t, x, x')s_1 s_2 + g_{xx'}(t, x, x')s'_1 s_2$$
$$+ g_{x'x}(t, x, x')s_1 s'_2 + g_{x'x'}(t, x, x')s'_1 s'_2\} dt$$

For an ordinary real function $f: \mathscr{R} \to \mathscr{R}$ we may improve on a local linear approximation by including the next term in the familiar

Taylor polynomial

$$f(\xi + \sigma) = f(\xi) + f'(\xi)\sigma + \tfrac{1}{2}f''(\xi)\sigma^2 + o[\sigma^2]^\dagger$$

provided the derivatives exist at ξ. What is the analogue of this result for an operator $T: \mathscr{V} \to \mathscr{U}$ which is twice Fréchet differentiable?

We may deduce what the likely form of the result might be by considering the familiar case of an ordinary function of two real variables $f: \mathscr{R}_2 \to \mathscr{R}$. Here we have

$$\begin{aligned}f(\xi_1 + \sigma_1, \xi_2 + \sigma_2) &= f(\xi_1, \xi_2) + (D_1 f(\xi_1, \xi_2)\sigma_1 + D_2 f(\xi_1, \xi_2)\sigma_2) \\&+ \tfrac{1}{2}(D_{11}f(\xi_1, \xi_2)\sigma_1^2 + 2D_{12}f(\xi_1, \xi_2)\sigma_1\sigma_2 \\&+ D_{22}f(\xi_1, \xi_2)\sigma_2^2) + o[(\sigma_1^2 + \sigma_2^2)].\end{aligned}$$

Now we saw in Example 4.2.2(a) that

$$f'(x)s = [D_j f(x)]\{\sigma_j\}, \qquad j = 1, 2,$$

and in Example 4.2.3 that

$$f''(x)st = \lfloor \sigma_i \rfloor [b_{ij}]\{\tau_j\}, \qquad i, j = 1, 2,$$

where

$$[b_{ij}] = \begin{bmatrix} D_{11}f(x) & D_{21}f(x) \\ D_{12}f(x) & D_{22}f(x) \end{bmatrix}$$

hence, in this example, we could write, with $x, s \in \mathscr{R}_2$,

$$f(x + s) = f(x) + f'(x)s + \frac{1}{2!}f''(x)s^2 + o[\|s\|_2^2],$$

where, by $f''(x)s^2$ we mean $(f''(x)s)s$, that is the restriction of the bilinear functional $f''(x)$ to the vectors $(s, s) \in \mathscr{R}_2 \times \mathscr{R}_2$.

For the general operator $T: \mathscr{V} \to \mathscr{U}$ the corresponding result is formally identical, viz.,

$$T(x + s) = T(x) + T'(x)s + \frac{1}{2!}T''(x)s^2 + o[\|s\|^2]$$

whenever the Fréchet derivatives exist. By means of induction this result can be extended to [64]:

Theorem 4.2.2 Let $T: \mathscr{V} \to \mathscr{U}$ have an nth Fréchet derivative $T^{(n)}(x)$ and Fréchet derivatives of order 1 to $(n-1)$ in an open ball

---

† The notation $o[\ \ ]$ is explained in Appendix 4.A.

$B(x;r) \in \mathscr{V}$; then for all $\|s\| < r$

$$T(x+s) = T(x) + T'(x)s + \frac{1}{2!}T''(x)s^2 + \ldots + \frac{1}{n!}T^{(n)}(x)s^n + o[\|s\|^n],$$

where $T^{(j)}(x)s^j$ denotes the restriction of the multilinear operator $T^{(j)}(x)$ to the vectors $(s, s, \ldots, s) \in \mathscr{V} \times \mathscr{V} \times \ldots \times \mathscr{V}$.

Exercises 4.2

1. (a) Let $f(\xi_1, \xi_2) = \xi_1 \xi_2/(\xi_1^2 + \xi_2^2)$ if $\xi_1, \xi_2 \neq 0$ and let $f(0,0) = 0$. Deduce that $D_1 f(0,0) = D_2 f(0,0) = 0$, but that f is not differentiable at $(0,0)$ (i.e., $df(\theta; s)$ does not exist).
   (b) If $T: \mathscr{V} \to \mathscr{U}$ then $f(\lambda) = T(x + \lambda s)$ is a function of the real variable λ with values in $\mathscr{U}$: deduce that

   $$dT(x; s) = \frac{d}{d\lambda} T(x + \lambda s)\big|_{\lambda=0}.$$

   (c) Let $f(\xi_1, \xi_2) = \xi_1^3/\xi_2$ if $\xi_1, \xi_2 \neq 0$ and let $f(0,0) = 0$: find $df(x; s)$. Show that $df(\theta; s)$ exists and is a continuous linear operator for all $s \in \mathscr{R}_2$: however, note that f is not continuous at $(0,0)$.
   (d) (i) Let $f: \mathscr{V} \to \mathscr{R}$ be a real-valued functional on $\mathscr{V}$ and let $f'^w(x)$ exist. Deduce that $f'^w(x) \in \mathscr{V}^*$, the dual space of $\mathscr{V}$, and hence that we may write

   $$df(x; s) = [f'^w(x), s], \qquad s \in \mathscr{V}.$$

   (ii) If, in (i), $\mathscr{V}$ is a Hilbert space $\mathscr{H}$ deduce that $f'^w(x) \in \mathscr{H}$ and hence that

   $$df(x; s) = \langle f'^w(x), s \rangle, \qquad s \in \mathscr{H}.$$

   In the context of (i) and (ii) $f'^w(x)$ is called *the weak gradient of f at x* and is written $\nabla^w f(x)$ or $\text{grad}^w f(x)$.
   (iii) Let $\mathscr{H}$ be a real Hilbert space. Show that if $f(x) = \langle x, x \rangle$ then $df(x; s) = \langle 2x, s \rangle$; that is to say,

   $$\nabla^w f(x) = 2x \in \mathscr{H}:$$

   hence deduce that $\nabla^w \|x\|_2 = x/\|x\|_2, x \neq \theta$.
   More generally if T is a positive definite, bounded real transformation on $\mathscr{H}$ and $\|x\|_T = \langle Tx, x \rangle^{1/2}$ show that

   $$\nabla^w \|x\|_T = Tx/\|x\|_T.$$

   (Note: results for $\nabla^w \|x\|_p, x \in \mathscr{L}_p$, can be found in [62].)

2. (a) Prove that the Fréchet derivative at a point is unique (Definition 4.2.2). (Hint: show that two different derivatives differ by $o[\|s\|]$.)
   (b) If $f: \mathscr{V} \to \mathscr{R}$ and $f'(x) \in \mathscr{V}^*$ exists then $f'(x)$ is often written $\nabla f(x)$ and called the *(strong) gradient of f at x*.

   (i) Prove that if $f: \mathscr{E}_2 \to \mathscr{R}$ is defined by $f(x) = \xi_1(\xi_2 + \xi_1)$ then
   $$\nabla f(x) = (\xi_2 + 2\xi_1, \xi_1).$$
   (You need to show that
   $$f(x+s) - f(x) - \langle \nabla f(x), s \rangle = o[\|s\|].)$$

   (ii) If $f: \mathscr{L}_2[0, 1] \to \mathscr{R}$ is defined by
   $$f(x) = \int_0^1 x^2(y)\,dt$$
   show that $\nabla f(x) = 2x$.

   (iii) Show that the weak gradients in Exercise 4.2,1(d)(iii) are also strong gradients.

   (iv) Let $T \in \mathscr{L}(\mathscr{V}, \mathscr{U})$; show that, for all $x \in \mathscr{V}$, $T'(x)$ exists and $T'(x) = T$.

3. (a) Prove that if $T: \mathscr{V} \to \mathscr{U}$ has a Gateaux derivative at every point of the line segment $L(x, x+s)$ then
   $$\|T(x+s) - T(x)\| \leq \sup_{0 < \alpha < 1} \|T'^w(x+\alpha s)\|\,\|s\|.$$
   (Hint: use the Hahn–Banach theorem as in the proof of Theorem 4.2.1.)
   (b) Let $S: \mathscr{W} \to \mathscr{V}$ be Fréchet differentiable at $x \in \mathscr{W}$ and let $R: \mathscr{V} \to \mathscr{U}$ be Fréchet differentiable at $y = Sx \in \mathscr{V}$. Let $T = RS$. Show that

   (i) $T(x+s) - T(x) = R(y+r) - R(y)$, $r = S(x+s) - S(x)$;
   (ii) $\|T(x+s) - T(x) - R'(y)r\| = o[\|r\|]$;
   (iii) $\|r - S'(x)s\| = o[\|s\|]$;
   (iv) $\|T(x+s) - T(x) - R'(y)S'(x)s\| = o[\|s\|]$ (use the fact that S is continuous at x).

   Hence deduce the chain rule (p. 293)

4. (a) Show that if $T: \mathscr{D}(T) \subset \mathscr{H} \to \mathscr{H}$ is a symmetric linear transformation with domain dense in $\mathscr{H}$,
   $$\left\langle x_1 - x_0, \int_{x_0}^{x_1} T(x)\,dx \right\rangle = \frac{1}{2}\langle x_1, Tx_1 \rangle - \frac{1}{2}\langle x_0, Tx_0 \rangle$$
   (cf. Exercise 4.2,1(d)(iii)).
   (b) Let $f(t)$ map the interval $[\alpha, \beta]$ to the Banach space $\mathscr{U}$, i.e., $f: \mathscr{R} \to \mathscr{U}$, and let $f'(t)$ (the Frechet derivative) be continuous in

$[\alpha, \beta]$. Show that

$$\int_\alpha^\beta f'(t)\,dt = f(\beta) - f(\alpha).$$

(Hint: let $f^* \in \mathscr{U}^*$ and show that

$$f^*\left(\int_\alpha^\beta f'(t)\,dt\right) = \int_\alpha^\beta \frac{d}{dt} f^*(f(t))\,dt.)$$

(c) Carry out the following steps to give a (partial) proof of the fundamental theorem of the calculus (p. 294). Let $S'^w(x) = T(x)$ on the line segment $L(x_0, x_1)$: show that

$$\int_{x_0}^{x_1} T(x)(x_1 - x_0)\,dx = \int_0^1 dS(x_0 + \lambda(x_1 - x_0); x_1 - x_0)\,d\lambda$$

$$= \int_0^1 \frac{d}{d\lambda} S(x_0 + \mu(x_1 - x_0))_{\mu=\lambda}\,d\lambda$$

$$= S(x_1) - S(x_0).$$

5. Find $f''(x)$ for the functionals given in Exercises 4.2, 2(b)(i) and (ii) and construct their Taylor polynomials to $o[\|s\|^2]$.

4.3 The variational calculus

It is a familiar result that, to locate a maximum or minimum of a (sufficiently smooth) function $f(x)$ of several variables, one equates the first partial derivatives to zero. This is clearly equivalent to asserting that the differential function

$$df(x; s) = \sum_{j=1}^n D_j f(x) \sigma_j$$

vanishes for all $s = (\sigma_1, \sigma_2, \ldots, \sigma_n)$.

The obvious extension of this result to more general functionals $f: \mathscr{V} \to \mathscr{R}$ is embodied in

Theorem 4.3.1 *Let the function $f: \mathscr{V} \to \mathscr{R}$ have a Gateaux differential on the vector space $\mathscr{V}$. A necessary condition for f to have a local minimum (maximum) at $\tilde{x} \in \mathscr{V}$ is that $df(\tilde{x}; s) = 0 \ \forall s \in \mathscr{V}$.*

Proof The point $\tilde{x}$ is a local minimum (maximum) of $f(x)$ if there is an open ball $B(\tilde{x}, r)$ such that $f(\tilde{x}) < f(x)$ ($f(\tilde{x}) > f(x)$) for all $x \in B(\tilde{x}, r), x \neq \tilde{x}$.

Let f have a local minimum (maximum) at $\tilde{x}$ and suppose that $df(\tilde{x};s) > 0$, $s \in \mathscr{V}$, then from Definition 4.2.1 we have

$$\frac{f(\tilde{x} + \lambda s) - f(\tilde{x})}{\lambda} > 0$$

for sufficiently small λ: if $\lambda > 0$, $f(\tilde{x} + \lambda s) > f(\tilde{x})$ while if $\lambda < 0$, $f(\tilde{x} + \lambda s) < f(\tilde{x})$. A similar argument holds if we suppose $df(\tilde{x};s) < 0$. In either case we obtain a contradiction and we conclude that it is necessary that $df(\tilde{x};s) = 0$. ∎

A point $\tilde{x}$ for which $df(\tilde{x};s) = 0$ is called a *stationary point*: it is important to notice that Theorem 4.3.1 only gives a necessary condition for the stationary point $\tilde{x}$ to be a minimum (maximum).[†] Notice also that the present discussion is concerned only with a local theory: the treatment of global minima or maxima requires a somewhat different approach and in its most useful form is based on the properties of convex functionals defined on convex sets (*see* Section 4.4, p. 351 *et seq.*).

Example 4.3.1

(a) For the functional $f: \mathscr{R}_n \to \mathscr{R}$ of Example 4.2.1(a) the condition $df(\tilde{x};s) = 0$ clearly reduces to

$$D_j f(\tilde{x}) = 0, \qquad j = 1, \ldots, n,$$

the familiar result for an ordinary function of n real variables.

(b) For the functional $f: \mathscr{C}[0, 1] \to \mathscr{R}$,

$$f(x) = \int_0^1 g(x(t), t) dt$$

of Example 4.2.1(b) the condition $df(\tilde{x};s) = 0$ leads to

$$\int_0^1 g_x(x, t) s(t) dt = 0.$$

Since $s \in \mathscr{C}[0, 1]$ is arbitrary, we may argue that the integral will vanish if, and only if,

$$g_x(x(t), t) = 0.$$

Part (b) of the above example is typical of the type of problem for which the methods of the variational calculus were developed. The classical problem is that of finding a function $x(t)$ defined on the

---

[†] The single word *extremum* is often used to denote either a maximum or a minimum.

interval $[0, 1]$ which minimizes the functional

$$J(x) = \int_0^1 g(x(t), x'(t), t) dt.^\dagger$$

In this context the function $x(t)$ is called an *extremal* or an *extremal arc*.

The question of finding sufficient conditions for the solution of a variational problem is one of the more difficult aspects of the variational calculus.

We have already met this type of difficulty in our discussion of completeness. In Example 3.6.2(b) we saw that the functional

$$J(u) = \int_0^1 \left(\frac{du}{dt}\right)^2 dt, \quad u(0) = 0, \quad u(1) = \beta,$$

has infinum equal to zero in the space of twice continuously differentiable functions but no 'solution' function exists which gives this value for J. Similarly, suppose we seek the continuous function $y(t)$ which minimizes the area between the curve $y(t), t \in [0, 1]$ with $y(0) = 0$ $y(1) = \beta$ and the t-axis, thus

$$J(y) = \int_0^1 y(t) dt.$$

This is a minimum norm problem in $\mathscr{L}_1$ and we know that the space of functions continuous on $[0, 1]$ is incomplete in $\mathscr{L}_1$; hence

$$\inf_y J = 0,$$

but no continuous function exists which gives this value.

In what follows, we shall be content to find the necessary conditions which a function must satisfy in order to be an extremal.

We return now to the classical problem of the variational calculus and specify that $x \in \mathscr{C}_0^{(1)}[0, 1]$, the space of continuously differentiable functions on $[0, 1]$, with $x(0) = x(1) = 0$ and we shall suppose that $g: \mathscr{R}_3 \to \mathscr{R}$ has continuous second partial derivatives with respect to x and x'. The set of functions x are often referred to as *admissible functions*.

The Gateaux differential of $J: \mathscr{C}_0^{(1)}[0, 1] \to \mathscr{R}$ is (cf., Example

---

† It is customary in the variational calculus to use the symbol J for a functional of this type.

4.2.2(c))

$$dJ(x;s) = \frac{d}{d\lambda}J(x+\lambda s)\bigg|_{\lambda=0}$$

$$= \frac{d}{d\lambda}\int_0^1 g(x+\lambda s, x'+\lambda s', t)dt\bigg|_{\lambda=0}$$

$$= \int_0^1 \{g_x(x, x', t)s(t) + g_{x'}(x, x', t)s'(t)\}\,dt.$$

In the context of the variational calculus the differential $dJ(x;s)$ is usually called *the first variation* and written δJ.[†]

To find a stationary point for J we set $dJ(x:s) = 0$ and integrate by parts to give

$$\int_0^1 \{g_x(x, x', t) - \frac{d}{dt}g_{x'}(x, x', t)\}s(t)dt = 0$$

upon using the boundary values $s(0) = s(1) = 0$.

Assuming that $\frac{d}{dt}g_{x'}$ is continuous, then the expression in braces is continuous in $[0, 1]$ and since $s \in \mathscr{C}_0^{(1)}[0, 1]$ is arbitrary we conclude that the extremal $x(t)$ must satisfy the *Euler–Lagrange equation*[‡]

$$\frac{d}{dt}g_{x'}(x, x', t) - g_x(x, x', t) = 0.$$

It is not necessary to assume continuity of the second partial derivatives of g, and the above equation can be derived without this assumption (Exercise 4.3,1(c)).

The Euler–Lagrange equation is a second-order differential equation for $x(t)$ with the boundary conditions $x(0) = x(1) = 0$.

If we wish to generalize the formulation of the problem so that $x'(t)$ is continuous in $[0, 1]$ but has inhomogeneous boundary values $x(0) = \alpha$, $x(1) = \beta$, we need only note that $x_1(t) - x_2(t) \in \mathscr{C}_0^{(1)}[0, 1]$,

---

[†] As a matter of fact, to discuss the variational calculus after the Gateaux derivative is to reverse the historical sequence, since the abstract concept of the differential had its origin in the first variation (so-called by Lagrange).

[‡] This follows from the so-called *fundamental lemma of the variational calculus*: if $y(t)$ is continuous on $[0, 1]$ and

$$\int_0^1 y(t)s(t)dt = 0 \quad \forall s \in \mathscr{C}^{(1)}[0, 1]$$

with $s(0) = s(1) = 0$, then $y(t) = 0$ on $[0, 1]$ (see Exercise 4.3,1).

$s(t) \in \mathscr{C}_0^{(1)}[0, 1]$ and the analysis leads again to the Euler–Lagrange equation, but now with inhomogeneous boundary conditions (*see* definition of coset, p. 25 *et seq.*).

By an obvious extension, if $x(t) \in \mathscr{R}_n$ the functional J takes the form

$$J(x) = \int_0^1 g(\xi_1(t), \ldots, \xi_n(t), \xi_1'(t), \ldots, \xi_n'(t), t) dt$$

and we obtain the set of Euler–Lagrange equations

$$\frac{d}{dt} g_{\xi_i'}(\xi_i, \xi_i', t) - g_{\xi_i}(\xi_i, \xi_i', t) = 0, \qquad i = 1, \ldots, n.$$

(Exercise 4.3,2(a)).

An extension of the foregoing analysis to functions of several variables is straightforward and one simply obtains one or a set of Euler–Lagrange equations with associated boundary values (Exercise 4.3,2(b)). For example, if

$$J(x) = \int_\Gamma g(x, D_1 x, D_2 x, t_1, t_2) dt_1 dt_2$$

the Euler–Lagrange equation is

$$D_1 g_{D_1 x} + D_2 g_{D_2 x} - g_x = 0.$$

Example 4.3.2

(a) A celebrated problem in the calculus of variations is the following. A mass slides on a frictionless wire connecting points P and Q in a vertical plane: for what shape of wire is the time of descent a minimum? In this example we shall use the customary Cartesian variables x and y.

During the motion energy is conserved, so that taking P as origin and measuring y vertically downwards we have $\frac{1}{2}mv^2 = mgy$, where v is the velocity of the mass. Hence, $v = \sqrt{2gy}$ and we wish to find the curve $y = y(x)$ which minimizes

$$J(y) = \int_P^Q dt = \int_P^Q \frac{ds}{v} = \int_0^\alpha \left[\frac{1 + (y')^2}{2gy} \right]^{1/2} dx,$$

with $y(0) = 0$ and $y(\alpha) = \beta$, say. The integrand $g(y, y')$ does not depend explicitly on x and the Euler–Lagrange equation can be integrated by noting that

$$\frac{d}{dx}[y' g_{y'} - g] = y'' g_{y'} + y'(g_{y'})' - g_y y' - g_{y'} y''$$
$$= y'((g_{y'})' - g_y)$$
$$= 0$$

since y satisfies the Euler–Lagrange equation; hence

$$y'g_{y'} - g = \text{const.}$$

In the specific case under consideration, this leads to

$$\frac{-1}{\sqrt{y(1+(y')^2)}} = \text{const.} = \frac{1}{\gamma},$$

say, or

$$(y')^2 = (\gamma^2 - y)/y$$

whence

$$dx = \sqrt{y/(\gamma^2 - y)}\,dy.$$

Using the substitution

$$y = \gamma^2 \sin^2 \eta = \tfrac{1}{2}\gamma^2(1 - \cos 2\eta)$$

the equation integrates to

$$x = \tfrac{1}{2}\gamma^2(2\eta - \sin 2\eta),$$

where γ is determined by the condition that $y(\alpha) = \beta$. These are the parametric equations of a cycloid.

(b) Let Γ be an open region in $\mathscr{R}_2$ and $\partial \Gamma$ its smooth boundary. Let $u(t_1, t_2)$ have continuous second partial derivatives in Γ and let $u = f$ on $\partial \Gamma$. Consider the functional

$$J(u) = \int_\Gamma (D_1 u)^2 + (D_2 u)^2 \, d\Gamma.$$

The Euler–Lagrange equation is (p. 305)

$$D_1(2D_1 u) + D_2(2D_2 u) = 0$$

or

$$\nabla^2 u = 0 \quad \text{in } \Gamma.$$

We saw in Example 3.6.2(b) that we could characterize the solution of $\nabla^2 u = 0$ in Γ with $u = f$ on $\partial \Gamma$ as a minimum norm problem in a suitable Hilbert space. We shall see in Section 4.4 that the above formulation and that of Example 3.6.2(b) are essentially equivalent.

(c) (Hamilton's principle and Lagrange's equations)[†] We consider a conservative mechanical system whose state can be described by n generalized coordinates q_i, $i = 1, \ldots, n$. That is to say, the position of each part of the system is fully specified in terms of the q_i, and furthermore, any kinematic constraints imposed on the system are automatically satisfied. The motion of such a system can be characterized by

[†] In this example we revert to the notation which is customarily used in this context.

Hamilton's principle. Assume that the kinetic energy of the system T is a function of the coordinates q_i, their rates of change $\dot{q}_i$ and t: in fact, T will be quadratic in the velocities $\dot{q}_i$ so that

$$T = \frac{1}{2} \sum_{i,j=1}^{n} t_{ij}(q_1, q_2, \ldots, q_n, t) \dot{q}_i \dot{q}_j, \qquad t_{ij} = t_{ji}.$$

The potential energy, U, is assumed to be a function only of the q_i and t, viz.,

$$U = U(q_1, q_2, \ldots, q_n, t).$$

For a conservative system these two functionals contain all the required information on the system and Newton's second law is conveniently embodied in Hamilton's principle: 'the motion of a conservative system between times t_1 and t_2 proceeds in such a way that the functional

$$J = \int_{t_1}^{t_2} (T - U) \, dt$$

is stationary'. We thus obtain the set of Euler–Lagrange equations

$$\frac{d}{dt} \frac{\partial T}{\partial \dot{q}_k} - \frac{\partial}{\partial q_k}(T - U) = 0, \qquad k = 1, \ldots, n,$$

which are usually simply referred to as *Lagrange's equations*. In the commonest instances neither T nor U depends explicitly on t and we have,

$$\sum_{j=1}^{n} \left\{ t_{kj} \ddot{q}_j - \sum_{i=1}^{n} \frac{1}{2} \frac{\partial t_{ij}}{\partial q_k} \dot{q}_i \dot{q}_j \right\} + \frac{\partial U}{\partial q_k} = 0.$$

For small motions about an equilibrium state which we may take to be the point $(0, 0, \ldots, 0)$ we obtain the set of constant coefficient, second-order differential equations

$$\sum_{j=1}^{n} (t_{kj}(\theta) \ddot{q}_j + k_{kj}(\theta) q_j) = 0,$$

where, for small q_i,

$$U = \frac{1}{2} \sum_{i,j=1}^{n} k_{ij}(\theta) q_i q_j.$$

For complicated mechanical systems it is almost always easier to deduce the equations of motion via Lagrange's equations rather than by a direct application of Newton's second law to each part of the system. The reason for this is that, through the variational formulation, all the internal reactions in the system and any external constraints which do no work do not need to be explicitly included in the formulation.

Treatments of dynamics based on Lagrange's equations are legion and we shall not pursue the matter any further here [49, 57, 66].

(d) (Canonical Euler equations) The Euler–Lagrange equations for the functional

$$J(x) = \int_0^1 g(\xi_1, \ldots, \xi_n, \xi_1', \ldots, \xi_n', t) \, dt$$

are the second-order equations

$$\frac{d}{dt} g_{\xi_i'} - g_{\xi_i} = 0, \qquad i = 1, \ldots, n.$$

We may formulate these equations in an alternative way as a set of $2n$ first-order equations by replacing the vector $x' = (\xi_1', \ldots, \xi_n')$ by the vector $p = (\pi_1, \ldots, \pi_n)$ through the transformation

$$\pi_i = g_{\xi_i'}, \qquad i = 1, \ldots, n.$$

We assume that the transformation can be inverted so that the ξ_i' can be expressed in terms of the ξ_i, π_i and t. This device in itself will replace the set of n second-order equations by the $2n$ first-order equations

$$\frac{d\pi_i}{dt} = g_{\xi_i}, \qquad \pi_i = g_{\xi_i'}, \qquad i = 1, \ldots, n.$$

However, we can arrive at a more satisfactory set of first-order equations exhibiting a symmetry which gives equal weight to the variables x and p by defining a new function $h(x, p, t)$ related to the function $g(x, x', t)$ by (Exercise 4.3,4(c))

$$h(\xi_1, \ldots, \xi_n, \pi_1, \ldots, \pi_n, t) = \sum_{i=1}^n \pi_i \xi_i' - g(\xi_1, \ldots, \xi_n, \xi_1', \ldots, \xi_n', t).$$

We have (cf., footnote on p. 285)

$$dh = -\frac{\partial g}{\partial t} dt + \sum_{i=1}^n \left(\xi_i' d\pi_i - \frac{\partial g}{\partial \xi_i} d\xi_i \right)$$

upon using $\pi_i = g_{\xi_i'}$; hence $h_{\pi_i} = \xi_i'$ and $h_{\xi_i} = -g_{\xi_i}$. In the second equation we write $g_{\xi_i} = \pi_i'$ and thus arrive at the *canonical Euler equations*

$$\frac{d\pi_i}{dt} = -h_{\xi_i}, \qquad \frac{d\xi_i}{dt} = h_{\pi_i}, \qquad i = 1, \ldots, n,$$

the solution of which renders stationary the functional

$$I(x, p) = \int_0^1 \left(\sum_{i=1}^n \pi_i \xi_i' - h(\xi_1, \ldots, \xi_n, \pi_1, \ldots, \pi_n, t) \right) dt,$$

wherein ξ_i' is to be regarded as a function of the ξ_i, π_i and t. The function $h(x, p, t)$ is called the *Hamiltonian*.

Let us illustrate the foregoing by reference to the equations of classical dynamics discussed in (c) of this example. The variable π_i is conventionally denoted by the symbol p_i and represents the generalized momentum of the system in the ith degree of freedom, viz.,

$$p_i = \sum_{j=1}^{n} t_{ij}\dot{q}_j.$$

(If the system consists only of interconnected massive particles, the ith particle having mass m_i, then $p_i = m_i \dot{q}_i$.)

The Hamiltonian is given by

$$h(q_1, \ldots, q_n, p_1, \ldots, p_n, t) = \sum_{i=1}^{n} \dot{q}_i \sum_{j=1}^{n} t_{ij}\dot{q}_j - (T - U) = T + U;$$

that is to say, it represents the sum of the kinetic and potential energies of the system.

It is not our intention in this book to deal at length with the classical variational calculus. For one thing, there are several excellent texts on the topic,[†] and for another, we wish to give an account of so-called direct methods for dealing with the minimization of functionals; these methods, which are more directly connected with functional analysis, will be taken up in later sections of this chapter. However, to round off this section, we introduce two further types of problem which are of interest in the variational calculus.

In the type of problem already discussed, the end points of the extremal arc have been specified. However, another class of problem that can be handled by the variational calculus is that in which the end points themselves are variable. For simplicity, suppose we wish to find an extremal arc for the functional

$$J(x) = \int_0^{t_r(x)} g(x, x', t) dt$$

where one end of the extremal arc is not fixed, but instead is constrained to lie on a given curve $r(x, t) = 0$. The Gateaux differential of J is again given by

$$dJ(x; s) = \frac{d}{d\lambda} J(x + \lambda s) \bigg|_{\lambda = 0}$$

$$= \frac{dt_r}{d\lambda}\bigg|_{\lambda=0} g(x(t_r), x'(t_r), t_r)$$

$$+ \int_0^{t_r} \{g_x(x, x', t) s(t) + g_{x'}(x, x', t) s'(t)\} dt,$$

[†] See, for example, [49, 57, 67–71].

and the customary integration by parts gives the Euler–Lagrange equation

$$\frac{d}{dt}g_{x'}(x, x', t) - g_x(x, x', t) = 0$$

together with the 'end point' expression:

$$\left.\frac{dt_r}{d\lambda}\right|_{\lambda=0} g(x(t_r), x'(t_r), t_r) + g_{x'}(x(t_r), x'(t_r), t_r)s(t_r).$$

We now find an expression for the derivative

$$\left.\frac{dt_r}{d\lambda}\right|_{\lambda=0}$$

involving the implicit equation of the termination curve, $r(x, t) = 0$. The function $t_r(\lambda)$ is, by definition, a solution of the equation

$$r((x(t_r) + \lambda s(t_r)), t_r) = 0$$

and, consequently,

$$dr = r_x(x'(t_r)dt_r + \lambda s'(t_r)dt_r + s(t_r)d\lambda) + r_t dt_r = 0,$$

giving

$$\left.\frac{dt_r}{d\lambda}\right|_{\lambda=0} = \frac{-r_x s(t_r)}{r_x x'(t_r) + r_t}.$$

The above 'end point' expression then becomes

$$\left(-\frac{r_x}{r_x x'(t_r) + r_t}g(x(t_r), x'(t_r), t_r) + g_{x'}(x(t_r), x'(t_r), t_r)\right)s(t_r)$$

and since $dJ(x; s)$ is required to vanish for arbitrary s, the expression in brackets must vanish, giving the so-called *transversality condition*:

$$[(g(x, x', t) - x'g_{x'}(x, x', t))r_x - g_{x'}(x, x', t)r_t]_{t=t_r} = 0.$$

Example 4.3.3 Consider the simple problem of finding the smooth curve of minimum arc length joining the origin to the given terminal curve $r(x, t) = 0$.
We have

$$J = \int_0^{t_r} \sqrt{1 + (x')^2}\, dt$$

with the Euler–Lagrange equation

$$\frac{d}{dt}\left(\frac{\partial}{\partial x'}\sqrt{1 + (x')^2}\right) = 0$$

and the transversality condition

$$\left(\sqrt{1+(x')^2} - \frac{x'^2}{\sqrt{1+(x')^2}}\right)r_x - \left(\frac{x'}{\sqrt{1+(x')^2}}\right)r_t\bigg|_{t=t_r} = 0.$$

The extremal arc is a straight line ($x' = $ const.) passing through the origin and meeting the terminal curve at right-angles. This latter conclusion follows from the transversality condition

$$x'(t_r) = r_x/r_t.$$

Lagrange multiplier rule

The last problem we discuss in this section is that of determining the stationary points of a functional $f(x)$ when the vector x is subject to a set of equality constraints.[†]

Let f and $c_j, j = 1, \ldots, m$, be real functionals defined on a (real) normed vector space $\mathscr{V}$. The functionals c_j constrain the extremal vector to lie in a set $\Gamma = \{x \in \mathscr{V} \mid c_j(x) = 0\} \subset \mathscr{V}$.

Let $\tilde{x} \in \Gamma$ be a point at which each of the weak derivatives $c_j'^w(x) \in \mathscr{L}(\mathscr{V}, \mathscr{R}) = \mathscr{V}^*$ exists (p. 289). The subspace of all $s \in \mathscr{V}$ such that

$$c_j'^w(\tilde{x})s = [c_j'^w(\tilde{x}), s] = 0, \qquad j = 1, \ldots, m,$$

is called the *tangent subspace* of Γ at $\tilde{x}$, and is denoted by Γ_t. The coset $\tilde{x} + \Gamma_t$ is called the *tangent manifold* (cf. p. 54); Γ_t is the intersection of m hyperplanes at $\tilde{x}$.

In the simple case $\mathscr{V} = \mathscr{R}_3, m = 1$ the constraint $c_1(x) = 0$ describes a surface in $\mathscr{R}_3$ and the tangent subspace is simply the tangent plane (i.e., the differential function) at any point of the surface which is smooth (i.e., differentiable). If $m = 2$, Γ is a curve in $\mathscr{R}_3$ and Γ_t is the tangent to the curve at $\tilde{x}$: looked at in another way, Γ is the intersection of the two surfaces $c_1(x) = c_2(x) = 0$ and Γ_t the intersection of their tangent planes at $\tilde{x}$.

Suppose $\tilde{x} \in \Gamma$ is a stationary point of f. This means that f is, to first order, constant for changes in $\tilde{x}$ along Γ. If Γ is smooth at $\tilde{x}$ then it would seem plausible that f is to first order also constant for changes in $\tilde{x}$ along Γ_t.

We shall not prove this result here, but merely outline the steps. We have first to assume that $\tilde{x}$ is a *regular point of* Γ, which is to say that the functionals $c_j'^w(\tilde{x})$ are linearly independent. The independence of the $c_j'^w(\tilde{x})$ guarantees the existence of a solution to the set of implicit

---

[†] The finding of stationary points of functionals subject to *inequality* constraints is a topic subsumed under the heading of linear or non-linear programming.

equations $c_j(x) = 0$ in the neighbourhood of $\tilde{x}$ (the *implicit function theorem* [9, 18]).

In the neighbourhood of $\tilde{x}$ one can show that the difference vector between the point $\tilde{x} + \varepsilon s$ on Γ_t and the point $\tilde{x} + \varepsilon s + d(\varepsilon)$ on Γ is at least second order in ε so that

$$df(\tilde{x}; s) = 0, \quad s \in \Gamma_t,$$

can replace the original condition [23]

$$df(\tilde{x}; s) = 0, \quad s \in \Gamma.$$

If we write the condition $df(\tilde{x}; s) = 0, s \in \Gamma_t$, as

$$[f'^w(\tilde{x}), s] = 0, \quad s \in \Gamma_t,$$

then we see that $f'^w(\tilde{x})$ is a vector in $\Gamma_t^0 \in \mathscr{V}^*$, the annihilator of $\Gamma_t \in \mathscr{V}$ (p. 54). But Γ_t^0 is spanned by the independent vectors $c_j'^w(\tilde{x}), j = 1, \ldots, m$, hence there exist scalars $\tilde{\lambda}^1, \tilde{\lambda}^2, \ldots, \tilde{\lambda}^m$ such that

$$f'^w(\tilde{x}) = - \sum_{j=1}^{m} \tilde{\lambda}^j c_j'^w(\tilde{x}).$$

If we now define a functional

$$L(x) = f(x) + \sum_{j=1}^{m} \lambda^j c_j(x)$$

and set $dL(\tilde{x}; s) = 0$ we obtain

$$f'^w(\tilde{x}) + \sum_{j=1}^{m} \tilde{\lambda}^j c_j'^w(\tilde{x}) = 0.$$

We have thus shown that if $\tilde{x}$ is a stationary point of the functional f subject to the constraints $c_j(x) = 0, j = 1, \ldots, m$, and $\tilde{x}$ is a regular point of the constraints, then there exists $(\tilde{\lambda}^1, \tilde{\lambda}^2, \ldots, \tilde{\lambda}^m) \in \mathscr{R}_m$ such that the (unconstrained) functional $L(x)$ is stationary at $\tilde{x}$. This is the *Lagrange multiplier rule*, the functional $L(x)$ is called the *Lagrangian* and the scalars λ^j *Lagrange multipliers*. We shall meet Lagrange multipliers again in subsequent sections in a somewhat different guise: meanwhile, we illustrate the theory with a few simple examples.

Example 4.3.4

(a) If f is a functional on $\mathscr{R}_n$ subject to $m < n$[†] algebraic constraints $c_j(x) = 0$

[†] Notice that, in this case, we must have $m < n$, whereas in the general case there is no upper limit on the number of constraint functionals. Roughly speaking, this is because the general case is set within the context of an infinite-dimensional rather than a finite-dimensional space.

then we obtain the familiar elementary form of the Lagrange multiplier rule. We form

$$L(\xi_1, \xi_2, \ldots, \xi_n; \lambda^1, \lambda^2, \ldots, \lambda^m) = f(\xi_1, \xi_2, \ldots, \xi_n)$$
$$+ \sum_{j=1}^{m} \lambda^j c_j(\xi_1, \xi_2, \ldots, \xi_n)$$

and then proceed as if dealing with an unconstrained extremum: namely, we set

$$L_{\xi_i} = 0, \quad i = 1, \ldots, n,$$

and

$$L_{\lambda^j} = 0, \quad j = 1, \ldots, m,$$

so obtaining $(m + n)$ equations for the scalars $(\tilde{\xi}_1, \tilde{\xi}_2, \ldots, \tilde{\xi}_n)$ and $(\tilde{\lambda}^1, \tilde{\lambda}^2, \ldots, \tilde{\lambda}^m)$.

(b) As we shall see in later sections, Lagrange multipliers often have physical significance in a specific problem. Let us return to the pin-jointed bar structure treated in Example 2.9.1 (p. 100). For reasons which will be made clear in a later section, we may analyse this problem by seeking a stationary point of the functional

$$U(e_i; d_j) = \frac{1}{2} \sum_{i=1}^{n} k_i e_i^2 + (f^1 d_1 + f^2 d_2)$$

subject to the (kinematic) constraint

$$\{e_i\} = [l_{ij}]\{d_j\}, \quad i = 1, \ldots, n, \quad j = 1, 2.$$

We form

$$L(e_i; d_j) = U(e_i, d_j) + \sum_{i=1}^{n} \lambda^i (e_i - \lfloor l_j \rfloor_i \{d_j\})$$

and set

$$\frac{\partial L}{\partial e_i} = \frac{\partial L}{\partial d_j} = \frac{\partial L}{\partial \lambda_i} = 0.$$

We obtain, respectively,

$$k_i e_i - \lambda^i = 0, \quad i = 1, \ldots, n,$$

$$f^j - \sum_{i=1}^{n} \lambda^i l_{ij} = 0, \quad j = 1, 2,$$

$$\{e_i\} - [l_{ij}]\{d_j\} = \{0\}, \quad i = 1, \ldots, n, \quad j = 1, 2,$$

a set of $2n + 2$ equations for the unknowns e_i, d_j, λ^i.

We may rewrite the second equation in the form

$$\{f^j\} = [l_{ij}]^T \{\lambda^i\}$$

and this, coupled with the equations

$$k_i e_i = \lambda^i,$$

shows that the Lagrange multipliers are the forces in the bars, t^i. As we saw in Example 2.9.1, the complete set of equations solves the framework problem.

(c) The classical problem of finding an extremum subject to constraints is the isoperimetric problem in the variational calculus: this is the class of problems wherein we seek curves of a given perimeter which enclose maximum area. The archetypal problem is to find the curve $x(t)$ with end points $(0, 0)$, $(1, 0)$ which is of fixed length $l > 1$ and encloses the maximum area between itself and the t-axis.

Hence

$$f(x) = \int_0^1 x(t)\,dt, \qquad x(0) = x(1) = 0,$$

with constraint

$$\int_0^1 \sqrt{1 + (x')^2}\,dt = l.$$

We form

$$L(x) = \int_0^1 x(t)\,dt + \lambda\left(\int_0^1 \sqrt{1 + (x')^2}\,dt - l\right).$$

$$= \int_0^1 (x + \lambda\sqrt{1 + (x')^2})\,dt - \lambda l,$$

giving the Euler–Lagrange equation

$$1 - \lambda \frac{d}{dt} \frac{x'}{\sqrt{1 + (x')^2}} = 0$$

or

$$\frac{x'}{\sqrt{1 + (x')^2}} = \frac{1}{\lambda}(t - C_1)$$

upon integrating. Solving for x' and integrating again gives

$$(x - C_2)^2 + (t - C_1)^2 = \lambda^2,$$

which is the equation for the arc of a circle with centre (C_1, C_2) and radius λ. The values of C_1, C_2 and λ are determined by the conditions $x(0) = x(1) = 0$ and

$$\int_0^1 \sqrt{1 + (x')^2}\,dt = l.$$

The Lagrange multiplier rule can be extended to equality constraints which take a more general form than those we have already discussed. Consider the problem of determining an extremum of the functional $f : \mathscr{V} \to \mathscr{R}$ subject to $Cx = \theta$, where C is a mapping from the Banach space $\mathscr{V}$ into a Banach space $\mathscr{U}$. The technique of solution is again to replace variations subject to $Cx = \theta$ by variations confined to the tangent manifold Γ_t of $\Gamma = \{x \in \mathscr{V} \mid Cx = \theta\}$. Specifically, if $\tilde{x} \in \mathscr{V}$ is a local extremum of f and f and C have continuous strong derivatives in an open neighbourhood of $\tilde{x}$ then $f'(\tilde{x})s = 0$ for all s satisfying $C'(\tilde{x})s = 0$.

As previously, we have to assume that $\tilde{x}$ is a *regular point of* Γ in the sense that in the neighbourhood of $\tilde{x}$ the operator C is invertible. (This is a generalized form of the classical implicit function theorem [23].) Similar reasoning to that used in the earlier analysis then leads to the conclusion that there exists $\tilde{\lambda} \in \mathscr{U}^*$ such that the Lagrangian

$$L(x) = f(x) + \lambda(Cx) = f(x) + [\lambda, Cx]$$

is stationary at $\tilde{x}$; that is to say,

$$f'(\tilde{x}) + [\tilde{\lambda}, C'(\tilde{x})] = 0.$$

A modified form of this result holds even if $\tilde{x}$ is not a regular point of Γ [36].

This extension of the multiplier rule allows constraints which may take the form of ordinary or partial differential equations, integral equations or difference equations. An example of considerable practical importance is the so-called optimal control problem wherein a dynamical system described by a set of non-linear ordinary differential equations is to be controlled in such a way that some objective or cost functional is to be minimized (or maximized). The following example illustrates a special case of this problem.

Example 4.3.5 On the interval $[0, T]$ of the real line, consider a dynamical system described by a set of linear differential equations

$$\{\dot{\xi}_i(t)\} = [d_{ij}]\{\xi_i(t)\} + [c_{ik}]\{\eta_k(t)\}, \qquad i, j = 1, \ldots, n, \quad k = 1, \ldots, m,$$

where $x(t) = (\xi_1(t), \ldots, \xi_n(t))$ is an n-tuple state vector and $y(t) = (\eta_1(t), \ldots, \eta_m(t))$ is an m-tuple control vector. The class of admissible control functions is taken to be $\mathscr{C}([0, T], \mathscr{E}_m)$, the continuous vector-valued functions on the Euclidean space $\mathscr{E}_m$, and similarly the class of admissible state functions is taken to be $\mathscr{C}([0, T], \mathscr{E}_n)$. The set of differential equations is a matrix representation of the operator equation

$$\dot{x}(t) = Dx(t) + Cu(t),$$

where $D: \mathscr{E}_n \to \mathscr{E}_n$ and $C: \mathscr{E}_m \to \mathscr{E}_n$.

The cost functional to be minimized is the quadratic form

$$f(x,u) = \frac{1}{2}\int_0^T (\lfloor x_i \rfloor [q_{ij}] \{x_j\} + \lfloor u_l \rfloor [r_{lk}] \{u_k\}) dt, \quad \begin{matrix} i,j = 1,\ldots,n, \\ l,k = 1,\ldots,m. \end{matrix}$$

$$= \frac{1}{2}\int_0^T (\langle x, Qx \rangle_{\mathscr{E}_n} + \langle u, Ru \rangle_{\mathscr{E}_m}) dt,$$

where $Q: \mathscr{E}_n \to \mathscr{E}_n$ is positive and $R: \mathscr{E}_m \to \mathscr{E}_m$ is positive definite (Definition 3.7.4). The initial value of the state vector is considered to be specified, say $x(0) = x_0$. Practically speaking, we seek to find a control, $u(t)$, which will keep the state of the system, $x(t)$, near zero without an unduly large cost in control during the time interval $[0, T]$.

The matrix differential equation describing the system is represented as a constraint in the form of the integral equation

$$x(t) = x_0 + \int_0^t (Dx(\tau) + Cu(\tau)) d\tau.$$

We form the Lagrangian

$$L(x,u) = \frac{1}{2}\int_0^T (\langle x, Qx \rangle_{\mathscr{E}_n} + \langle u, Ru \rangle_{\mathscr{E}_m}) dt$$

$$+ \left[\lambda(t), x(t) - x_0 - \int_0^t (Dx(\tau) + Cu(\tau)) d\tau \right],$$

where $\lambda(t) \in \mathscr{C}^*([0,T], \mathscr{E}_n)$. Now the dual space of $\mathscr{C}[0,T]$ is the space $\mathscr{B}_N[0,T]$ of (normalized) functions of bounded variation on $[0,T]$ (see p. 15), so that the second term in the Lagrangian may be written

$$\int_0^T \left\langle d\lambda(t), x(t) - x_0 - \int_0^t (Dx(\tau) + Cu(\tau)) d\tau \right\rangle_{\mathscr{E}_n}.$$

Stationarity of L at $\tilde{x}, \tilde{u}$ gives

$$\int_0^T \left(\langle Q\tilde{x}, s \rangle_{\mathscr{E}_n} dt + \left\langle d\tilde{\lambda}, s - \int_0^t Ds(\tau) d\tau \right\rangle_{\mathscr{E}_n} \right) = 0$$

and

$$\int_0^T \langle R\tilde{u}, h \rangle_{\mathscr{E}_m} dt - \left\langle d\tilde{\lambda}, \int_0^t Ch(\tau) d\tau \right\rangle_{\mathscr{E}_n} = 0$$

$\forall s \in \mathscr{C}([0,T], \mathscr{E}_n), s(0) = \theta$ and $h \in \mathscr{C}([0,T], \mathscr{E}_m), h(0) = \theta$.

Integrating the first equation by parts gives

$$\int_0^T \{\langle Q\tilde{x}, s\rangle_{\mathscr{E}_n} + \langle D^*\tilde{\lambda}, s\rangle_{\mathscr{E}_n} - \langle \tilde{\lambda}, \dot{s}\rangle_{\mathscr{E}_n}\} dt = 0$$

with $\tilde{\lambda}(T) = 0$. Applying the extended form of the fundamental lemma of the calculus of variations (Exercise 4.3,1(c)) we deduce that

(i) $\tilde{\lambda}(t)$ is differentiable in $[0, T]$;
(ii) $-\dot{\tilde{\lambda}} = Q\tilde{x} + D^*\tilde{\lambda}$.

From the second of the stationarity equations we obtain, again by an integration by parts:

(iii) $R\tilde{u} + C^*\tilde{\lambda} = \theta$.

Since R is positive definite,

$$\tilde{u}(t) = -R^{-1}C^*\tilde{\lambda}(t)$$

and replacing u in the system equation by this expression gives

$$\dot{\tilde{x}}(t) = D\tilde{x}(t) - CR^{-1}C^*\tilde{\lambda}(t),$$

and this expression, together with

$$-\dot{\tilde{\lambda}}(t) = Q\tilde{x}(t) + D^*\tilde{\lambda}(t),$$

represents a set of $2n$ linear differential equations for the vectors $\tilde{x}, \tilde{\lambda}$. The set of equations is subject to the two-point boundary values $\tilde{x}(0) = x_0$, $\tilde{\lambda}(T) = 0$, and this makes an analytical (or numerical) solution very difficult to obtain. If we postulate that $\tilde{\lambda}$ and $\tilde{x}$ are linearly related by

$$\tilde{\lambda}(t) = P(t)\tilde{x}(t), \qquad P: \mathscr{E}_n \to \mathscr{E}_n,$$

then we find, by direct substitution, that $P(t)$ must satisfy the matrix Ricatti equation[†]

$$\dot{P}(t) = -P(t)D - D^*P(t) + P(t)CR^{-1}C^*P(t) - Q$$

[†] The classical scalar Ricatti equation is

$$\dot{\phi} = a(t)\phi^2 + b(t)\phi + c(t).$$

This can be transformed to the second-order linear equation

$$\ddot{\psi} = (\dot{a} + b)\dot{\psi} - ca\psi$$

by the substitution

$$\phi(t) = -\dot{\psi}(t)/a\psi(t).$$

The Ricatti equation has been studied intensively by mathematicians and many properties of its solutions are known, both for the scalar and matrix cases [72, 73].

with $P(T) = 0$. If we take the adjoint (in $\mathscr{E}_n$) of this equation, we find

$$P^*(t) = -P^*(t)D - D^*P^*(t) + P^*(t)CR^{-1}C^*P^*(t) - Q$$

with $P^*(T) = 0$, showing that P is a symmetric matrix. The matrix Ricatti equation thus represents a set of $\frac{1}{2}n(n+1)$ first-order non-linear differential equations.

It can be shown that the solution $P(t)$ is unique on $[0, T]$ [72]: knowing $P(t)$ the optimal control vector $\tilde{u}(t)$ is given *in terms of the state vector* $\tilde{x}(t)$ by

$$\tilde{u}(t) = -R^{-1}C^*P(t)\tilde{x}(t).$$

This is the required feedback law for implementation of the optimal control. By substituting the control law into the expression for $f(x, u)$, we find (Exercise 4.3,6(d)) that the minimum value is given by

$$f(\tilde{x}, \tilde{u}) = \tfrac{1}{2}\langle x_0, P(0)x_0 \rangle_{\mathscr{E}_n}.$$

Now since Q is positive and R is positive definite, $f(x, u) > 0$ provided $u(t) \neq 0$: this implies that $P(0)$ is positive definite. However, so far as the Ricatti equation is concerned the initial instant $t = 0$ is arbitrary and we would obtain the same solution irrespective of the initial instant $t_0 < T$. This implies that the solution $P(t)$ of the matrix Ricatti equation is positive definite in $[0, T]$.

We should remark here that the methods of the variational calculus are not always the most effective in optimal control problems. This is particularly true when the control is constrained to lie in a bounded region or when the functional $L(x, u)$ is not sufficiently smooth: for these problems Pontryagin's maximum principle is more suitable [67, 72].

Necessary and sufficient conditions for an extremum

After an extremal curve has been determined one needs to investigate whether the functional $f(x)$ attains a maximum or minimum value.

While it is very difficult to apply a test which is both necessary and sufficient [68, 71], it is not too difficult to check only necessity. Such a test, when combined with physical reasoning and perhaps with a numerical calculation of $f(x)$ on neighbouring curves, is generally able to resolve the question satisfactorily.

From ordinary calculus we know that the sign of the second derivative of a scalar function at a stationary point tells us whether we have a minimum or maximum (or if it is zero, a point of inflexion). An analogous result holds for the more general case if we replace the ordinary second derivative by the second strong derivative; we quote the following result without proof [18, 67].

We know that an extremum of $f: \mathscr{V} \to \mathscr{R}$ subject to the constraints $c_j(x) = 0$ can be found from

$$f'^w(\tilde{x}) + \sum_{j=1}^{n} \tilde{\lambda}^j c_j'^w(\tilde{x}) = 0.$$

In order to be able to find a necessary condition for a maximum (or minimum) of f we need to insist on a reasonable degree of smoothness in the neighbourhood of the stationary point $\tilde{x}$. We require that the strong derivatives $f'(\tilde{x})$, $c_j'(\tilde{x})$ exist in a neighbourhood of $\tilde{x}$ and further that the strong second derivatives $f''(\tilde{x})$, $c_j''(\tilde{x})$ exist.

In that case, we can assert that if $\tilde{x}$ is a local minimum (maximum) of f subject to $x \in \Gamma = \{x | c_j(x) = 0\}$ then, for all s in the tangent subspace Γ_t,

$$L''(\tilde{x})s^2 \geq 0 \quad (\leq 0),$$

where

$$L(x) = f(x) + \sum_{j=1}^{m} \lambda^j c_j(x).$$

Example 4.3.6

(a) (See Example 4.3.4(a).) If f is a functional on $\mathscr{R}_n$ subject to $m < n$ algebraic constraints $c_j(x) = 0$, then the above condition becomes (Example 4.2.3(a))

$$\lfloor \sigma_i \rfloor [b_{ik}] \{\sigma_k\} = \sum_{i=1}^{n} \sum_{k=1}^{n} D_{ki} L(x) \sigma_i \sigma_k \geq 0 \quad (\leq 0)$$

for all vectors $(\sigma_1, \sigma_2, \sigma_3, \ldots, \sigma_n)$ such that

$$\lfloor D_i c_j(x) \rfloor \{\sigma_i\} = \sum_{i=1}^{n} D_i c_j(x) \sigma_i = 0, \quad j = 1, \ldots, m.$$

(b) In the isoperimetric problem of Example 4.3.4(c),

$$L(x) = \int_0^1 x(t) dt + \lambda \left(\int_0^1 \sqrt{1 + (x')^2} \, dt - l \right)$$

with (see Example 4.2.3(b))

$$L''(\tilde{x})s^2 = \lambda \int_0^1 \frac{1}{(1 + (\tilde{x}')^2)^{3/2}} (s')^2 \, dt$$

and

$$\Gamma_t = \left\{ s \left| \int_0^1 \frac{\tilde{x}' s'}{\sqrt{1 + (\tilde{x}')^2}} \, dt = 0 \right. \right\}.$$

If we restrict ourselves to extremal arcs for which $\tilde{x}(t) \geq 0$, we can deduce from the first integral of the Euler equation of Example 4.3.4(c) that $\tilde{\lambda} < 0$ and hence that $L''(\tilde{x})s^2 \leq 0$, consistent with the fact that the circular extremal arcs maximize the 'positive' area. (If $\tilde{x}(t) < 0$ then a 'negative' area is minimized.) In this particular case we did not need to use the fact that $s \in \Gamma_t$.

(c) There is one important case in which the necessary condition $L''(\tilde{x})s^2 \geq 0$ for a local minimum is also a sufficient condition. This is the case of a *quadratic functional* on a Hilbert space when there are no constraint conditions.

Take $f: \mathcal{H} \to \mathcal{R}$ to be

$$f(x) = \tfrac{1}{2} \langle Tx, x \rangle - \langle b, x \rangle,$$

where b is a fixed vector in $\mathcal{H}$ and T is a self-adjoint transformation (Definition 3.7.3).

We easily find

$$f'(\tilde{x}) = T\tilde{x} - b,$$

so that the stationary point is a solution of the linear equation $T\tilde{x} = b$; also

$$f''(\tilde{x}) = T$$

and the higher derivatives of f are null. Hence (Theorem 4.2.2)

$$f(\tilde{x} + s) - f(\tilde{x}) = \langle T\tilde{x} - b, s \rangle + \tfrac{1}{2} \langle Ts, s \rangle$$
$$= \tfrac{1}{2} \langle Ts, s \rangle, \qquad s \in \mathcal{H},$$

since $\tilde{x}$ is the stationary point. If $\tilde{x}$ is a local minimum then, necessarily, $\langle Ts, s \rangle \geq 0$, implying that T is a positive transformation (Definition 3.7.4). Conversely, if $\tilde{x}$ is a solution of $T\tilde{x} = b$ and $\langle Ts, s \rangle \geq 0$ then $\tilde{x}$ is a minimum of f. Hence the condition that T is a positive transformation is a necessary and sufficient condition for f to have a minimum at $\tilde{x}$. If T is positive definite then the minimum is attained only at the point $\tilde{x}$.

Exercises 4.3

1. (a) Prove the fundamental lemma of the variational calculus, namely that if $y \in \mathscr{C}[0, 1]$ and

    $$\int_0^1 y(t) s(t) \, dt = 0 \qquad \forall \, s \in \mathscr{C}_0^{(1)}[0, 1]$$

    then $y(t) = 0$ on $[0, 1]$. (Hint: assume $y(t) > 0$ for $t \in [t_1, t_2]$ and take $s(t) = (t - t_1)^2 (t - t_2)^2, t \in [t_1, t_2]$.)

(b) If $y \in \mathscr{C}[0, 1]$ and

$$\int_0^1 y(t)\,ds(t) = 0 \qquad \forall\, s \in \mathscr{C}_0^{(1)}[0, 1]$$

show that $y(t) = \alpha$, a constant, in $[0, 1]$. (Hint: choose

$$\int_0^t s(t) = (y(\tau) - \alpha)\,d\tau, \qquad \alpha = \int_0^1 y(t)\,dt.)$$

(c) If $y(t), z(t) \in \mathscr{C}[0, 1]$ and

$$\int_0^1 \left(y(t)s(t) + z(t)\frac{ds(t)}{dt} \right) dt = 0 \qquad \forall\, s \in \mathscr{C}_0^{(1)}[0, 1]$$

show that z is differentiable and $dz/dt = y$ in $[0, 1]$.

2. (a) If $x(t) = (\xi_1(t), \xi_2(t), \ldots, \xi_n(t)) \in \mathscr{R}_n[0, 1]$ and

$$J(x) = \int_0^1 g(\xi_1(t), \ldots, \xi_n(t), \xi_1'(t), \ldots, \xi_n'(t), t)\,dt$$

$$= \int_0^1 g(\xi_i(t), \xi_i'(t), t)\,dt,$$

show that the Euler–Lagrange equations are

$$\frac{d}{dt}g_{\xi_i'}(\xi_i, \xi_i', t) - g_{\xi_i}(\xi_i, \xi_i', t) = 0, \qquad i = 1, \ldots, n.$$

(b) If

$$J(x) = \int_\Gamma g(x, D_1 x, D_2 x, t_1, t_2)\,dt_1\,dt_2,$$

where $x \in \mathscr{C}^{(1)}(\Gamma)$, show that the Euler–Lagrange equation is

$$D_1 g_{D_1 x} + D_2 g_{D_2 x} - g_x = 0.$$

(Hint: use the divergence theorem and assume that the result of Exercise 4.3, 1(a) extends to multiple integrals.)

3. (a) Find the extremal arc for the functionals

(i) $J(x) = \int_0^1 xx'\,dt$, $x(0) = 0, x(1) = 1$;

(ii) $J(x) = \int_\alpha^\beta \frac{\sqrt{1 + (x')^2}}{x}\,dt$, $x(\alpha) = x(\beta) = 1$ (cf. example 4.3.2(a));

(iii) $J(x) = \int_\alpha^\beta t^2 x'^2\,dt$ $\begin{cases} \alpha = 1, \beta = 2, x(1) = 1, x(2) = \frac{1}{2}, \\ \alpha = -2, \beta = 1, x(1) = 1, x(-2) = -\frac{1}{2}. \end{cases}$

(b) Write down the Euler–Lagrange equation for the functionals

(i) $J(x) = \int_\Gamma (D_1 x)^2 - (D_2 x)^2 \, dt_1 \, dt_2$,

(ii) $J(x) = \int_\Gamma (D_1 x)^2 + (D_2 x)^2 + 2xf(t_1, t_2) \, dt_1 \, dt_2$

where $f(t_1, t_2)$ is a given function.

4. (a) Derive the canonical Euler equations (Example 4.3.2(d))

$$\frac{d\pi_i}{dt} = -h_{\xi_i}, \quad \frac{d\xi_i}{dt} = h_{\pi_i}, \quad i = 1, \ldots, n,$$

as the Euler–Lagrange equations for the functional

$$I(x, p) = \int_0^1 \left(\sum_{i=1}^n \pi_i \xi_i' - h(\xi_1, \ldots, \xi_n, \pi_1, \ldots, \pi_n, t) \right) dt.$$

(b) If the kinetic energy of an n-degree of freedom mechanical system is given by

$$T = \frac{1}{2} \sum_{i=1}^n \sum_{j=1}^n t_{ij} \dot{q}_i \dot{q}_j$$

(Example 4.3.2(c)) show directly the equivalence of Lagrange's and Hamilton's dynamical equations for the system.

(c) Given a function $f(x)$, $x = (\xi_1, \xi_2, \xi_3, \ldots, \xi_n)$ define a new variable $y = (\eta_1, \eta_2, \ldots, \eta_n)$ by the transformation $\eta_i = f_{\xi_i}$.

If the Hessian (Example 4.2.3(a)) of f is non-zero then the transformation is invertible, thus giving the ξ_i as functions of the η_i. Define

$$g(y) = \sum_{i=1}^n \xi_i \eta_i - f(x)$$

and show that

$$\xi_i = g_{\eta_i}.$$

This transformation, called a *Legendre* or *contact* transformation, is hence symmetrical in its treatment of the two sets of variables. When f is also a function of other variables which play no part in the transformation, those variables which are transformed are called active and those which are not passive. Thus in Example 4.3.2(d) the ξ_i and t are passive variables, the ξ_i' and π_i active.

(d) Consider a string stretched between two points a distance l apart. Let the tension in the string be τ and the (constant) mass per unit length of the string be ρ. Let the string suffer a small transverse displacement $\phi(x, t)$, small in the sense that higher powers than the first in ϕ and its derivatives are taken to be negligible. Show that

(i) The kinetic energy, $T = \dfrac{\rho}{2}\int_0^l \phi_t^2\,dx$;

(ii) the potential energy, $U = \dfrac{\tau}{2}\int_0^l \phi_x^2\,dx$ (hint: U is proportional to change in length of the string).

Hence deduce that the Euler–Lagrange equation for the transverse motion of the string is

$$\phi_{tt} - c^2\phi_{xx} = 0, \qquad c^2 = \tau/\rho.$$

(Hint: use the result of Exercise 4.3, 2(b).)

5. (a) Show that, in terms of the canonical variable $\pi = g_{x'}$, the transversality condition (p. 310) becomes

$$[hr_x + \pi r_t] = 0,$$

where $h = \pi x' - g$ is the Hamiltonian.

(b) Show that if the terminal curve $r(x, t) = 0$ is a line parallel to the x-axis, i.e., $r(x, t) = t - \alpha = 0$, the tranversality condition reduces to

$$\pi\big|_{t=\alpha} = g_{x'}\big|_{t=\alpha} = 0.$$

Deduce the same result from the vanishing of $dJ(x; s)$, where

$$J(x) = \int_0^\alpha g(x, x', t)\,dt,$$

taking $x(0) = 0$. (When no boundary condition is specified for x at one end of the interval the vanishing of $dJ(x;s)$ generates a so-called *natural boundary condition* which usually involves the derivative of x; this point is discussed in Section 4.4.)

(c) Find an extremal arc for the functional

$$J(x) = \int_0^{t_r} \frac{\sqrt{1 + (x')^2}}{x}\,dt,$$

where $x(0) = 0$ and $x(t_r)$ lies on the circle $(t - 6)^2 + x^2 = 9, x \geqslant 0$. (Hint: (i) see Exercise 4.3, 3(a) (ii)—extermal arc is a circle containing an unknown constant; (ii) in the transversality condition x' is evaluated *on the terminal curve*; (iii) the extremal arc, the terminal curve and the transversality condition give three equations for the coordinates t_r, $x_r = x(t_r)$ and the unknown constant.)

6. (a) By finding the extremal arc for the functional

$$J(x) = \frac{1}{l}\int_\alpha^\beta x\sqrt{1 + (x')^2}\,dt$$

subject to

$$l = \int_\alpha^\beta \sqrt{1+(x')^2}\,dt$$

deduce that the shape adopted by a heavy uniform cable suspended between the points (α, γ) and (β, μ) is a catenary curve

$$x + \lambda = k_1 \cosh\left(\frac{t+k_2}{k_1}\right).$$

Discuss how to find λ, k_1 and k_2. Justify the given physical interpretation.

(b) (i) Deduce directly from $dJ(x;s) = 0$ that the Euler–Lagrange equation for the functional

$$J(x) = \int_0^1 g(x, x', x'', t)\,dt, \qquad x \in \mathscr{C}^{(2)}[0,1],$$

is

$$\frac{d^2}{dt^2} g_{x''} - \frac{d}{dt} g_{x'} + g_x = 0.$$

(ii) Derive an equivalent result by seeking a stationary value of

$$J(x, y) = \int_0^1 g(x, x', y', t)\,dt$$

subject to the differential constraint $y = x'$.

(c) (i) Show that the functional

$$J(x) = \int_0^l \tfrac{1}{2}(EIx'')^2 - px\,dt$$

yields the Euler–Lagrange equation

$$(EIx'')'' = p,$$

where $p(t)$ is a given function, together with the boundary conditions

$$\begin{cases} x' = 0 \\ \text{or} \\ EIx'' = 0 \end{cases} \quad \text{and} \quad \begin{cases} x = 0 \\ \text{or} \\ (EIx'')' = 0 \end{cases}$$

at $t = 0, l$. (Here $x(t)$ represents the transverse deflection of a beam of length l and bending stiffness $EI(t)$ while $p(t)$ is the applied lateral load.) Which of the boundary conditions are natural (Exercise 4.3,5(b))? Give a physical interpretation for each set of possible boundary conditions.

(ii) Let $m/EI = -x''$ and deduce that the Euler–Lagrange equation for the functional

$$J(x,m) = \int_0^l \left(\frac{1}{2}\frac{m^2}{EI} - px\right)dt$$

subject to $m/EI = -x''$ is $\lambda'' = p$ with

$$\begin{Bmatrix} x' = 0 \\ \text{or} \\ \lambda = 0 \end{Bmatrix} \text{ and } \begin{Bmatrix} x = 0 \\ \text{or} \\ \lambda' = 0 \end{Bmatrix}$$

at $t = 0, l$, where $\lambda(t)$ is a Lagrange multiplier. Show also that $\lambda = -m$. ($m(t)$ represents the bending moment in the beam, hence we see that the multiplier has a definite physical interpretation—*see also* Example 4.7.4.)

(iii) Show that, if $x = 0$ on $\partial\Gamma$ the functional

$$J(x) = \int_\Gamma K/2 \{(\nabla^2 x)^2 - 2(1-\mu)(D_{11}xD_{22}x - (D_{12}x)^2)\} - px \, dt_1 \, dt_2,$$

μ constant, yields the Euler–Lagrange equation

$$K\nabla^2(\nabla^2 x) = p \quad \text{in} \quad \Gamma$$

together with the natural boundary condition

$$\mu\nabla^2 x + (1-\mu)D_{\nu\nu} x = 0,$$

where ν denotes the outward normal to $\partial\Gamma$. (These are the equations of equilibrium and boundary conditions for a transversely loaded thin plate having bending stiffness K and which is simply supported along its edge $\partial\Gamma$; μ is Poisson's ratio [74].) (Hint: in $dJ(x;s)$ the required boundary condition appears as the multiplier of $D_\nu s$.)

(d) (i) In Example 4.3.5 verify that the minimum value of the cost function (p. 318) is

$$f(\tilde{x}, \tilde{u}) = \tfrac{1}{2}\langle x_0, P(0)x_0\rangle_{\mathscr{E}_n}.$$

(ii) Apply the analysis of Example 4.3.5 to the scalar system $\dot{x} = u$ with cost functional

$$f(x,u) = \frac{1}{2}\int_0^T x^2 + u^2 \, dt$$

to show that $P(t) = \tanh(T-t)$, and thence

$$\tilde{u}(t) = -\tilde{x}(t)\tanh(T-t).$$

Deduce also that

$$\tilde{x}(t) = x_0 - \frac{\cosh(t-T)}{\cosh T} \quad \text{and} \quad \tilde{u}(t) = x_0 \frac{\sinh(t-T)}{\cosh T};$$

hence verify that

$$f(\tilde{x}, \tilde{u}) = \tfrac{1}{2} x_0^2 P(0) = \tfrac{1}{2} x_0^2 \tanh T.$$

4.4 Direct solution of positive operator equations

Example 4.3.6(c) showed us that the solution of the operator equation

$$Tx = y,$$

where T is a positive definite self-adjoint transformation is also a (global) minimum of the quadratic functional

$$f(x) = \tfrac{1}{2} \langle Tx, x \rangle - \langle y, x \rangle.$$

This suggests that we might be able to compute the solution, or an approximation to it, by seeking to minimize the functional $f(x)$ rather than by solving the operator equation. This process is the reverse of that followed in the variational calculus where, beginning with a given functional, we deduce the Euler–Lagrange equation and then attempt to solve it. Since the step of forming the Euler–Lagrange equation is omitted and instead a direct attack is made on minimizing the functional this type of solution is, for historical reasons, referred to as 'direct'.

For our present purposes the result embodied in Example 4.3.6(c) has two shortcomings:

(i) we assumed the existence of a minimum $\tilde{x}$;
(ii) we assumed that T was a bounded transformation.

We shall deal with the second point by relaxing the condition that T is self-adjoint and assuming that T is merely symmetric (Definition 3.7.3). At the same time we shall somewhat strengthen the condition of positive definiteness by insisting that T is *positive-bounded-below*: as we shall show, this settles the question of the existence of a minimum.

Definition 4.4.1 *A symmetric transformation T on a Hilbert space $\mathcal{H}$ is said to be* positive-bounded-below *if, for any $x \in \mathcal{D}(T)$,*

$$\langle Tx, x \rangle \geq \beta^2 \langle x, x \rangle, \qquad \beta > 0.$$

If T is positive-bounded-below then clearly T is also positive definite.

From the Schwartz inequality we have,

$$\|Tx\|\|x\| \geq \langle Tx, x \rangle \geq \beta^2 \|x\|^2$$

or

$$\|Tx\| \geq \beta^2 \|x\|, \quad x \in \mathcal{D}(T).$$

It follows from Theorem 3.4.2 that T^{-1} exists and is bounded. It is also important to note that $\langle Tx, z \rangle$ is a coercive bilinear functional (Exercise 3.7.3).

Theorem 4.4.1 *Let T be a symmetric transformation with domain $\mathcal{D}(T)$, which is dense in a Hilbert space $\mathcal{H}$, and let T be positive-bounded-below. Then the solution $\tilde{x}$ of $Tx = y$ minimizes the functional*

$$f(x) = \tfrac{1}{2}\langle Tx, x \rangle - \langle y, x \rangle$$

and, conversely, the vector in $\mathcal{D}(T)$ which minimizes $f(x)$ is the solution of the operator equation.

Proof (Cf., Example 4.3.6(c).) For any $s \in \mathcal{D}(T)$, we have

$$f(\tilde{x} + s) - f(\tilde{x}) = \langle T\tilde{x} - y, s \rangle + \tfrac{1}{2}\langle Ts, s \rangle$$
$$\geq \langle T\tilde{x} - y, s \rangle + \tfrac{1}{2}\beta^2 \|s\|^2.$$

If $\tilde{x}$ satisfies $T\tilde{x} = y$ then clearly $\tilde{x}$ is a minimum of $f(x)$. Conversely, if $\tilde{x}$ is a minimum point of $f(x)$ we require that

$$\langle T\tilde{x} - y, s \rangle = 0 \quad \forall s \in \mathcal{D}(T).$$

But $\mathcal{D}(T)$ is dense in $\mathcal{H}$, hence $T\tilde{x} - y = \theta$ and $\tilde{x}$ is a solution of $Tx = y$. ∎

The theorem holds if T is merely positive (Exercise 4.4,3(a))

It appears to be implicit in the statement of the above theorem that we should search for the minimum of the functional $f(x)$ in the subspace $\mathcal{D}(T)$. However, it turns out that this restriction is not necessary and we can, in our search for the minimum point, use a class of function considerably larger than $\mathcal{D}(T)$; this is a result of great practical significance. To see why this might be so, we shall consider a simple example.

Example 4.4.1 The differential operator defined in $[0, 1]$,

$$Tx = -\frac{d^2x}{dt^2}, \quad x(0) = x'(1) = 0,$$

is positive definite (cf., Example 3.7.5(a)). That it is also bounded below can be shown as follows. Since $x(0) = 0$,

$$x(t) = \int_0^t \frac{dx}{dt} dt$$

and applying the Schwarz inequality (Theorem 3.6.1) to the right-hand side we obtain

$$x^2(t) \leq \int_0^t 1^2 dt \int_0^t \left(\frac{dx}{dt}\right)^2 dt = t \int_0^t \left(\frac{dx}{dt}\right)^2 dt \leq t \int_0^1 \left(\frac{dx}{dt}\right)^2 dt.$$

Integrating the last inequality we have

$$\int_0^1 x^2(t) dt \leq \frac{1}{2} \int_0^1 \left(\frac{dx}{dt}\right)^2 dt$$

or (Example 3.7.5(a))

$$\|x\|^2 \leq \tfrac{1}{2}\langle Tx, x \rangle;$$

that is to say,

$$\langle Tx, x \rangle \geq \beta^2 \|x\|^2$$

with $\beta = \sqrt{2}$.

The functional $f(x)$ is given by

$$f(x) = \frac{1}{2} \int_0^1 \left(\frac{dx}{dt}\right)^2 dt - \int_0^1 xy\, dt.$$

In this case the functions in $\mathscr{D}(T)$ are the twice differentiable functions defined on $[0, 1]$ with $x(0) = x'(1) = 0$ and having norm

$$\|x\|^{(2)} = \left(\int_0^1 \left(\frac{d^2x}{dt^2}\right)^2 + \left(\frac{dx}{dt}\right)^2 + x^2\, dt \right)^{1/2},$$

a subspace of $\mathscr{H}^{(2)}[0, 1]$.[†] However, it is striking that the highest derivative to appear in the expression for $f(x)$ is the first, and hence the functional $f(x)$ is defined for all functions in $[0, 1]$ having a first derivative which is merely piecewise continuous and for which

$$\|x\|_T^2 = \int_0^1 \left(\frac{dx}{dt}\right)^2 < \infty;$$

let us call this space $\mathscr{H}^T[0, 1]$. Could we then seek for a minimum using functions in the larger space $\mathscr{H}^T$? If $\tilde{x} \in \mathscr{D}(T)$ is the solution, then we know that

$$f(\tilde{x} + s) - f(\tilde{x}) \geq 0,$$

[†] See Example 3.6.1(f)—spaces of this type will be more fully described in Chapter 5.

but we shall now allow s to be an arbitrary vector in $\mathscr{H}^T$. This gives, directly from the expression for $f(x)$,

$$f(\tilde{x}+s) - f(\tilde{x}) = \int_0^1 \left(\left(\frac{d\tilde{x}}{dt}\right)\left(\frac{ds}{dt}\right) - ys\right) dt + \frac{1}{2}\int_0^1 \left(\frac{ds}{dt}\right)^2 dt.$$

As we already know, the first term (representing the first derivative of f) must vanish for all s, giving

$$\int_0^1 \left(\left(\frac{d\tilde{x}}{dt}\right)\left(\frac{ds}{dt}\right) - ys\right) dt = 0.$$

Now $\tilde{x} \in \mathscr{D}(T)$, so we may integrate by parts to give

$$-\int_0^1 \left(\frac{d^2x}{dt^2} + y\right) s\, dt + \left[\frac{d\tilde{x}}{dt} s\right]_0^1 = 0$$

and the integral term yields

$$-\frac{d^2\tilde{x}}{dt^2} = y.$$

The boundary term will vanish at $t = 0$ if we require the functions in $\mathscr{H}^T[0, 1]$ to satisfy the single boundary condition $s(0) = 0$. The vanishing of the boundary term at $t = 1$ then implies $\tilde{x}'(1) = 0$. Hence, by enlarging the class of functions within which we may search for a minimum, we satisfy not only the differential equation but also, automatically, the derivative boundary condition.

This in turn, of course, implies that we can find a sequence of points $s^{(n)}$ in $\mathscr{H}^T \supset \mathscr{D}(T)$ with $s^{(n)}(0) = 0$ such that the limit of the sequence is the solution point $\tilde{x} \in \mathscr{D}(T)$ (with $\tilde{x}'(1) = 0$).

To see why we need to impose the condition $s(0) = 0$ on the functions in $\mathscr{H}^T$ but not the condition $s'(1) = 0$ we should note that, in $\mathscr{H}^T$, we demand only mean convergence of the first derivative, which can clearly be satisfied by the limit function without satisfaction of the pointwise condition $\lim_{n \to \infty} s^{(n)'}(1) = 0$. However, using the Schwarz inequality together with $s^{(n)}(0) = 0$, we have

$$|s^{(n)}(t) - s^{(m)}(t)|^2 = \left|\int_0^t \left(\frac{ds^{(n)}}{dt} - \frac{ds^{(m)}}{dt}\right) dt\right|^2$$

$$\leq \int_0^t 1^2\, dt \int_0^t \left(\frac{ds^{(n)}}{dt} - \frac{ds^{(m)}}{dt}\right)^2 dt,$$

so that convergence (in the mean) in $\mathscr{H}^T$ implies uniform convergence of $s^{(n)}$ to $\tilde{x}$. Thus it is essential to impose the boundary condition $s(0) = 0$ on the trial functions in $\mathscr{H}^T$.

In this context the boundary condition $x(0) = 0$ is termed *essential* (or *principal*), while the boundary condition $x'(1) = 0$ is termed *natural*: we shall pursue this matter in detail presently.

The foregoing example raises two important questions that need further investigation:

1. What is the nature of the space $\mathcal{H}^T$?
2. If we seek for a minimum of $f(x)$ in the enlarged space $\mathcal{H}^T$, does a minimum always exist, and is this minimum attained at the solution point $\tilde{x} \in \mathcal{D}(T)$?

We shall now investigate these questions and at the same time generalize the ideas contained in Example 4.4.1.

We begin by introducing the energy product $\langle Tx, z \rangle$ (p. 225) associated with the positive-bounded-below (and therefore positive definite) transformation T into the domain space $\mathcal{D}(T) \subset \mathcal{H}$: in this way we make $\mathcal{D}(T)$ an inner product space with norm $\langle Tx, x \rangle^{1/2}$. This space will, in general, be incomplete, but we may complete it in the usual way by adjoining elements which are the limits of Cauchy sequences in $\mathcal{D}(T)$ (Theorem 3.2.1): let us denote this Hilbert space by $\mathcal{H}^T$.[†] Since $\mathcal{H}^T$ contains points which are not in $\mathcal{D}(T)$, the inner product $\langle Tx, z \rangle$ is not defined on $\mathcal{H}^T$. We define the inner product $\langle x, z \rangle_T$ on the Hilbert space $\mathcal{H}^T$ by

$$\langle x, z \rangle_T = \lim_{n \to \infty} \langle Tx^{(n)}, z^{(n)} \rangle,$$

where $x^{(n)}, z^{(n)}$ are (Cauchy) sequences in $\mathcal{D}(T)$. In the same way the *energy norm* in $\mathcal{H}^T$ is defined as

$$\|x\|_T^2 = \lim_{n \to \infty} \langle Tx^{(n)}, x^{(n)} \rangle.[‡]$$

Does the Hilbert space $\mathcal{H}^T$ contain points which are not in the parent space $\mathcal{H}$? The answer is, 'No' (Exercise 4.4,3(b)). That is to say, the space with inner product $\langle Tx, y \rangle$, $x, y \in \mathcal{D}(T)$, can be completed using elements from the parent space $\mathcal{H}$ (i.e., $\mathcal{H}^T \subset \mathcal{H}$). It is clear that $\mathcal{D}(T)$ is dense in both $\mathcal{H}^T$ and $\mathcal{H}$.

Since the transformation T is positive-bounded-below, we have,

---

[†] Elements of $\mathcal{H}^T$ are often referred to as 'functions with finite energy'.
[‡] The name 'energy norm' has its origins in the early application of variational methods to problems in elastic structures. In this context the equation $Tx = y$ relates the 'displacement' x to the 'force' y through an 'elastic' operator T: the product of force × displacement, i.e., $\langle y, x \rangle = \langle Tx, x \rangle$, is a kind of energy or work function directly related to the strain energy function in structural analysis.

for any Cauchy sequence $x^{(n)} \in \mathscr{D}(T)$,
$$\beta^2 \|x^{(n)}\|^2 \leqslant \langle Tx^{(n)}, x^{(n)} \rangle$$
and since every point in $\mathscr{H}^T$ is also in $\mathscr{H}$ we obtain, in the limit,
$$\|x\| \leqslant \frac{1}{\beta} \|x\|_T.$$

Hence 'convergence in energy' implies strong convergence.

We have now elucidated the first question posed on p. 330 and can turn our attention to the second. First of all, we extend the functional
$$f(x) = \tfrac{1}{2} \langle Tx, x \rangle - \langle y, x \rangle$$
defined on $\mathscr{D}(T)$ to the whole of $\mathscr{H}^T$. The first term on the right-hand side is replaced by the energy product $\langle x, x \rangle_T$, $x \in \mathscr{H}^T$. The second term is, for fixed y and any $x \in \mathscr{H}^T$, a bounded linear functional, $y(x)$ on $\mathscr{H}^T$ since, using the Schwarz inequality, we have
$$|y(x)| = |\langle y, x \rangle| \leqslant \|y\| \|x\| \leqslant \frac{\|y\|}{\beta} \|x\|_T.$$

Now to every bounded linear functional $l(x)$ on a Hilbert space $\mathscr{H}$ there corresponds a unique element $z \in \mathscr{H}$ such that (Example 3.6.2(c))
$$l(x) = \langle z, x \rangle;$$
hence there exists $\hat{x} \in \mathscr{H}^T$, say, such that
$$\langle y, x \rangle = \langle \hat{x}, x \rangle_T.$$

We can now write the functional $f(x)$ extended to $\mathscr{H}^T$ as
$$\begin{aligned} f(x) &= \tfrac{1}{2} \langle x, x \rangle_T - \langle \hat{x}, x \rangle_T, \quad x \in \mathscr{H}^T, \\ &= \tfrac{1}{2}(\langle x - \hat{x}, x - \hat{x} \rangle_T - \langle \hat{x}, \hat{x} \rangle_T) \\ &= \tfrac{1}{2}(\|x - \hat{x}\|_T^2 - \|\hat{x}\|_T^2) \end{aligned}$$
so that $f(x)$ assumes its minimum value at $x = \hat{x}$.

Now suppose $\hat{x}^{(n)}$ is a sequence of points in $\mathscr{H}^T$ such that
$$\lim_n f(\hat{x}^{(n)}) = -\tfrac{1}{2} \|\hat{x}\|_T^2.$$

This implies that $\|\hat{x}^{(n)} - \hat{x}\|_T \to 0$ and that $\|\hat{x}^{(n)} - \hat{x}\| \to 0$; hence there exists a point $\hat{x} \in \mathscr{H}^T \subset \mathscr{H}$ such that
$$\lim_n \hat{x}^{(n)} = \hat{x}.$$

Since $\mathscr{D}(T)$ is dense in $\mathscr{H}^T$ there must exist a sequence $x^{(n)} \in \mathscr{D}(T)$ such that $\lim_n x^{(n)} = \hat{x}$.

The statement above,

$$\langle y, x \rangle = \langle \hat{x}, x \rangle_T,$$

can be written

$$\lim_n \langle y - Tx^{(n)}, x \rangle = 0 \qquad \forall x \in \mathscr{D}(T).$$

If $\hat{x} \in \mathscr{D}(T)$ then clearly

$$\lim_n x^{(n)} = \hat{x} = \tilde{x},$$

so that minimization of the functional $f(x)$ in $\mathscr{H}^T$ leads to the same minimum point as minimization in $\mathscr{D}(T)$ (Theorem 4.4.1).

If $\hat{x} \notin \mathscr{D}(T)$ there is no point in $\mathscr{D}(T)$ such that $T\hat{x} = y$. In that case, we say that $\hat{x}$ is a *generalized solution* of the operator equation characterized by the weak convergence of $x^{(n)}$ (*see* Example 3.6.2(c)),

$$\lim_n \langle x^{(n)}, x \rangle_T = \langle y, x \rangle.$$

We shall pursue this question further in Chapter 5 when we discuss generalized functions or distributions (*see also* Example 3.6.2(b)).

In conclusion, we should note the crucial role played throughout by the condition that the operator is positive-bounded-below, which implies convergence in $\mathscr{H}$ from convergence in $\mathscr{H}^T$.

When T is a differential operator together with an associated set of (homogeneous) boundary conditions, the functions of the energy space $\mathscr{H}^T$ are obtained from $\mathscr{D}(T)$ by adjoining functions which

1. do not have the required degree of differentiability;
2. do not satisfy all the boundary conditions.

A systematic way of delineating those functions which are admisible in $\mathscr{H}^T$ is the following.

In Example 3.7.4(c) and (d) we verified the symmetry of two differential operators by integration by parts or its analogue in higher dimensions, the divergence theorem.

Let $D^{(2s)}$ be a symmetric differential operator containing (partial) derivatives up to and including order $2s$. The energy inner product takes the form

$$\langle D^{(2s)} x, z \rangle = \int_\Gamma z D^{(2s)} x \, d\Gamma$$

for some open region Γ with boundary $\partial \Gamma$.

Integration by parts will yield the expression

$$\langle D^{(2s)}x, z \rangle = \int_\Gamma D^{(s)}z \, D^{(s)}x \, d\Gamma + \int_{\partial \Gamma} b(z, x) \, dS,$$

where $D^{(s)}$ is a differential operator of maximum order s defined in Γ, and b is a bilinear functional involving the (partial) derivatives of z up to order $(s-1)$ and of x from order s to $2s-1$. Since the operator $D^{(2s)}$ is symmetric it must be possible, by using the prescribed boundary conditions, to express the boundary integrand in a form $c(z, x)$ which is symmetric in z and x. We now form the functional

$$f(x) = \tfrac{1}{2}\langle D^{(2s)}x, x \rangle - \langle y, x \rangle$$
$$= \int_\Gamma \tfrac{1}{2}(D^{(s)}x)^2 - yx \, d\Gamma + \int_{\partial \Gamma} c(x, x) \, dS$$

and consider it defined for all functions which are $2s$ times differentiable in Γ but satisfy no boundary conditions whatsoever.

Setting the first derivative of f equal to zero we may, again using integration by parts, generate the Euler–Lagrange equation

$$D^{(2s)}x = y$$

to be satisfied in Γ, together with a boundary term. A prescribed boundary condition which is automatically satisfied by virtue of the vanishing of the first derivative of f is a *natural boundary condition*: all others are *essential*. From these considerations the following rule emerges: a boundary condition is natural if it contains derivatives of x of order s and higher and essential if of order $s - 1$ or less.

We conclude that functions in $\mathcal{H}^T$ need possess only sth-order derivatives in Γ and satisfy only the essential boundary conditions on $\partial \Gamma$. It is important to note that we can, by adding different boundary terms to a basic functional, modify the natural boundary conditions while still yielding the same Euler–Lagrange equation.

Example 4.4.2

(a) We saw in Example 4.4.1 that the vanishing of the first derivative of

$$f(x) = \tfrac{1}{2}\int_0^1 \left(\frac{dx}{dt}\right)^2 dt - \int_0^1 xy \, dt$$

led to the Euler–Lagrange equation

$$-\frac{d^2x}{dt^2} = y$$

and to the boundary condition $x'(1) = 0$. Hence the boundary condition $x'(1) = 0$ is natural while $x(0) = 0$ is essential.

(b) Consider the equation (Example 3.7.4(d))
$$-\nabla^2 x = y \quad \text{in} \quad \Gamma$$
with boundary condition $\partial x/\partial v + kx = 0$ on $\partial \Gamma$, $k > 0$. This differential operator is positive-bounded-below (Exercise 4.4, 1(a)). We have
$$\langle Tx, z \rangle = \int_\Gamma \sum_{i=1}^3 \frac{\partial x}{\partial t_i} \frac{\partial z}{\partial t_i} \, d\Gamma - \int_{\partial \Gamma} \frac{\partial x}{\partial v} z \, dS,$$
$$= \int_\Gamma \sum_{i=1}^3 \frac{\partial x}{\partial t_i} \frac{\partial z}{\partial t_i} \, d\Gamma + k \int_{\partial \Gamma} xz \, dS,$$
a symmetric form in x and z. Hence
$$f(x) = \int_\Gamma \tfrac{1}{2} \sum_{i=1}^3 \left(\frac{\partial x}{\partial t_i}\right)^2 - yx \, d\Gamma + \frac{k}{2} \int_{\partial \Gamma} x^2 \, dS,$$
and putting $f'(x)s = 0$ gives (after integration by parts)
$$-\int_\Gamma (\nabla^2 x + y)s \, d\Gamma + \int_{\partial \Gamma} \left(\frac{\partial x}{\partial v} + kx\right) s \, dS = 0.$$
Thus the condition $\partial x/\partial v + kx = 0$ on $\partial \Gamma$ is natural.

Minimizing sequences—the Rayleigh–Ritz method

We have seen that the solution point of $Tx = y$ is also the minimum of the functional
$$f(x) = \tfrac{1}{2}\langle x, x \rangle_T - \langle y, x \rangle$$
on $\mathcal{H}^T$. Hence by seeking an approximation to the minimum point of $f(x)$ we can, in some sense to be made clear, approximate to the solution of $Tx = y$.

A sequence of points $x^{(n)} \in \mathcal{H}^T$ such that
$$\lim_{n \to \infty} f(x^{(n)}) = \inf_{x \in \mathcal{H}_T} f(x)$$
is called a *minimizing sequence for f*. As we have seen, if T is positive-bounded-below the minimum is actually attained by an element of $\mathcal{H}^T$.

The Rayleigh–Ritz method gives a systematic way of constructing a minimizing sequence for f which yields an approximation to the minimum point in a finite-dimensional subspace of $\mathcal{H}^T$. Take $\mathcal{H}$ to be separable: then it can be shown [79] that this implies that $\mathscr{H}^T$ is also separable.

Let $\{z^i\}$ be a basis for $\mathscr{H}^T$ (Definition 3.3.2) and let $\mathscr{H}_n^T$ be the finite-dimensional subspace spanned by the n vectors $\{z^1, z^2, \ldots, z^n\}$. Let

$$x^{(n)} = \sum_{i=1}^{n} \xi_i^{(n)} z^i,$$

then

$$f(x^{(n)}) = \frac{1}{2} \sum_{i=1}^{n} \sum_{j=1}^{n} \xi_i^{(n)} \xi_j^{(n)} \langle z^j, z^i \rangle_T - \sum_{i=1}^{n} \xi_i^{(n)} \langle y, z^i \rangle$$

and we select the coordinates $\{\xi_j^{(n)}\}$ so that $x^{(n)}$ is a minimum point of f on $\mathscr{H}_n^T$. This is an ordinary minimum problem in n variables and leads to the set of linear equations (the *Ritz equations*).

$$\sum_{j=1}^{n} \langle z^j, z^i \rangle_T \xi_j^{(n)} = \langle y, z^i \rangle, \qquad i = 1, 2, \ldots, n.$$

Because the z^i are linearly independent, the Gram determinant,

$$\det[\langle z^j, z^i \rangle_T] \neq 0$$

(Example 3.6.2(a)) and the set of equations has a unique solution.

By increasing the dimension of the finite subspaces $\mathscr{H}_n^T$ we obtain a sequence $\{x^{(n)}\}$ which tends to the minimum $\hat{x}$ of f on $\mathscr{H}^T$; the sequence of finite-dimensional spaces $\mathscr{H}_n^T$ of increasing dimension is often referred to as *a nesting sequence of spaces*, since each succeeding member contains all its predecessors. To show that $x^{(n)} \to \hat{x}$ we generate (e.g., by the Gram–Schmidt process, p. 194) another set of basis vectors in $\mathscr{H}_n^T$, $\{v^i\}$, which are orthonormal with respect to the inner product $\langle \cdot, \cdot \rangle_T$. Then, with respect to the basis $\{v^i\}$ the above set of equations becomes diagonal and

$$x^{(n)} = \sum_{i=1}^{n} \langle y, v^i \rangle v^i.$$

Now $\hat{x}$, the minimum point of f in $\mathscr{H}^T$, can be expanded in the Fourier series

$$\hat{x} = \sum_{i=1}^{\infty} \langle \hat{x}, v^i \rangle_T v^i$$

relative to the orthonormal basis $\{v^i\}$ for $\mathscr{H}^T$ (p. 198). But $\langle \hat{x}, v^i \rangle_T = \langle y, v^i \rangle$ (p. 331) so we see that the approximation $x^{(n)}$ is the nth partial sum of the Fourier series for $\hat{x}$: hence $\|\hat{x} - x^{(n)}\|_T \to 0$ and consequently $\|\hat{x} - x^{(n)}\| \to 0$.

Let us embody the foregoing results in

Theorem 4.4.2 *Let T be a positive-bounded-below operator with domain $\mathcal{D}(T)$ dense in a Hilbert space $\mathcal{H}$. Let $\mathcal{H}^T \subset \mathcal{H}$ be the completion of the space with energy inner product*

$$\langle x, z \rangle_T = \langle Tx, z \rangle, \qquad x, z \in \mathcal{D}(T),$$

and let f be the functional

$$f(x) = \tfrac{1}{2}\langle x, x \rangle_T - \langle y, x \rangle \qquad \forall x \in \mathcal{H}^T.$$

Then the minimizing sequence of vectors $\{x^{(n)}\}$ generated by the Rayleigh–Ritz method from a basis in $\mathcal{H}^T$ converges in energy and strongly to the solution of $Tx = y$.

Corollary *Writing $f(x)$ in the form (p. 331)*

$$f(x) = \tfrac{1}{2}(\|x - \hat{x}\|_T^2 - \|\hat{x}\|_T^2)$$

we see that the minimum of f over $\mathcal{H}_n^T$, i.e.,

$$\min_{x \in \mathcal{H}_n^T}(\|x - \hat{x}\|_T^2 - \|\hat{x}\|_T^2),$$

is attained by that point $x^{(n)} \in \mathcal{H}_n^T$ which is closest to $\hat{x}$ in the energy norm sense.

Theorem 3.6.2 tells us that the point giving the minimum distance from $\hat{x} \in \mathcal{H}_T$ to the finite-dimensional subspace $\mathcal{H}_n^T$ is given by the orthogonal projection of $\hat{x}$ onto $\mathcal{H}_n^T$. Hence if $x^{(n)}$ is the minimum point of f in $\mathcal{H}_n^T$,

$$\langle \hat{x} - x^{(n)}, z \rangle_T = 0 \qquad \forall z \in \mathcal{H}_n^T,$$

In particular, this is true for $z = x^{(n)}$, so that from the Pythagorean theorem (p. 185) we have

$$\|\hat{x}\|_T^2 = \|\hat{x} - x^{(n)}\|_T^2 + \|x^{(n)}\|_T^2$$

or

$$\|\hat{x} - x^{(n)}\|_T^2 = \|\hat{x}\|_T^2 - \|x^{(n)}\|_T^2,$$

which is to say that 'the energy in the error is equal to the error in the energies' and the energy in the error (of the approximate solution) decreases monotonically as the dimension of the spaces $\mathcal{H}_n^T$ is increased.

We also see that

$$\|x^{(n)}\|_T \leq \|\hat{x}\|_T,$$

so that the energy of the approximate solution is always less than the energy of the true solution.

Recalling the earlier discussion on natural boundary conditions for a differential operator of order 2s we see that, for this type of operator, the *admissible basis functions* for $\mathscr{H}_n^T$ need only be s times differentiable and satisfy only the essential boundary conditions.

Example 4.4.3

(a) Let us return to the operator of Examples 4.4.1 and 4.4.2(a) and attempt to find a solution of the equation

$$-\frac{d^2x}{dt^2} = 1, \quad x(0) = x'(1) = 0$$

using the Rayleigh–Ritz method.

Let us adopt the basis functions $\{t^i\}, i = 1, 2, 3, \ldots$, which satisfy the essential boundary condition $x(0) = 0$. We transform these into an orthonormal set $\{v^i\}$ with respect to the energy inner product,

$$\langle x, z \rangle_T = \int_0^1 \frac{dx}{dt}\frac{dz}{dt} dt,$$

giving

$$v^1 = t, \quad v^2 = \sqrt{3}(t^2 - t), \quad \text{etc.}$$

The inner products $\alpha_i = \langle y, v^i \rangle$ are $\alpha_1 = \frac{1}{2}$, $\alpha_2 = -\sqrt{3}/6$ and $\alpha_i = 0$ for $i \geq 3$ by virtue of the orthonormalization. Hence the Ritz solution is

$$x(t) = \tfrac{1}{2}t - \tfrac{1}{2}(t^2 - t) = t - t^2/2,$$

which is, of course, the exact solution.

(b) Let us seek the Ritz solution of the operator equation $-\nabla^2 x = 1$ with $x = 0$ on $\partial \Gamma$ (the Dirichlet problem), where Γ is the plane rectangular region $0 \leq t_1 \leq a$, $0 \leq t_2 \leq b$. The energy inner product is (Example 4.4.2(b))

$$\langle x, z \rangle_T = \int_\Gamma \sum_{i=1}^2 \frac{\partial x}{\partial t_i}\frac{\partial z}{\partial t_i} d\Gamma.$$

The functions

$$v^{ij} = \sin\frac{i\pi t_1}{a} \sin\frac{j\pi t_2}{b}$$

are in $\mathscr{H}^T$ (in fact in the domain of the operator): they are also orthogonal with respect to the energy inner product since we have

$$\langle v^{ij}, v^{rs} \rangle_T = \int_0^a \int_0^b \frac{ir\pi^2}{a^2} \left(\cos\frac{i\pi t_1}{a} \cos\frac{r\pi t_1}{a} \right) \left(\sin\frac{j\pi t_2}{b} \sin\frac{s\pi t_2}{b} \right) dt_1 dt_2$$

$$+ \int_0^a \int_0^b \frac{js\pi^2}{b^2} \left(\sin\frac{i\pi t_1}{a} \sin\frac{r\pi t_2}{a} \right)$$

$$\times \left(\cos\frac{j\pi t_2}{b} \cos\frac{s\pi t_2}{b} \right) dt_1 dt_2.$$

If $i \neq r$ or $j \neq s$ then $\langle v^{ij}, v^{rs} \rangle = 0$ while if $i = r$ and $j = s$ we have

$$\langle v^{ij}, v^{ij} \rangle_T = \|v_{ij}\|_T^2 = \frac{\pi^2(i^2 b^2 + j^2 a^2)}{4ab}.$$

The inner products

$$\langle 1, v^{ij} \rangle = \frac{ab}{ij\pi^2}(1 - (-1)^i)(1 - (-1)^j)$$

$$= \begin{cases} 0, & \text{if } i \text{ or } j \text{ is even,} \\ 4ab/ij\pi^2, & \text{if } i \text{ and } j \text{ are odd;} \end{cases}$$

hence the Ritz solution is

$$x(t_1, t_2) = \frac{16 a^2 b^2}{\pi^4} \sum_{i,j} \frac{\sin\frac{i\pi t_1}{a} \sin\frac{j\pi t_2}{b}}{ij(i^2 b^2 + j^2 a^2)}, \quad i, j = 1, 3, 5, 7, \ldots,$$

and

$$\|x\|_T^2 = \left(\frac{4ab}{\pi^2} \right)^3 \sum_{i,j} \frac{1}{i^2 j^2 (i^2 b^2 + j^2 a^2)}.$$

Practical implementation of the Rayleigh–Ritz method depends crucially on the availability of a basis in $\mathcal{H}^T$: while these are not difficult to find for regular domains in two- and three-dimensional problems, the situation is not straightforward for the irregular domains which arise in practical problems. We shall see how to overcome this technical difficulty when we discuss the so-called finite element method in Section 4.6.

For the moment we shall show how the Rayleigh–Ritz method can be applied to the solution of a certain class of non-linear operator equations. The resulting Ritz equations are of course themselves non-linear.

Potential operators

Let f be a functional on the normed vector space $\mathscr{V}$ and suppose that f is Fréchet differentiable at $\tilde{x} \in \mathscr{V}$ (Definition 4.2.2). The Fréchet derivative $f'(\tilde{x})$ is an element of the space $\mathscr{L}(\mathscr{V}, \mathscr{R})$, which is to say that $f'(\tilde{x}) \in \mathscr{V}^*$, the dual of $\mathscr{V}$. The differential of f at $\tilde{x}$ is given by

$$df(\tilde{x}; s) = f'(\tilde{x})s$$

and since $f'(\tilde{x}) \in \mathscr{V}^*$ we can write this in the form

$$df(\tilde{x}; s) = [f'(\tilde{x}), s].$$

Considered in this way the Fréchet derivative of the functional f at $\tilde{x}$ is often referred to as the *gradient of f at $\tilde{x}$*. If f is Fréchet differentiable at every point of $\mathscr{U} \subset \mathscr{V}$ then the mapping $f': \mathscr{U} \to \mathscr{V}^*$ (see p. 295) is called the *gradient of f*.

We are now ready for

Definition 4.4.2 *A mapping $P: \mathscr{U} \to \mathscr{V}^*$, $\mathscr{U} \subset \mathscr{V}$, is called a* potential operator *(or* gradient mapping) *if there exists a differentiable functional $f: \mathscr{U} \to \mathscr{R}$ such that $f' = P$ on $\mathscr{U}$.*

Given a non-linear operator we would like to have a test which will tell us whether the operator is potential or not. It can be shown [62, 75] (*see also* Exercise 4.4,6) that a necessary and sufficient condition for an operator $P: \mathscr{U} \to \mathscr{V}^*$ to be potential on $\mathscr{U}$ is that the bilinear functional (Exercise 3.5,9) $[P'(x)s_1, s_2]$ is symmetric; that is,

$$[P'(x)s_1, s_2] = [s_1, P'(x)s_2] \qquad \forall x \in \mathscr{U}, \quad s_1, s_2 \in \mathscr{V}.$$

The importance of potential operators in the present context is that the non-linear operator equation $Px = y$ can be solved using the Ritz method under conditions which are very similar to those for positive linear operators. We shall not give proofs of the following results, but simply state them as a theorem: for comparison with our previous work, a Hilbert space setting is adopted [62, 75].

Theorem 4.4.3 (i) *Let P be a non-linear operator defined on a dense subspace $\mathscr{D}(P)$ of a Hilbert space $\mathscr{H}$. Assume $P\theta = \theta$ and let P be a potential operator whose derivative P' is positive definite for all $x \in \mathscr{D}(P)$, i.e.,*

$$\langle P'(x)s, s \rangle > 0 \qquad \forall s \in \mathscr{D}(P).$$

If the operator equation $Px = y$ has a solution $\tilde{x}$ then the solution is unique and minimizes the functional

$$f(x) = \int_0^1 \langle P\lambda x, x \rangle \, d\lambda - \langle y, x \rangle.$$

(ii) *If the derivative P' is positive-bounded-below, i.e.,*

$$\langle P'(x)s, s \rangle \geq \beta^2 \langle s, s \rangle \qquad \forall x, s \in \mathcal{D}(P),$$

then any minimizing sequence for the functional f converges strongly to some limit in $\mathcal{H}$. The limit is a (generalized) solution of the operator equation (p. 332).

(iii) *If, in addition, there exist scalars γ, β such that*

$$\langle P'(x)s, s \rangle \geq \gamma^2 \langle P'(\theta)s, s \rangle \geq \beta^2 \langle s, s \rangle \qquad \forall x, s \in \mathcal{D}(P),$$

then the minimum of $f(x)$ is the limit of a sequence in the space $\mathcal{H}^{P'(\theta)}$, the energy space with norm $\langle P'(\theta)s, s \rangle^{1/2}$.

Example 4.4.4

(a) Let T be a linear positive definite operator. In this case (Exercise 4.2,4(a))

$$f(x) = \int_0^1 \langle T\lambda x, x \rangle \, d\lambda - \langle y, x \rangle$$

$$= \int_0^1 \lambda \langle Tx, x \rangle \, d\lambda - \langle y, x \rangle$$

$$= \frac{1}{2} \langle Tx, x \rangle - \langle y, x \rangle.$$

(b) Consider the steady-state diffusion equation in two space dimensions augmented by a non-linear contribution viz.,

$$Px = -\nabla^2 x + r(x) = 0 \quad \text{in} \quad \Gamma,$$

with $x = 0$ on $\partial \Gamma$: here r is a scalar function of $x(t_1, t_2)$. The differential of P is

$$P'(x)s = -\nabla^2 s + r'(x)s$$

and the bilinear functional

$$\int_\Gamma (-\nabla^2 s_1 + r'(x)s_1)s_2 \, d\Gamma$$

is clearly symmetric; P' is positive definite if $r'(x) > 0$ for all x. The associated functional $f(x)$ is

$$f(x) = \int_\Gamma \left(\frac{1}{2} \sum_{i=1}^{2} \left(\frac{\partial x}{\partial t_i} \right)^2 + x \int_0^1 r(\lambda x) \, d\lambda \right) d\Gamma,$$

which can be rewritten in the form

$$f(x) = \int_\Gamma \left(\frac{1}{2} \sum_{i=1}^{2} \left(\frac{\partial x}{\partial t_i}\right)^2 + \int_0^x r(\lambda) d\lambda \right) d\Gamma.$$

Further examples of potential operators, including the application of the Rayleigh–Ritz method, are given in [65] and [75].

Application of the Rayleigh–Ritz method to the computation of eigenvalues

In this section we assume T to be a symmetric positive-bounded-below transformation with compact resolvent; in Section 3.9 we saw that such transformations have a spectrum which consists entirely of a finite or countable set of real eigenvalues. The eigenvalues can be written as an ordered increasing sequence $\lambda_1, \lambda_2, \ldots, \lambda_n, \ldots$; since T is positive definite, all the eigenvalues are positive.

Because T is assumed to be positive-bounded-below, we have

$$\langle Tx, x \rangle / \langle x, x \rangle \geq \beta^2,$$

so that the left-hand side has an infimum; we now show that the infimum is λ_1.

Theorem 4.4.4 *If the domain $\mathscr{D}(T)$ of T is dense in $\mathscr{H}$ and there exists $x^{(1)} \in \mathscr{D}(T)$ such that the infimum*

$$\lambda_1 = \inf_{\|x^{(1)}\| = 1} \langle Tx^{(1)}, x^{(1)} \rangle$$

is actually attained, then λ_1 is the smallest eigenvalue of T and $x^{(1)}$ is the corresponding eigenvector.[†]

Proof The proof is almost identical to that used in Example 3.8.3(a), relating to the symmetric matrix transformation S. The denseness of $\mathscr{D}(T)$ needs to be used to conclude that $Tx^{(1)} = \lambda_1 x^{(1)}$ from the statement

$$\langle (\lambda_1 I - T)x^{(1)}, x \rangle = 0 \quad \forall x \in \mathscr{D}(T). \blacksquare$$

We see from this theorem that we can characterize the smallest eigenvalue of T as the minimum of the functional $\langle Tx, x \rangle$ subject

[†] The ratio $\langle Tx, x \rangle / \langle x, x \rangle$ is often called *Rayleigh's quotient* and the fact that λ_1 is given by the minimum of this quotient is called *Rayleigh's principle*.

to the constraint $\|x\| = 1$. We can find successive eigenvalues in the following way. Suppose we have found the first r orthonormal eigenvectors of T: then λ_{r+1} is the infimum attained by the functional $\langle Tx, x \rangle$ subject to the constraint $\|x\| = 1$ and the orthogonality constraints $\langle x, x^{(s)} \rangle = 0$, $s = 1, 2, \ldots, r$ (Exercise 4.4,7(a)).

Let us now apply the Rayleigh–Ritz method to find approximations to the eigenvalues and eigenvectors of T by seeking minima of the functional $\langle Tx, x \rangle$ (subject to the above constraints), not in $\mathscr{H}$ but in a sequence of finite-dimensional subspaces $\mathscr{H}_n$. We shall begin by seeking an approximation to the smallest eigenvalue λ_1 by finding the minimum of the functional

$$f(x) = \langle Tx, x \rangle - \lambda \langle x, x \rangle$$

in the finite-dimensional subspace $\mathscr{H}_n \subset \mathscr{D}(T)$, spanned by the vectors $\{z^1, z^2, \ldots, z^n\}$. If the minimum in $\mathscr{H}_n$ is

$$\tilde{x}^{(1)} = \sum_{i=1}^{n} \xi_i^{(1)} z^i$$

then by steps which are virtually identical to those followed on p. 335 we find the set of linear homogeneous equations

$$\sum_{j=1}^{n} (\langle Tz^j, z^i \rangle - \lambda \langle z^j, z^i \rangle) \xi_j^{(1)} = 0, \qquad i = 1, 2, \ldots, n,$$

for the $\xi_j^{(1)}$. This is a finite-dimensional (matrix) eigenvalue problem; the smallest eigenvalue, $\tilde{\lambda}_1$, is an approximation to the smallest eigenvalue λ_1 of T and $\tilde{x}^{(1)}$ is an approximation (in the mean square sense) to the eigenvector $x^{(1)}$.

In fact, the approximation $\tilde{\lambda}_1$ is an overestimate or upper bound for the eigenvalue λ_1 since the minimum of $\langle Tx, x \rangle$ on $\mathscr{H}_n$ is clearly greater than (more strictly, not less than) the minimum of $\langle Tx, x \rangle$ on $\mathscr{H}$.

Now let us move on to the second smallest eigenvalue: we construct the Lagrangian

$$L(x) = \langle Tx, x \rangle - \lambda \langle x, x \rangle - 2\mu \langle x, \tilde{x}^{(1)} \rangle$$

to be minimized on $\mathscr{H}_n$. Notice that we can only demand orthogonality of x to the already found approximate eigenvector $\tilde{x}^{(1)} \in \mathscr{H}_n$.

Let

$$x^{(2)} = \sum_{i=1}^{n} \xi_i^{(2)} z^i$$

then we find
$$\sum_{j=1}^{n} (\langle Tz^j, z^i \rangle - \lambda \langle z^j, z^i \rangle)\xi_j^{(2)} - \mu \langle z^j, z^i \rangle \xi_j^{(1)} = 0.$$

Multiplying both terms by $\xi_i^{(1)}$, summing over i and rearranging leads to
$$\sum_{i=1}^{n} \xi_i^{(2)} \sum_{j=1}^{n} (\langle Tz^j, z^i \rangle - \lambda \langle z^j, z^i \rangle)\xi_j^{(1)} - \mu = 0.$$

But
$$\sum_{j=1}^{n} \langle Tz^j, z^i \rangle \xi_j^{(1)} = \tilde{\lambda}_1 \sum_{j=1}^{n} \langle z^j, z^i \rangle \xi_j^{(1)};$$

hence
$$\sum_{i=1}^{n} \xi_i^{(2)} (\tilde{\lambda}_1 - \lambda) \sum_{j=1}^{n} \langle z^j, z^i \rangle \xi_j^{(1)} - \mu = 0$$

or
$$(\tilde{\lambda}_1 - \lambda)\langle \tilde{x}^{(1)}, \tilde{x}^{(2)} \rangle - \mu = 0,$$

implying that the value of the Lagrange multiplier μ is zero. Thus we obtain the same set of equations as we had before but now we approximate the second smallest eigenvalue of T by the second smallest eigenvalue of the finite system,
$$\sum_{j=1}^{n} (\langle Tz^j, z^i \rangle - \lambda \langle z^j, z^i \rangle)\xi_j^{(2)} = 0.$$

A similar analysis applies to successive eigenvalues, so that we may say, in summary, that we can obtain approximations to the first n eigenvalues and eigenvectors of T by finding the eigenvalues and eigenvectors of the matrix equation
$$([k_{ij}] - \lambda[m_{ij}])\{\xi_j\} = 0,$$

where
$$k_{ij} = \langle Tz^j, z^i \rangle, \quad m_{ij} = \langle z^j, z^i \rangle.$$

It can readily be shown [81] that each approximate eigenvalue is an upper bound to the exact eigenvalue and that as the dimension of the approximating space $\mathcal{H}_n$ is increased each upper bound is lowered (at the same time, of course, we begin to approximate more eigenvalues of T), giving monotonic convergence of the eigenvalues to the exact values.

The foregoing discussion has assumed that the approximating spaces are subspaces of $\mathscr{D}(T)$, but clearly all the above results hold for sequences generated from nesting subspaces $\mathscr{H}_n^T$ of the energy space $\mathscr{H}^T$. In that case, we should rewrite the approximating matrix equation as

$$\sum_{j=1}^{n} (\langle z^j, z^i \rangle_T - \lambda \langle z^j, z^i \rangle) \xi_j = 0.$$

Before concluding this discussion of the Rayleigh–Ritz method we should remark that in all the foregoing considerations perfect arithmetic has been assumed. If the dimension of the approximating subspaces is small there is little cause for concern, but when spaces of large dimension are used then one must begin to consider the numerical stability of the resulting set of equations. This question relates to the specification of the basis functions in the spaces $\mathscr{H}_n^T$. The question is fully discussed in [78] and we shall not pursue it here, although we shall touch on it again when we discuss the finite element method.

The method of orthogonal projections

The corollary to Theorem 4.4.2 (p. 336) shows that the energy of an approximate solution of the operator equation $Tx = y$ is always less than the energy of the exact solution; unfortunately we have no way of estimating the actual difference between the two, since we do not know the exact solution. We can only say that $\|x^{(n)}\|_T$ provides a lower bound for $\|\hat{x}\|_T$: if we could by some means also find a quantity which provided an upper bound to $\|\hat{x}\|_T$ then we would have a direct estimate of the error of the approximate solution (in the sense of the energy norm, of course). For a certain class of positive-bounded-below operators it is possible to find an upper bound and it is this method which we now describe.

We shall assume that the positive-bounded-below transformation $T: \mathscr{H}_1 \to \mathscr{H}_1$ can be represented as the composition S^*S of the linear operator $S: \mathscr{H}_1 \to \mathscr{H}_2$ and its adjoint $S^*: \mathscr{H}_2 \to \mathscr{H}_1$. We have already given an instance of this in Example 3.7.4(d) where we showed that S^*S is necessarily symmetric (*see* Fig. 4.2).

Putting $Sx = w, w \in \mathscr{H}_2$, the operator equation $Tx = y$ becomes $S^*w = y$, so that now we have to find

(a) $\hat{w} \in \mathscr{H}_2$ such that $S^*\hat{w} = y$;
(b) $\hat{x} \in \mathscr{H}_1$ such that $S\hat{x} = \hat{w}$.

DIRECT SOLUTION OF POSITIVE OPERATOR EQUATIONS

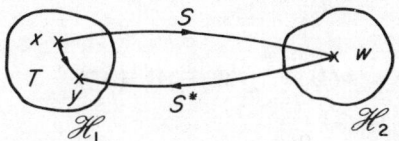

Fig. 4.2

In most practical instances it is sufficient to find $\hat{w}$ only, since the quantity of interest (and the quantity we are trying to bound) is the energy norm $\langle x, x \rangle_T$.

Now

$$\langle x, x \rangle_T = \langle Tx, x \rangle_{\mathcal{H}_1}$$
$$= \langle S^*Sx, x \rangle_{\mathcal{H}_1}$$
$$= \langle Sx, Sx \rangle_{\mathcal{H}_2}$$
$$= \langle w, w \rangle_{\mathcal{H}_2},$$

so that a knowledge of w enables us to know $\langle x, x \rangle_T$ (since $\mathcal{D}(T)$ is dense in $\mathcal{H}_1^T$ the equality is true for all $x \in \mathcal{H}_1^T$).

In what follows we shall assume that $\mathcal{N}(S^*) \neq \mathcal{O}$. In that case we can write

$$\mathcal{H}_2 = \mathcal{N}(S^*) \oplus \mathcal{N}^\perp(S^*),$$

and since $\mathcal{R}(S) \subset \mathcal{N}^\perp(S^*)$ (p. 214) then it follows that $\hat{w} \in \mathcal{N}^\perp(S^*)$: the situation is illustrated diagrammatically in Fig. 4.3.

Let $w_0 \in \mathcal{H}_2$ be some vector satisfying the equation

$$S^*w_0 = y$$

and let M be the coset $w_0 + \mathcal{N}(S^*)$. Then it is clear that the vector we seek, $\hat{w}$, lies at the intersection of the coset M and the subspace $\mathcal{N}^\perp(S^*)$: that is to say, $\hat{w}$ is the (orthogonal) projection of w_0 onto $\mathcal{N}^\perp(S^*)$.

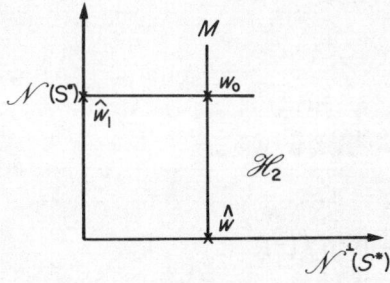

Fig. 4.3

This means that
$$\langle \hat{w} - w_0, w_2 \rangle_{\mathcal{H}_2} = 0 \quad \forall w_2 \in \mathcal{N}^\perp(S^*)$$
or
$$\langle S\hat{x} - w_0, Sx \rangle_{\mathcal{H}_2} = 0$$
or
$$\langle S^*S\hat{x} - S^*w_0, x \rangle_{\mathcal{H}_1} = 0 \quad \forall x \in \mathcal{D}(T)$$
or
$$\langle T\hat{x} - y, x \rangle_{\mathcal{H}_1} = 0,$$
which result is equivalent to our former method for characterizing the solution of $Tx = y$ by minimizing the functional
$$f(x) = \tfrac{1}{2}\langle x, x \rangle_T - \langle y, x \rangle_{\mathcal{H}_1}$$
(*see* Corollary to Theorem 4.4.2).

However, we have another orthogonal projection at our disposal, namely the projection of w_0 onto $\mathcal{N}(S^*)$. This vector, which we shall call $\hat{w}_1$, is characterized by
$$\langle \hat{w}_1 - w_0, w_1 \rangle_{\mathcal{H}_2} = 0 \quad \forall w_1 \in \mathcal{N}(S^*)$$
or equivalently as the minimum of the functional
$$g(w_1) = \tfrac{1}{2}\langle w_1, w_1 \rangle_{\mathcal{H}_2} - \langle w_0, w_1 \rangle_{\mathcal{H}_2}$$
over the subspace $\mathcal{N}(S^*)$. If we apply the Rayleigh–Ritz method to calculate a sequence of approximations for the minimum point of $g(w)$, admitting as trial functions elements of $\mathcal{N}(S^*)$, then we shall obtain in this way a sequence of upper bounds for the functional $g(w_1)$. This will clearly lead to a sequence of lower bounds for the functional
$$h(w_1) = -\tfrac{1}{2}\langle w_0, w_0 \rangle_{\mathcal{H}_2} - g(w_1),$$
$$= -\tfrac{1}{2}\langle w_0 - w_1, w_0 - w_1 \rangle_{\mathcal{H}_2}.$$

Now clearly (Fig. 4.3)
$$w_0 = \hat{w}_1 + \hat{w}$$
so that,
$$\max_{w_1 \in \mathcal{N}(S^*)} h(w_1) = -\tfrac{1}{2}\langle \hat{w}, \hat{w} \rangle_{\mathcal{H}_2}$$
$$= -\tfrac{1}{2}\langle \hat{x}, \hat{x} \rangle_T$$
$$= \min_{x \in \mathcal{H}^T} f(x);$$

hence at the solution point the maximum and minimum values of

DIRECT SOLUTION OF POSITIVE OPERATOR EQUATIONS 347

the functionals h and f are equal. It follows that if $x^{(n)} \in \mathcal{H}^T$, $w^{(m)} \in \mathcal{N}(S^*)$ are Rayleigh–Ritz approximations for $\hat{x}$ and $\hat{w}_1$, then

$$f(x^{(n)}) \geqslant -\tfrac{1}{2}\langle \hat{x}, \hat{x}\rangle_T \geqslant h(w^{(m)})$$

or (p. 336)

$$\|x^{(n)}\|_T^2 \leqslant \|\hat{x}\|_T^2 \leqslant \|w_0\|_{\mathcal{H}_2}^2 - \|w^{(m)}\|_{\mathcal{H}_2}^2,$$

thus we have provided simultaneous upper and lower bounds for $\|\hat{x}\|_T^2$. The error estimate can be put in the alternative form (Exercise 4.4,9(b))

$$\|\hat{x} - x^{(n)}\|_T^2 \leqslant (\|w_0\|_{\mathcal{H}_2}^2 - \|w^{(m)}\|_{\mathcal{H}_2}^2) - \|x^{(n)}\|_T^2.$$

If $\{w^i\}$ is a basis for an m-dimensional subspace of $\mathcal{N}(S^*)$ then the Ritz equations for

$$w^{(m)} = \sum_{i=1}^{m} \phi_i^{(m)} w^i$$

take the form

$$\sum_{j=1}^{m} \langle w^j, w^i \rangle \phi_j^{(m)} = \langle w_0, w^i \rangle, \qquad 1 \leqslant i \leqslant m.$$

The principal difficulty with the method of orthogonal projections lies in the generation of suitable basis functions in the subspace $\mathcal{N}(S^*)$. When the operator S^* is a partial differential operator, for example, defined on an irregular domain then it may simply be impossible to find solutions of $S^*w = \theta$. We give now a straightforward example of the method.

Example 4.4.5 In Example 4.4.3(b) we applied the Rayleigh–Ritz method to the solution of $-\nabla^2 x = 1$ in the rectangular region $0 \leqslant t_1 \leqslant a, 0 \leqslant t_2 \leqslant b$ with x vanishing on the boundary. Let us find bounds on the quantity $\|\hat{x}\|_T$, using the method of orthogonal projections for the special case $a = b = 1$.

In this instance the operators S and S^* are (see Example 3.7.3(f))

$$S \equiv \text{grad}, \qquad S^* \equiv -\text{div}$$

with inner products

$$\langle x, z \rangle_T = \int_\Gamma \sum_{i=1}^{2} \frac{\partial x}{\partial t_i} \frac{\partial z}{\partial t_i} d\Gamma$$

and

$$\langle w, u \rangle_{\mathcal{H}_2} = \int_\Gamma \sum_{i=1}^{2} \phi_i v_i \, d\Gamma,$$

where w, u are the two-dimensional vectors with coordinates (ϕ_1, ϕ_2), (v_1, v_2), respectively.

We need to choose the vector w_0 and a basis set for a finite-dimensional subspace of $\mathcal{N}(S^*)$, that is, a set of vectors with vanishing divergence. We shall choose $w_0 = (\frac{1}{2} - t_1, 0)$ and

$$w^{ij} = \pi(j \cos i\pi t_1 \sin j\pi t_2, -i \sin i\pi t_1 \cos j\pi t_2), \qquad i, j = 1, 3, 5, 7, \ldots.$$

Here we are taking advantage of the fact that the solution $x(t_1, t_2)$ will be symmetrical about the point $(\frac{1}{2}, \frac{1}{2})$; hence, relative to this point, the vectors $w \in \mathcal{R}(S)$ will have the properties

$$(v_1, v_2) = (\text{odd in } t_1/\text{even in } t_2, \text{ even in } t_1/\text{odd in } t_2).$$

Taking just one function for each projection we find (Exercise 4.4, 9(c))

$$0.182 \leqslant \|\hat{x}\|_T \leqslant 0.223$$

and taking four functions (i.e., $i, j = 1, 3$) we find

$$0.187 \leqslant \|\hat{x}\|_T \leqslant 0.208,$$

giving relative errors in $\|\hat{x} - x^{(n)}\|$ of 20 per cent and 10 per cent, respectively.

An alternative method for estimating the error of an approximate solution is due to Trefftz: the reader will find the method explained and exemplified in [34]. As has been pointed out, the Rayleigh–Ritz method gives upper bound estimates for the eigenvalues of a positive operator. The question of obtaining lower bound estimates is still largely unsolved, although a method due to Weinstein and fully described in [81] is available. This method requires the construction of a sequence of operators whose eigenvalues represent lower bounds for the eigenvalues of the given operator (*see also* [80]).

A method for finding upper and lower bounds on the lowest eigenvalue is due to Temple [86].

Complementary variational principles

The idea of obtaining upper and lower bound estimates for a functional of the solution of an operator equation can be extended somewhat beyond the situation covered by the method of orthogonal projections. The theory, which leads to complementary variational principles (or dual extremum principles), has been fully developed only within the last decade, but its origins lie in earlier work, including of course the method of orthogonal projections, and in the duality which exists between the Lagrangian and Hamiltonian formulations of the classical variational principles of mechanics.

We begin with a pair of operator equations which represent a generalization of the canonical Euler equations of Example 4.3.2(d), viz.,

$$\left.\begin{array}{l} Sx = a'_w(x,w) \\ S^*w = a'_x(x,w) \end{array}\right\} \quad x \in \mathscr{D}(S) \in \mathscr{H}_1, \quad w \in \mathscr{D}(S^*) \in \mathscr{H}_2.$$

Here $S: \mathscr{H}_1 \to \mathscr{H}_2$ is a linear operator and $S^*: \mathscr{H}_2 \to \mathscr{H}_1$ its adjoint (cf., p. 344 and Fig. 4.2); $a(x,w)$ is a Fréchet differentiable functional defined on the product space $\mathscr{H}_1 \times \mathscr{H}_2$ and $a'_x(x,w), a'_w(x,w)$ denote its partial Fréchet derivatives.[†]

We recall that $a'_w \in \mathscr{H}_2^* = \mathscr{H}_2$, so that both sides of the first equation are (as they must be) elements of $\mathscr{H}_2$; similarly both sides of the second equation are elements of $\mathscr{H}_1$.

Now consider the functional

$$\begin{aligned} l(x,w) &= \langle x, S^*w \rangle_{\mathscr{H}_1} - a(x,w) \\ &= \langle Sx, w \rangle_{\mathscr{H}_2} - a(x,w). \end{aligned}$$

Let $(\hat{x}, \hat{w})$ be a stationary point of l, then

$$dl(\hat{x};s,\hat{w};r) = \frac{d}{d\lambda} l(x+\lambda s)\Big|_{\lambda=0} + \frac{d}{d\mu} l(\hat{w}+\mu r)\Big|_{\mu=0} = 0$$

$$\forall s \in \mathscr{H}_1, \quad r \in \mathscr{H}_2.$$

This implies (Exercise 4.4,10(a)) that $(\hat{x}, \hat{w})$ is the solution of the equations

$$Sx = a'_w, \quad S^*w = a'_x;$$

that is to say, these are the (canonical) Euler equations corresponding to the functional $l(x,w)$.

Suppose now that in $l(x,w)$ we choose w so that it is a solution of the first Euler equation,

$$Sx = a'_w.$$

[†] The partial Fréchet derivative [64] (cf., Definition 4.2.2) is given by

$$a(x+s,w) - a(x,w) = a'_x(x,w)s + \varepsilon(x,w;s), \quad s \in \mathscr{H}_1$$

with

$$\lim_{\|s\|_{\mathscr{H}_1} \to 0} \frac{|\varepsilon(x,w;s)|}{\|s\|_{\mathscr{H}_1}} = 0.$$

If we denote this solution by $\tilde{w}(x)$ then we may write

$$l(x, \tilde{w}(x)) = l_1(x) = \langle Sx, \tilde{w}(x) \rangle_{\mathscr{H}_2} - a(x, \tilde{w}(x))$$

and clearly $l_1(x)$ will be stationary at the solution point $(\hat{x}, \tilde{w}(\hat{x}))$ of both Euler equations.

Similarly if $\tilde{x}(w)$ denotes a solution of the second Euler equation

$$S^*w = a'_x$$

then the functional

$$l(\tilde{x}(w), w) = l_2(w) = \langle \tilde{x}(w), S^*w \rangle_{\mathscr{H}_1} - a(\tilde{x}(w), w)$$

will be stationary at the solution point $(\tilde{x}(\hat{w}), \hat{w})$ of both Euler equations. Furthermore, at the point $(\hat{x}, \hat{w})$ we clearly have

$$l_1(\hat{x}) = l_2(\hat{w}).$$

If we could show that $\hat{x}$ was a minimum point for the functional $l_1(x)$ and that $\hat{w}$ was a maximum point for the functional $l_2(w)$ then we could make the statement

$$l_1(x) = l(x, \tilde{w}(x)) \geqslant l(\hat{x}, \hat{w}) \geqslant l(\tilde{x}(w), w) = l_2(w),$$

so providing upper and lower bounds on the functional $l(\hat{x}, \hat{w})$.[†]

We can ensure that this desirable situation holds by placing suitable conditions on the functional $a(x, w)$: in effect we require the functional $l(x, w)$ to exhibit, near $(\hat{x}, \hat{w})$, the characteristic illustrated in Fig. 4.4. Clearly this will be ensured if the second (Fréchet) derivatives of $l_1(x)$ and $l_2(w)$ satisfy

$$l_1''(\hat{x})s^2 \geqslant 0, \quad s \in B(\hat{x}; r_1) \in \mathscr{H}_1;$$
$$l_2''(\hat{w})r^2 \leqslant 0, \quad r \in B(\hat{w}; r_2) \in \mathscr{H}_2.$$

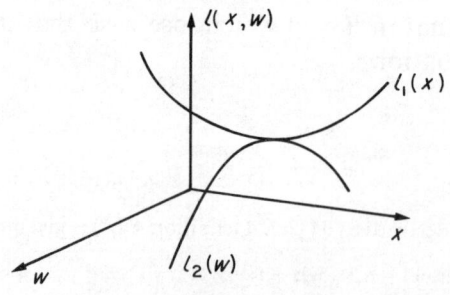

Fig. 4.4

[†] Clearly minimum and maximum can be interchanged throughout—this is simply equivalent to interchanging the variable names.

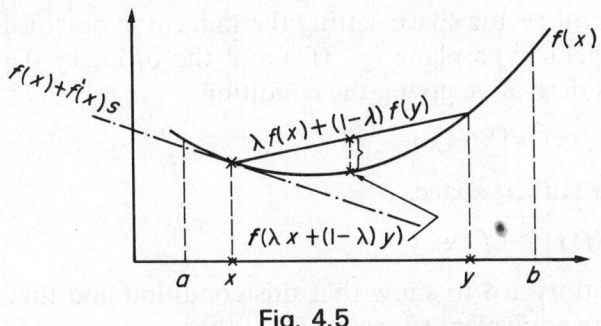

Fig. 4.5

This follows directly from Taylor's theorem (Theorem 4.2.2).

However, a global rather than merely a local approach to the desired condition is to require that, in some region, $a(x, w)$ is a saddle functional, concave in x and convex in w. This approach has the added advantage that it can be applied to situations wherein the Euler equations are replaced by inequalities [88]. Let us now explain what we mean by these terms.

We first recall what we mean by a convex set (Definition 2.3.6). For any two points x, y in a convex set $S \subset \mathscr{V}$, the line segment $\{\lambda x + (1 - \lambda)y \mid 0 < \lambda < 1\} \in S$. An ordinary function of one variable is said to be *convex* if it has the characteristic shape illustrated in Fig. 4.5. We can embody this characteristic into an analytic statement by observing that the value of the function at the point $\lambda x + (1 - \lambda)y$ in the convex set (a, b) lies below the corresponding point on the chord (line-segment) joining $f(x)$ and $f(y)$; that is to say,

$$f(\lambda x + (1 - \lambda)y) \leq \lambda f(x) + (1 - \lambda)f(y), \quad 0 < \lambda < 1.$$

This leads to

Definition 4.4.3 *The real functional $f(x)$ defined on a convex subset S of a vector space $\mathscr{V}$ is said to be* **convex** *if*

$$f(\lambda x + (1 - \lambda)y) \leq \lambda f(x) + (1 - \lambda)f(y), \quad 0 < \lambda < 1, \quad \forall x, y \in S.$$

The functional is said to be concave *if the inequality is reversed.*

Another way of characterizing convexity for an ordinary function is to note (Fig. 4.5) that the function lies above its tangent at any point. We may embody this in the statement

$$f(y) - f(x) \geq f'(x)(y - x) \quad \forall x, y \in (a, b).$$

In a general vector space setting the tangent is obviously replaced by the tangent hyperplane (p. 311) and the ordinary derivative by the Fréchet derivative giving the condition

$$f(y) - f(x) \geq [f'(x), y - x]$$

or, if $\mathscr{V}$ is a Hilbert space,

$$f(y) - f(x) \geq \langle f'(x), y - x \rangle.$$

It is straightforward to show that this condition and that of Definition 4.4.3 are equivalent (Exercise 4.4,10(b)).

The functional $f(x, w)$ is said to be a *saddle functional*, concave in x and convex in w, if for (x, w), (y, z) in some convex subset of $\mathscr{H}_1 \times \mathscr{H}_2$,

$$f(y, z) - f(x, w) \geq \langle f'_x(y, z), y - x \rangle_{\mathscr{H}_1} + \langle f'_w(x, w), z - w \rangle_{\mathscr{H}_2}$$

If $a(x, w)$ is assumed to have this property then we have

$$\begin{aligned}
l_1(x) - l_1(\hat{x}) &= \langle Sx, \tilde{w} \rangle_{\mathscr{H}_2} - \langle S\hat{x}, \hat{w} \rangle_{\mathscr{H}_2} + (a(\hat{x}, \hat{w}) - a(x, \tilde{w})) \\
&\geq \langle Sx, \tilde{w} \rangle_{\mathscr{H}_2} - \langle S\hat{x}, \hat{w} \rangle_{\mathscr{H}_2} + \langle a'_x(\hat{x}, \hat{w}), \hat{x} - x \rangle_{\mathscr{H}_1} \\
&\quad + \langle a'_w(x, \tilde{w}), \hat{w} - \tilde{w} \rangle_{\mathscr{H}_2} \\
&= \langle Sx, \tilde{w} \rangle_{\mathscr{H}_2} - \langle \hat{x}, S^*\hat{w} \rangle_{\mathscr{H}_1} + \langle S^*\hat{w}, \hat{x} - x \rangle_{\mathscr{H}_1} \\
&\quad + \langle Sx, \hat{w} - \tilde{w} \rangle_{\mathscr{H}_2} \\
&= \langle Sx, \hat{w} \rangle_{\mathscr{H}_2} - \langle S^*\hat{w}, x \rangle_{\mathscr{H}_1} = 0,
\end{aligned}$$

hence

$$l_1(x) \geq l_1(\hat{x}) = l(\hat{x}, \hat{w});$$

a similar argument shows that

$$l_2(w) \leq l_2(\hat{w}) = l(\hat{x}, \hat{w}).$$

We have thus shown that the condition that $a(x, w)$ be a saddle functional is sufficient to ensure the existence of complementary upper and lower bounds on $l(x, w)$. The reader should note that the functional $a(x, w)$ can be generated from a given pair of operator equations by direct integration.

When the operators S, S^* are differential operators over a region Γ then the vectors x and w must clearly be restricted to their respective domains, and this will involve satisfaction of the boundary conditions on $\partial \Gamma$. Inhomogeneous boundary equations can be formally included in the functional $a(x, w)$ by the addition of appropriate boundary integrals: this approach is extensively adopted in [87, 88] wherein the forms of the boundary terms are set out for a variety of cases. We shall merely illustrate the procedure through the examples which follow. The role of complementary variational principles in a range of physical problems is to be found in [89, 90].

Example 4.4.6

(a) Consider the Sturm–Liouville differential equation on [0, 1],

$$-\frac{d}{dt}\left(p(t)\frac{dx}{dt}\right) + q(t)x = y(t), \qquad x(0) = x(1) = 0,$$

with $p(t) > 0$, $q(t) \geq 0$ (see Example 3.10.5).

We rewrite this second-order differential equation as the first-order pair

$$\frac{dx}{dt} = \frac{w}{p}, \qquad -\frac{dw}{dt} = -qx + y.$$

We identify S with the operator d/dt, $x(0) = x(1) = 0$ and S^* with the adjoint $-d/dt$ subject to no boundary conditions: note that $\mathscr{H}_1 = \mathscr{H}_2$. Then we have

$$a'_w = w/p, \qquad a'_x = -qx + y;$$

and by direct integration we find that

$$a(x, w) = \tfrac{1}{2}\langle w/p, w\rangle - \tfrac{1}{2}\langle qx, x\rangle + \langle y, x\rangle.$$

It is readily verified that a is convex in w and concave in x, and explicitly, since the inner product on $\mathscr{H}_1 = \mathscr{H}_2$ is $\int_0^1 fg\,dt$,

$$a(x, w) = \int_0^1 \left(\tfrac{1}{2}\frac{w^2}{p} - \tfrac{1}{2}qx^2 + yx\right)dt.$$

The functional $l(x, w)$ is given by

$$l(x, w) = \left\langle x, -\frac{dw}{dt}\right\rangle - a(x, w)$$

$$= \left\langle \frac{dx}{dt}, w\right\rangle - a(x, w),$$

so that

$$l_1(x) = \tfrac{1}{2}\int_0^1 \left(p\left(\frac{dx}{dt}\right)^2 + qx^2 - 2yx\right)dt$$

is to be minimized among the once differentiable functions on [0, 1] with $x(0) = x(1) = 0$ while

$$l_2(w) = -\tfrac{1}{2}\int_0^1 \left(\frac{1}{q}\left(\frac{dw}{dt} + y\right)^2 + \frac{w^2}{p}\right)dt$$

is to be maximized among the once differentiable functions on $[0,1]$.

If the boundary values are changed to, say, $x(0) = 0$, $x'(1) = w(1) = 0$ then these conditions should be applied in extremizing l_1 and l_2, respectively: notice that these conditions are consistent with the adjoint relation between d/dt and $-d/dt$ on $[0,1]$.

The stationary value of $l(x, w)$ is

$$l(\hat{x}, \hat{w}) = -\tfrac{1}{2}\langle y, \hat{x}\rangle = -\tfrac{1}{2}\int_0^1 y\hat{x}\,dt$$

and it is this quantity for which we are able to provide upper and lower bounds.

Suppose we now replace the homogeneous boundary conditions by the inhomogeneous set $x(0) = x_0$, $x(1) = x_1$. We add to the functional $a(x, w)$ the boundary term $[wx]_0^1 = w(1)x_1 - w(0)x_0$, so giving

$$l(x, w) = \int_0^1 \left(-x\frac{dw}{dt} - \frac{1}{2}\frac{w^2}{p} + \frac{1}{2}qx^2 - yx\right)dt + (w(1)x_1 - w(0)x_0).$$

It is easy to verify that the vanishing of $dl(x; s, w; r)$ implies the Euler equations

$$\frac{dx}{dt} = \frac{w}{p}, \qquad -\frac{dw}{dt} = -qx + y$$

together with the boundary conditions $x(0) = x_0$, $x(1) = x_1$. Likewise we find

$$l_1(x) = \frac{1}{2}\int_0^1 \left(p\left(\frac{dx}{dt}\right)^2 + qx^2 - 2yx\right)dt,$$

it being understood that x satisfies the (essential) boundary conditions, while

$$l_2(w) = -\frac{1}{2}\int_0^1 \left(\frac{1}{q}\left(\frac{dw}{dt} + y\right)^2 + \frac{w^2}{p}\right)dt + (w(1)x_1 - w(0)x_0).$$

(b) We have pointed out that the Rayleigh–Ritz method could be applied to non-linear potential operator equations (pp. 339–341). We can bring this class of operator equation within the ambit of the present theory by considering the operator equation

$$(T + P)x = \theta,$$

where the positive-bounded-below operator T is representable as the composition S^*S and P is a potential operator. This gives us the pair of equations

$$Sx = w, \qquad S^*w + Px = \theta,$$

and clearly

$$a(x, w) = \frac{1}{2}\langle w, w\rangle_{\mathscr{H}_2} - \int_0^1 \langle P\lambda x, x\rangle_{\mathscr{H}_1} d\lambda$$

$$= \frac{1}{2}\langle w, w\rangle_{\mathscr{H}_2} - f(x),$$

say. This will be a saddle functional if $-f(x)$ is concave. The functionals $l_1(x), l_2(w)$ are given by

$$l_1(x) = \tfrac{1}{2}\langle Sx, Sx\rangle_{\mathscr{H}_2} + f(x) = \tfrac{1}{2}\langle Tx, x\rangle_{\mathscr{H}_1} + f(x)$$

and

$$\begin{aligned}l_2(w) &= -\langle x, Px\rangle_{\mathscr{H}_1} - \tfrac{1}{2}\langle w, w\rangle_{\mathscr{H}_2} + f(x) \\ &= -\langle P^{-1}S^*w, S^*w\rangle_{\mathscr{H}_1} - \tfrac{1}{2}\langle w, w\rangle_{\mathscr{H}_2} + f(P^{-1}S^*w),\end{aligned}$$

assuming P to be invertible.

It is often convenient to consider the function w to be generated from a function $z \in \mathscr{H}_1$, via the first Euler equation, $Sz = w$: in that case

$$l_2(Sz) = -\langle P^{-1}Tz, Tz\rangle_{\mathscr{H}_1} - \tfrac{1}{2}\langle Tz, z\rangle_{\mathscr{H}_1} + f(P^{-1}Tz).$$

Let us consider the following specific example in one variable,

$$\frac{d^2 x}{dt^2} = 2 \sinh x \quad \text{in} \quad [0, 1]$$

with $x(0) = 0, x(1) = 1$. This is a degenerate case of the diffusion equation discussed in Example 4.4.4(b). We identify S with d/dt, S^* with $-d/dt$ and P with the scalar operator $2\sinh$. The functional $f(x)$ is given by

$$f(x) = 2\int_0^1 \int_0^1 x \sinh \lambda x \, dt \, d\lambda$$

$$= 2\int_0^1 (\cosh x - 1) dt.$$

Adding the boundary term $w(1)$ to the functional $a(x, w)$ we obtain

$$l_1(x) = \int_0^1 \left(\frac{1}{2}\left(\frac{dx}{dt}\right)^2 + 2(\cosh x - 1)\right) dt, \quad x(0) = 0, \quad x(1) = 1,$$

$$\begin{aligned}l_2(Sz) = \int_0^1 &\left(-\frac{1}{2}\left(\frac{dz}{dt}\right)^2 - \frac{dz^2}{dt^2}\sinh^{-1}\left(\frac{1}{2}\frac{d^2 z}{dt^2}\right)\right. \\ &\left. + 2\left(\cosh\left(\sinh^{-1}\left(\frac{1}{2}\frac{d^2 z}{dt^2}\right)\right) - 1\right)\right) dt + \left.\frac{dz}{dt}\right|_{t=1}.\end{aligned}$$

Arthurs [86] has used the trial functions
$$x = \frac{\sinh \alpha t}{\sinh \alpha}, \quad z = \frac{\sinh \beta t}{\sinh \beta},$$
obtaining $\alpha = 1.46$, $l_1 = 0.3083$ and $\beta = 1.48$, $l_2 = 0.3079$.

Exercises 4.4

1. (a) The energy norm for the Sturm–Liouville operator
$$Tx \equiv -\frac{d}{dt}\left(p(t)\frac{dx}{dt}\right) + q(t)x,$$
where $p(t) > 0$, $q(t) \geq 0$ and $x(0) = x'(1) = 0$, is
$$\|x\|_T = \left\{\int_0^1 \left(p\left(\frac{dx}{dt}\right)^2 + qx^2\right)dt\right\}^{1/2}.$$

The norm $\|x\|^{(1)}$ of functions $x(t)$ whose derivatives are square integrable in $[0, 1]$ is
$$\|x\|^{(1)} = \left\{\int_0^1 \left(\left(\frac{dx}{dt}\right)^2 + x^2\right)dt\right\}^{1/2}.$$

Show that the norms $\|x\|_T$ and $\|x\|^{(1)}$ are equivalent (Exercise 3.3,4) by showing that there exist constants γ, δ such that
$$\gamma\|x\|^{(1)} \leq \|x\|_T \leq \delta\|x\|^{(1)}.$$
(Hint: it is easy to show that $\|x\|_T \leq \delta\|x\|^{(1)}$; to obtain the other inequality use an argument similar to that used in Example 4.4.1.)

(b) Repeat (a) when
$$Tx \equiv -D_1(pD_1x) - D_2(pD_2x) + rx$$
in $\Gamma \subset \mathcal{R}_2$ with $D_\nu x + kx = 0$ on $\partial \Gamma$, $k > 0$, $p(t)$ strictly positive and continuous and $r(t)$ non-negative. Deduce that T is positive-bounded-below (Example 4.4.2(b)). (Hint: use the Freidrichs inequality (Exercise 5.4,3(e))
$$\int_\Gamma x^2 \, d\Gamma \leq K\left\{\int_\Gamma (\nabla x)^2 \, d\Gamma + \int_{\partial \Gamma} x^2 \, ds\right\}.$$

The equivalence of the norms $\|x\|_T$ and $\|x\|^{(1)}$ can be proved for a more general class of second-order operator, namely those which are elliptic (Example 4.5.1 and p. 465).

2. Show that if $x(0) = x(1) = 0$ then on $\mathscr{H}^{(1)}[0,1]$ the norm

$$\|x\|^{(1)} = \left\{\int_0^1 \left(\left(\frac{dx}{dt}\right)^2 + x^2\right)dt\right\}^{1/2}$$

is equivalent to the norm

$$\|x\|_0^{(1)} = \left\{\int_0^1 \left(\frac{dx}{dt}\right)^2 dt\right\}^{1/2}.$$

Similarly, using the Friedrichs inequality, show that, for $\Gamma \in \mathscr{R}_2$,

$$\|x\|^{(1)} = \left\{\int_\Gamma ((\nabla x)^2 + x^2) d\Gamma\right\}^{1/2}$$

is equivalent to

$$\|x\|_0^{(1)} = \left\{\int_0^1 (\nabla x)^2\right\}^{1/2}$$

if $x = 0$ on $\partial\Gamma$.

3. (a) Show that theorem 4.4.1 holds if T is merely positive provided we assume the existence of $\tilde{x}$.
   (b) Prove that $\mathscr{H}^T \subset \mathscr{H}$ (Hint: use the inequality $\|x\| \leq \beta^2 \|x\|_T$ to show that a Cauchy sequence in $\mathscr{H}^T$ is also Cauchy in $\mathscr{H}$.)

4. (a) Show, using the method of Example 4.4.2, that for the equation $d^4x/dt^4 = y$, $t \in [0,1]$, with the boundary conditions $x(0) = x(1) = x'(0) = 0$ and $x''(1) = -kx'(1)$, $k > 0$, the last boundary condition is natural. (Here $x(t)$ represents the lateral deflection of a uniform beam built-in at $t=0$, supported at $t=1$ and there restrained against rotation by a linear spring.)
   (b) (i) Show that the solution of

$$-\nabla^2 x = 0 \quad \text{in} \quad \Gamma$$

with $\partial x/\partial \nu = b(s)$ on $\partial\Gamma$, where b is a given function of the boundary coordinate s, minimizes the functional

$$f(x) = \tfrac{1}{2}\int_\Gamma (\nabla x)^2 d\Gamma - \int_{\partial\Gamma} xb\,ds$$

amongst those functions for which

$$\int_\Gamma (\nabla x)^2 d\Gamma < \infty.$$

Note that the inhomogeneous boundary condition is natural —what is $f(x)$ if the boundary condition is essential, i.e., $x = b(s)$ on $\partial\Gamma$?

(ii) Let $d^4x/dt^4 = 0$, $t \in [0, 1]$, with $x(0) = x'(0) = 0$ and $x''(1) = M$, $x'''(1) = S$. Show that x minimizes the functional

$$f(x) = \frac{1}{2}\int_0^1 \left(\frac{d^2x}{dt^2}\right)^2 dt - M\left.\frac{dx}{dt}\right|_{t=1} + Sx\Big|_{t=1}$$

amongst those functions x with $x(0) = x'(0) = 0$ and for which

$$\int_0^1 \left(\frac{d^2x}{dt^2}\right)^2 dt < \infty.$$

5. (a) Show that if $x^{(n)}$ is the Rayleigh–Ritz approximation to the minimum of the functional

$$f(x) = \tfrac{1}{2}\langle x, x\rangle_T - \langle y, x\rangle$$

then

$$\min_{x \in \mathscr{H}_n^T} f(x) = f(x^{(n)}) = -\tfrac{1}{2}\langle y, x^{(n)}\rangle.$$

(b) Determine the exact solution for the function minimizing the functional

$$f(x) = \int_0^{\pi/2} \left((x')^2 - x^2 - \frac{4tx}{\pi}\right) dt, \qquad x(0) = x\left(\frac{\pi}{2}\right) = 0,$$

and hence find the minimum value of $f(x)$.

Compare this value with the minimum attained by using the Rayleigh–Ritz method with the single basis function $t(2t - \pi)$.

(c) Let the region Γ be enclosed by the semicircular curve $t_1^2 + t_2^2 = a^2$, $t_2 \geq 0$ and the t_1-axis, $-a \leq t_1 \leq a$.

Using the single basis function $t_2(a^2 - t_1^2 - t_2^2)$, obtain a lower bound on the quantity $2\int_\Gamma x \, d\Gamma$, where $-\nabla^2 x = 2$ in Γ with $x = 0$ on $\partial \Gamma$. (The equation $-\nabla^2 x = 2$ in Γ with $x = 0$ on $\partial \Gamma$ yields a solution of the St Venant torsion problem for a prismatic bar having cross-section Γ—the quantity $2\int_\Gamma x \, d\Gamma$ is proportional to the torsional rigidity—see, for example, [52]).

(d) (i) Verify that if Γ is the elongated rectangular region $-l \leq t_1 \leq l$, $-d \leq t_2 \leq d$ with $l \gg d$ then, except near the ends of the rectangle, the solution of $-\nabla^2 x = 2$ in Γ with $x = 0$ on $\partial \Gamma$ will take the form

$$x(t_1, t_2) = \text{const.}(d^2 - t_2^2)$$

(ii) It would seem reasonable that a better approximation to x will be obtained if we take

$$x(t_1, t_2) = \phi(t_1)(d^2 - t_2^2), \qquad \phi(-l) = \phi(l) = 0,$$

where $\phi(t_1)$ is to be determined by minimizing the functional

$$f(x) = \int_\Gamma (\tfrac{1}{2}(\nabla x)^2 - 2x)\,d\Gamma.$$

(iii) Show that the condition $df(x;s) = 0$ with $s = \sigma(t_1)(d^2 - t_2^2)$, $\sigma(-l) = \sigma(l) = 0$ leads to the differential equation

$$\frac{d^2\phi}{dt_1^2} - \frac{5}{2d^2}\phi = -\frac{5}{2d^2}.$$

(iv) Hence show that

$$2\int x\,d\Gamma = \frac{16}{3}ld^3\left[1 - \frac{\tanh\beta}{\beta}\right], \qquad \beta^2 = \frac{5\,l^2}{2\,d^2}$$

gives a lower bound on the torsional rigidity of a prismatic bar having a slender rectangular cross-section. (The result is in fact rather accurate for nearly square sections!)

(Note: this variant of the Rayleigh–Ritz method whereby the unknown function is expressed in the form

$$x(t_1, t_2) = \sum_i \psi^i(t_1)\phi^i(t_2),$$

where the ψ^i, say, are prescribed and the ϕ^i are to be determined by solving a set of *ordinary* differential equations is due to Kantorovich [85].)

(e) A cantilever beam of length l has bending stiffness $EI(t) = (EI)_0(1 - \alpha t)$, $1/l > \alpha > 0$. Find an estimate for the bending deflection at the tip of the beam ($t = l$) due to a uniformly distributed load of magnitude p force units per unit length (*see* Exercise 4.3.6(c)). (Hint: use one, two or three appropriate basis functions in the Rayleigh–Ritz method with $\alpha = 1/2l$.)

(f) Let $\nabla^2 x = 0$ in the square $[0,1] \times [0,1]$ with $x(t_1, 0) = 1$, $D_1 x(0, t_2) = D_2 x(t_1, 1) = 0$ and $D_1 x(1, t_2) + kx(1, t_2) = 0, k > 0$. Find the functional whose minimum yields the solution to this problem and use the basis function $t_2 t_1^2(2 - t_2)$ to estimate the quantity

$$\int_0^1 D_1 x(1, t_2)\,dt_2.$$

(Here x represents the temperature in a square of conducting material with one edge at unit temperature, two edges insulated and the fourth having a heat flux proportional to edge temperature; the estimated quantity is the net heat flux along this side.)

6. (a) Let a general non-linear differential operator T defined on $[0,1]$

be represented by

$$Tx \equiv h(x, x', x''), \qquad x(0) = x(1) = 0,$$

where h is a given real function.

(i) Show that the Fréchet derivative of T at x is given by

$$T'(x)s = h_x s + h_{x'} s' + h_{x''} s''$$

(ii) Assuming the inner product $\int_0^1 xy \, dt$, show that the adjoint (Example 3.7.3(e)) of the linear operator $T'(x)$ is given by

$$T'^*(x)s = h_x s - (h_{x'} s)' + (h_{x''} s)''$$

and hence that T is potential if and only if

$$h_x - (h_{x'})' + (h_{x''})'' = h_x$$

and

$$-h_{x'} + 2(h_{x''})' = h_{x'}.$$

Show that these conditions are subsumed under the single condition

$$h_{x'} - (h_{x''})' = 0.$$

(iii) Hence verify that the operator

$$Tx = -2xx'' - (x')^2, \qquad x(0) = x(1) = 0,$$

is potential and that it is the gradient of the functional

$$f(x) = \int_0^1 x(x')^2 \, d\Gamma.$$

(b) In a similar fashion to (a) show that the operator on $\mathcal{R}_2$ defined by

$$-2x\nabla^2 x - (D_1 x)^2 - (D_2 x)^2 \qquad \text{in } \Gamma$$

with $x = 0$ on $\partial \Gamma$ is potential and is the gradient of the functional

$$f(x) = \int_\Gamma x(\nabla x)^2 \, dt.$$

Note: a useful tabular presentation of the conditions to be satisfied in order that a class of partial differential operators be potential is given in [80].

7. (a) Show that $\lambda_{r+1} = \inf_{\|x\|=1} \langle Tx, x \rangle$ subject to the constraints $\langle x, x^{(s)} \rangle = 0, \; s = 1, 2, \ldots, r$.

(b) Show that the Rayleigh quotient

$$f_R(x) = \frac{\langle Tx, x \rangle}{\langle x, x \rangle}$$

has a stationary point $(df_R(x;s)=0)$ at an eigenvector $x^{(n)}$ of the positive operator T. Hence show that, in the neighbourhood of $x^{(n)}$,

$$f_R(x^{(n)}+s) = \lambda_n + O[\|s\|^2].$$

(Note: this shows that although the 'error' in the eigenfunction is $O[\|s\|]$ the 'error' in the eigenvalue is $O[\|s\|^2]$.)

(c) Verify that the eigensolutions of the operator $Tx \equiv -x''$ in $[0, 1]$ with $x(0) = x(1) = 0$ are $(n^2\pi^2; \sin n\pi t)$, $n = 1, 2, 3, \ldots$.
  (i) Using the basis functions $z^i(t) = t^i(1-t)$ find approximations to λ_1 using one, two and three functions. How do the approximations to λ_2, λ_3 compare to the exact values?
  (ii) Repeat (i) by first generating an orthonormal basis from the set z^i.

(d) (i) Verify that the eigensolutions of the operator $Tx = -\nabla^2 x$ in the region $\Gamma = [0, 1] \times [0, 1]$ with $x = 0$ on $\partial\Gamma$ are $(\pi^2(m^2 + n^2); \sin m\pi t_1 \sin n\pi t_2)$, $m, n = 1, 2, 3, \ldots$. Using the basis function $t_1(1-t_1)t_2(1-t_2)$ find an estimate for the lowest eigenvalue and compare it with the exact value.
  (ii) Estimate the lowest eigenvalue for the operator in (i) if the region Γ is an equilateral triangle of unit side length. (The exact value is $16\pi^2/3$.)

8. (a) Consider the *generalized eigenvalue problem*

$$Tx - \lambda Sx = 0,$$

where T and S are both positive-bounded-below and $\mathscr{D}(T) \subseteq \mathscr{D}(S)$. Show that
  (i) the eigenvalues are positive;
  (ii) the eigenfunctions are orthogonal with respect to the inner product $\langle Sx, x \rangle$ and with respect to the inner product $\langle Tx, x \rangle$;
  (iii) the rth eigenvalue λ_r is given by

$$\lambda_r = \inf \frac{\langle Tx, x \rangle}{\langle Sx, x \rangle}$$

subject to $\langle Sx, x^{(s)} \rangle = 0$, $s = 1, 2, \ldots, r-1$;
  (iv) the Ritz equations for the eigenvalues are (p. 343)

$$([k_{ij}] - \lambda[m_{ij}])\{\xi_j\} = 0,$$

where $k_{ij} = \langle Tz^i, z^j \rangle$, $m_{ij} = \langle Sz^i, z^j \rangle$. What form do these equations take if the basis functions z^i are orthogonal with respect to the inner product $\langle x, x \rangle_S$?

(b) (i) Interpret the results in (a) in the case of the equation
$$-D_1(pD_1x) - D_2(pD_2x) + rx - \lambda sx = 0 \quad \text{in} \quad \Gamma$$
with $D_\nu x + kx = 0$ on $\partial \Gamma, p, r, s > 0$ in $\bar{\Gamma}$ (Exercise 4.4,1(b)).

(ii) Estimate the lowest eigenvalue of the equation
$$(tx')' + \lambda tx = 0, \qquad t \in [0, 1],$$
with $x(0) = x(1) = 0$ by using the single basis function $t(t-1)$. Note: if p vanishes at $x = 0$, say, it can be shown [34] that $-(px')' + rx - \lambda sx = 0$ has a discrete spectrum (i.e., has an integral inverse operator which is compact) if
$$\int_0^1 \frac{t\,dt}{p(t)} < \infty.$$

(iii) If the mass per unit length of the cantilever beam of Exercise 4.4,5(e) is $m(t) = m_0(1 - \alpha t)^{1/3}$ find an estimate for the lowest frequency of transverse vibration when $\alpha = 1/2l$; sketch the corresponding eigenfunction (mode shape). (Hint: the equation of motion upon assuming harmonic motion of frequency ω is $(EIx'')'' - \omega^2 mx = 0$).

(iv) Estimate the lowest frequency of transverse vibration ω of a square, uniform elastic plate of side l, having bending stiffness K and mass per unit area m which is cantilevered along one edge. (Hint: use the functional of Exercise 4.3,6(c)(iii) with an 'effective' inertia loading $p = \omega^2 mx$; boundary conditions on the free edges are natural and need not be satisfied by the basis functions—use as basis functions products of the form $z^i(t_1)y^i(t_2)$, where z^i, y^i are suitable trigonometric or polynomial functions—for extensive numerical calculations see [83]—take $\mu = 0.3$.)

9. (a) Show that if $w^{(m)}$ is the Rayleigh–Ritz approximation to the maximum of the functional (p. 346)
$$h(w) = -\tfrac{1}{2}\|w_0 - w_1\|_{\mathscr{H}_2}^2$$
then
$$\max_{w \in \mathscr{N}_m(S^*)} h(w) = \tfrac{1}{2}\|w^{(m)}\|_{\mathscr{H}_2}^2 - \tfrac{1}{2}\|w_0\|_{\mathscr{H}_2}^2$$
$$= \tfrac{1}{2}\langle w_0, w^{(m)} - w_0\rangle_{\mathscr{H}_2}.$$

(b) Show that the method of orthogonal projections leads to the error estimate
$$\|\hat{x} - x^{(n)}\|_T^2 \leq (\|w_0\|_{\mathscr{H}_2}^2 - \|w^{(m)}\|_{\mathscr{H}_2}^2) - \|x^{(n)}\|_T^2.$$
(Hint: begin with $\|\hat{x} - x^{(n)}\|_T^2 = \|\hat{x}\|_T^2 - \|x^{(n)}\|_T^2$, p. 336).

(c) Verify the numerical error bounds given in Example 4.4.5.

(d) Formulate the method of orthogonal projections for the operator equation $-\nabla^2 x = y$ in Γ with the boundary conditions $x = 0$ on $\partial \Gamma_1$, $D_\nu x = 0$ on $\partial \Gamma_2$, $\partial \Gamma = \partial \Gamma_1 \cup \partial \Gamma_2$.

(e) Consider the application of the method of orthogonal projections to the elliptic equation (Example 4.5.1)

$$-\sum_{i,j=1}^{m} D_i(\alpha_{ij} D_j x) = y \quad \text{in} \quad \Gamma \subset \mathscr{R}_m$$

with $x = 0$ on $\partial \Gamma$; taking

$$\langle x, z \rangle_{\mathscr{H}_2} \equiv \int_\Gamma \lfloor \xi_i \rfloor [\alpha_{ij}]^{-1} \{\zeta_j\} \, d\Gamma$$

deduce the form of the orthogonal subspaces $\mathscr{N}(S^*), \mathscr{N}^\perp(S^*)$ (p. 345).

10. (a) Show that the Euler equations for the functional (p. 349)

$$l(x, w) = \langle x, S^* w \rangle_{\mathscr{H}_1} - a(x, w) = \langle Sx, w \rangle_{\mathscr{H}_2} - a(x, w)$$

are $Sx = a'_w(x, w)$, $S^* w = a'_x(x, w)$.

(b) Show that the convexity conditions

$$f(\lambda x + (1 - \lambda)y)) \leq \lambda f(x) + (1 - \lambda)f(y), \quad 0 < \lambda < 1,$$

and

$$f(y) - f(x) \geq \langle f'(x), y - x \rangle$$

for the real functional $f(x)$ are equivalent. Note: If equality is excluded we speak of *strict convexity*. (Hint: replacing λ by $1 - \lambda$ the first condition gives

$$f(y) - f(x) \geq \frac{f(x + \lambda(y - x)) - f(x)}{\lambda}$$

—now use the definition of Fréchet derivative; in the other direction begin with
$$\lambda f(x) + (1 - \lambda)f(y) - f(\lambda x + (1 - \lambda)y)$$
$$= \lambda(f(x) - f(z)) + (1 - \lambda)(f(y) - f(z)), \text{ where}$$
$$z = \lambda x + (1 - \lambda)y).$$

(c) (i) Show that the functional $f(x) = \langle x, x \rangle$ is convex.

(ii) Show that the functional $f(x)$ on $\mathscr{L}_2[0, 1]$ given by

$$f(x) = \int_0^1 x^2 \, dt + ax(1) - bx(0),$$

where a, b are constants, is convex.

11. (a) Deduce suitable forms for $a(x, w), l(x, w), l_1(x)$ and $l_2(w)$ for the

linear operator equation in $\mathcal{H}_1$,

$$S^*Sx + Rx = y,$$

where $R: \mathcal{H}_1 \to \mathcal{H}_1$ is a positive linear operator with inverse R^{-1}. Is $a(x, w)$ a saddle functional? (Hint: begin with $Sx = w$.)

What form does $l_2(w)$ take if we choose to approximate by using basis functions of the form $w^i = Sx^i$, $x^i \in \mathcal{H}_1$? (cf., Example 4.4.6(b)).

(b) Rework (a) when $R = 0$ (the zero operator) and reconcile your results with the method of orthogonal projections. Apply your results to the equation

$$-\frac{d}{dt}\left((2+t)\frac{dx}{dt}\right) = -1 \quad \text{on} \quad \mathcal{L}_2[-1, 1]$$

with $x(-1) = x(1) = 0$, and hence find upper and lower bounds on the functional

$$f(x) = \frac{1}{2}\int_{-1}^{1} x \, dt.$$

In this case the lower bound is exact—why? (Hint: try $x = \beta \cos(\pi t/2)$ to find an upper bound.)

(c) Deduce suitable forms for $a(x, w)$, $l(x, w)$, $l_1(x)$ and $l_2(w)$ for the matrix equations

$$[s_{ij}]\{\xi_j\} = \{c_i(\phi_i)\}, \qquad [s_{ji}]\{\phi_i\} = \{b_j\},$$

where $[s_{ij}]: \mathcal{E}_m \to \mathcal{E}_n$, $\{b_j\}$ is a given m-tuple of constants and each c_i, $i = 1, \ldots, n$, is a given (non-linear) convex function of the scalar ϕ_i only.

What form do the functionals take if the f_i are linear, that is $c_i(\phi_i) = r_i \phi_i$, say?

The reader should compare these equations with the framework equations of Example 2.9.1, identifying the matrix $[l_{ij}]$ there with the matrix $[s_{ij}]$ here, the d_j with the ξ_j, and the t^i with the ϕ_i—the b_j are the imposed external force components f^j and the functions c_i represent the non-linear characteristics of the rods of the framework giving the strain in the ith rod, e_i, in terms of the tension t^i.

4.5 Projection solution of operator equations

The corollary to Theorem 4.4.2 points out that the nth member of the Rayleigh–Ritz sequence $\{x^{(n)}\}$ for the approximate solution of $Tx = y$ can be considered to be the orthogonal projection of the exact solution $\hat{x}$ onto the finite-dimensional subspace $\mathcal{H}_n^T$, viz.,

$$\langle \hat{x} - x^{(n)}, z \rangle_T = 0 \qquad \forall z \in \mathcal{H}_n^T.$$

Assuming $Tx = y$ to have the solution $\tilde{x} \in \mathscr{D}(T)$, then in terms of the inner product on $\mathcal{H}$ this is equivalent to

$$\langle T\hat{x} - Tx^{(n)}, z \rangle = 0$$

or

$$\langle y - Tx^{(n)}, z \rangle = 0 \qquad \forall z \in \mathcal{H}_n \subset \mathscr{D}(T)$$

with $x^{(n)} \in \mathcal{H}_n$. Now if we regard the vector $(y - Tx^{(n)})$ as the 'error' in failing to satisfy the operator equation exactly, we could interpret the above condition as saying that we require the projection of the error onto the subspace $\mathcal{H}_n$ to vanish. In an intuitive sense, as the spaces $\mathcal{H}_n$ get larger, more and more components of the error are required to vanish and we should expect that $Tx^{(n)} - y \to \theta$ (see p. 192).

In this development we have not made any reference to the symmetry of T nor to the associated problem of minimizing an energy functional: hence we have, in principle, a method for finding a sequence of approximate solutions of the general operator equation $Tx = y$. The question of the convergence of such a sequence is expected to be more difficult than that met with in the Rayleigh–Ritz method, since we no longer have at our disposal the 'nice' properties of positive-bounded-below operators.

Let $\mathscr{V}, \mathscr{U}$ be Banach spaces and let $T : \mathscr{V} \to \mathscr{U}$ have domain $D(T) \subset \mathscr{V}$. Let $\{\mathscr{V}_n\}$, $\{\mathscr{U}_n\}$ be two given sequences of nesting subspaces with $\mathscr{V}_n \subset D(T) \subset \mathscr{V}, \mathscr{U}_n \subset \mathscr{U}$, and let $\{P_n\}$ be a sequence of (linear) projections of $\mathscr{U}$ onto $\mathscr{U}_n$ (see Definition 2.5.3, Example 2.6.5 and Example 3.5.8(c)). Suppose we seek an approximate solution $x^{(n)}$ of $Tx = y$, where $x^{(n)} \in \mathscr{V}_n$: in general $(Tx^{(n)} - y)$ will not be zero, but we can at least insist that the projection of this vector (the 'error') onto $\mathscr{U}_n$ should vanish, i.e.,

$$P_n(Tx^{(n)} - y) = \theta.$$

If we rewrite this equation in the form

$$P_n Tx^{(n)} = P_n y,$$

then we could take the view that we have replaced the original equation $Tx = y$ by the approximating equation $P_n Tx^{(n)} = P_n y$. A method for finding approximate solutions of operator equations which is based on the foregoing idea is generally referred to as a *projection method*. We should note that any such method may lead

only to a weak solution of $Tx = y$ in the sense that (p. 169 et seq.)
$$[Tx - y, z] = 0 \quad \forall z \in \mathcal{U}^*.$$

Let $\mathcal{U}_n \subset \mathcal{U}$ be the finite-dimensional subspace spanned by $\{u^1, u^2, \ldots, u^n\}$; then the projection P_n clearly takes the form
$$P_n \cdot = \sum_{j=1}^{n} l_j(\cdot) u^j, \quad l_j \in \mathcal{U}^*.$$

Indeed, we know (Example 3.5.8(c)), that the direct decomposition $\mathcal{U} = \mathcal{U}_n \oplus \mathcal{U}'_n$ induces the decomposition $\mathcal{U}^* = \mathcal{U}_n^0 \oplus (\mathcal{U}'_n)^0$: hence if $\{u_1, u_2, \ldots, u_n\}$ is a basis for $(\mathcal{U}'_n)^0$ which is reciprocal to $u^1, u^2, \ldots, u^n$, then we have explicitly (Example 2.6.5).
$$P_n \cdot = \sum_{j=1}^{n} [\cdot, u_j] u^j.$$

If $\{v^1, v^2, \ldots, v^n\}$ is a basis for $\mathcal{V}_n \subset \mathcal{D}(T) \subset \mathcal{V}$, then we may write
$$x^{(n)} = \sum_{i=1}^{n} \xi_i^{(n)} v^i$$
and the projection method leads to the finite set of equations
$$\left[T\left(\sum_{i=1}^{n} \xi_i^{(n)} v^i \right), u_j \right] = [y, u_j], \quad j = 1, \ldots, n,$$
or, if T is a linear operator, to
$$\sum_{i=1}^{n} \xi_i^{(n)} [T v^i, u_j] = [y, u_j], \quad j = 1, \ldots, n.$$

It is not at all easy to determine the conditions for convergence of the projection method in a general Banach space setting. One of the principal difficulties concerns the fact that, by allowing an arbitrary choice of basis sets, we have no way of controlling the angle between basis vectors—in other words, there may be a countable number of vectors which are 'nearly linearly dependent': this can result in the projection operators P_n being unbounded [77].

If we restrict consideration to a Hilbert space setting, and use only orthogonal projections, then this particular difficulty is absent, hence many of the concrete forms of the projection method use the machinery of orthogonal projections.[†]

[†] If we write the projection equations in a Hilbert space with inner product
$$\int_{\Gamma} uv \, d\Gamma$$

We shall now list most of the well-known and widely used forms of the projection method, restricting the discussion, for the moment, to linear operators.

The Galerkin–Petrov method [77] With $\mathscr{U} = \mathscr{V} = \mathscr{H}$ we choose two sets of linearly independent vectors $\{u^1, u^2, \ldots, u^n\}$, $\{v^1, v^2, \ldots, v^n\}$. Taking

$$x^{(n)} = \sum_{i=1}^{n} \xi_i^{(n)} v^i$$

we make the error $Tx^{(n)} - y$ orthogonal to the subspace spanned by $\{u^i\}$, hence the equations take the form

$$\sum_{i=1}^{n} \xi_i^{(n)} \langle Tv^i, u^j \rangle = \langle y, u^j \rangle, \qquad j = 1, \ldots, n.$$

The *method of moments* is obtained if we choose $u^i = Sv^i$, where S is some suitably chosen operator. A particularly important choice is $S = T$, which gives:

The method of least squares [76, 92] Taking $u^i = Tv^i$ in the Galerkin–Petrov equations gives

$$\sum_{i=1}^{n} \xi_i^{(n)} \langle Tv^i, Tv^j \rangle = \langle y, Tv^j \rangle.$$

Now if we minimize the 'error' functional $\| Tx^{(n)} - y \|^2$ when

$$x^{(n)} = \sum_{i=1}^{n} \xi_i^{(n)} v^i,$$

we obtain the same set of equations. Hence we see that this choice for the basis functions $\{u^i\}$ is equivalent to minimizing the 'mean square error'.

The Bubnov–Galerkin method [77] In this method, more commonly known in the western hemisphere simply as Galerkin's method, the

---

defined over some physical region Γ then we obtain

$$\int_{\Gamma} \left(\sum_{i=1}^{n} \xi_i^{(n)} Tv^i - y \right) u^j d\Gamma = 0.$$

The term in parentheses is the residual or error, while u^j can be viewed as a type of weighting function. In this context the projection method is often referred to as the *method of weighted residuals* (MWR) [80].

same subspaces of $\mathscr{H}$ are chosen for approximating the solution and for projection. Using the single basis set $\{v^i\}$ we obtain the equations

$$\sum_{i=1}^{n} \xi_i^{(n)} \langle Tv^i, v^j \rangle = \langle y, v^j \rangle, \qquad j = 1, \ldots, n.$$

We now return to a general Banach space setting for:

The collocation method [35] Suppose T is a differential or integral operator defined on a physical region Γ. Once again, we seek an approximate solution for $x^{(n)}$ in the form

$$x^{(n)} = \sum_{i=1}^{n} \xi_i^{(n)} v^i,$$

where the basis functions v^i are defined on Γ and satisfy, where necessary, boundary conditions on $\partial\Gamma$. We now specify that the operator equation is to be satisfied at a set of points $(t^1, t^2, \ldots, t^n)$ within Γ, giving the equations

$$\sum_{i=1}^{n} \xi_i^{(n)} (Tv^i)_{t=t^k} = (y)_{t=t^k}.$$

This method can apparently be brought into the general scheme by defining the projection operators P_n by

$$P_n u = \sum_{k=1}^{n} u(t^k) u^k.$$

However, the situation is not as simple as it seems, since functionals with the property $l_k(u) = u(t^k)$ are elements of spaces of distributions or generalized functions, hence we shall defer detailed consideration of the collocation method to Chapter 5, Example 5.5.1 (*see also* Exercise 4.5,4).

We shall now show that there is a very close connection between the method of least squares and the Rayleigh–Ritz method when $T: \mathscr{H} \to \mathscr{H}$ has a bounded inverse. Consider the equation

$$T^*Tx = T^*y,$$

where T^* is the adjoint of T. The solution of $Tx = y$ is also the solution of $T^*Tx = T^*y$. Furthermore, the operator T^*T is positive-bounded-below since

$$\langle T^*Tx, x\rangle = \langle Tx, Tx\rangle$$
$$= \|Tx\|^2$$
$$\geq \alpha^2 \|x\|^2,$$

the last line being a consequence of the assumption that T is invertible (Theorem 3.4.2).

Now, finding the solution of $T^*Tx = T^*y$ is equivalent to finding the minimum point of the functional

$$f(x) = \tfrac{1}{2}\langle T^*Tx, x\rangle - \langle T^*y, x\rangle$$
$$= \tfrac{1}{2}\langle Tx, Tx\rangle - \langle y, Tx\rangle$$
$$= \tfrac{1}{2}\langle Tx - y, Tx - y\rangle - \tfrac{1}{2}\langle y, y\rangle$$
$$= \tfrac{1}{2}(\|Tx - y\|^2 - \|y\|^2).$$

The functionals $2f(x)$ and $\|Tx - y\|^2$ differ only by a constant and hence have the same minimum point. We can therefore conclude that the application of the method of least squares to $Tx = y$ is equivalent to an application of the Rayleigh–Ritz method to $T^*Tx = T^*y$.

If T is not invertible then a solution of $Tx = y$ may not exist and correspondingly a solution of the Galerkin–Petrov equations,

$$\sum_{i=1}^{n} \xi_i^{(n)} \langle Tv^i, u^j\rangle = \langle y, u^j\rangle,$$

may not exist. However, if $\mathcal{R}(T) \subset \mathcal{H}$ is closed a least squares solution of $Tx = y$ always exists. This follows from the projection theorem (Theorem 3.6.2) since if we wish to minimize

$$\|Tx - y\|^2$$

we are in effect seeking a point in the closed subspace $\mathcal{R}(T)$ which is closest to the fixed vector y. Furthermore, if $\tilde{y}$ is such a point, we know that $(y - \tilde{y}) \in \mathcal{R}^\perp(T) = \mathcal{N}(T^*)$ (p. 214); hence

$$T^*(y - \tilde{y}) = \theta$$

or
$$T^*T\tilde{x} - T^*y = \theta,$$

showing that the least squares solution of $Tx = y$ is a solution of the equation $T^*Tx = T^*y$.

Convergence of projection methods

As we have already pointed out, the general question of the conditions for convergence of the various forms of projection method is one of extreme difficulty. Even for specific methods, different convergence theorems are available depending on which assumptions one is prepared to make about the underlying spaces, the projection operators and the operator T itself. There still remains, of course, the practical difficulty of establishing that the necessary conditions are satisfied for a given concrete operator equation. Very extensive consideration is given to these questions in [35, 76, 77]: here we shall be content to state theorems on the convergence of the method of least squares and the (Bubnov) Galerkin method in a Hilbert space setting.

Convergence of the method of least squares can be dealt with in a straightforward manner; the result is embodied in

Theorem 4.5.1 *The method of least squares gives a sequence of approximate solutions which converges to the exact solution $\tilde{x}$ of $Tx = y$ if*:

1. *the equation is soluble;*
2. *T has a bounded inverse;*
3. *the sequence $\{Tv^i\}$ is a basis for $\mathscr{R}(T)$.*

Proof Since the sequence $\{Tv^i\}$ is a basis or spanning set for $\mathscr{R}(T) \subset \mathscr{H}$, for any $\varepsilon > 0$ there exists N and $(\xi_1, \xi_2, \ldots, \xi_N)$ such that (p. 139).

$$\left\| T\tilde{x} - \sum_{i=1}^{N} \xi_i Tv^i \right\| < \varepsilon.$$

In particular, this inequality is true for the N-tuple $(\xi_1^{(N)}, \xi_2^{(N)}, \ldots, \xi_N^{(N)})$, where

$$x^{(N)} = \sum_{i=1}^{N} \xi_i^{(N)} v^i$$

is the least squares approximation to $\tilde{x}$ in the subspace with basis $\{v^i\}, i = 1, \ldots, N$; this particular choice minimizes the left-hand side of the inequality. Furthermore, the left-hand side does not increase with increase in N, so that for $n \geq N$ we have

$$\| T\tilde{x} - Tx^{(n)} \| < \varepsilon.$$

But T has a bounded inverse and therefore satisfies an inequality of the form

$$\| x \| \leq \frac{1}{\alpha} \| Tx \|;$$

hence it follows that, for any $\varepsilon > 0$, there exists $n \geq N$ such that $\|\tilde{x} - x^{(n)}\| < \varepsilon/\alpha$: this implies that $\lim_n x^{(n)} = \tilde{x}$. ∎

Corollary *An estimate of the error $\|\tilde{x} - x^{(n)}\|$ is given by*

$$\|\tilde{x} - x^{(n)}\| \leq \frac{1}{\alpha} \|Tx^{(n)} - y\|$$

(*see also* Exercise 4.5,1(c)).

Proof We have

$$\|\tilde{x} - x^{(n)}\| \leq \frac{1}{\alpha} \|T(\tilde{x} - x^{(n)})\|$$

$$= \frac{1}{\alpha} \|T\tilde{x} - Tx^{(n)}\|$$

$$= \frac{1}{\alpha} \|Tx^{(n)} - y\|. \quad \blacksquare$$

Now we turn to the question of convergence of the Bubnov–Galerkin method and this turns out to be much less straightforward. First of all, we shall assume that we can write the operator T as the sum

$$T = R + S,$$

where R is symmetric positive-bounded-below and $\mathcal{D}(R) \subset \mathcal{D}(S)$. Let us say at once that this is an eminently practical approach to adopt. Many operators which arise from physical situations will have a symmetric part which we can extract. We can then allow convergence of the method to depend on the 'nice' properties of this part of the operator; in particular, we again have at our disposal the energy norm $\langle Rx, x \rangle^{1/2}$ which proved to be so useful in connection with the Rayleigh–Ritz method.

The proof of convergence falls into two parts. First, one rewrites the operator equation

$$Tx = (R + S)x = y$$

as

$$x + R^{-1}Sx = R^{-1}y:$$

one then considers the application of the Galerkin method to the latter equation. We shall not give the proof in full since it is rather extensive, but rather indicate the main steps and leave the sceptical reader to consult, for example, [77]. As previously, we denote the completed space with inner product $\langle Rx, z \rangle$ by $\mathcal{H}^R$.

Theorem 4.5.2 *The Bubnov–Galerkin method gives a sequence of approximate solutions which converges to the solution of $Tx = (R + S)x = y$ in the energy norm $\langle x, x \rangle_R^{1/2}$ if:*

1. $Tx = y$ has at most one solution;
2. the operator $R^{-1}S$ is compact in $\mathcal{H}^R$.

Proof First, we show the equivalence between the Galerkin approximations to $(R + S)x = y$ and $x + R^{-1}Sx = R^{-1}y$. Let $x^{(n)} \in \mathcal{H}_n \subset \mathcal{H}^R$: then the Galerkin equation for $x^{(n)}$ is

$$P_n(Rx^{(n)} + Sx^{(n)} - y) = 0,$$

where P_n is the orthogonal projection of $\mathcal{H}$ onto $\mathcal{H}_n$. Let P_n^R be the orthogonal projection of $\mathcal{H}^R$ onto $\mathcal{H}_n$ in the sense of the energy inner product $\langle x, z \rangle_R$. Now $P_n x = 0$ if, and only if, $\langle x, z^{(n)} \rangle = 0$ $\forall z^{(n)} \in \mathcal{H}_n$: but

$$\langle x, z^{(n)} \rangle = \langle RR^{-1}x, z^{(n)} \rangle = \langle R^{-1}x, z^{(n)} \rangle_R,$$

so that $P_n x = 0$ also implies that $P_n^R R^{-1} x = 0$.

This means that we can write the Galerkin equation above as

$$P_n^R(x^{(n)} + R^{-1}Sx^{(n)} - R^{-1}y) = 0, \qquad x^{(n)} \in \mathcal{H}_n,$$

or

$$x^{(n)} + P_n^R R^{-1} S x^{(n)} - P_n^R R^{-1} y = 0$$

provided the (closure of the) range space of $R^{-1}S$ is contained in $\mathcal{H}^R$; this requires that we assume that $R^{-1}S$ is bounded (in $\mathcal{H}^R$). But this last equation is the Galerkin equation for the operator equation

$$x + R^{-1}Sx = R^{-1}y$$

in $\mathcal{H}^R$.

Now we can consider the convergence of the Galerkin method applied to the equation

$$(I + Q)x = w,$$

where $Q = R^{-1}S$ and $w = R^{-1}y$. The main assumption here is that the operator Q is compact (Definition 3.4.3). Now it was pointed out in Section 3.9, p. 254, that a compact operator has a pure point spectrum and the only limit point is zero. This means that the Fredholm alternative applies, so that either $x + Qx = \theta$ has a nontrivial solution (the eigenvector(s) belonging to -1) or $(I + Q)^{-1}$ exists and is bounded, in which case $x + Qx = w$ has a unique solution. We are concerned only with the latter case. We should also recall that the range of a compact operator is compact and therefore there exists a finite-dimensional subspace of $\mathscr{R}(Q)$ such that every point in $\mathscr{R}(Q)$ is near some point in that subspace (p. 152). Suppose then that we write

$$Q = P_n^R Q + (I - P_n^R)Q$$
$$= Q_1 + Q_2,$$

say.

In that case $\mathscr{R}(Q_1)$ is a finite-dimensional subspace of $\mathscr{R}(Q)$ and we can make

$$\|Qx - Q_1 x\| = \|Q_2 x\|$$

as small as we please by taking n large enough. But this implies that, for any given $\varepsilon > 0$, there exists n such that

$$\|Q_2\| = \|(I - P_n^R)Q\| < \varepsilon.$$

Now by hypothesis the operator $(I + Q)^{-1}$ exists and since $\|(I + Q)\|$ can be made to differ by as little as we please from $\|I + Q_1\|$ then we may assert that $(I + Q_1)^{-1}$ exists for n sufficiently large; hence the Galerkin equation has the unique solution

$$x^{(n)} = (I + Q_1)^{-1} w^{(n)}$$
$$= (I + P_n^R Q)^{-1} P_n^R w.$$

But $w^{(n)} \to w$ and $(I + P_n^R Q)^{-1} \to (I + Q)^{-1}$ as $n \to \infty$, so $x^{(n)} \to x$, where x is the solution of $(I + Q)x = w$. ∎

We have assumed in the foregoing that $Tx = y$ has a solution. A solution of $(R + S)x = y$ is clearly a solution of $x + Qx = w$, but the converse is only true if the solution of the latter equation is an element of $\mathscr{D}(T)$. In general, the limits of the Galerkin approximations are in $\mathscr{H}^R$ and in that case we must regard them as generalized solutions. We should note that the Galerkin equations should be written

$$\sum_{i=1}^{n} \xi_i^{(n)}(\langle v^i, v^j \rangle_R + \langle Sv^i, v^j \rangle) = \langle y, v^j \rangle, \quad j = 1, \ldots, n.$$

As for the Rayleigh–Ritz method, if T is a differential operator it is not necessary that the basis functions v^i should satisfy any natural boundary conditions.

A useful point to note is that $R^{-1}S$ is compact in $\mathcal{H}^R$ if R^{-1} is compact in $\mathcal{H}$ and S is bounded in $\mathcal{H}$ (Exercise 4.5,2(a)). As was pointed out in Section 3.9, many differential operators have inverses which are compact integral operators.

A common application of the Bubnov–Galerkin method is to equations of evolutionary type, for example, a partial differential equation of the form

$$\frac{\partial^2 x}{\partial t^2} - Mx = y,$$

where t is a time-like variable and x, y are functions of position and time; M is a partial differential operator involving only space derivatives. In applying the Bubnov–Galerkin method the coordinates $\xi_i^{(n)}$ are taken to be functions of the time variable t with the result that the Galerkin equations are ordinary differential equations rather than algebraic equations [100]. This approach is reminiscent of the separation of variables method so familiar in mathematical physics; alternatively, one may consider the method to be based on an integral transformation of the operator equation from the time domain to a parameter domain (*see*, for example, [80, 93, 94]).

A similar modification to the Galerkin method for problems which are not time dependent was suggested and exploited by Kantorovich [85]. Suppose the operator equation applies to a two-dimensional physical region Γ. A trial function expansion of the form

$$x(t_1, t_2) = \sum_i u_i(t_1) v_i(t_2)$$

is employed, where the functions $u_i(t_1)$, say, are chosen while the functions $v_i(t_2)$ are to be determined. Application of the Galerkin method or one of the projection methods in the coordinate t_1 then leads to a coupled set of ordinary differential equations with associated boundary conditions for the functions v_i (cf., Exercise 4.4,5(d)).

The eigenvalue problem

Projection methods may be used for the determination of approximations to the eigenvalues (and eigenvectors) of the operator equation

$$Qx = \lambda x.$$

Almost all known results [77] assume that Q is compact, but unbounded operators can also be dealt with in the manner of Theorem 4.5.2. Indeed, we can state a result for the eigenvalue problem which is completely analogous to Theorem 4.5.2 for the Bubnov–Galerkin method: namely, if the operator $R^{-1}S$ is compact in $\mathscr{H}^R$ then the Galerkin method gives sequences of approximate eigenvalues which converge to the eigenvalues of the equation

$$Sx = \lambda Rx.$$

The Galerkin equations are

$$\sum_{i=1}^{n} \xi_i^{(n)} \langle Sv^i, v^j \rangle = \lambda \sum_{i=1}^{n} \xi_i^{(n)} \langle v^i, v^j \rangle_R \quad j = 1, \ldots, n:$$

the solution vectors of this set of homogeneous equations are approximations (in $\mathscr{H}_n^R$) to the eigenvectors of $Sx = \lambda Rx$ provided that the eigenvalues have multiplicity one [34].

We now consider the applicability of Galerkin's method to some differential equations.

Example 4.5.1 Let us see whether, according to Theorem 4.5.2, we can apply the Bubnov–Galerkin method to the ordinary differential equation

$$-x'' + \alpha_1(t)x' + \alpha_2(t)x = y$$

in $[0, 1]$ with $x(0) = x(1) = 0$. We assume that $\alpha_1(t)$, $\alpha_2(t)$ are continuous functions.

For the operator

$$Tx = -x'' + \alpha_1(t)x' + \alpha_2(t)x$$

we shall identify Rx with the positive-bounded-below operator $-x''$ (Example 4.4.1) leaving Sx to be identified with $\alpha_1(t)x' + \alpha_2(t)x$. Now the inverse of R is the integral operator

$$R^{-1}x = \int_0^1 k(t,s) x(s) \, ds,$$

where (Exercise 3.4, 5)

$$k(t,s) = \begin{cases} -s(t-1), & 0 \leq s \leq t \leq 1, \\ -t(s-1), & 0 \leq t \leq s \leq 1. \end{cases}$$

The energy inner product on $\mathscr{D}(R) \subset \mathscr{H}$ is

$$\langle Rx, z \rangle = \int_0^1 -x'' z \, dt = \int_0^1 x' z' \, dt$$

upon integration by parts and use of the boundary conditions. Hence in $\mathcal{H}^R$,

$$\langle x, z \rangle_R = \int_0^1 x'z' \, dt \quad \text{and} \quad \|x\|_R^2 = \int_0^1 (x')^2 \, dt.$$

Consider now the operator

$$R^{-1}Sx = \int_0^1 k(t,s)(\alpha_1(s)x'(s) + \alpha_2(s)x(s)) \, ds$$

and in the right-hand side express $x(t)$ in terms of $x'(t)$ by

$$x(t) = \int_0^1 k_1(t,s)x'(s) \, ds,$$

say, where (Example 2.5.6)

$$k_1(t,s) = \begin{cases} 1, & 0 \leq s \leq t \leq 1, \\ 0, & 0 \leq t \leq s \leq 1. \end{cases}$$

Now changing the order of integration where necessary we obtain a representation of the operator $R^{-1}S$ considered as operating on the function $x'(t) \in \mathcal{H}_0^{(1)}$ (Exercise 4.4,2; *see also* Section 5.4), namely,

$$R^{-1}Sx' = \int_0^1 g(t,s)x'(s) \, ds,$$

where

$$g(t,s) = k(t,s)\alpha_1(s) + \int_0^1 k(t,s_1)k_1(s_1,s)\alpha_2(s_1) \, ds_1$$

is a continuous function satisfying

$$\int_0^1 dt \int_0^1 g^2(t,s) \, ds < \infty.$$

Hence (Example 3.4.5(e)) the operator $R^{-1}S$ with argument $x'(t)$ is compact in $\mathcal{H}_0^{(1)}$ and hence compact in $\mathcal{H}^R$, since the norms on $\mathcal{H}_0^{(1)}$ and $\mathcal{H}^R$ are equivalent.

The method used in this example can be extended to ordinary differential equations of higher order provided the highest derivative is of even order [34]. An analogous result also holds for partial differential equations of the type

$$-\sum_{i,j=1}^n D_i(\alpha_{ij} D_j x) + \sum_{i=1}^n \beta_i D_i x + \gamma x = y$$

defined over some domain Γ with $x = 0$ on $\partial \Gamma$, provided the operator

$$Rx = -\sum_{i,j=1}^n D_i(\alpha_{ij} D_j x)$$

PROJECTION SOLUTION OF OPERATOR EQUATIONS

is elliptic in $\bar{\Gamma}$. That is to say, for all points in $\bar{\Gamma}$, the matrix operator $[\alpha_{ij}]$ is positive-bounded-below in the n-dimensional inner product space of n-tuples; this, with the given boundary condition, ensures that R is positive-bounded-below in $\mathscr{H}$ (cf., Exercise 4.4,1(b)).

Exercises 4.5

1. (a) Show that if T^{-1} exists then the least squares equations (p. 367)
$$\sum_{i=1}^{n} \xi_i^{(n)} \langle Tv^i, Tv^j \rangle = \langle y, Tv^j \rangle$$
have a unique solution.

   (b) Let $T: \mathscr{H} \to \mathscr{H}$ be positive-bounded-below (see Definition 4.4.1). If $x^{(n)}$ is the least squares approximation in $\mathscr{H}_n$ to the solution $\tilde{x}$ of $Tx = y$, show that
$$\|\tilde{x} - x^{(n)}\|_T \leq \frac{1}{\beta} \|Tx^{(n)} - y\|$$
and hence that the sequence $x^{(n)}$ is a minimizing sequence for the functional $f(x) = \frac{1}{2}\langle x, x \rangle_T - \langle y, x \rangle$. (Note however that convergence in $\mathscr{H}^T$ is slower for least squares than for the Ritz sequence.)

   (c) If $x^{(n)} \in \mathscr{H}_n$ is the least squares approximation to $\tilde{x}$, show that
$$\|Tx^{(n)} - y\|^2 = \|y\|^2 - \langle Tx^{(n)}, y \rangle.$$

   (d) Let $T: \mathscr{H}_1 \to \mathscr{H}_2$ and suppose $y \notin \mathscr{R}(T) \subset \mathscr{H}_2$, then $Tx = y$ has no solution: however (p. 369), a 'least squares solution' $\tilde{x}$ exists such that $T\tilde{x} = \tilde{y}$ and $\|y - \tilde{y}\|_{\mathscr{H}_2}$ is a minimum. Whilst $\tilde{y}$ is unique (assuming $\mathscr{R}(T)$ to be closed) $\tilde{x}$ need not be unique since T may not be invertible. Suppose we choose as our 'solution' the vector $\tilde{x}$ having minimum norm (in $\mathscr{H}_1$). Show that this vector, say $\tilde{x}_m$, is given by
$$\tilde{x}_m = T^\dagger y$$
where $T^\dagger: \mathscr{H}_2 \to \mathscr{H}_1$ is the transformation having the properties
   (i) $T^\dagger T x = x \quad \forall x \in \mathscr{N}^\perp(T) = \mathscr{R}(T^*) \subset \mathscr{H}_1$;
   (ii) $T^\dagger z = \theta \quad \forall z \in \mathscr{R}^\perp(T) = \mathscr{N}(T^*) \subset \mathscr{H}_2$.
   Note that we have $\mathscr{H}_2 = \mathscr{R}(T) \oplus \mathscr{N}(T^*)$, hence $T^\dagger$ is uniquely defined on $\mathscr{H}_2$ with $\mathscr{R}(T^\dagger) = \mathscr{R}(T^*)$. The transformation $T^\dagger$ is called a *pseudo-inverse* of T: the one given is only a particular case of a family of possible pseudo-inverses—see [91].

   If $T: \mathscr{E}_2 \to \mathscr{E}_3$ has the matrix representation
$$[t_{ij}] = \begin{bmatrix} 1 & 2 \\ 0 & 1 \\ 3 & 2 \end{bmatrix}$$

verify that the pseudo-inverse $T^\dagger : \mathscr{E}_3 \to \mathscr{E}_2$ given by

$$[t_{ij}]^\dagger = \frac{1}{26}\begin{bmatrix} -7 & -8 & 11 \\ 12 & 10 & -4 \end{bmatrix}$$

has the required properties.

(e) Apply the method of least squares to the equation of Exercise 4.4,11 (b). Evaluate

$$\frac{1}{2}\int_{-1}^{1} x\, dt \quad \text{and} \quad \|Tx^{(n)} - y\|.$$

(f) Apply the method of least squares to find an approximate solution of the equation

$$t^2 x'' + tx' + x = 1 \quad \text{in} \quad [1, 2]$$

with $x(1) = x(2) = 0$: evaluate $\|Tx^{(n)} - y\|$. (Note that this equation can be solved analytically by making the independent variable change $t = e^s$).

(g) Apply the method of least squares to find an approximate solution of the equation

$$xx'' + (x')^2 = \tfrac{1}{2} \quad \text{in} \quad [0, 1]$$

with $x'(0) = 0$, $x(1) = 1$. (Hint: use a coset with one simple polynomial basis function.) Compare your solution with the exact solution of this equation.

(h) Let $\mathscr{H} = \mathscr{L}_2[0, 1]$ and let T be a linear transformation $\mathscr{H} \to \mathscr{H}$. Show that the functional

$$f(x) = \int_0^1 w(t)(Tx - y)^2\, dt,$$

where $w(t) > 0$, yields the Euler equation

$$T^*(wTx) = T^*wy.$$

(This is an instance of the weighted least squares method—see, for example, [92].)

2. (a) Let the operator $R: \mathscr{H} \to \mathscr{H}$ be positive-bounded-below ($\|x\|_R \geq \beta \|x\|$), let R^{-1} be compact in $\mathscr{H}$ and let $S: \mathscr{H} \to \mathscr{H}$ be bounded in $\mathscr{H}$. If $T = R^{-1}S$ show that

$$\|Tx - Ty\|_R^2 \leq \|S\|\, \|x - y\|\, \|Tx - Ty\|.$$

If A is a bounded set in $\mathscr{H}^R$, say $\|x\|_R < \lambda$ if $x \in A$, then show further that

$$\|Tx - Ty\|_R^2 < \frac{2\lambda \|S\|}{\beta} \|Tx - Ty\|,$$

hence by choosing any sequence $\{x^{(n)}\}$ from A deduce that $\{Tx^{(n)}\}$ is Cauchy in $\mathscr{H}^R$ and hence that T is compact in $\mathscr{H}^R$ (see p. 152).

(b) Let $T: \mathscr{H}_1 \to \mathscr{H}_2$ be a linear operator and T^* its adjoint. Show that if

$$f(x, z) = \langle x, T^*z - w \rangle_{\mathscr{H}_1} - \langle z, y \rangle_{\mathscr{H}_2}$$

then the vanishing of the differential $df(x, z; s, r)$ at $(\tilde{x}, \tilde{z})$ implies
(i) $\langle T^*\tilde{z} - w, s \rangle_{\mathscr{H}_1} = 0 \ \forall s \in \mathscr{H}_1$;
(ii) $\langle T\tilde{x} - y, r \rangle_{\mathscr{H}_2} = 0 \ \forall r \in \mathscr{H}_2$.

(This formal manipulation thus leads to projection (or weak) solutions of the equations $Tx = y$ and $T^*z = w$. In this sense any unsymmetric operator can be associated with a variational principle [80] but the functional $f(x, z)$ does not of course have the properties characterized by the coercive bilinear functional associated with a positive-bounded-below operator—see remark on p. 327).

3. Apply the Galerkin (or Galerkin–Petrov) method of approximation to each of the following equations. In many cases it is convenient, in the Galerkin–Petrov method, to allow the basis vectors $\{u^i\}$ (p. 367) to satisfy the adjoint boundary conditions.

(a) $x'' + tx = 0$ in $[0, 1]$ with $x(0) = 0, x(1) = 1$. (Try polynomial functions: the exact solution is $ct^{1/2} J_{1/3}(\tfrac{2}{3} t^{3/2})$, where c is a constant and $J_{1/3}$ is a Bessel function.)

(b) $(2 - t^2)x'' + \lambda x = 0$ in $[0, 1]$ with $x(0) = x(1) = 0$. Estimate the lowest eigenvalue λ_1 using the polynomial $v^1 = t - 2t^3 + t^4$,
(i) in the Galerkin method;
(ii) in the Galerkin–Petrov method with $u^1 = (v^1)''$.
Extend (i) and (ii) to two basis functions.

(c) $D_{11}x + D_{22}x = 0$ in the semi-infinite strip $t_1 \in [0, 1]$, $t_2 \in [0, \infty)$ with $x(0, t_2) = x(1, t_2) = x(t_1, \infty) = 0$ and $x(t_1, 0) = t_1(1 - t_1)$. Use the Kantrovich method (Exercise 4.4,5(d)): the function $x(t_1, t_2) = t_1(1 - t_1)\phi_1(t_2) + \psi(t_1)\phi_2(t_2)$ with $\phi_1(0) = 1$, $\phi_2(0) = 0$ will satisfy the (inhomogeneous) boundary condition. Begin with $\phi_2 = 0$ and then try $\phi_2 \neq 0$.

(d) $(EIx')'' - kV^2 x = 0$ in $[0, l]$ with $(EIx')' = EIx' = 0$ at $t = l$ and $x(0) = 0$; k is a given constant. Estimate the lowest eigenvalue V^2 if

$$EI(t) = (EI)_0(1 - \alpha t), \quad 1/l > \alpha > 0.$$

(This is a model of a wing leading edge in an airstream of speed V: the leading edge is assumed to act as a cantilever beam and the air loading is proportional to the local slope of the deflection curve (cf., Exercise, 4.3,6(c)): the leading edge will 'roll-up' if V exceeds the 'critical speed', V_c.)

Make use of functions satisfying the adjoint boundary conditions and show that, using only one pair of functions, we obtain the estimate

$$V_c^2 = \frac{(EI)_0 \int_0^l \frac{EI}{(EI)_0} (v^1)'(u^1)'' dt}{k \int_0^l v^1 u^1 dt}.$$

Extend the analysis to a two-dimensional approximating space. (For $EI = $ const. there is, of course, a simple analytical solution to this problem.)

(e) $EIx'''' + \gamma x'' - \omega^2 mx = 0$ in $[0, l]$ with $x(0) = x'(0) = 0$ and
 (i) $x''(l) = 0$, $EIx'''(l) + \gamma x'(l) = 0$ or
 (ii) $x''(l) = x'''(l) = 0$;
γ is a known positive parameter and $EI = $ const.

Show that in case (i) the differential operator $EI\,x'''' + \gamma x''$ with the given boundary conditions is symmetric whereas in case (ii) it is not. Deduce that in case (i) ω^2 is always real.

(This equation represents a cantilevered beam (or column) carrying an axial load γ at the tip; if $\gamma > 0$ the load is compressive. In case (i) the load is always aligned along the t-axis while in case (ii) the load is tangential to the deflection curve at the tip. Case (i) corresponds to classical Euler buckling. The beam is assumed to be executing a harmonic motion (in bending) of frequency ω.)

Use approximation in a two-dimensional space to deduce, in both cases, the variation of ω as γ is increased and deduce what you can about the motion of the system [96].

(f) $(EIx'')'' + \Omega^2 [mtx' - x'' \int_t^l m(\tau)\tau d\tau] - \omega^2 mx = 0$ with $x(0) = x''(0) = x''(l) = (EIx'')'_{t=l} = 0$; Ω is a given parameter and EI, m are given functions of t.

Show that the projection (or weak) form of this equation (p. 366) is

$$\Omega^2 \int_0^l \int_t^l m(\tau)\tau d\tau\, x'z'\, dt + \int_0^l EIx''z''\, dt - \omega^2 \int_0^l mxz\, dt = 0.$$

and hence deduce that the operator is symmetrical; verify this formally. (This equation will yield the normal vibration modes and natural transverse vibration frequencies of a beam which is hinged at $t = 0$ and rotates about that point with angular frequency Ω (e.g., a helicopter blade [97]).) The eigenfunctions will clearly be orthogonal.

Taking m, EI to be constant, estimate the first two eigenvalues (ω_1/Ω) and (ω_2/Ω) when

$$\frac{m\Omega^2 l^4}{EI} = 150.$$

(Note: it is necessary to include a rigid 'flapping' mode.)

(g) Interpret the virtual work identity in Example 3.10.1 as a weak or projection form of the (Cauchy) equation of equilibrium

$$\sum_{j=1}^{3} \frac{\partial \tau^{ij}}{\partial t_j} = 0 \quad \text{in} \quad \Gamma$$

with

$$\sum_{j=1}^{3} \tau^{ij} v_j = f^i \quad \text{on} \quad \partial \Gamma_1 \subset \partial \Gamma.$$

Take $_s e_i$ to vanish on $\partial \Gamma - \partial \Gamma_1$.

4. Suppose the Galerkin equations (p. 368) are expressed in terms of the $\mathscr{L}_2(\Gamma)$ inner product, viz.,

$$\sum_{i=1}^{n} \xi_i^{(n)} \int_\Gamma T v^i v^j \, d\Gamma = \int_\Gamma y v^j \, d\Gamma \qquad j = 1, \ldots n.$$

In many cases it is not possible to evaluate the integrals analytically and hence numerical quadrature must be used based on a set of M points $\{t^k\} \subset \Gamma$. If the quadrature formula is (cf., Example 3.5.7)

$$\int_\Gamma x(t) \, d\Gamma = \sum_{k=1}^{M} a_k x(t^k),$$

where a_k are the weights, show that the resulting form of the Galerkin equations is

$$\sum_{k=1}^{M} a_k v^j(t^k) \sum_{i=1}^{n} \xi_i^{(n)} T v^i(t^k) = \sum_{k=1}^{M} a_k v^j(t^k) y(t^k), \qquad j = 1, \ldots, n.$$

If we choose $M = n$ and the matrix $[v^j(t^k)]$ is non-singular, then deduce that the Galerkin equations are equivalent to the collocation equations (p. 368)

$$\sum_{i=1}^{n} \xi_i^{(n)} (T v^i)_{t=t^k} = (y)_{t=t^k}$$

(*see also* Example 5.5.1)

4.6 The finite element method

As we have already pointed out, the practical implementation of the Rayleigh–Ritz method (or the Galerkin method) is crucially dependent on constructing a suitable basis for the appropriate Hilbert space. If we take the Rayleigh–Ritz method as an example we may identify the steps involved in finding an approximate solution of an operator equation as follows:

1. Choose a suitable basis $\{z^i\}$ for $\mathcal{H}^T$.
2. Form the Ritz equations,

$$\sum_{j=1}^{n} \langle z^j, z^i \rangle_T \xi_j^{(n)} = \langle y, z^i \rangle, \qquad i = 1, 2, 3, \ldots, n.$$

3. Solve the Ritz equations.

Assuming step 1 to be completed for the moment, then in implementing step 2 we will, in the general case, be faced with the numerical evaluation of the integrals $\langle z^j, z^i \rangle_T$ and $\langle y, z^i \rangle$: the Ritz equations we obtain will be an approximation to the true Ritz equations in $\mathcal{H}_n^T$. Step 3 involves the solution of a set of algebraic equations. If the operator equation is non-linear, some type of iterative solution will probably be necessary; the body of results pertaining to the solution of such systems of equations is, to date, very sparse [78]. If the operator equation is linear, then we are, with regard to both methods of solution and error analysis, on very firm ground indeed. However, the practical computation of the solution of large systems of linear equations is fraught with difficulties, not the least of which is the cost of computer time; the choice of basis functions plays a most significant part in the solution in this respect.

The modern computer allows one to use, in the Rayleigh–Ritz and similar methods, approximating spaces whose dimension is of the order of several hundreds (or even thousands!) rather than the small spaces used in classical applications. This has focused attention on the dependence of steps 1, 2 and 3 on the choice of basis and the associated questions of rate of convergence, error propagation and cost. The finite element method has emerged as the front runner in best satisfying these requirements and at the same time giving great flexibility in application.

The finite element method originated in early computer analysis of structures, particularly aircraft structures. The structure could be considered to be made up of different physical elements like shear webs, tension elements, bending elements, etc., which were, in themselves, very simple to specify: these were then connected up or joined together at a *finite number of junctions*. That is to say, compatibility and equilibrium were enforced only at certain locations throughout the structure: distributed loadings were 'lumped' at these junctions in a more or less consistent way. The result was to replace the partial differential equations of three-dimensional elasticity by a set of algebraic equations. The basis of the method was the principal of virtual work (Example 4.7.4), which is completely analogous to the Ritz equation. It turns out that the structural elements are assembled to satisfy compatibility requirements and that the equilibrium at each joint is satisfied only in terms of 'forces' which are weighted integrals of the stress resultants in the element.

From these beginnings the finite element method developed extremely rapidly, for the most part in the hands of engineers. From structural analysis the method was extended to other problems, but generally with an actual physical subdivision of the region of interest as the underlying theme. However, with the recognition that the finite element method was really an application of the classical Rayleigh–Ritz method with a particular choice of *local basis functions*, the mathematical justification of the method for a large class of field problems was in sight. Active work in this direction is still being pursued.

As we shall see, the finite element method

1. makes available basis sets which may be catalogued and are readily applied to irregular domains;
2. allows the Ritz equations to be formed automatically;
3. leads to Ritz equations which are sparse, banded and well-conditioned.

These advantages make the method ideal for automatic implementation on computers and the role of the analyst is reduced to a minimum, both in formulating and solving the problem.

There is now a vast literature on the finite element method covering both basic theory and a multitude of specific applications, e.g. [65, 95, 98–118]. Here we will be content to outline the basic ideas and formulation of the method and for this purpose we will restrict attention to plane physical domains: extension to three dimensions is of course more complex, but not different in principle.

The finite element method in two dimensions

Suppose the linear operator equation $Tx = y$, with T positive-bounded-below, is defined on a two-dimensional region Γ. For example, T might be a partial differential operator with associated boundary conditions specified on $\partial \Gamma$; it is assumed for the moment that $\partial\Gamma$ is smooth. For ease of notation we shall represent the coordinate pair (t_1, t_2) of a point in the plane by the single symbol t.

In $\bar{\Gamma} = \Gamma + \partial\Gamma$ we choose a set of points $t^1, t^2, \ldots, t^G$ which are called *global nodes* or *global nodal points*. The functions of the approximating subspace $\mathscr{H}_N^T$ are then chosen such that their values or their derivatives are unity at one nodal point and zero at every other nodal point, that is to say, we choose an interpolation basis for $\mathscr{H}_N^T$. Let $\{v_m^i\}, i = 1, \ldots, I; m = 1, \ldots, M; MI = N$, be this basis and let $L_n v_m^i(t^j)$ denote the value of a specified derivative of v_m^i at the node t_j; for example, L_n could denote the value of the function or one of the operators $\partial/\partial t_1$, $\partial/\partial t_2$, $\partial^2/\partial t_1 \partial t_2$ or some form of directed derivative. Then the fact that $\{v_m^i\}$ is an interpolation basis is embodied in the statement

$$L_n v_m^i(t^j) = \delta_i^j \delta_m^n.$$

In terms of this basis a vector $x^{(N)} \in \mathscr{H}_N^T$ is given by

$$x^{(N)}(t) = \sum_{m=1}^M \sum_{i=1}^I \mu_i^m v_m^i(t),$$

where the μ_i^m, $i = 1, \ldots, I; m = 1, \ldots, M$, are the global values of the function $x^{(N)}$ or its specified derivatives at the nodes. To see this we only need to operate on both sides with L_n and evaluate the result at t^j, giving

$$L_n x^{(N)}(t^j) = \sum_{m=1}^M \sum_{i=1}^I \mu_i^m L_n v_m^i(t^j) = \mu_j^n.$$

It is convenient, in what follows, to reindex the basis set $\{v_m^i\}$ into a single sequence $\{z^r\}, r = 1, \ldots, N$, by assuming that a one-to-one correspondence has been set up between the indices r and the pairs (i, m): we shall then write the coordinates of $x^{(N)}(t)$ as ξ_r, giving

$$x^{(N)}(t) = \sum_{r=1}^N \xi_r z^r(t).$$

If no derivatives of the interpolation functions are involved then $M = 1$ and $N = G$, otherwise the number of nodes is less than the number of coordinates; that is to say, some of the nodes are *multiple nodes*.

So far we have done nothing more than make a choice of interpolation functions as basis functions, for $\mathscr{H}_N^T$. Now we come to the core of the finite element method: we choose these interpolation functions to have *local support*. By this we mean that the functions z^i

THE FINITE ELEMENT METHOD

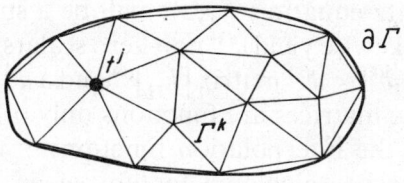

Fig. 4.6

vanish outside a subregion of Γ which is adjacent to the nodal point t^i. We do this by dividing the region Γ into subregions or elements $\Gamma^{(k)}$ whose intersection is either empty or consists only of a common boundary curve or point. In Fig. 4.6 the region is subdivided into triangular subregions with the global nodes situated at their vertices. Nodes need not all be on the boundaries of elements and those that are not are referred to as *internal nodes*. Let there be E such regions. We shall denote the number of nodes in the kth element by $n^{(k)}$ and by $d^{(k)}$ the number of *degrees of freedom*; this is the number obtained by allotting to each node its multiplicity and then summing.

We then choose the interpolation functions in the following way. For the global node t^j the function z^j (or its specified derivative) takes the value 1 and *vanishes identically outside the subregion consisting of the union of those $\Gamma^{(k)}$ which are incident on the node t^j*.

We still need to specify the actual form of the local function we wish to use and we need to ensure that it lies in $\mathcal{H}_N^T \subset \mathcal{H}^T$. The finite element method gives a beautifully systematic way of doing this by first of all *disassembling* the elements (or subregions), choosing a local basis in each element, computing the inner products required for the Ritz equations over each element and then reassembling the whole so that the full Ritz equations are given by a simple summation.

Consider the term $\langle z^i, z^j \rangle_T$ in the Ritz equations. This scalar is an integral of z^i, z^j and their derivatives over Γ: but, given certain conditions which we shall discuss below, the integral over Γ may be replaced by the sum of integrals over the subregions $\Gamma^{(k)}$: that is, we could write — with obvious notation —

$$\langle z^i, z^j \rangle_T = \sum_{k=1}^{E} \langle z^i, z^j \rangle_T^{(k)}.$$

But since the z^i are functions with local support, most of the scalars $\langle z^i, z^j \rangle_T^{(k)}$ will be zero. In fact, it is easy to see that if the element $\Gamma^{(k)}$ has $d^{(k)}$ degrees of freedom then the number of non-vanishing scalars $\langle z^i, z^j \rangle_T^{(k)}$ is simply $d^{(k)} \times d^{(k)}$. In a similar way, the right-hand side term in the Ritz equation $\langle y, z^j \rangle$ will be a summation of terms $\langle y, z^j \rangle^{(k)}$ and will likewise yield $d^{(k)}$ non-zero scalars. We may assemble these scalars into a $d^{(k)} \times d^{(k)}$ matrix $[k_{ij}]^{(k)}$ and a $d^{(k)}$ vector $\{y_i\}^{(k)}$, respectively, and these matrices are functions only of the element shape and the nature of the interpolation functions z^i *within the element*. They can therefore be calculated for any given element as if that element were isolated from its neighbours using a local coordinate system. For a standard element shape and a fixed class of interpolation functions, the matrix $[k_{ij}]^{(k)}$ can be computed once for all and it then only remains, knowing the position and orientation of the element in a global coordinate system, to inject the elements of this matrix into their proper positions in a global left-hand side $N \times N$ matrix $[k_{ij}] = [\langle z^i, z^j \rangle_T]$ and a global right-hand side vector $\{y_i\} = \{\langle y, z^i \rangle\}$. The structural analysis background to the finite element method has left a permanent imprint by virtue of the fact that we call $[k_{ij}]^{(k)}$ and $[k_{ij}]$, respectively, element and global stiffness matrices, and likewise $\{y_i\}^{(k)}$ and $\{y_i\}$ element and global load vectors.

Given a correspondence between a global numbering system for nodes and a local numbering system for nodes in each element, the assembly of a global stiffness matrix from the element stiffness matrices can be performed by the computer. Because of the local nature of the basis functions, the resulting matrix is sparse and banded; the bandwidth is dependent on the numbering of the global nodes, but roughly speaking will be of order $d^{(k)}$. In practical applications of the method $d^{(k)}$ will usually be the same for all elements.

So far we have said nothing explicit about the basis functions having to satisfy any essential homogeneous boundary conditions on $\partial \Gamma$. First, we should note that our subdivision of Γ yields a region $\Gamma^{(E)}$ which is only an approximation to Γ and we shall consider at a later point the effect of this on the solution. For the moment let us simply apply essential conditions on $\Gamma^{(E)}$. This means that for nodes which lie on $\partial \Gamma^{(E)}$ certain nodal values are specified to be zero. Rather than exclude the corresponding basis functions from the analysis the usual procedure is to assemble $[k_{ij}]$ and $\{y_i\}$, treating all nodes as unconstrained and then to eliminate those nodal unknowns which are specified. Inhomogeneous essential boundary conditions can be incorporated in a similar way (Exercise 4.6,1).

Now let us return to the point we skimmed over earlier, namely the assumption that the integral $\langle z^i, z^j \rangle_T$ over Γ could be replaced by the sum of corresponding integrals over the subregions or elements $\Gamma^{(k)}$. For this to be true it is necessary that the functions being integrated have only simple jump discontinuities over interelement boundaries. Thus if, for example, the highest (partial) derivative appearing in $\langle z^i, z^j \rangle_T$ is of order m, then the functions z^i must have continuous derivatives of order $(m-1)$ across interelement boundaries, that is to say, $z^i \in \mathscr{C}^{(m-1)}$. This condition is known as *conformability*,[†] and elements which satisfy the condition for the specific energy inner product under consideration are said to be *conforming*.

Let us now look at a few of the many finite elements which are in current use. All of these use polynomials as interpolation functions.

Triangular and rectangular elements

Consider the triangular element shown in Fig. 4.7.

The node numbering system is local but a global Cartesian coordinate system is used.

The simplest form of interpolation we can choose is linear, and to implement this choice we express x within the element as

$$x(t_1, t_2) = \alpha_1 + \alpha_2 t_1 + \alpha_3 t_2.$$

Since we have three nodes and three coefficients α_i we are able to define three linear interpolation functions within the element by relating the α_i to the nodal values ξ_i of x, viz.,

$$\begin{bmatrix} 1 & t_1^1 & t_2^1 \\ 1 & t_1^2 & t_2^2 \\ 1 & t_1^3 & t_2^3 \end{bmatrix} \begin{bmatrix} \alpha_1 \\ \alpha_2 \\ \alpha_3 \end{bmatrix} = \begin{bmatrix} \xi_1 \\ \xi_2 \\ \xi_3 \end{bmatrix},$$

where t_i^j is the ith coordinate of the jth node.

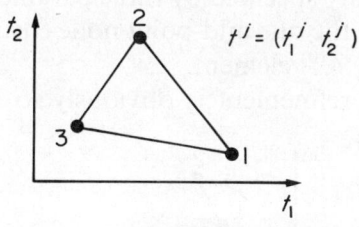

Fig. 4.7

[†] The conformability condition need not always be satisfied for the finite element method to converge, but certain other conditions need to be met. A test exists called the 'patch test' which will identify those non-conforming elements which will allow summation of element integrals [99, 109].

Solving the equations gives

$$\begin{bmatrix} \alpha_1 \\ \alpha_2 \\ \alpha_3 \end{bmatrix} = \frac{1}{2A} \begin{bmatrix} t_1^2 t_2^3 - t_1^3 t_2^2 & t_1^3 t_2^1 - t_1^1 t_2^3 & t_1^1 t_2^2 - t_1^2 t_2^1 \\ t_2^2 - t_2^3 & t_2^3 - t_2^1 & t_2^1 - t_2^2 \\ t_1^3 - t_1^2 & t_1^1 - t_1^3 & t_1^2 - t_1^1 \end{bmatrix} \begin{bmatrix} \xi_1 \\ \xi_2 \\ \xi_3 \end{bmatrix}$$

or

$$\{\alpha_i\} = \frac{1}{2A}[g_{ij}]\{\xi_j\},$$

where A is the area of the element. The interpolation function corresponding to a unit nodal value at vertex 1 is, for example,

$$\frac{1}{2A}((t_1^2 t_2^3 - t_1^3 t_2^2) + (t_2^2 - t_2^3)t_1 + (t_1^3 - t_1^2)t_2).$$

Along any edge of the triangle the variation of x is linear and is determined by the two end nodal values and completely unaffected by the third; hence we see immediately that x will be continuous across element boundaries and therefore these elements will be conforming whenever the highest derivative in $\langle z^i, z^j \rangle_T$ is the first. We call this a $\mathscr{C}^{(0)}$ element.

We may achieve higher accuracy both overall and within an element by including quadratic terms, viz.,

$$x(t_1, t_2) = \alpha_1 + \alpha_2 t_1 + \alpha_3 t_2 + \alpha_4 (t_1)^2 + \alpha_5 (t_1)(t_2) + \alpha_6 (t_2)^2.$$

Now we have six coefficients α_i and so require six nodes: we may introduce three extra nodes at the mid-points of the sides (Fig. 4.8). Since the variation of x along any side is quadratic in one variable and a quadratic is fully specified by three parameters, then continuity at the end nodes and the one mid-point node ensures continuity along the edge, so this is a $\mathscr{C}^{(0)}$ element.

The next stage of refinement is obviously to include cubic terms,

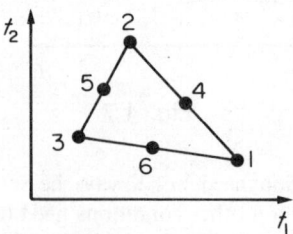

Fig. 4.8

and this requires a total of ten nodes; standard elements have either vertex plus two nodes per side plus a centroidal node, or the vertex nodes are triple nodes at which the value and two derivatives are specified plus the value at a centroidal node[†] (see, for example, [99]). The generation of interpolation functions for triangular elements is considerably simplified if so-called area coordinates are used—these are explained and developed in Exercise 4.6,2.

Similar interpolation functions can be constructed on rectangular elements with sides parallel to the global coordinate axes. The simplest element stems from a bilinear expression for x, viz.,

$$x(t_1, t_2) = \beta_1 + \beta_2 t_1 + \beta_3 t_2 + \beta_4 t_1 t_2.$$

The four coefficients β_1 are determined by the values of the function at the corners of the rectangle. This is a $\mathscr{C}^{(0)}$ element. Higher-order Lagrange-type elements can of course be constructed. In the construction of these a local coordinate system is used and the interpolation functions are constructed directly using the product of one-dimensional Lagrange interpolation polynomials [98]. Alternatively, we may use Hermite-type interpolation and make some of the nodes multiple. For example, if we adopt a cubic representation involving 12 coefficients then we can adopt four corner nodes plus two nodes per side or make the four corner nodes triple nodes at which the value and the partial derivatives $\partial/\partial t_1, \partial/\partial t_2$ are specified. These are $\mathscr{C}^{(0)}$ elements.

Triangular or rectangular elements which possess $\mathscr{C}^{(1)}$ continuity are obviously more complicated than the $\mathscr{C}^{(0)}$ type but many ingenious $\mathscr{C}^{(1)}$ elements exist. The simplest triangular element involves a quintic representation for x with 21 coefficients. When the function value and all the first- and second-order derivatives are specified at the vertices, there are only 18 conditions and another three are needed: these are taken as the normal derivative at each side mid-point. The normal slope along each side is a quartic which needs five conditions to be fully specified; these are the two end values of the function, the two end normal derivatives and the mid-point normal derivative [99].

Isoparametric elements

An important limitation of straight-sided elements is that they

[†] Interpolation involving only function values is usually called *Lagrange interpolation*, while if derivatives are also involved the allusion is to *Hermite interpolation*.

replace the smooth boundary $\partial \Gamma$ by a polygon. Clearly elements with curved sides will give a better representation. The idea of the isoparametric element is to approximate the boundary (locally) by a polynomial—indeed to use the same type of polynomial as is used to represent the variation of the function within the element.

We shall illustrate the basic idea using quadratic representation (p. 388) in a triangle with six nodes. For simplicity we take only one side of the required element to be curved. We first transform the triangle to a local axis system (s_1, s_2) in which the angle between the two straight sides is $\pi/2$ (Fig. 4.9).

This is achieved by the linear mapping
$$s_1 = \alpha_1 + \alpha_2 t_1 + \alpha_3 t_2,$$
$$s_2 = \beta_1 + \beta_2 t_1 + \beta_3 t_2,$$

where
$$\{\alpha_i\} = \frac{1}{2A}[g_{ij}]\begin{bmatrix}1\\0\\0\end{bmatrix}$$

and
$$\{\beta_i\} = \frac{1}{2A}[g_{ij}]\begin{bmatrix}0\\1\\0\end{bmatrix}.$$

Next we transform the curved side to a straight side using a quadratic mapping so that there is a fit at vertices 1 and 2 and at some intermediate point—let this point be the mid-point of the side 12 (Fig. 4.9).

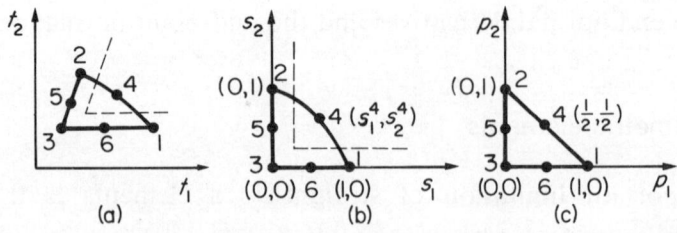

Fig. 4.9

This is achieved with the bilinear map

$$s_1 = p_1 + 2(2s_1^4 - 1)p_1 p_2,$$
$$s_2 = p_2 + 2(2s_2^4 - 1)p_1 p_2.$$

The straight sides lying along the coordinate axes are mapped identically while the point (s_1^4, s_2^4) goes to $(\frac{1}{2}, \frac{1}{2})$. The Jacobian of the transformation is

$$J(p_1, p_2) = \det \begin{bmatrix} 1 + 2(2s_1^4 - 1)p_2 & 2(2s_1^4 - 1)p_1 \\ 2(2s_2^4 - 1)p_2 & 1 + 2(2s_2^4 - 1)p_1 \end{bmatrix}$$
$$= 1 + 2(2s_1^4 - 1)p_2 + 2(2s_2^4 - 1)p_1.$$

The Jacobian equals 1 at $(0, 0)$ and is linear; it will not vanish in the triangle if it is positive at vertices 1 and 2. This requires $s_1^4 > \frac{1}{4}, s_2^4 > \frac{1}{4}$ so that the node on the curved side must lie in the sectors shown in Fig. 4.9(a) and (b).

Any transformation of the above type must clearly satisfy continuity conditions between elements. By using interpolation functions which are already known to satisfy continuity conditions this presents little difficulty. It is perhaps worth noting that the integrals $\langle z^i, z^j \rangle_T^{(k)}$ no longer have simple polynomial integrands in the variables (p_1, p_2). For a typical integral of the form

$$\int_{T^{(k)}} f\left(x, \frac{\partial x}{\partial t_1}, \frac{\partial x}{\partial t_2}\right) dt_1 \, dt_2$$

the derivatives $\partial/\partial t_1, \partial/\partial t_2$ are expressed as

$$\begin{bmatrix} \dfrac{\partial}{\partial t_1} \\ \dfrac{\partial}{\partial t_2} \end{bmatrix} = [J]^{-1} \begin{bmatrix} \dfrac{\partial}{\partial p_1} \\ \dfrac{\partial}{\partial p_2} \end{bmatrix},$$

where $J(p_1, p_2)$ is the Jacobian matrix of the transformation and the integral becomes

$$\int_{\triangle} h(p_1, p_2) \det [J] dp_1 \, dp_2.$$

This integral generally requires numerical evaluation (*see*, for example, [105, 111]).

The other common type of isoparametric element is based on the rectangle. If a linear mapping is used, a quadrilateral is obtained, while for quadratic (or higher-order) mappings a 'quadrilateral-like' curved-sided element is obtained.†

Convergence of the finite element method

So far we have not considered the implication of the word 'basis' in the first step of the finite element method (p. 382), viz., 'choose a suitable basis $\{z^i\}$ for $\mathcal{H}^T$'. If $\{z^i\}$ is a basis for $\mathcal{H}^T$ then we require that, given any $\varepsilon > 0$, there exists N such that, for $n > N$,

$$\|x - \sum_{i=1}^{n} \xi_i z^i\|_T < \varepsilon \qquad \forall x \in \mathcal{H}^T.$$

This condition is often called 'completeness in energy' of the set $\{z^i\}$ (cf., footnote on p. 139).

The piecewise polynomial functions of the finite element method do not automatically provide a basis for $\mathcal{H}^T$ even though the elements are conforming. The conforming condition relates to the degree of conformity of $x^{(N)}$ across interelement boundaries; the 'completeness in energy' condition relates to the degree of smoothness of $x^{(N)}$ *within* elements as $N \to \infty$, that is to say, as the element size tends to zero. Roughly speaking we should ensure, for example, that—within the element—the function $x^{(N)}$ is capable of representing a constant value or a linear variation with a constant superimposed; in other words, we should use polynomials in which we have not omitted terms of low degree. In the context of structural analysis, where x is the displacement field, we must ensure that, within each element, we can represent a motion of the element as a rigid body.

There is a simple rule for choosing element polynomials that lead to global basis functions for $\mathcal{H}^T$ [106]. First of all, a polynomial of degree r is called *complete* (this over-used word again!) if it contains all terms whose total degree (involving products of one, two or three variables) is r or less. A complete cubic polynomial in two variables, for example, will contain 10 terms. If the energy inner product $\langle z^j, z^i \rangle_T$ contains derivatives of order not exceeding m, then the condition we need to satisfy is that our polynomials have at least $\mathcal{C}^{(m)}$ continuity within the element. This will be ensured if the element polynomial contains a complete polynomial of degree m whose

† The number of functions used to represent the element shape and the unknown $x(t)$ need not be the same: in this way we can generate families of *subparametric* and *superparametric* elements.

coefficients can be independently varied by varying a subset of the nodal values of the element: terms of higher degree, if present, can be made to vanish independently. This means that $x^{(N)}$ or any of its partial derivatives up to and including those of order m can take on arbitrary constant values within the element.

The isoparametric element formulation offers the attractive feature that if completeness is satisfied in the standard element then it is automatically satisfied in the curved element. This is intimately connected with the fact that the Jacobian of the transformation is bounded away from zero and, for example, all angles in a transformed triangle exceed the same lower bound; that is to say, the triangle cannot become degenerate [98].

Satisfaction of the conformity and completeness requirements ensures convergence of the finite element method as element size decreases: let us now turn to the question of estimating the rate of convergence. We know (p. 331) that the approximation $x^{(N)} \in \mathscr{H}_N^T$ to the solution $\hat{x} \in \mathscr{H}^T$ of $Tx = y$ minimizes the distance $\|x^{(N)} - \hat{x}\|_T$. Hence if z is any point in $\mathscr{H}_n^T$ we have

$$\|\hat{x} - x^{(N)}\|_T \leq \|\hat{x} - z\|_T, \qquad z \in \mathscr{H}_N^T,$$

with equality if, and only if, $z = x^{(N)}$. In particular, the inequality holds when z is an interpolation function in $\mathscr{H}_N^T$. If we can estimate the energy distance between $\hat{x}$ and its interpolation function $\hat{z}^{(N)}$ then we know that the Ritz solution $x^{(N)}$ will be nearer $\hat{x}$ or, at the least, no further away. But the estimation of this distance is a standard problem in approximation theory and has been studied extensively. Since

$$\|\hat{x} - \hat{z}^{(N)}\|_T^2 = \sum_{k=1}^{E} (\|\hat{x} - \hat{z}^{(N)}\|_T^2 : t \in \Gamma^{(k)})$$

it is necessary only to estimate the interpolation error $\|\hat{x} - \hat{z}^{(N)}\|$ within a typical element, where the function $\hat{x}$ has the same nodal values as the interpolation function $\hat{z}^{(N)} \in \mathscr{H}_N^T$. Convergence rate is usually expressed in terms of a parameter h_k which is a measure of the size of the kth finite element, say the maximum side length or diameter of an element. An analysis along these lines leads to an error bound of the form

$$\|\hat{x} - x^{(N)}\| \leq K h^s |\hat{x}|,$$

where, on the left-hand side, $\|\cdot\|$ denotes some chosen norm of the error; on the right-hand-side $|\cdot|$ denotes a norm or seminorm

(Exercise 3.3,1(b)) of the (unknown) solution, K is a constant, $h = \max_k h_k$ and s is a positive integer. This type of estimate gives an asymptotic rate of convergence. In almost all cases the constant K is not known and the norm of $\hat{x}$ may be difficult to estimate, although in the case of elliptic differential equations $\|\hat{x}\|$ can be related to a norm of the right-hand term y in $Tx = y$ (p. 465). However, such error bounds are of use in comparing different finite element formulations since presumably the 'best' is that which leads to the largest value of s.

For elliptic differential operators of order $2m$ (cf., Example 4.5.1) the energy norm is equivalent (Exercise 3.3,4) to the norm

$$\|x\|^{(m)} = \left\{ \sum_{k \leq m} \int_\Gamma (D^k x)^2 d\Gamma \right\}^{1/2} :$$

if all polynomials in Γ of degree r are in the span of the (global) basis functions $\{z^i\}, i = 1, \ldots, N$, then for this class of operator a typical result for the error bound is

$$\|\hat{x} - x^{(N)}\|^{(m)} \leq K_m h^{r+1-m} |\hat{x}|^{(r+1)}.$$

As we know, the error $\|\hat{x} - x^{(N)}\|^{(m)}$ is natural for the Ritz method, but similar formulae apply for the errors $\|\hat{x} - x^{(N)}\|^{(s)}$, $s < m$, and take the form (Exercise 4.6,3)

$$\|\hat{x} - x^{(N)}\|^{(s)} \leq K_s h^{r+1-s} |\hat{x}|^{(r+1)}.$$

For example, if $T \equiv \nabla^2$ ($m = 1$) and if we use linear interpolation functions in each element ($r = 1$), then

$$\|\hat{x} - x^{(N)}\|^{(1)} = O[h],$$
$$\|\hat{x} - x^{(N)}\| = O[h^2].$$

Many other error bounds of the same type have been derived for specific operators and for specific finite elements based on the use of piecewise polynomial interpolation [99, 108, 114–116].

There are two practical aspects of the finite element method which might be expected to affect the rate of convergence:

1. replacement of Γ by a polygonal region implying only the approximate satisfaction of essential boundary conditions;
2. the use of numerical integration to evaluate the energy inner product within elements.

The first question is involved and has only been worked out for particular differential operators [113]. The difficulty can largely be avoided, of course, by the use of isoparametric elements.

The second question obviously involves a choice of quadrature formula which will not degrade the theoretical order of convergence; at the same time it should not be over-accurate, for then the cost of computation would be unnecessarily high. Most commonly used quadrature formulae integrate polynomials of a certain order (or less) exactly.

The quantity to be calculated is, of course, $\langle z^i, z^j \rangle_T^{(k)}$, and if T is an elliptic operator of order $2m$ this apparently involves the accurate integration of the product of the mth derivatives of the functions z^i, z^j. However, it turns out that this accuracy is not required and it is sufficient for convergence of the finite element method that the quadrature formula should integrate the mth derivatives of each global, piecewise polynomial, basis function z^i exactly. For example, using Lagrange or Hermite interpolation polynomials of degree r, the quadrature formula should be exact for polynomials of degree $r - m$ or less. In effect, rules such as this lead to the specification of a minimum number of quadrature (usually Gaussian) points in a given element: these matters are fully discussed in [98, 99, 112].

Finally, we should note that although we have here discussed the finite element method entirely within the context of linear operator equations, its use is by no means confined to these. The subdivision of the region of interest and the adoption of a localized basis, the ideas which form the backbone of the finite element method, can be extended to any of the projection methods discussed in Section 4.5. The analysis is, of course, more complicated, but the summation of the inner product integrals over elements still holds given conforming elements; the result is that the analysis leads to a finite set of non-linear equations with the nodal values as unknowns. We may formally write the resulting equations as

$$\{k_i(x)\} = \{y_i\}, \quad i = 1, \ldots, N,$$

where the k_i are non-linear functions of the N-tuple $x = (\xi_1, \xi_2, \ldots, \xi_N)$ (*see*, for example, [65, 104]).

Example 4.6.1 We can illustrate many aspects of the finite element method by using a very simple example in one space dimension. Let us solve the operator equation

$$-\frac{d^2 x}{dt^2} = 1 \quad \text{in} \quad [0, 1]$$

with $x(0) = 0$, $x'(1) = 0$ using interpolation polynomials of first degree.

We subdivide the region $[0, 1]$ into E subintervals or finite elements each of length h (see Fig. 4.10 (a)) and in each element we choose as a local basis set the interpolation polynomials $\tau, 1 - \tau$, where τ is a local t-like coordinate with $\tau = 0$ at $t = t^j$, $\tau = 1$ at $t = t^{j+1}$ (Fig. 4.10(b)). Since the energy inner product for this case is

$$\langle x, z \rangle_T = \int_0^1 \frac{dx}{dt} \frac{dz}{dt} dt$$

we see that the elements are conforming and the interpolation functions are complete in energy.

The energy inner products for a typical element are given by

$$k_{11}^{(k)} = h \int_0^1 \frac{1}{h} \frac{1}{h} d\tau = \frac{1}{h},$$

$$k_{12}^{(k)} = h \int_0^1 \frac{1}{h} \times \frac{-1}{h} d\tau = -\frac{1}{h},$$

etc., giving a local stiffness matrix

$$[k_{ij}]^{(k)} = \frac{1}{h} \begin{bmatrix} 1 & -1 \\ -1 & 1 \end{bmatrix}.$$

We now assemble the local stiffness matrices into a global stiffness matrix by identifying the global and local numbering systems for the nodes. The

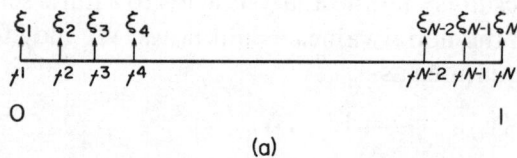

(a)

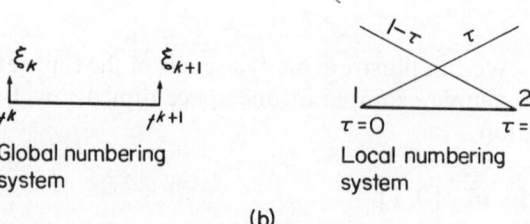

(b)

Fig. 4.10

correspondence is

$$\begin{array}{llllllll} \text{Global} & 1 & 2 & 3 & 4 & \ldots & N-1 & N \\ \text{Local} & 1 & 2 & & & & & \\ & & 1 & 2 & & & & \\ & & & 1 & 2 & & & \\ & & & & & \ddots & & \\ & & & & & & 1 & 2 \end{array}$$

and a schematic of the assembled stiffness matrix is

$$\frac{1}{h}\begin{bmatrix} 1+\begin{bmatrix} 1 & -1 \\ -1 & 1 \end{bmatrix} + \begin{bmatrix} 1 & -1 \\ -1 & 1 \end{bmatrix} \\ \ddots \end{bmatrix}$$

$$=\frac{1}{h}\begin{bmatrix} 2 & -1 & 0 & 0 & 0 & \ldots & 0 & 0 \\ -1 & 2 & -1 & 0 & 0 & \ldots & 0 & 0 \\ 0 & -1 & 2 & -1 & 0 & \ldots & 0 & 0 \\ 0 & 0 & -1 & 2 & -1 & \ldots & 0 & 0 \\ \vdots & \vdots & \vdots & \vdots & \vdots & \vdots & \vdots & \vdots \\ 0 & 0 & 0 & 0 & 0 & \ldots & 2 & -1 \\ 0 & 0 & 0 & 0 & 0 & \ldots & -1 & 1 \end{bmatrix}$$

the nodal value ξ_1 is zero by virtue of the essential boundary value $x(0) = 0$ and is omitted.

In a similar way the local element loading vector has components given by

$$y_1^{(k)} = h\int_0^1 \tau d\tau = \frac{h}{2} = y_2^{(k)}$$

and upon assembly we obtain the global load vector

$$\frac{h}{2}\begin{bmatrix} 1+\begin{bmatrix} 1 \\ 1 \end{bmatrix} + \begin{bmatrix} 1 \\ 1 \end{bmatrix} \\ \ddots \end{bmatrix} = h\begin{bmatrix} 1 \\ 1 \\ \vdots \\ \frac{1}{2} \end{bmatrix}.$$

For example, if we take three elements we obtain the global Ritz equation

$$\begin{bmatrix} 2 & -1 & 0 \\ -1 & 2 & -1 \\ 0 & -1 & 1 \end{bmatrix}\begin{bmatrix} \xi_2 \\ \xi_3 \\ \xi_4 \end{bmatrix} = \frac{1}{9}\begin{bmatrix} 1 \\ 1 \\ \frac{1}{2} \end{bmatrix}$$

with solution

$$\begin{bmatrix} \xi_2 \\ \xi_3 \\ \xi_4 \end{bmatrix} = \frac{1}{9} \begin{bmatrix} 1 & 1 & 1 \\ 1 & 2 & 2 \\ 1 & 2 & 3 \end{bmatrix} \begin{bmatrix} 1 \\ 1 \\ 1 \end{bmatrix} = \frac{1}{18} \begin{bmatrix} 5 \\ 8 \\ 9 \end{bmatrix}.$$

The exact solution of the equation is, of course, $x = t(1 - t/2)$ and it is readily seen that the nodal values coincide with the values of the exact solution at the nodal points.[†]

We also have

$$\|\hat{x}\|_T^2 = \int_0^1 \left(\frac{d\hat{x}}{dt}\right)^2 dt = \int_0^1 1 \hat{x}\, dt = \frac{1}{3}$$

and

$$\|x^{(4)}\|_T^2 = \lfloor \xi_i \rfloor [k_{ij}] \{\xi_j\} = \lfloor y_j \rfloor \{\xi_j\} = 0.3240$$

so that (p. 336)

$$\|\hat{x} - x^{(4)}\|_T = 0.0964.$$

The asymptotic rate of convergence for $\|\hat{x} - x^{(N)}\|_T$ in this problem is $O(h)$ (p. 394), $r = 1$, $m = 1$: the computed error for increasing N agrees closely with this estimate (Exercise 4.6, 4(a)).

For an arbitrary right-hand side the integrals $\langle y, z^j \rangle^{(k)}$ would have to be evaluated numerically: in this case a one-point Gaussian quadrature formula will suffice (this integrates first-degree polynomials exactly and is therefore adequate for the interpolation polynomials used).

It is instructive to derive the error bound (p. 394):

$$\|\hat{x} - x^{(N)}\|_T \leqslant Kh \|\hat{x}\|^{(2)}$$

for this particular example. Within a typical finite element the 'error' function

$$\varepsilon(t) = \hat{x}(t) - x^{(N)}(t)$$

may be represented as a Fourier sine series since $\varepsilon(t)$ vanishes at each end of the element, viz.,

$$\varepsilon(t) = \sum_{i=1}^\infty \alpha_i \sin \frac{i\pi t}{h}.$$

[†] The reason for agreement of the nodal values with the exact solution lies in the facts that (a) the Green's function for this operator is piecewise linear; (b) in the right-hand side the 'constant' function can be effectively replaced by concentrated 'loads' at the nodal points. The above situation will also apply if y is piecewise linear, but not for higher-order polynomials or other functions.

Now
$$\int_0^h (\varepsilon')^2 \, dt = \frac{h}{2} \sum_{i=1}^{\infty} \left(\frac{i\pi}{h}\right)^2 \alpha_i^2$$
and
$$\int_0^h (\varepsilon'')^2 \, dt = \frac{h}{2} \sum_{i=1}^{\infty} \left(\frac{i\pi}{h}\right)^4 \alpha_i^2;$$
also
$$\left(\frac{i\pi}{h}\right)^2 \alpha_i^2 \leqslant \frac{h^2}{\pi^2}\left(\frac{i\pi}{h}\right)^4 \alpha_i^2,$$
so that
$$\int_0^h (\varepsilon')^2 \, dt \leqslant \frac{h^2}{\pi^2} \int_0^h (\varepsilon'')^2 \, dt.$$
But within each element $\varepsilon'' = \hat{x}''$, hence
$$\int_0^h (\varepsilon')^2 \, dt \leqslant \frac{h^2}{\pi^2} \int_0^h (\hat{x}'')^2 \, dt.$$

Summing this inequality over all elements (the elements are conforming and $\hat{x}''$ is continuous) and taking the square root, we have

$$\langle \varepsilon', \varepsilon' \rangle^{1/2} = \| \hat{x} - x^{(N)} \|_T \leqslant \frac{h}{\pi} \langle \hat{x}'', \hat{x}'' \rangle^{1/2} = \frac{h}{\pi} |\hat{x}|^{(2)}.$$

There has been no attempt in this section to acquaint the reader with the practical computational and organizational intricacies of the finite element method, nor are the exercises which follow designed as practice in the method. There are many excellent texts (and a flood of papers) which deal in detail with these important matters, most of them written by engineers for engineers ([65, 101–105, 110, 111, 118] represent a small sample).

Exercises 4.6

1. Let the equations to be solved for a finite element analysis be
$$[k_{ij}]\{\xi_j\} = \{y_i\}, \quad i,j = 1, \ldots, N$$
and suppose that $\xi_r = \alpha_r$ for $r = \{r_1, r_2, \ldots, r_k\}$ a subset of $\{1, \ldots, N\}$.
   (a) We may proceed formally, renumber the nodes and so produce the

partitioned system

$$\left[\begin{array}{c|c} [k_{ij}]_{11} & [k_{ij}]_{12} \\ \hline [k_{ij}]_{21} & [k_{ij}]_{22} \end{array}\right] \left[\begin{array}{c} \{\xi_j\}_1 \\ \hline \{\alpha_j\} \end{array}\right] = \left[\begin{array}{c} \{y_i\}_1 \\ \hline \{y_i\}_2 \end{array}\right].$$

We then have

$$[k_{ij}]_{11}\{\xi_j\}_1 = \{y_i\}_1 - [k_{ij}]_{12}\{\alpha_j\}.$$

(b) A renumbering of nodes will destroy the banded property of $[k_{ij}]$: verify that the following procedure will give the result of part (a) without renumbering and without altering the order of the matrix equation to be solved. Begin with the typical prescribed nodal value $\xi_r = \alpha_r$.
  (i) Set $k_{rj} = 0, j = 1, \ldots, N$ except $j = r$.
  (ii) Replace y_r by $k_{rr}\alpha_r$.
  (iii) Subtract $k_{ir}\alpha_r$ from $y_i, i \neq r$.
  (iv) Set $k_{ir} = 0, i = 1, \ldots, N$ except $i = r$.
The same procedure is carried through for each prescribed nodal value.

(c) Show how to find the unknown 'reactions' $\{y_i\}_2$ in parts (a) and (b) above.

2. Area or natural coordinates.
  (a) Let a plane triangular element have vertices t^j (Fig. 4.7) and let $P(t)$ be any interior point of the triangle. The vectors $(t - t^j) \in \mathcal{R}_2$, $j = 1, 2, 3$, must be linearly dependent (Exercise 2.4,2); hence show that there exist scalars $\sigma_j, j = 1, 2, 3$, with $\sigma_1 + \sigma_2 + \sigma_3 = 1$ such that

  $$t = \sigma_1 t^1 + \sigma_2 t^2 + \sigma_3 t^3.$$

  (b) Show that the transformation between the coordinates $(\sigma_1, \sigma_2, \sigma_3)$ and (t_1, t_2) of P is

  $$\{\sigma_i\} = \frac{1}{2A}[g_{ij}]^T \begin{bmatrix} 1 \\ t_1 \\ t_2 \end{bmatrix}$$

  where $[g_{ij}]$ is the matrix given on p. 388.
  Show that the lines $\sigma_1 = \alpha, 0 \leqslant \alpha \leqslant 1$, are parallel to side 2–3 of the triangle with corresponding cyclic results for $\sigma_2, \sigma_3 = $ const. What does α represent?

  (c) Show that the scalar σ_1 is the ratio A_{P23}/A, where A_{P23} is the area of the triangle with vertices P, 2, 3. This is the justification for the name 'area coordinates'; note that $\sigma_1, \sigma_2, \sigma_3$ are not independent.

  (d) Show that the linear interpolation functions for the triangle derived on p. 388 are simply the functions σ_j in terms of the area coordinates.

(e) Verify that quadratic interpolation functions for the six-node triangle of Fig. 4.8 are, in terms of the area coordinates,

$$\sigma_j(2\sigma_j - 1), \quad j = 1, 2, 3,$$

and

$$4\sigma_j\sigma_k, \quad j = 1, 2, 3, \quad k = 2, 3, 1.$$

(f) Find an expression for

$$\langle z^i, z^j \rangle_T^{(k)} = \int_{\Delta^k} D_1 z^i D_1 z^j + D_2 z^i D_2 z^j \, dt_1 \, dt_2$$

using area coordinates when z^i, z^j are (i) linear, (ii) quadratic interpolation functions on a triangular element $\Delta^{(k)}$ (cf. p. 391).

Note: natural or area coordinates also exist for tetrahedral elements. Analytical integration formulae exist for triangular (and tetrahedral) elements, in terms of area coordinates: for example, for triangles [105]

$$\int_{\Gamma = \Delta} (\sigma_1)^\alpha (\sigma_2)^\beta (\sigma_3)^\gamma \, d\Gamma = 2A \frac{\alpha! \beta! \gamma!}{(\alpha + \beta + \gamma + 2)!}.$$

3. For a second-order elliptic differential operator the energy norm is equivalent to the $\mathcal{H}^{(1)}$-norm and the natural measure of error of an N-dimensional finite-element solution based on linear interpolation functions appears in the form (p. 394)

$$\|\hat{x} - x^{(N)}\|^{(1)} \leq K_1 h \|\hat{x}\|^{(2)}.$$

To obtain an estimate of the mean-square error in x rather than in its first derivatives one can proceed in the following way:

(i) Show that the weak solution $\hat{z}$ of the operator equation $Tz = \hat{x} - x^{(N)}$ is characterized by

$$\langle \hat{z}, w \rangle_T = \langle \hat{x} - x^{(N)}, w \rangle \quad \forall w \in \mathcal{H}^T.$$

(ii) Using the fact that the result in (i) holds for $w = \hat{x} - x^{(N)}$ and that the Ritz solution of $Tx = y$ is characterized by

$$\langle \hat{x} - x^{(N)}, w^{(N)} \rangle_T = 0 \quad \forall w^{(N)} \in \mathcal{H}_N^T,$$

show that

$$\langle \hat{z} - w^{(N)}, \hat{x} - x^{(N)} \rangle_T = \|\hat{x} - x^{(N)}\|^2.$$

(iii) Now use the Schwarz inequality (in $\mathcal{H}^T$) to show that

$$\|\hat{x} - x^{(N)}\|^2 \leq \|\hat{z} - w^{(N)}\|_T \|\hat{x} - x^{(N)}\|_T.$$

(iv) Choosing $w^{(N)}$ as the Ritz approximation to $\hat{z}$ we have, from the basic error estimate in $\mathcal{H}^T$,

$$\|\hat{z} - w^{(N)}\|^{(1)} \leq K_1 h \|\hat{z}\|^{(2)}.$$

Now from the theory of elliptic operators (Section 5.4, p. 465 $\|\hat{z}\|^{(2)}$ can be estimated in terms of the right-hand side, $\hat{x} - x^{(N)}$, viz.,

$$\|\hat{z}\|^{(2)} \leq M \|\hat{x} - x^{(N)}\|;$$

hence show that

$$\|\hat{x} - x^{(N)}\|^2 \leq Ch^2 \|\hat{x} - x^{(N)}\| \|\hat{x}\|^{(2)}$$

or

$$\|\hat{x} - x^{(N)}\| \leq Ch^2 \|\hat{x}\|^{(2)} \leq Dh^2 \|y\|.$$

4. (a) Repeat Example 4.6.1 with two, four and six elements and compare the computed errors with the predicted asymptotic rate of convergence. (Hint: nodal values are exact!)
   (b) Solve the equation

   $$-\nabla^2 x = 1 \quad \text{in} \quad \Gamma = [0, 1] \times [0, 1]$$

   with $x = 0$ on $\partial \Gamma$ using triangular finite elements. Notice that the solution has fourfold symmetry and hence only one eighth of the region needs to be considered namely, an isosceles triangle.
   (i) Divide this triangle into four triangular elements and use linear interpolation functions (three nodal points per element—Exercise 4.6,2(f)). Assemble the global stiffness matrix as described on p. 386 and impose the boundary conditions only at the final stage with the technique of Exercise 4.6,1(b).
   (ii) Treat this triangle as a single element and use quadratic interpolation functions (six nodal points per element—using Exercise 4.6,2(f)).
   Compare your results for $\|x^{(N)}\|_T$ with those of Example 4.4.5.
   (c) In problems of plane stress (or strain) or in two-dimensional flow problems, the quantity of interest, x, is a vector rather than a scalar. In that case interpolation functions are used to approximate to each of the (Cartesian) components of $x = (\xi_1, \xi_2)$
   (i) Construct a (plane) triangular element in which each displacement component ξ_1, ξ_2 is interpolated linearly with vertex nodal values (six degrees of freedom).
   (ii) The two-dimensional strain–displacement relations are

   $$\varepsilon_{11} = D_1 \xi_1, \qquad \varepsilon_{22} = D_2 \xi_2, \qquad \varepsilon_{12} = D_1 \xi_2 + D_2 \xi_1$$

   while, in the case of plane stress, the stress–strain relations are

   $$\tau_{11} = \frac{E}{1-\mu^2}(\varepsilon_{11} + \mu \varepsilon_{22}), \qquad \tau_{22} = \frac{E}{1-\mu^2}(\mu \varepsilon_{11} + \varepsilon_{22}),$$

   $$\tau_{12} = \frac{E}{2(1+\mu)} \varepsilon_{12},$$

where μ and E are material constants [52]. This scheme, although not consistent with the tensorial properties of strain and stress, is convenient in that it allows the introduction of the strain and stress 'vectors', $(\varepsilon_{11}, \varepsilon_{22}, \varepsilon_{12})$ and $(\tau_{11}, \tau_{22}, \tau_{12})$.

(iii) Thence evaluate the (6×6) element stiffness matrix (cf. Example 4.7.4)

$$[k_{ij}]^{(k)} = \langle z^i, z^j \rangle_T^{(k)} = \int_{\Gamma^{(k)}} \varepsilon_{11}^i \tau_{11}^j + \varepsilon_{22}^i \tau_{22}^j + \varepsilon_{12}^i \tau_{12}^j \, d\Gamma$$

in the form

$$[k_{ij}]^{(k)} = [b_{ij}]^T [d_{ij}] [b_{ij}],$$

where $[d_{ij}]$ is a (3×3) 'material property' matrix and $[b_{ij}]$ is a (3×6) 'displacement–gradient' matrix. We are actually dealing here with stiffness per unit thickness of material assumed homogeneous and isotropic.

(iv) A boundary element may be subject to a 'stress' boundary condition along one of its sides. Assuming the side joining nodes i and j is subjected to the surface traction $f = (f_1, f_2)$, evaluate the element load vector

$$y_i^{(k)} = \int_{i \to j} f_1 \xi_1^i + f_2 \xi_2^i \, dS$$

(see Example 4.7.4) and deduce that the total traction on the side is effectively replaced by equal concentrated forces at each of the end nodes.

4.7 Application examples

Example 4.7.1 (Newton's method) The reader is no doubt familiar with Newton's (or the Newton–Raphson) method for finding a zero of the real function $x(t)$. Starting with an approximation $t^{(0)}$ to a root of $x(t) = 0$ we compute the sequence

$$t^{(m+1)} = t^{(m)} - \frac{x(t^{(m)})}{x'(t^{(m)})}.$$

If the sequence $t^{(m)}$ converges to a point $\tilde{t}$ and if x' exists and is continuous at $\tilde{t}$ and $x'(\tilde{t}) \neq 0$ then, passing to the limit in the above expression, we obtain

$$\tilde{t} = \tilde{t} - \frac{x(\tilde{t})}{x'(\tilde{t})},$$

which implies that $x(\tilde{t}) = 0$. This also shows that $\tilde{t}$ is a fixed point of the function

$$z(t) = t - \frac{x(t)}{x'(t)}$$

(Example 3.10.6) and that $z'(\tilde{t}) = 0$. It follows from a simple application of Taylor's formula that for $t^{(0)}$ sufficiently near $\tilde{t}, z$ is a contraction mapping and therefore that the sequence of iterates converges provided $x'(t)$ is bounded away from zero in a neighbourhood of $\tilde{t}$. Furthermore, if x'' is continuous and x''' exists and is bounded in a neighbourhood of $\tilde{t}$ then we have

$$z(t^{(m)}) - z(\tilde{t}) = \tfrac{1}{2} z''(\tau)(t^{(m)} - \tilde{t})^2, \qquad \tau \in (\tilde{t}, t^{(m)})$$

or

$$\left| t^{(m+1)} - \tilde{t} \right| = \tfrac{1}{2} \left| z''(\tau) \right| \left| t^{(m)} - \tilde{t} \right|^2 \leqslant \text{const.} \left| t^{(m)} - \tilde{t} \right|^2$$

so that the 'error' at the $(m+1)$th iteration behaves like the square of the 'error' at the mth iteration. Newton's method thus has *quadratic convergence* (or is a *second-order method*) in contrast to the basic iteration method which is a *first-order method* (p. 269).

Geometrically we may view Newton's method as the replacement of the function locally by its tangent as depicted schematically in Fig. 4.11.

This suggests that the natural extension of Newton's method to the iterative solution of the non-linear operator equation $Tx = \theta$ lies in the local replacement of T by its linear part via the Fréchet derivative. Thus if $\mathscr{V}, \mathscr{U}$ are normed vector spaces and $T: \mathscr{V} \to \mathscr{U}$ is Fréchet differentiable in some ball $B(\tilde{x}, r) \in \mathscr{V}$ then Newton's method for finding the solution point $\tilde{x}$ of $Tx = \theta$ is

$$x^{(m+1)} = x^{(m)} - (T'(x^{(m)}))^{-1} T(x^{(m)}).$$

Again we may consider Newton's method as yielding a fixed point of the operator

$$S(x) = x - (T'(x))^{-1} T(x).$$

Notice that at every iteration it is necessary to solve the *linear* operator

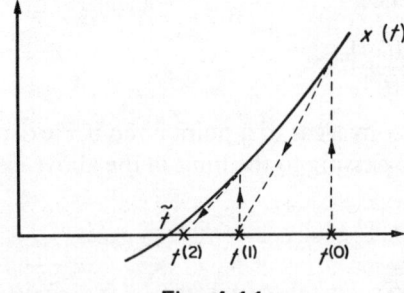

Fig. 4.11

equation

$$T'(x^{(m)})(x^{(m+1)} - x^{(m)}) = - T(x^{(m)}),$$

hence Newton's method is considerably more complex computationally than basic iteration. The compensation for increased complexity lies, of course, in the quadratic rate of convergence

$$\| x^{(m+1)} - \tilde{x} \|_{\mathscr{V}} \leqslant \text{const.} \, \| x^{(m)} - \tilde{x} \|_{\mathscr{V}}^2.$$

The conditions for convergence of Newton's method require that, in a neighbourhood of $\tilde{x}$, $\| T''(x) \| \, \| T'(x)^{-1} \|$ is sufficiently small (*see*, for example, [18, 32]), but in practice these conditions are difficult to check and it is usually simpler to carry out the computations and verify convergence *a posteriori*. This does, however, require the analyst to have some knowledge (possibly arising from a physical model) of the location and nature of the solutions which are of interest to him.

Generally speaking, Newton's method will converge (and converge quadratically) provided that the initial approximation is sufficiently near the solution. If *a priori* information on the location of the solution is not known with any accuracy then a slower iterative method can be used initially with a subsequent change to Newton's method.

The need to evaluate the derivative at each step in Newton's method leads to a computational difficulty. Two principal methods for overcoming this are [18, 32]:

1. Replace the derivative by a difference formula (the method of false position).
2. Keep the value of the derivative constant over successive iterates (the chord method).

Let us now look at two specific applications.

(a) (Cf., Example 4.2.2 (a).) If $f: \mathscr{R}_n \to \mathscr{R}_n$ is a vector-valued function $(f_1(x), f_2(x), \ldots, f_n(x))$ of $x = (\xi_1, \xi_2, \ldots, \xi_n)$ then Newton's method for a root of $f(x) = \theta$ leads to the sequence of linear equations

$$[D_j f_i(x^{(m)})] \{\delta_j\}^{(m)} = - \{f_i(x^{(m)})\},$$

where

$$\{\delta_i\}^{(m)} = \{\xi_i\}^{(m+1)} - \{\xi_i\}^{(m)}.$$

An important practical application of the Newton method for this type of non-linear operator equation arises in the solution of finite element equations which arise from non-linear operator equations. As an illustration, consider the finite element method applied to the large deflection analysis of an elastic structure. In this type of situation the strains throughout the structure are small enough to allow the use of a linear stress–strain relation, but the resulting deflections are not small enough to allow us to ignore the gross changes in geometry which take place due to applied load. The typical situation is that of a thin rod subjected to an initial axial load (the *elastica*): beyond a certain load the straight form of equilibrium becomes unstable and the stable form becomes one involving a large lateral displacement. This second equilibrium state can only be calculated if we allow for large translation and rotation of elements in the analysis. The same type of situation arises in the deflection analysis of thin shell structures [65, 121].

Let us write the finite element equations as (p. 395)

$$\{k_i(x)\} = \{y_i\}, \qquad x = (\xi_1, \xi_2, \ldots, \xi_N), \quad i = 1, \ldots, N.$$

The basic method for dealing with the solution of this equation is the *incremental load method*. We assume that the loading can be specified in terms of a single magnitude parameter p: we can then consider that $x = x(p)$ describes a continuous mapping of an interval of the real line to $\mathcal{R}_n$. This mapping describes the (nodal vector) displacement of the structure as a fixed loading system is increased in magnitude.

Beginning with, say, $p = 0$, $x = 0$, we apply an increment of load $\Delta p^{(0)}$ and replace the non-linear operator by its linear part at $x = 0$, viz.,

$$[D_j k_i(0)]\{\delta_i\}^{(0)} = \Delta p^{(0)}\{r_i\},$$

where $\{r_i\}$ is a fixed vector.

This set of linear equations is solved to give $\{\xi_i\}^{(1)} = \{\delta_i\}^{(0)}$. We now replace the operator by the linear part at $\{\xi_i\}^{(1)}$ and so on, giving the recursion

$$[D_j k_i(x^{(r)})]\{\delta_i\}^{(r)} = \Delta p^{(r)}\{r_i\},$$

$$\{\xi_i\}^{(r+1)} = \{\delta_i\}^{(r)} + \{\xi_i\}^{(r)}.$$

Now this is really a predictor type of method using a tangent approximation appropriate to the beginning of an interval: hence after a sequence of steps, without correction, an error will build up (Fig. 4.12).

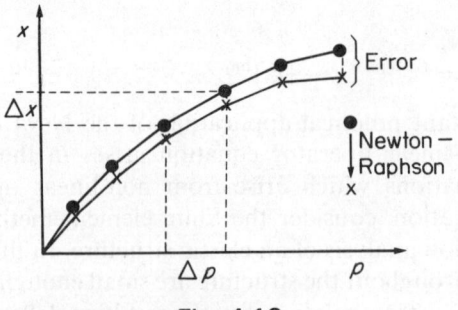

Fig. 4.12

The error can be reduced at each step by using Newton's method. In this case Newton's recursion is, at the rth step,

$$[D_j k_i(x^{(r,m)})]\{\delta_i\}^{(r,m)} = p^{(r)}\{r_i\} - \{k_i(x^{(r,m)})\},$$

$$\{\delta_i\}^{(r,m)} = \{\xi_i\}^{(r,m+1)} - \{\xi_i\}^{(r,m)},$$

$$p^{(r)} = \sum_{k=0}^{r-1} \Delta p^{(k)}.$$

Notice that the left-hand side matrix for the incremental load equations and for Newton's equations are the same.

Along a loading curve $x(p)$ a point may be reached at which the current equilibrium state becomes unstable—a point of bifurcation. Such a point is characterized by the vanishing of

$$\det[D_j k_i(x)].$$

Near such points the deflection clearly varies very rapidly with load increment and it is generally necessary either to

1. jump to a higher load and seek a new equilibrium path, or
2. increment the deflection rather than the load, so producing a family of curves $p(x)$ [122].

(b) Newton's method can be used to find a solution of a non-linear two-point boundary-value problem in the following way. For simplicity we shall outline the method for two dependent variables.

Suppose we wish to solve the pair of non-linear equations

$$\{\xi_i'\} = \{f_i(\xi_1, \xi_2, t)\}, \quad i = 1, 2,$$

on $[0, 1]$ subject to $\xi_1(0) = \alpha$, $\xi_2(1) = \beta$.

The central difficulty here lies in the fact that the boundary values are specified at both ends of the interval; if we know $\xi_1(0)$ and $\xi_2(0)$, say, then numerical integration of the equations would be straightforward (using a Runge–Kutta method, for example).

A simple way to proceed would be to guess values of $\xi_2(0)$, compute the trajectories, hence $\xi_2(1)$, and hope that we can obtain a match (a so-called 'shooting method'). In fact, we can use Newton's method to give rapid (quadratic) convergence beginning with a guessed value for $\xi_2(0)$.

We embed this particular problem in a class of problems by recognizing that the trajectory is an implicit function of the initial values, $\xi_1(0) = \alpha$, $\xi_2(0) = \gamma$, viz.,

$$\xi_i = \xi_i(\alpha, \gamma, t).$$

We are essentially interested in knowing how $\xi_i(t)$ changes with changes in $\xi_2(0) = \gamma$. To this end we introduce the functions $\eta_i, i = 1, 2,$

defined by

$$\eta_i = \frac{\partial \xi_i}{\partial \gamma}.$$

Differentiating the differential equations with respect to γ we obtain

$$\{\eta'_i\} = [D_j f_i(\xi_1, \xi_2, t)]\{\eta_j\}, \qquad i, j = 1, 2,$$

a set of linear differential equations subject to the initial values $\eta_1(0) = 0$, $\eta_2(0) = 1$.

Now we want to find that value of γ which will satisfy

$$\xi_2(\alpha, \gamma, 1) - \beta = 0.$$

Applying Newton's method to this equation yields the recursion

$$\gamma^{(m+1)} = \gamma^{(m)} - \frac{\xi_2(\alpha, \gamma^{(m)}, 1) - \beta}{D_\gamma \xi_2(\alpha, \gamma^{(m)}, 1)}$$

$$= \gamma^{(m)} - \frac{\xi_2(\alpha, \gamma^{(m)}, 1) - \beta}{\eta_2(1)}.$$

Thus the procedure is as follows: with the initial values $(\alpha, \gamma^{(0)})$ we integrate both the original differential equations and the differential equations for η_1, η_2. From these we find $\xi_2(\alpha, \gamma^{(0)}, 1)$ and $\eta_2(1)$ and hence a new estimate of the initial value $\gamma^{(1)}$—the cycle is repeated.

Example 4.7.2 (Linear stability) Consider the set of ordinary differential equations

$$\{\dot{\xi}_i\} = \{f_i(\xi_i; t)\}, \qquad t \in [0, \infty),$$

which might be the equations describing the motion of a dynamic system.

Suppose $\{\tilde{\xi}_i(t)\}$ is a solution (trajectory) of these equations: the question arises as to whether this solution is stable to small disturbances. This question can be decided by considering the motion of the system in the neighbourhood of the trajectory $\{\tilde{\xi}_i(t)\}$, that is, by studying the local linear part of the system equations. Let the motion in a neighbouring trajectory be

$$\{\xi_i(t)\} = \{\tilde{\xi}_i(t)\} + \{\sigma_i(t)\}$$

then, using a Taylor expansion (Theorem 4.2.2) of $\{f_i\}$ about $\{\tilde{\xi}_i(t)\}$ and neglecting terms of second order, we obtain the linear equations[†]

$$\{\dot{\sigma}_i(t)\} = [D_j f_i(\tilde{\xi}_i; t)]\{\sigma_j(t)\}.$$

If all solutions of these equations are such that $\sigma_i(t) \to 0$ as $t \to \infty$ when $\sigma_i(0)$ is sufficiently small, then we say that the trajectory $\{\tilde{\xi}_i(t)\}$ is *asymptotically stable in the small in the sense of Lyapunov* [42].

[†] These equations are sometimes called the *equations of first variation*.

A particularly important case arises when the equations of motion are autonomous so that f_i is not explicitly a function of time. The equations will then have stationary solutions or equilibrium points $\{\tilde{\xi}_i\}$ whose stability will be decided by the behaviour of the solutions of the constant coefficient linear equations

$$\{\dot{\sigma}_i(t)\} = [D_j f_i(\tilde{\xi}_i)]\{\sigma_j(t)\}.$$

These equations have exponential solutions whose indices are the eigenvalues of the Jacobian matrix $[D_j f_i(\tilde{\xi}_i)]$, so that the question of asymptotic stability is settled if we can show that none of the eigenvalues of this matrix have real parts which lie in the (closed) right half of the complex plane (for the basic theorem *see* [42]). Roughly speaking, we can use this technique to examine stability in the small provided the linear terms in the Taylor expansion of $\{f_i(\tilde{\xi}_i)\}$ are dominant; this is precisely the condition required for $\{f_i\}$ to have a strong derivative at $\{\tilde{\xi}_i\}$ (*see* Definition 4.2.2).

Example 4.7.3 (Steepest descent) Consider the problem of minimizing a functional $f(x)$ defined on a Hilbert space $\mathcal{H}$; there are no constraints on x. If we can construct a minimizing sequence $\{x^{(n)}\}$ for f (p. 334) then we know that (provided f is bounded below)

$$f(x^{(n)}) \to \inf_{x \in \mathcal{H}} f(x).$$

It is not necessary that the sequence $\{x^{(n)}\}$ converges in $\mathcal{H}$. In practice, of course, one continues the sequence only to a point at which successive values of $f(x^{(n)})$ differ by an acceptably small amount.

One way to generate a minimizing sequence is the following. Suppose f has a strong derivative $f'(x^{(n)})$ then (Definition 4.2.2)

$$f(x^{(n)} + s^{(n)}) - f(x^{(n)}) = f'(x^{(n)})s^{(n)} + \varepsilon(x^{(n)}; s^{(n)})$$
$$= \langle f'(x^{(n)}), s^{(n)} \rangle + \varepsilon(x^{(n)}; s^{(n)}),$$

where, in this context, $f'(x^{(n)}) \in \mathcal{L}(\mathcal{H}, \mathcal{R}) = \mathcal{H}$ is the gradient (vector) of f at $x^{(n)}$.

Our object is to make the left-hand side negative and this is guaranteed for sufficiently small $\|s^{(n)}\|$ if we choose

$$s^{(n)} = -\sigma_n f'(x^{(n)})$$

for then

$$f(x^{(n)} - \sigma_n f'(x^{(n)})) - f(x^{(n)}) = -\sigma_n \|f'(x^{(n)})\|^2 + \varepsilon(x^{(n)}; s^{(n)}).$$

This suggests that we use

$$x^{(n+1)} = x^{(n)} - \sigma_n f'(x^{(n)})$$

to generate a minimizing sequence for f.

In a Hilbert space $f'(x^{(n)})$ is the gradient vector of the functional f and it is clear that at each step we move in the negative direction of the gradient. How far should we move in this direction at each step? It would clearly be sensible to choose σ_n so as to achieve as large a reduction in f as possible, that is, we choose σ_n to minimize

$$g^{(n)}(x^{(n)}; \sigma) = f(x^{(n)} - \sigma f'(x^{(n)})).$$

This means that σ_n is a root of the equation

$$\frac{d}{d\sigma} g^{(n)}(x^{(n)}; \sigma) = 0,$$

although in practice it is not usually possible to find σ_n in this way. What is usually done is that, at each step, $g^{(n)}(x^{(n)}; \sigma)$ is computed for several values of σ until $g^{(n)}$ increases and σ_n is then found by interpolation. The above method for generating a minimizing sequence for f is called the *method of steepest descent*.

An explicit expression for σ_n can be found if f is a quadratic form, say, $f(x) = \frac{1}{2}\langle Tx, x \rangle - \langle b, x \rangle$, where T is positive definite: in this case $f'(x^{(n)}) = Tx^{(n)} - b$ and

$$\sigma_n = \frac{\|Tx^{(n)} - b\|^2}{\|Tx^{(n)} - b\|_T^2}.$$

As we have seen (Example 4.3.6(c)) the minimum point of f is the solution of $Tx = b$.

Convergence of the method of steepest descent requires [18] that $f'(x)$ is continuous on the convex hull $S = \{x : f(x) < f(x^{(0)})\}$ and that S should be compact. In practice the method rarely fails to converge to at least a local minimum. The method of steepest descent and other gradient methods are slowly convergent, not exceeding a linear rate in general.

Many functionals $f : \mathscr{E}_n \to \mathscr{R}$ behave like quadratic functionals in a sufficiently small neighbourhood of a minimum point, hence there is a particular interest in the problem of minimizing

$$f(x) = \tfrac{1}{2}\langle Tx, x \rangle - \langle b, x \rangle, \qquad x \in \mathscr{E}_n, \quad T \text{ positive definite},$$

and there are methods which will converge to the minimum in $\leq n$ iterations. One of these is the *conjugate gradient method* which is essentially based on the construction of an orthonormal basis for $\mathscr{E}_n$ with respect to the inner product $\langle x, z \rangle_T$.

In this context the basis vectors are said to define *conjugate directions* for f and the conjugate directions are generated iteratively by adding one dimension to the problem at each step (for further details *see* [18, 36]).

We should add that the foregoing methods are available for use in those constrained problems which may be converted to unconstrained problems through the use of Lagrange multipliers.

Example 4.7.4 (Variational principles in structural analysis) Let us apply the method of orthogonal projections (p. 344) to the equations of linear elasticity: we have already gone some way along this path in Example 3.10.1 and the reader should refer to that example for the notation used here. However, since we shall frame the analysis in a Hilbert space context there is no need to retain the subscript and superscript notation of that example.

Specifically we wish to solve

$$\sum_{j=1}^{3} \frac{\partial \tau_{ij}}{\partial t_j} = 0 \quad \text{in} \quad \Gamma,^{\dagger}$$

$$\sum_{j=1}^{3} \tau_{ij} v_j = f_i \quad \text{on} \quad \partial \Gamma_2,$$

with

$$e_i = 0 \quad \text{on} \quad \partial \Gamma_1 = \partial \Gamma - \partial \Gamma_2,$$

subject to the material constitutive law

$$\varepsilon_{ij} = \sum_{k,l=1}^{3} \gamma_{ijkl} \tau_{kl}; \qquad \tau_{ij} = \sum_{k,l=1}^{3} c_{ijkl} \varepsilon_{kl}.$$

The spaces $\mathscr{H}_1$, $\mathscr{H}_2$ used in describing the method of orthogonal projections are, in the present context:

$\mathscr{H}_1$: Hilbert space of vector functions with components in $\mathscr{C}^{(1)}(\Gamma)$ and with inner product

$$\langle u, v \rangle_{\mathscr{H}_1} = \sum_{i=1}^{3} \int_{\Gamma} u_i v_i \, d\Gamma;$$

$\mathscr{H}_2$: Hilbert space of symmetric stress tensors with components in $\mathscr{C}^{(1)}(\Gamma)$ and with inner product

$$\langle \bar{\tau}, \bar{\bar{\tau}} \rangle_{\mathscr{H}_2} = \sum_{i,j=1}^{3} \int_{\Gamma} \tau_{ij} \bar{\bar{\varepsilon}}_{ij} \, d\Gamma,$$

where $\bar{\bar{\varepsilon}}$ is the strain tensor associated with the stress τ through the material constitutive law.[‡]

The operator $S: \mathscr{H}_1 \to \mathscr{H}_2$ is the displacement–stress relation

$$Se = \sum_{k,l=1}^{3} \frac{1}{2} c_{ijkl} \left(\frac{\partial e_k}{\partial t_l} + \frac{\partial e_l}{\partial t_k} \right), \qquad e_i = 0 \quad \text{on} \quad \partial \Gamma_1$$

[†] There is no difficulty in including a non-zero body force in the analysis.
[‡] It is a consequence of thermodynamic principles that $\langle \bar{\tau}, \bar{\bar{\tau}} \rangle_{\mathscr{H}_2}$ as defined is an inner product, i.e., is a positive definite, bilinear, symmetric form. It is twice the strain energy [65].

while $S^*: \mathcal{H}_2 \to \mathcal{H}_1$ is the 'equilibrium' operator

$$S^*\tau \equiv \begin{cases} \sum_{j=1}^{3} \dfrac{\partial \tau_{ij}}{\partial t_j} & \text{in } \Gamma, \\ \sum_{j=1}^{3} \tau_{ij} v_j & \text{on } \partial \Gamma_2. \end{cases}$$

The null space of S^*, $\mathcal{N}(S^*)$, consists of those tensors satisfying

$$\sum_{j=1}^{3} \frac{\partial \tau_{ij}}{\partial t_j} = 0 \quad \text{in } \Gamma$$

and

$$\sum_{j=1}^{3} \tau_{ij} v_j = 0 \quad \text{on } \partial \Gamma_2.$$

Let $\tau^0 \in \mathcal{H}_2$ be such that

$$\sum_{j=1}^{3} \frac{\partial \tau_{ij}^0}{\partial t_j} = 0 \quad \text{in } \Gamma,$$

$$\sum_{j=1}^{3} \tau_{ij}^0 v_j = f_i \quad \text{on } \partial \Gamma_2.$$

Then the primary projection (onto $\mathcal{N}^\perp(S^*)$) gives

$$\langle \hat{\tau} - \tau^0, \bar{\tau} \rangle_{\mathcal{H}_2} = 0 \qquad \forall \bar{\tau} \in \mathcal{R}(S)$$

or, using the virtual work identity (p. 258),

$$\sum_{i,j=1}^{3} \int_\Gamma \hat{\tau}_{ij} \bar{\varepsilon}_{ij} \, d\Gamma = \sum_{i=1}^{3} \int_{\partial \Gamma_2} f_{i,s} \bar{e}_i \, dS,$$

where

$$\bar{\varepsilon}_{ij} = \frac{1}{2}\left(\frac{\partial \bar{e}_i}{\partial t_j} + \frac{\partial \bar{e}_j}{\partial t_i} \right), \qquad \bar{e}_i = 0 \quad \text{on } \partial \Gamma_1.$$

This projection is generally called the *principle of virtual work*. In using it to find an approximate solution of the given problem via the Rayleigh–Ritz method the steps are as follows:

1. Choose a finite set of displacement functions $\{e^k\}$ which vanish on $\partial \Gamma_1$, $k = 1, \ldots, n$.
2. Compute the tensors ε^k and τ^k.
3. Form the Ritz equations

$$\sum_{l=1}^{n} \xi_l \left(\sum_{i,j=1}^{3} \int_\Gamma \tau_{ij}^l \varepsilon_{ij}^k \, d\Gamma \right) = \sum_{i=1}^{3} \int_{\partial \Gamma_2} f_{i,s} e_i^k \, dS, \qquad k = 1, \ldots, n,$$

where the approximate solution for the displacement field is given by $\sum_{k=1}^{n} \xi_k e^k$.

Notice that in this approximation, since strain is found from displacement, compatibility is exactly satisfied whereas equilibrium is satisfied only in a projectional sense.

The orthogonal projection (onto $\mathcal{N}(S^*)$) gives

$$\langle \hat{\bar{\tau}} - \tau^0, \bar{\tau} \rangle_{\mathcal{H}_2} = 0 \quad \forall \bar{\tau} \in \mathcal{N}(S^*)$$

or

$$\sum_{i,j=1}^{3} \int_{\Gamma} \hat{\bar{\tau}}_{ij} \bar{\varepsilon}_{ij} \, d\Gamma = \sum_{i,j=1}^{3} \int_{\Gamma} \tau^0_{ij} \bar{\varepsilon}_{ij} \, d\Gamma,$$

where

$$\sum_{j=1}^{3} \frac{\partial \bar{\tau}_{ij}}{\partial t_j} = 0 \quad \text{in} \quad \Gamma, \quad \sum_{j=1}^{3} \bar{\tau}_{ij} v_j = 0 \quad \text{on} \quad \partial \Gamma_2.$$

This projection is called the *principle of complementary virtual work* (or the *principle of least work*). In using it to find an approximate solution of the given problem we have to:

1. find a particular tensor, τ^0;
2. find a set of stress tensors $\bar{\tau}^k$ which are elements of $\mathcal{N}(S^*)$.
3. Form the Ritz equations

$$\sum_{l=1}^{n} \eta_l \left(\sum_{i,j=1}^{3} \int_{\Gamma} \bar{\tau}^l_{ij} \bar{\varepsilon}^k_{ij} \, d\Gamma \right) = \sum_{i,j=1}^{3} \int_{\Gamma} \tau^0_{ij} \bar{\varepsilon}^k_{ij} \, d\Gamma, \quad k=1,\ldots,n,$$

where the approximate solution for the stress field is given by

$$\tau^0 - \sum_{k=1}^{n} \eta_k \bar{\tau}^k.$$

In this approximation we satisfy equilibrium exactly whereas the kinematic conditions are satisfied only in a projectional sense. Indeed, there is no guarantee that the approximate strain field can be integrated to give a connected displacement field: we can, however, obtain a generalized displacement field by an application of the so-called *unit load equation* [84]. The calculation of the energy norm (in this case twice the strain energy) needs only a knowledge of the element $\hat{\bar{\tau}} \in \mathcal{H}_2$, as we showed on p. 345.

The principle of virtual work (*the displacement method*) has the great advantage over the principle of complementary virtual work (*the force method*) that there is no requirement to use trial tensors which have to satisfy a partial differential equation (i.e. be self-equilibrating). Instead we only need to choose displacement vectors which satisfy the given kinematic condition.

In the early days of automated methods for structural analysis many structures were physically subdivided into simple elements within which self-equilibrating stress resultant systems were easy to find, and in this era the force method was the more popular; it has the added advantage of directly approximating stress, which is the designer's main target. However, the extension of automatic methods to more general continuum structures has led to the dominance of the displacement method.

The displacement method also has the great advantage (*vide* the finite element method) that a large part of the actual setting up of the problem (forming the Ritz equations, for example) can be automated; this cannot be done in the same way with the force method, which requires active participation by the analyst.

Let us illustrate the two projections by applying them to the bending of a prismatic beam within the confines of the so-called *engineer's theory of bending*. The operator T is, in this case,

$$Tx = \frac{d^2}{dt^2}\left(EI\frac{d^2x}{dt^2}\right) \quad \text{in} \quad [0, 1],$$

where $EI > 0$ is the bending stiffness. T is clearly symmetric and is positive-bounded-below provided sufficient kinematic conditions are imposed at the end points 0, 1. We express T as the composition S^*S where

$$S^* \equiv -\frac{d^2}{dt^2} \quad \text{and} \quad S \equiv -EI\frac{d^2}{dt^2}.$$

Now the quantity $m = Sx = -EI(d^2x/dt^2)$ has a perfectly definite physical interpretation: it is the bending moment at any section of the beam due to the transverse displacement x (see Fig. 4.13).

Thus the stress resultant $m(t)$ replaces stress τ in this special case. The

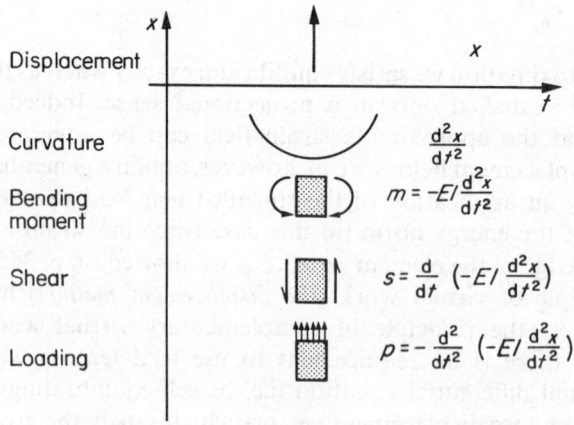

Fig. 4.13

APPLICATION EXAMPLES 415

equation of equilibrium is represented by the operator S^*, viz.,

$$-\frac{d^2 m}{dt^2} = p,$$

where p is the transverse loading per unit length.

The inner product in $\mathcal{H}_1$ is simply

$$\langle x, y \rangle_{\mathcal{H}_1} = \int_0^1 xy \, dt$$

while in $\mathcal{H}_2$ we take

$$\langle \bar{m}, \bar{\bar{m}} \rangle_{\mathcal{H}_2} = \int_0^1 \frac{\bar{m}\bar{\bar{m}}}{EI} \, dt.$$

This inner product is of the form

$$\int_0^1 (\text{Bending moment}) \times (\text{Curvature}) \, dt$$

and is the form we would expect for the stress resultant m combined with a strain resultant which is simply the curvature. In this form we are taking account only of the 'bending strain energy' in the beam, which is the dominant contribution when depth of section $\ll$ span. We could, if we wished, add a term to take approximate account of 'shear strain energy'. The virtual work identity is

$$\int_0^1 \frac{\bar{m}\bar{\bar{m}}}{EI} dt = \int_0^1 \bar{p}\bar{\bar{x}} \, dt = \int_0^1 EI \frac{d^2\bar{x}}{dt^2}\frac{d^2\bar{\bar{x}}}{dt^2} dt$$

provided that, at $t = 0, 1$,

$$\begin{pmatrix} m = 0 \\ \text{or} \\ x' = 0 \end{pmatrix} \quad \text{and} \quad \begin{pmatrix} m' = s = 0 \\ \text{or} \\ x = 0 \end{pmatrix}$$

The principle of virtual work takes the form

$$\int_0^1 EI \frac{d^2\hat{x}}{dt^2}\frac{d^2 x}{dt^2} dt = \int_0^1 px \, dt,$$

where x satisfies the essential (kinematic) boundary conditions while the principle of complementary virtual work takes the form

$$\int_0^1 \frac{\hat{\bar{m}}\bar{m}}{EI} dt = \int_0^1 \frac{m^0 \bar{m}}{EI} dt,$$

where m^0 satisfies the natural (statical) boundary conditions and the equation

$$-\frac{d^2 m^0}{dt^2} = p,$$

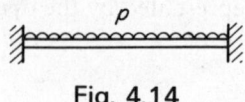

Fig. 4.14

while the functions $\bar{m}$ satisfy homogeneous statical boundary conditions and the equation

$$-\frac{d^2\bar{m}}{dt^2} = 0.$$

Let us apply these equations to the solution of a simple statically indeterminate problem, namely to find the bending moment distribution in an *encastré* beam of constant cross-section carrying a uniformly distributed load p per unit length (Fig. 4.14); $x(0) = x'(0) = x(1) = x'(1) = 0$.

There are no statical boundary conditions so we can take any solution of

$$-\frac{d^2 m^0}{dt^2} = p$$

we like. Let us take $m^0(0) = m^0(1) = 0$, then this gives

$$m^0 = \frac{p}{2} t(1-t),$$

the bending moment distribution for a beam simply supported at its ends. As for the functions $\bar{m}$, since they are required to satisfy

$$-\frac{d^2\bar{m}}{dt^2} = 0$$

they can only be linear functions of t. Indeed, the solution space of this equation is two-dimensional and we can choose any basis we like. We shall choose the orthogonal vectors $\{1, 1 - 2t\}$. If we apply the Rayleigh–Ritz method, then, because the space $\mathcal{N}(S^*)$ is finite-dimensional, we shall obtain an exact solution of the problem. The Ritz equations are

$$\eta_1 \int_0^1 dt = \int_0^1 \frac{p}{2} t(1-t) dt,$$

$$\eta_2 \int_0^1 (1-2t)^2 dt = \int_0^1 \frac{p}{2} t(1-t)(1-2t) dt,$$

with solution $\eta_1 = p/12$, $\eta_2 = 0$. The basis vectors for $\mathcal{N}(S^*)$ have a clear physical interpretation, being the bending moment distributions for the self-equilibriating load systems shown in Fig. 4.15.

It should be clear that there is no difficulty in extending the method to a continuous beam over many supports leading to a 'finite element' form of the force method. Having found the total bending moment distribution, $m^0 - \bar{m}$

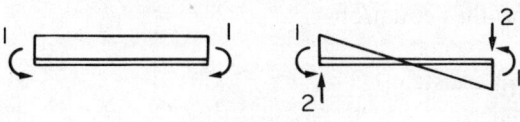

Fig. 4.15

(p. 413) we can find the displacement if required by integration of

$$-EI\frac{d^2x}{dt^2} = p\left(\frac{1}{2}t(1-t) - \frac{1}{12}\right)$$

with $x(0) = x'(0) = 0$. (Symmetry ensures satisfaction of $x(1) = x'(1) = 0$.)

The virtual work solution of the problem is not quite so neat. If we note that the basic operator equation

$$EI\frac{d^4x}{dt^4} = \text{const.}$$

needs to be satisfied then we can see that a trial space of polynomials up to and including order four will yield an exact solution. However, if the loading is of a more complicated form and/or the bending stiffness is not constant, then the solution will have to be approximate. By contrast the basis for $\mathcal{N}(S^*)$ is *always* two-dimensional and, provided we can evaluate the Ritz integrals, we can still obtain an 'exact' solution for the bending moment distribution.

This simple example should show the advantages of the force method provided self-equilibrating stress systems can be found: unfortunately, this problem is too difficult in the general case and applications are restricted to structures built up from simple elements and some plate and thin shell problems, e.g. [123].

Example 4.7.5 Variational principles exist for certain problems in fluid mechanics but have not yet been developed for the most general case. Projection methods, on the other hand, can always be used if the describing equations are available, but of course they do not necessarily lead to extremum principles and even less so to complementary bounds.

One situation in which complementary variational principles have been shown to exist is that of steady, inviscid, irrotational subsonic flow.

For such a flow a velocity potential ϕ exists such that

$$\text{grad } \phi = v,$$

where v is the fluid velocity vector in the Eulerian sense. The equation of motion,

$$(v \cdot \text{grad})v = -\frac{1}{\rho}\text{grad } p,$$

where p is the pressure and ρ the density, may be integrated along a streamline to give Bernoulli's equation

$$\frac{1}{2}v^2 + \int \frac{dp}{\rho} = \text{const.}$$

when p is a function only of ρ (the fluid is barotropic). Such a flow is also homentropic, so that the internal energy u is a function only of the density, $u = u(\rho)$, with

$$du = -p\,d\left(\frac{1}{\rho}\right) = \frac{p}{\rho^2}d\rho.$$

Now

$$d\left(\frac{p}{\rho}\right) = \frac{dp}{\rho} - \frac{p\,d\rho}{\rho^2} = \frac{dp}{\rho} - du$$

so that we may rewrite Bernoulli's equation as,

$$\frac{1}{2}v^2 + u + \frac{p}{\rho} = 0$$

or

$$\frac{1}{2}v^2 + \frac{d}{d\rho}(\rho u) = 0.$$

Finally we have the conservation of mass equation

$$-\text{div}(\rho v) = 0.$$

Let us introduce the momentum vector $m = \rho v$ as a variable in place of v: then our governing equations are

$$\text{grad}\,\phi = m/\rho, \qquad -\text{div}\,m = 0,$$

with ρ and m connected by Bernoulli's equation,

$$\frac{1}{2}m^2 + \rho^2 \frac{d}{d\rho}(\rho u) = 0,$$

where the internal energy,

$$u = \int_{\rho_0}^{\rho} \frac{p\,d\rho}{\rho^2},$$

is a known function of ρ.

The first two equations are in a form appropriate for deriving complementary bounds since the operators grad and $-$ div are, as we know, adjoint. Hence, taking ϕ and m as our principal unknowns, we seek a functional $a(\phi, m)$ such that (p. 349)

$$S\phi = \text{grad}\,\phi = m/\rho = a'_m$$

and $S^*m = -\text{div}\,m = 0 = a'_\phi$: clearly the functional a should not involve ϕ.

Now, by virtue of Bernoulli's equation, ρ is a function of m^2, hence we have, by direct integration (p. 294),

$$a(m) = \int_0^1 \left\langle m, \frac{m\lambda}{\rho(m^2 \lambda^2)} \right\rangle_{\mathscr{H}_2} d\lambda,$$

where

$$\langle m, n \rangle_{\mathscr{H}_2} = \sum_{i=1}^3 \int_\Gamma m_i n_i \, d\Gamma.$$

Hence,

$$\begin{aligned} a(m) &= \int_\Gamma \int_0^1 \frac{m^2 \lambda \, d\lambda}{\rho(m^2 \lambda^2)} d\Gamma \\ &= \int_\Gamma \frac{1}{2} \int_0^{m^2} \frac{dr}{\rho(r)} d\Gamma \\ &= \int_\Gamma \left(\frac{1}{2} \frac{m^2}{\rho} + \int^\rho \frac{1}{2} \frac{m^2}{\rho^2} d\rho \right) d\Gamma \\ &= \int_\Gamma \frac{1}{2} \frac{m^2}{\rho} - \rho u \, d\Gamma \end{aligned}$$

upon using Bernoulli's equation.

The functional $l(\phi, m)$ is given by (p. 349)

$$\begin{aligned} l(\phi, m) &= \int_\Gamma -\phi \operatorname{div} m + \rho u - \frac{1}{2} \frac{m^2}{\rho} d\Gamma \\ &= \int_\Gamma m \cdot \operatorname{grad} \phi + \rho u - \frac{1}{2} \frac{m^2}{\rho} d\Gamma, \end{aligned}$$

leading to

$$\begin{aligned} l_1(\phi) &= \int_\Gamma \rho (\operatorname{grad} \phi)^2 + \rho u - \frac{1}{2} \rho v^2 \, d\Gamma \\ &= \int_\Gamma \frac{1}{2} \rho v^2 + \rho u \, d\Gamma = -\int_\Gamma p(\phi) \, d\Gamma. \end{aligned}$$

Here p is a function of ϕ via the barotropic relation and Bernoulli's equation.

The complementary functional is

$$l_2(m) = \int_\Gamma \rho u - \frac{1}{2} \frac{m^2}{\rho} d\Gamma = \int_\Gamma p + \frac{m^2}{\rho} d\Gamma.$$

Here p is a function of m via the barotropic relation and Bernoulli's equation:

any trial momentum vector must satisfy div $m = 0$ together with any homogeneous boundary conditions for v on $\partial \Gamma$.

Now consider inhomogeneous boundary values. If on $\partial \Gamma$ we specify a normal momentum flux condition

$$\sum_{i=1}^{3} v_i m_i = f_i,$$

for example, then we add a term

$$\int_{\partial \Gamma} \phi f \, dS$$

to $a(m, \phi)$

A sufficient condition for $a(\phi, m)$ to have the required saddle property of being convex in m and degenerately concave in ϕ is that the flow be subsonic; that is $v^2 < a^2$, where $a = (dp/d\rho)^{1/2}$ is the local sonic speed [88].

Appendix 4.A. Order of magnitude notation

The *order symbols* O and o are conventionally defined in the following way. Let f and g be real-valued functions defined on R and suppose $g(x) > 0$.

We write

$$f = O[g] \quad \text{as} \quad x \to x_0$$

if $|f| < Mg$ for some positive constant M as $x \to x_0$. Here g has the character of a comparison function (usually of a simple type) and the order statement asserts that, near x_0, $|f|/g$ does not exceed some finite value, say M. In most cases x_0 is 0 or ∞; for example

$$3x^2 + 2x = \begin{cases} O[x^2] & \text{as} \quad x \to \infty, \\ O[x] & \text{as} \quad x \to 0. \end{cases}$$

We write

$$f = o[g] \quad \text{as} \quad x \to x_0$$

if $|f| < \varepsilon g$ for any positive ε, however small, as $x \to x_0$. We are usually concerned with the situation wherein $f(x) \to 0$ as $x \to x_0$ and we wish to provide information on 'how fast' $f(x)$ tends to zero. The order statement says that $f(x)$ tends to zero at least as fast as $g(x)$ tends to zero, where again $g(x)$ has the character of a simple comparison function. For example

$$\sin x = x + o[x] \quad \text{as} \quad x \to 0$$

since

$$\frac{\sin x - x}{x} = -\frac{x^2}{3!} + \frac{x^4}{5!} - \cdots,$$

and the right-hand side can be made as small as we please by taking x sufficiently small.

5 Distributions

5.1 Introduction

A very familiar tool of applied mathematics is the so-called Dirac delta function $\delta(t)$. Its properties are usually described by

$$\delta(t) = 0, \quad t \neq 0,$$

$$\int_{-\infty}^{\infty} \delta(t)\mathrm{d}t = 1,$$

and if $g(t)$ is a function which is continuous at $t = 0$,

$$\int_{-\infty}^{\infty} \delta(t)g(t)\mathrm{d}t = g(0).$$

Unfortunately, no function exists which has these properties, since the integral of a function which vanishes everywhere except at the isolated point $t = 0$ is zero. But perhaps what we might mean by $\delta(t)$ is the following: imagine a sequence of functions each of which has its maximum value at $t = 0$ and, as we move along the sequence, the maximum value increases while the graph of the function gets narrower, so as to make its integral unity for each member of the sequence. One can see that such a sequence of functions $\{f^{(n)}\}$ might have the property

$$\lim_{n} \int_{-\infty}^{\infty} f^{(n)}(t)g(t) = g(0).$$

Indeed, sequences of this type have been known for a very long time: for example, Hermite in 1891 discovered the sequence

$$f^{(n)}(t) = \frac{1}{\pi} \frac{n}{n^2 t^2 + 1}$$

while Kirchoff in 1882 used the sequence

$$f^{(n)}(t) = \frac{n}{\sqrt{\pi}} e^{-n^2 t^2}.$$

For an obvious reason these are examples of what are called *delta sequences* (Exercise 5.2,1).

So what is the delta function? Is it, in some sense, a limit of such sequences, or can it be given another interpretation? The answer to both these questions is, 'Yes', and we shall show why in Section 5.2. It turns out that the delta function is only one representative, albeit the most commonly known, of a whole class of entities called singular distributions (or generalized functions). We shall see that we can interpret these entities as the ideal elements obtained by completing certain spaces of linear functionals. It is only within the framework of functional analysis that the theory of distributions can be properly understood.

The Fourier transform is an extremely powerful tool in the field of linear analysis, but in its classical form it suffers from the defect that the class of functions which is Fourier transformable is rather limited. This is because such functions must tend to zero at both $+\infty$ and $-\infty$ fast enough to ensure the existence of the integral. The Fourier transform of the simple step function

$$1_+(t) = \begin{cases} 1, & t \geq 0, \\ 0, & t < 0, \end{cases}$$

for example, does not exist in the classical sense. Many attempts have been made to circumvent this difficulty and applied mathematics texts abound with treatments of linear systems by transform methods in which convergence factors are introduced to make the Fourier integrals converge, only to be removed at some later point in the analysis. By extending the Fourier transform to include distributions, which we do in Section 5.3, all these intricate and doubtful arguments become unnecessary. The class of functions which is transformable is greatly enlarged and many apparently obscure results are made understandable.

One property of distributions which is startlingly different from the situation pertaining to ordinary functions is that they are infinitely differentiable. This means, for example, that a convergent Fourier series can be differentiated term by term as many times as desired and the resulting series will always have a meaning in the sense of distributions. A very significant feature is introduced into normed and inner product spaces by the availability of distributional differentiation. If we construct, say, an inner product space whose norm involves not only values of the member functions but also the values of those derivatives which are square integrable we find that the space is incomplete for the simple reason that the limit of a sequence of differentiable functions is not necessarily differentiable. However, if we allow distributional differentiation, completion gives us a class of important Hilbert spaces called Sobolev spaces. We give a short account of these spaces in Section 5.4 and discuss very briefly why they are of importance in the abstract theory of partial differential equations.

5.2 Distributions

We have seen that there are a number of function sequences $\{f^{(n)}\}$ which are delta sequences, that is to say,

$$\lim_{n} \int_I f^{(n)}(t)g(t)dt = g(0),$$

where I is an interval of the real line containing zero. However, although this limit exists we cannot write the result in the form

$$\int_I \lim_n f^{(n)}(t)g(t)dt = g(0)$$

since $\lim_n f^{(n)}(t)$ does not exist as an ordinary function. But we have met this type of situation before when we discussed completion of a metric space. We proceed by treating the sequence $\{f^{(n)}\}$ as a type of weak fundamental sequence and the 'delta function' is understood to be a weak limit of the sequence. In this sense the function space in which we generate the sequence $\{f^{(n)}\}$ is incomplete and we complete it by adjoining ideal elements. Two delta sequences, for example, are equivalent weak fundamental sequences (p. 124).

To carry through this reasoning we need to define what we mean by convergence of the sequence $\{f^{(n)}\}$. This we do by introducing convergence with respect to testing functions, so that we say that the sequence $\{f^{(n)}\}$ has a limit if

$$\lim_{n \to \infty} \int_I f^{(n)}(t)g(t)dt$$

exists where g belongs to a class of 'testing functions'. We can then define the distribution (or generalized function) f as an ideal element by the formal statement

$$\int_I f(t)g(t)dt = \lim_{n \to \infty} \int_I f^{(n)}(t)g(t)dt,$$

where of course the left-hand side has no meaning other than that assigned to it by the limit. The ideal f represents the class of equivalent fundamental sequences, and rules for its manipulation can be deduced from the definition. This is somewhat analogous to the deduction of the rules of arithmetic for the irrational numbers from the rules governing the elements of their fundamental sequences, namely the rationals.

This approach to a theory of distributions is a very natural one and was developed and presented by Temple [125]; a later book by Lighthill also adopts this line of approach [126]. However, there is another way to develop a theory of distributions. In this theory a distribution is considered to be a linear functional on a suitably chosen vector space. In this context the delta function is represented by a linear functional δ with the property

$$\delta(g(t)) = g(0).$$

This second theory is the one we shall pursue here. Both theories lead, of course, to largely the same set of results (although there are minor technical differences) and both viewpoints can be useful.[†] However, the interpretation of distributions as linear functionals leads, on the whole, to simpler and more direct proofs of the basic results. We recall that a linear functional on a vector space $\mathscr{V}$ is characterized solely by its evaluation on the elements of $\mathscr{V}$.

In our treatment of distributions we shall, for the most part, deal with functions defined on $\mathscr{R}$: an extension to functions defined on $\mathscr{R}_n$ does not present major difficulties (Section 5.4).

In order to define a distribution as a linear functional we need first to choose a vector space of *testing functions*. We shall initially choose this space to be $\mathscr{C}_0^{(\infty)}$, the space of infinitely differentiable functions on $\mathscr{R}$ with compact support (*see* footnote on p. 159). In order to be able to discuss the continuity of linear functionals we need to introduce a notion of convergence on this space; we shall demand uniform convergence of a sequence of functions and of all the derivatives. Let us denote this *space of testing functions* by $\mathscr{D}$: then $\phi \in \mathscr{D}$ if

(i) ϕ vanishes outside some finite interval;
(ii) ϕ has continuous derivatives of all orders.

If the sequence $\{\phi^{(n)}\}$ has support in some fixed finite interval then $\phi^{(n)} \underset{\mathscr{D}}{\to} \phi$ if, for every k, $D^k \phi^{(n)}(t) \to D^k \phi$ uniformly.[‡] Under these

---
[†] There is a third approach based on an 'algebraic' operator theory due to Mikusinski [127].
[‡] Here $D^k \phi(t)$ denotes the kth derivative of $\phi(t)$, not to be confused with $D_j \phi(t)$ which denotes the (first) partial derivative of $\phi(t_1, t_2, \ldots, t_n)$ with respect to t_j.
This notation is consistent with that used in Example 3.6.1(f) and later in Section 5.4.

conditions the limit of such sequences is in $\mathscr{D}$, that is to say, $\mathscr{D}$ is closed under convergence.

As an example of a testing function of the above type consider

$$\phi(t) = \begin{cases} 0, & |t| \geq 1, \\ \exp(1/(t^2 - 1)), & |t| < 1. \end{cases}$$

It is straightforward to show that $\phi(t)$ has continuous derivatives of all orders.

Why should we choose testing functions which are required to satisfy such stringent conditions? Suffice it to say at this stage that the more restrictions we put on the testing functions the larger the class of distributions we can define.

Definition 5.2.1 *A continuous linear functional on the space $\mathscr{D}$ is a distribution.*

The space of distributions on $\mathscr{D}$ is, of course, the dual space, $\mathscr{D}^*$; the value assigned to the testing function ϕ by the distribution $f \in \mathscr{D}^*$ is denoted in the usual notation by $f(\phi)$. It is important to notice that we have not introduced a norm into $\mathscr{D}$ or $\mathscr{D}^*$. There is no 'natural' norm in the space $\mathscr{D}$ and this space can only be given a topology by means of a limit process [143].

One way to generate distributions in $\mathscr{D}^*$ is to associate them with locally integrable functions, that is functions which are integrable (perhaps in the Lebesgue sense) over every finite interval. If f is such a function then the *distribution* $f \in \mathscr{D}^*$ corresponding to the function f is given by

$$f(\phi) = \int_{-\infty}^{\infty} f(t)\phi(t)dt.$$

Distributions generated in this way are called *regular distributions*. One can show (Exercise 5.2,2(a)) that two continuous functions which produce the same regular distribution are equal. However, if the function f has discontinuities then the distribution f corresponds to the equivalence class of functions which differ on a set of measure zero. For example, the set of functions

$$s_\alpha(t) = \begin{cases} 0, & t < 0, \\ \alpha, & t = 0, \\ 1, & t > 0, \end{cases}$$

all lead to the same distribution

$$1_+(\phi) = \int_0^\infty \phi(t)dt$$

which we shall call the *Heaviside* or *unit step distribution*. Clearly, wherever a function is continuous we may effectively identify the function with its regular distribution. In this spirit one sometimes writes the Heaviside distribution, for example as $1_+(t)$.

Elements of $\mathscr{D}^*$ which cannot be associated in this way with an integrable function are called *singular distributions*. For example, the delta distribution δ is defined as having the (sifting) property,

$$\delta(\phi) = \phi(0).$$

This functional is clearly linear and continuous on $\mathscr{D}$. Similar delta-like distributions $D^k\delta$ can be defined by

$$D^k\delta(\phi) = (-1)^k D^k\phi(0).$$

We shall see later that, in a sense, these distributions are the derivatives of the delta distribution.

The *support of a distribution* f is defined as the smallest closed interval (or set) outside of which $f(\phi) = 0$. For example, the support of the delta distribution is the isolated point $t = 0$. Since $f(\phi)$ vanishes outside the support of f it can be seen that the distribution f can be characterized by its evaluation on only those functions ϕ which do not vanish over its support. In fact, it is necessary that ϕ does not vanish over a neighbourhood of the support of f, that is, any open set which contains the support of f. For example, the delta distribution has $t = 0$ as its support and hence may be characterized by its evaluation on functions which do not vanish in an arbitrarily small open interval containing 0. This is an extremely important practical point for it allows us to assign a meaning to $\delta(g)$ when $g \notin \mathscr{D}$ since in applications we cannot always be fortunate enough to work solely with suitable testing functions. Provided g is continuous at 0 we can find a testing function which coincides with g in the neighbourhood of 0, hence we can write

$$\delta(g) = g(0).$$

More generally we can assign a meaning to $f(g)$ provided f has bounded support and g is sufficiently smooth over a neighbourhood of the support of f [128].

Another type of singular distribution is obtained from certain types of divergent integral. Consider the integral

$$f(\phi) = \int_0^\beta t^{-3/2} \phi(t) dt, \qquad \phi \in \mathcal{D},$$

which we may write as

$$f(\phi) = \int_0^\beta t^{-3/2}(\phi(t) - \phi(0)) dt + \phi(0) \lim_{\varepsilon \to 0} \int_\varepsilon^\beta t^{-3/2} dt.$$

Since $\phi(t)$ has a Taylor expansion about 0 the first integral is convergent while the second term leads to

$$-\lim_{\varepsilon \to 0} 2\phi(0) \left[\frac{1}{\sqrt{\beta}} - \frac{1}{\sqrt{\varepsilon}} \right].$$

We obtain *Hadamard's finite part of a divergent integral*,[†] denoted Fp, by simply discarding the term $1/\sqrt{\varepsilon}$. We may note in addition that since the testing functions ϕ vanish outside finite intervals we may let $\beta \to \infty$ so obtaining

$$\text{Fp} \int_0^\infty t^{-3/2} \phi(t) dt = \int_0^\infty t^{-3/2}(\phi(t) - \phi(0)) dt.$$

Now the right-hand side is a continuous linear functional on $\mathcal{D}$ (Exercise 5.2,3(a)) and hence defines a distribution in $\mathcal{D}^*$ which is denoted by $\text{Pf } 1_+(t) t^{-3/2}$, that is to say,

$$\text{Pf} 1_+(t) t^{-3/2}(\phi) = \int_0^\infty t^{-3/2}(\phi(t) - \phi(0)) dt.$$

This type of singular distribution is called a *pseudofunction* and is denoted by the prefix Pf. This pseudofunction coincides with the ordinary function $t^{-3/2}$ except at $t = 0$, where the function is not defined.

Other types of pseudofunction can be constructed in a similar fashion (Exercise 5.2,3(c)), but we shall gain a clearer idea of their natural occurrence when we later come to deal with the differentiation of distributions.

The notation used to represent the pseudofunction '$1_+(t) t^{-3/2}$'

[†] Hadamard, 1923. In many problems in applied mathematics (particularly in wave theory) the operations of differentiation and integration are illegally, but conveniently, interchanged. This can be corrected later in the analysis by evaluating the resulting divergent integrals using the concept of finite part (*see*, for example, [131] and Exercise 5.2,3(b)).

highlights a possible source of confusion in the use and manipulation of distributions. It is extremely convenient, as in this case, to use a formula like $t^{-3/2}$ to represent a distribution since over any interval excluding 0 it coincides with the ordinary function $t^{-3/2}$. In the same way we may write $\delta(t), 1_+(t)$ knowing that there are functions (namely 0 and 1) which coincide with these distributions in certain intervals. In this sense it is common to represent the distribution f by $f(t)$, but it should always be remembered that this does not imply that $f(t)$ has 'values' for every value of t.

We may introduce a notion of convergence in $\mathscr{D}^*$ in the following way. The sequence $\{f^{(n)}\} \in \mathscr{D}^*$ converges to $f \in \mathscr{D}^*$ if, for every $\phi \in \mathscr{D}$,

$$f^{(n)}(\phi) \to f(\phi).$$

This is clearly a type of weak-star convergence (Definition 3.5.4).

For example, the sequence of (regular) distributions $\{\sin nt\}$ converges to the zero distribution since

$$\sin nt(\phi) = \int_{-\infty}^{\infty} \sin nt\, \phi(t)\, dt = \frac{1}{n} \int_{-\infty}^{\infty} \cos nt\, \phi'(t)\, dt$$

or

$$|\sin nt(\phi)| \leq \frac{1}{n} \int_{-\infty}^{\infty} |\phi'(t)|\, dt,$$

which tends to zero for all ϕ as $n \to \infty$.

Operations on distributions

Apart from addition and scalar multiplication, there are three important operations which we may apply to the elements of $\mathscr{D}^*$. In each case the form of the operation is generalized from that appropriate to a regular distribution.

(i) The *translation* f_τ or $f(t-\tau)$ of the distribution f is given by

$$f_\tau(\phi(t)) = f(\phi(t+\tau)),$$

which is obviously consistent with the result for a regular distribution,

$$f_\tau(\phi) = \int_{-\infty}^{\infty} f(t-\tau)\phi(t)\, dt = \int_{-\infty}^{\infty} f(t)\phi(t+\tau)\, dt.$$

For example, the translated delta distribution δ_τ has the property

$$\delta_\tau(\phi) = \phi(\tau).$$

(ii) The effect of a scaling of the variable t can be seen by examining the result for a regular distribution f. Suppose α is a scalar, then

$$\int_{-\infty}^{\infty} f(\alpha t)\phi(t)dt = \begin{cases} \dfrac{1}{\alpha} \int_{-\infty}^{\infty} f(t)\phi\left(\dfrac{t}{\alpha}\right)dt, & \alpha > 0, \\ \dfrac{1}{-\alpha} \int_{-\infty}^{\infty} f(t)\phi\left(\dfrac{t}{\alpha}\right)dt, & \alpha < 0, \end{cases}$$

which suggests the definition

$$f(\alpha t)(\phi(t)) = \frac{1}{|\alpha|} f\left(\phi\left(\frac{t}{\alpha}\right)\right)$$

of the distribution $f(\alpha t)$.

(iii) We may multiply a distribution by an element of $\mathscr{D}$. Suppose $\psi \in \mathscr{D}$, then

$$f\psi(\phi) = f(\psi\phi), \qquad \psi\phi \in \mathscr{D}.$$

It is not, in general, possible to define the product of distributions: this is one of the basic deficiencies of distribution theory and has not as yet been successfully overcome. As a simple instance, we cannot give a meaning to $\delta^2(t)$; the product of two delta sequences is certainly not a delta sequence or for that matter anything to which we can attach a meaning in distribution theory. We shall see later that we can define another 'product of distributions' which is of considerable practical importance.

Calculus of distributions

The principal reason for insisting that the testing functions $\phi \in \mathscr{D}$ are infinitely differentiable is to give us the nice property that *every distribution is infinitely differentiable*; this is, of course, in stark contrast to differentiation of ordinary functions.

As usual, the definition of distributional differentiation is suggested by considering the situation for regular distributions. Let $f(t)$ be a function which is everywhere continuously differentiable; the derivative $Df(t)$ generates a regular distribution Df through

$$Df(\phi) = \int_{-\infty}^{\infty} Df(t)\phi(t)dt.$$

Integration by parts leads to

$$Df(\phi) = -\int_{-\infty}^{\infty} f(t)D\phi(t)\,\mathrm{d}t = -f(D\phi)$$

and we adopt this as a general definition with an obvious extension to higher derivatives.

Definition 5.2.2 *The kth derivative of the distribution $f \in \mathscr{D}^*$ is the distribution $D^k f \in \mathscr{D}^*$ given by*

$$D^k f(\phi) = (-1)^k f(D^k \phi).$$

Differentiation is a continuous linear operator in the space $\mathscr{D}^*$ (Exercise 5.2,5(a)).

Example 5.2.1

(a) (i) The distribution $D^k \delta$ introduced earlier by the definition

$$D^k \delta(\phi) = (-1)^k D^k \phi(0)$$

is the kth distributional derivative of δ since

$$D^k \delta(\phi) = (-1)^k \delta(D^k \phi) = (-1)^k D^k \phi(0).$$

(ii) The function $\sin \alpha t$ is infinitely smooth in a neighbourhood of the support of $D\delta$, hence $\sin \alpha t \, D\delta(t)$ is a distribution. We have

$$\sin \alpha t \, D\delta(\phi) = D\delta(\phi(t) \sin \alpha t)$$
$$= -\delta(D\phi \sin \alpha t) - \alpha\delta(\phi \cos \alpha t)$$
$$= -\alpha\phi(0),$$

hence, in the sense of distributions, we may write

$$\sin \alpha t \, D\delta(t) = -\alpha\delta(t).$$

(iii) Consider the distribution $D^k \delta(\alpha t - \beta)$. By a combination of the translation and scaling operations (Exercise 5.2,4(a)) and the definition of differentiation, we have

$$D^k \delta(\alpha t - \beta)(\phi) = \frac{1}{|\alpha|} D^k \delta\left(\phi\left(\frac{t+\beta}{\alpha}\right)\right)$$

$$= \frac{(-1)^k}{|\alpha|} \delta\left(D^k \phi\left(\frac{t+\beta}{\alpha}\right)\right)$$

$$= \frac{(-1)^k}{|\alpha|\alpha^k} D^k \phi\left(\frac{\beta}{\alpha}\right)$$

$$= \frac{(-1)^k}{|\alpha|\alpha^k} \delta\left(t - \frac{\beta}{\alpha}\right)(D^k\phi(t))$$

$$= \frac{1}{|\alpha|\alpha^k} D^k \delta\left(t - \frac{\beta}{\alpha}\right)(\phi),$$

hence we may write

$$D^k \delta(\alpha t - \beta) = \frac{1}{|\alpha|\alpha^k} D^k \delta\left(t - \frac{\beta}{\alpha}\right).$$

(b) (i) Consider the regular distribution $1_+(t)t^\lambda$, where $-1 < \lambda < 0$. By distributional differentiation we have

$$D\,1_+(t)t^\lambda(\phi) = -1_+(t)t^\lambda(D\phi)$$

$$= -\lim_{\varepsilon \to 0} \int_\varepsilon^\infty t^\lambda \phi'(t)\,dt$$

$$= \lim_{\varepsilon \to 0}\left(\lambda \int_\varepsilon^\infty t^{\lambda-1}\phi(t)\,dt + \varepsilon^\lambda \phi(\varepsilon)\right)$$

upon integrating by parts. Now $\phi(\varepsilon) = \phi(0) + O[\varepsilon]$; hence we may write the last expression as

$$\lim_{\varepsilon \to 0}\left(\lambda \int_\varepsilon^\infty t^{\lambda-1}\phi(t)\,dt + \varepsilon^\lambda \phi(0)\right),$$

and this is simply the definition of the pseudofunction $\text{Pf}\,1_+(t)\lambda t^{\lambda-1}$ (Exercise 5.2,3(c) (iii)): thus, in the sense of distributions,

$$D\,1_+(t)t^\lambda = \text{Pf}\,1_+(t)\lambda t^{\lambda-1}.$$

By an extension of this argument we can show that this rule holds for all negative λ provided λ is *not a negative integer*. (Of course, in the above formula both sides are then pseudofunctions.)

This shows that we may consider pseudofunctions to be generated by applying the formal rules of differentiation to regular distributions.

(ii) When λ is a negative integer, derivatives of the delta distribution appear. To illustrate this we consider the pseudofunction $\text{Pf}(1_+(t)/t^2)$. Now (cf., Exercise 5.2,3(c)(ii))

$$\text{Pf}\frac{-1_+(t)}{t^2}(\phi) = \lim_{\varepsilon \to 0}\left(-\int_\varepsilon^\infty \frac{\phi(t)}{t^2}\,dt + \frac{\phi(0)}{\varepsilon} - \phi'(0)\ln \varepsilon\right),$$

and upon integrating by parts and noting that $\phi(\varepsilon) = \phi(0) +$

$\varepsilon\phi'(0) + O[\varepsilon^2]$ we may write the right-hand side as

$$-D\phi(0) - \mathrm{Pf}\frac{1_+(t)}{t}(D\phi) = D\delta(\phi) + D\,\mathrm{Pf}\frac{1_+(t)}{t}(\phi);$$

hence

$$D\,\mathrm{Pf}\frac{1_+(t)}{t} = \mathrm{Pf}\frac{-1_+(t)}{t^2} - D\delta.$$

(iii) The special behaviour of the pseudofunctions $\mathrm{Pf}\,1_+(t)t^\lambda$ when λ is a negative integer can be clarified by considering λ to be a complex number.

For $\mathrm{Re}\,\lambda > -1$, $1_+(t)t^\lambda$ is the regular distribution

$$1_+(t)t^\lambda(\phi) = \int_0^\infty t^\lambda \phi(t)\,dt = F(\lambda),$$

say. Considered as a function of the complex variable λ, $F(\lambda)$ is analytic in the half-plane $\mathrm{Re}\,\lambda > -1$ since

$$\frac{dF}{d\lambda} = \int_0^\infty t^\lambda \ln t\,\phi(t)\,dt$$

exists for $\mathrm{Re}\,\lambda > -1$. If we write $F(\lambda)$ as

$$F(\lambda) = \int_0^1 t^\lambda(\phi(t) - \phi(0))\,dt + \int_1^\infty t^\lambda \phi(t)\,dt + \frac{\phi(0)}{1+\lambda}$$

then we see that the first term exists for $\mathrm{Re}\,\lambda > -2$, the second for all λ and the third provided $\lambda \neq -1$. Thus we have analytically continued $F(\lambda)$ into the strip $-2 < \mathrm{Re}\,\lambda \leqslant -1$ except for the isolated point $\lambda = -1$: at this point $F(\lambda)$ has a simple pole with residue $\phi(0)$. This means that the distribution $1_+(t)t^\lambda$ has a singularity at $\lambda = -1$ with residue $\delta(t)$.

The analytic continuation of $F(\lambda)$ can clearly be carried into successive strips of the left-hand half-plane with poles at $\lambda = -1$, $-2, -3, \ldots$: at the pole $\lambda = -(k+1)$ the residue is

$$\frac{(-1)^k D^k \delta(t)}{k!}.$$

In this example we began with a regular distribution and by successive differentiations generated singular distributions. This is, in fact, a general result. It can be shown that every distribution with compact support can be generated by a finite number of differentiations[†] from a regular distribution (or for that matter,

---

[†] A distribution which can be represented as the rth derivative of a regular distribution is called a distribution of *rank r*.

by employing more differentiations, from a continuous function) [128].

(c) Convergence in $\mathscr{D}^*$ is insensitive to differentiation which is, of course, in marked contrast to the corresponding situation for ordinary function sequences. Thus suppose $f^{(n)} \to f$ in $\mathscr{D}^*$, then

$$D^k f^{(n)}(\phi) = (-1)^k f^{(n)}(D^k \phi) \to (-1)^k f(D^k \phi) = (-1)^k D^k f(\phi)$$

so that

$$D^k f^{(n)} \to D^k f$$

in $\mathscr{D}^*$.

This means, for example, that term-by-term differentiation of a convergent series is always possible provided the resulting series is interpreted in the sense of distributions. The following result [128] is of considerable practical utility in the (distributional) theory of Fourier series. The series

$$\sum_{n=-\infty}^{\infty} c_n e^{int},$$

where $|c_n| < M|n|^\gamma, n \neq 0$, and γ is a real constant, converges in $\mathscr{D}^*$; its sum equals

$$c_0 + D^k g(t),$$

where $k = [\text{least positive integer} \geq \gamma + 2]$ and $g(t)$ is the continuous periodic function given by

$$g(t) = \sum_{\substack{n=-\infty \\ n \neq 0}}^{\infty} \frac{c_n}{(in)^k} e^{int}.$$

For example, the series $\sum_{n=-\infty}^{\infty} e^{int}$ can be given no meaning in the sense of ordinary functions, but according to the above result

$$\sum_{n=-\infty}^{\infty} e^{int} = 1 + D^2 g(t),$$

where

$$g(t) = -\sum_{\substack{n=-\infty \\ n \neq 0}}^{\infty} \frac{1}{n^2} e^{int}$$

is the Fourier series for a periodic function consisting of parabolic arcs.

As Fig. 5.1 shows, the second distributional derivative of g is the sum of the distribution -1 and a series of delta distributions at the points....

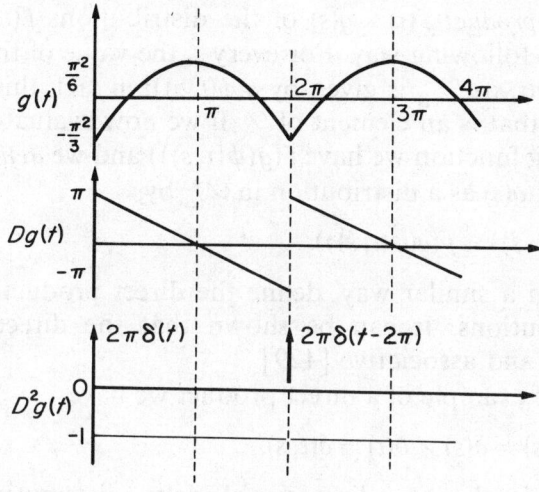

Fig. 5.1

$-4\pi, -2\pi, 0, 2\pi, 4\pi, \ldots$. Hence we can write

$$\sum_{n=-\infty}^{\infty} e^{int} = 2\pi \sum_{n=-\infty}^{\infty} \delta(t - 2n\pi),$$

the meaning of this being that, for any $\phi \in \mathscr{D}$,

$$\sum_{n=-\infty}^{\infty} e^{int}(\phi) = 2\pi \sum_{n=-\infty}^{\infty} \phi(t - 2n\pi).$$

Note that, since every ϕ vanishes outside a finite interval, the summation on the right-hand side is always finite.

Convolution of distributions

We have already seen that it is not, in general, possible to define the product of distributions as we would the product of functions. However, we can define another type of product, called the *direct product of distributions*, by using a space of testing functions in two variables, say s and t.

First, a regular distribution f of two variables is defined quite obviously by

$$f(\phi(t,s)) = \int_{-\infty}^{\infty} \int_{-\infty}^{\infty} f(t,s)\phi(t,s) \, dt \, ds,$$

where $\phi(t, s)$ is an element of the space $\mathscr{D}_{t,s}$ of testing functions which vanish outside bounded regions of $\mathscr{R}_2$.

The *direct product* $f(t) \times g(s)$ of the distributions $f(t), g(s) \in \mathscr{D}^*$ is defined in the following way. For every t, the value of the functional $g(s) \in \mathscr{D}^*$ on $\phi(t,s) \in \mathscr{D}_{t,s}$ is given by $g(\phi(t,s))$: in fact, this is a testing function in t, that is an element of $\mathscr{D}$. If we now evaluate $f(t)$ on this class of testing function we have $f(g(\phi(t,s)))$ and we *define* the direct product $f(t) \times g(s)$ as a distribution in $\mathscr{D}_{t,s}^*$ by,

$$f \times g(\phi(t,s)) = f(g(\phi(t,s))).$$

One can, in a similar way, define the direct product of three (or more) distributions. It can be shown that the direct product is commutative and associative [129].

As a simple example of a direct product we have

$$\delta(t) \times \delta(s) = \delta(s) \times \delta(t) = \delta(t,s),$$

where $\delta(t,s)$ is the two-dimensional delta distribution with the property

$$\delta(\phi(t,s)) = \phi(0,0).$$

To see this we simply note that

$$\begin{aligned}
\delta \times \delta(\phi(t,s)) &= \delta(\delta(\phi(t,s)) \\
&= \delta(\phi(t,0)) \\
&= \phi(0,0) = \delta(\phi(t,s)).
\end{aligned}$$

We now turn to the relation between the direct product of two distributions and their convolution. For two continuous functions $f(t), g(t)$ their convolution $h(t)$ is defined by

$$h(t) = f(t) * g(t) = \int_{-\infty}^{\infty} f(s) g(t-s) ds.$$

It is not immediately obvious how we might extend this definition to distributions. However, suppose we think of h as being a regular distribution, then

$$h(\phi) = f * g(\phi) = \int_{-\infty}^{\infty} dt \int_{-\infty}^{\infty} f(s) g(t-s) \phi(t) ds.$$

If we treat the iterated integral as a double integral and apply the change of variable $s = \sigma, t = \sigma + \tau$, we obtain

$$f * g(\phi) = \int_{-\infty}^{\infty} \int_{-\infty}^{\infty} f(\sigma) g(\tau) \phi(\sigma + \tau) d\sigma d\tau,$$

which is in a form reminiscent of the direct product of distributions.

If $f(t), g(t)$ are two distributions in $\mathscr{D}^*$, this suggests that we *define* their *convolution* $f*g$ by

$$f*g(\phi) = f \times g(\phi(t+s))$$
$$= f(g(\phi(t+s))), \qquad \phi \in \mathscr{D}.$$

There is, however, one difficulty with this definition. Even though the function $\phi(t+s)$ is infinitely smooth it is not a testing function in $\mathscr{D}_{t,s}$ since its support is not bounded in the (t, s) plane. (t and s may be unboundedly large while $t+s$ remains finite.) We should ensure that the intersection of the supports of $f(t) \times g(s)$ and $\phi(t+s)$ is a bounded set. With this restriction it can be shown that $f*g$ exists as a distribution (in $\mathscr{D}^*$) if one or other of the following conditions is fulfilled:

(i) f or g has bounded support;
(ii) both f and g have supports bounded on the left (on the right).[†]

A simple sketch in the (t, s) plane will convince the reader of the validity of these conditions.

Convolution can be defined for distributions under other conditions, but the behaviour of f and/or g must be restricted in some way as $t \to \infty$.

The operation of convolution is:

(a) commutative, $f*g = g*f$;
(b) linear, $f*(\alpha g + \beta h) = \alpha f*g + \beta f*h$;
(c) under the conditions given above the support of $f*g$ is (i) bounded; (ii) bounded on the left (on the right).

Example 5.2.2

(a) If f and g are regular distributions with support bounded on the left at $t=0$ then

$$h(t) = f(t)*g(t) = \int_0^t f(s)g(t-s)\,\mathrm{d}s.$$

This is the familiar form of convolution integral one meets in connection with the analysis of linear stationary systems which are assumed to be quiescent for $t < 0$ (Examples 2.9.4, 3.10.4 and 5.5.3).

(b) The support of the delta distribution is bounded so we are able to take the convolution of δ with *any* distribution f.

[†] The support of a distribution is bounded on the left at t_0 if $f(\phi) = 0$ for all $\phi \in \mathscr{D}$ which vanish for $t > t_0$.

We have
$$\delta * f(\phi) = \delta \times f(\phi(t+s))$$
$$= f(\delta(\phi(t+s)))$$
$$= f(\phi),$$
whence
$$\delta * f = f.$$
We may show in a similar way (Exercise 5.2,8(a)) that
$$D^k \delta * f = D^k f.$$

(c) Let Df be a distribution with support bounded on the left at $t=0$, then
$$1_+ * Df(\phi) = 1_+(Df(\phi(t+s)))$$
$$= -1_+(f(D\phi(t+s)))$$
$$= -f(1_+(D\phi(t+s)))$$
$$= f(\phi);$$
hence convolution of a right-sided distribution with the unit step distribution is equivalent to integration (Exercise 5.2,8(b)).

(d) (Associativity) The convolution of three or more distributions can be built up provided that, at every stage, the appropriate conditions are fulfilled. However, the process is *not*, in general, associative.

For example, let 1 be the constant distribution, then
$$1*(D\delta*1_+) = 1*\delta = 1$$
and
$$(1*D\delta)*1_+ = 0*1_+ = 0.$$
while $(1*1_+)*D\delta$ does not exist since the first convolution is not defined. However, if each distribution has support bounded on the left then repeated convolutions always exist and furthermore the operation is associative.

(e) (Differentiation of a convolution) Since $f*g$ is a distribution, we may consider its (distributional) derivative $D(f*g)$. We have
$$D(f*g)(\phi) = f*g(-D\phi)$$
$$= f(g(-D\phi(t+s)))$$
$$= f(Dg(\phi(t+s)))$$
$$= f*Dg(\phi),$$
hence
$$D(f*g) = f*Dg = g*Df$$

upon interchanging the roles of f and g. More generally we can show that

$$D^k(f*g) = D^{k_1}f * D^{k_2}g, \qquad k = k_1 + k_2.$$

The previous part, (c), of this example is an instance of this rule, viz.,

$$D(1_+ * f) = \delta * f = f = 1_+ * Df.$$

(f) (*Regularization of a distribution*) The convolution of a distribution f with a testing function ϕ converts the distribution into a function which is infinitely smooth. This follows from the rule for differentiation of a convolution, since

$$D^k(f*\phi) = f * D^k\phi.$$

In a sense, convolution of a distribution with a testing function gives a smooth version of the distribution; this smooth version is called a *regularization of the distribution*.

A family of regularizations of the pseudofunction $\text{Pf}(1_+(t)/t)$ is illustrated in Fig. 5.2. These are generated from the results of Exercise 5.2,9.

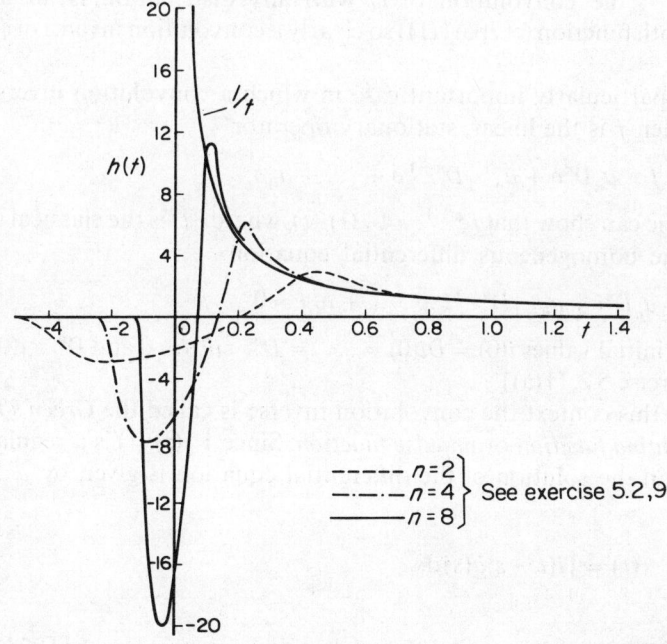

Fig. 5.2

(g) (A convolution algebra) Part (b) of this example showed that, for the operation of convolution, the delta distribution acts like an identity element. Furthermore, provided we confine ourselves to right-sided distributions, convolution is associative (part (d)). These considerations, together with the commutativity of convolution suggest that we have a *commutative algebra* with *unit element* δ for right (left)-sided distributions.

This *convolution algebra* does *not* contain divisors of zero, that is to say, $f*g = 0$ necessarily implies that either $f = 0$ or $g = 0$ [127].

The context of a convolution algebra allows us to consider convolution equations and convolution inverses. A convolution equation is an operator equation of the form

$$f*x = g,$$

where f and g are given and we require to find the distribution x. If we can find a distribution f^{*-1}, the convolution inverse of f, such that

$$f^{*-1}*f = \delta,$$

then a formal solution of the equation is

$$x = f^{*-1}*g.$$

However, it must not be assumed that every distribution has an inverse. Suppose we seek the convolution inverse of a testing function $\phi \in \mathscr{D}^*$; the convolution of ϕ with any distribution is an infinitely smooth function (cf., part (f)) so clearly a convolution inverse of ϕ cannot exist.

A particularly important case in which a convolution inverse exists is when f is the linear, stationary operator

$$f \equiv a_n D^n \delta + a_{n-1} D^{n-1} \delta + \ldots + a_0 \delta.$$

One can show that $f^{*-1} = 1_+(t)i(t)$, where $i(t)$ is the classical solution of the homogeneous differential equation

$$a_n D^n i + a_{n-1} D^{n-1} i + \ldots + a_0 i = 0$$

with initial values $i(0) = Di(0) = \ldots = D^{n-2}i(0) = 0$ and $D^{n-1}i(0) = 1/a_n$ (Exercise 5.2, 11(a))

In this context the convolution inverse is called the *Green's function*, *weighting function* or *impulse function*. Since $1_+(t)i(t)$ is a regular distribution the solution of the differential equation is given by

$$x(t) = \int_0^t i(t-s)g(s)ds$$

whenever g is also a regular distribution (*see*, for example, [82, 129, 132, 135]).

Exercises 5.2

1. (a) Let $\{f^{(n)}\}$ be a sequence of locally integrable functions satisfying the following conditions:
   (1) $\lim\limits_{n} \int_{-\alpha}^{\alpha} f^{(n)}(t)\,dt = 1$ for some constant α;
   (2) $f^{(n)}(t) \to 0$ uniformly for $|t| \geq \beta$ for any $\beta > 0$, however small;
   (3) $f^{(n)}(t) \geq 0 \quad \forall\, t, n$;
   then $\{f^{(n)}\}$ is called a *positive delta sequence* [31, 124].
   (i) Show that the Hermite and Kirchoff sequences (p. 422) are positive delta sequences.
   (ii) Let
   $$f^{(n)}(t) = \begin{cases} \dfrac{1}{\gamma_n}(1-t^2)^n, & |t| \leq 1, \\ 0, & |t| > 1, \end{cases}$$
   where
   $$\gamma_n = \int_{-1}^{1} (1-t^2)^n\,dt.$$
   Show that this is a positive delta sequence.
   (Note: there are other types of delta sequences such as $f_\lambda(t) = \sin \lambda t / \pi t$, which is a delta sequence of Dirichlet type [31].)
   (b) Let $\{f^{(n)}\}$ be a positive delta sequence consisting of even functions. Let $g(t)$ be an integrable function having a jump discontinuity at 0. Show that
   $$\lim_{n} \int_{-\infty}^{\infty} f^{(n)}(t)g(t)\,dt = \frac{1}{2}(g(-0) + g(+0))$$
   (Hint: express g as the sum of an even and odd function.)

2. (a) Show that two continuous functions which produce the same regular distribution are equal. (Hint: assume, say, $f(t) - g(t) \neq 0$ in some interval—using a testing function of the type given on p. 426 obtain a contradiction.)
   (b) Deduce from (a) that $\mathscr{D} \subset \mathscr{D}^*$.

3. (a) Verify that
   $$\int_{0}^{\infty} t^{-3/2}(\phi(t) - \phi(0))\,dt$$
   is a continuous linear functional on $\mathscr{D}$. (Hint: use the fact that,

for some $\alpha(\phi)$,

$$\left|\frac{\phi(t) - \phi(0)}{t}\right| \leq \sup_{t \in (0,\alpha)} |D\phi|.)$$

(b) Consider the simple result

$$\frac{d}{dt}\int_\alpha^t \frac{d\tau}{\sqrt{t-\tau}} = \frac{1}{\sqrt{t-\alpha}}.$$

Show that, on the left-hand side, we can 'interchange the operations of integration and differentiation' if we employ the interpretation

$$\frac{d}{dt}\int_\alpha^t \frac{d\tau}{\sqrt{t-\tau}} = \mathrm{Fp}\int_\alpha^t \frac{\partial}{\partial t}\left(\frac{1}{\sqrt{t-\tau}}\right)d\tau.$$

Note that we are not properly differentiating the integral but rather interchanging operations. If $g(t) \in \mathscr{C}^{(1)}[\alpha, t]$ show that

$$\frac{d}{dt}\int_\alpha^t \frac{g(\tau)d\tau}{\sqrt{t-\tau}} = -\frac{1}{2}\mathrm{Fp}\int_\alpha^t \frac{g(\tau)d\tau}{(t-\tau)^{3/2}}.$$

(c) (i) Why cannot we define a pseudofunction $\mathrm{Pf}\, 1_+(t)/t$ by (cf., p. 428)

$$\mathrm{Pf}\frac{1_+(t)}{t}(\phi) = \int_0^\infty \frac{\phi(t) - \phi(0)}{t}dt?$$

(ii) Show that we can define a pseudofunction (i.e., a continuous linear functional on $\mathscr{D}$) corresponding to 'the function $1_+(t)/t$' by the definition

$$\mathrm{Pf}\frac{1_+(t)}{t}(\phi) = \lim_{\varepsilon \to 0}\left\{\int_\varepsilon^\infty \frac{\phi(t)}{t}dt + \phi(0)\ln\varepsilon\right\}$$

$$= \int_0^1 \frac{\phi(t) - \phi(0)}{t}dt + \int_1^\infty \frac{\phi(t)}{t}dt$$

(iii) Show that if $-1 < \lambda < 0, k = 1, 2, 3, \ldots$,

$$\mathrm{Pf}\, 1_+(t)t^{\lambda-k}(\phi) = \int_0^\infty t^{\lambda-k}\left(\phi(t) - \sum_{m=0}^{k-1}\frac{D^m\phi(0)}{m!}t^m\right)dt.$$

(Hint: to prove continuity use Taylor's theorem with remainder.)

What form will the definition take if $\lambda = 0$? (Hint: look at Example 5.2.1(b)(ii).)

(iv) Let $\mathrm{Pv}(1/t)$ denote the distribution defined by

$$\mathrm{Pv}\frac{1}{t}(\phi) = \lim_{\varepsilon \to 0}\left\{\int_{-\infty}^{-\varepsilon}\frac{\phi(t)}{t}dt + \int_\varepsilon^\infty \frac{\phi(t)}{t}dt\right\}$$

This is of course the familiar *Cauchy principal value* of a divergent integral.

Show that
$$\text{Pv}\frac{1}{t} = \text{Pf}\frac{1_+(t)}{t} - \text{Pf}\frac{1_+(-t)}{|t|}$$

4. (a) Show that if $f \in \mathcal{D}^*$
$$f(\alpha t + \beta)(\phi(t)) = \frac{1}{|\alpha|} f\left(\phi\left(\frac{t-\beta}{\alpha}\right)\right).$$

   (b) Using (a), deduce that we may write,
$$\delta(\alpha t + \beta) = \frac{1}{|\alpha|} \delta\left(t + \frac{\beta}{\alpha}\right).$$

   (c) Let $p(t)$ be a smooth function with $p'(0) \neq 0$ and $p(0) = 0$; then deduce from (b) that we may write
$$\delta(p(t)) = \frac{1}{|p'(0)|} \delta(t).$$

   (d) Extend the result (c) to find $\delta(q(t))$ where the smooth function $q(t)$ has zeros at $t = t_i$ at each of which $q'(t_i) \neq 0$. Hence deduce the (formal) expression
$$\delta(t^2 - \alpha^2) = \frac{1}{2\alpha}(\delta(t-\alpha) + \delta(t+\alpha)).$$

5. (a) Show that distributional differentiation (Definition 5.2.2) is a continuous operation on $\mathcal{D}$.

   (b) Let $f(t)$ be a function which is piecewise continuous, having jump discontinuities $\Delta_i = f(t_i + 0) - f(t_i - 0)$ at the points t_i and let $f(t)$ be differentiable except at the points t_i.

   Show that we can construct a continuous function
$$f_c(t) = f(t) - \sum_i \Delta_i 1_+(t - t_i)$$
   and hence deduce that the distributional derivative of (the regular) functional f is given by
$$Df(t) = f'_c(t) + \sum_i \Delta_i \delta(t - t_i)$$

   (c) Show that, in the sense of distributions:
   (i) $D^2[1_+(t)\sin \alpha t] = \alpha \delta(t) - \alpha^2 1_+(t) \sin \alpha t$;

   (ii) $t^n D^m \delta(t) = \begin{cases} 0, & m < n, \\ (-1)^n n! \, \delta(t), & m = n, \\ (-1)^n \dfrac{m!}{(m-n)!} D^{m-n} \delta(t), & m > n; \end{cases}$

(iii) $\psi D\delta = \psi(0)D\delta - \psi'(0)\delta$, $\psi \in \mathscr{D}$;

(iv) $D(\psi f) = \psi' f + \psi Df$, $\psi \in \mathscr{D}, f \in \mathscr{D}^*$;

(v) $D(\operatorname{sgn} t) = 2\delta(t)$, where the *signum function* is given by
$$\operatorname{sgn} t = \begin{cases} 1, & t > 0, \\ -1, & t < 0; \end{cases}$$

(vi) $D^2|\sin t| = 2\delta(t) - |\sin t|$.

(d) (i) Show that, in the sense of distributions,
$$\lim_{\tau \to +0} \ln(t + i\tau) = \ln|t| + i\pi 1_+(-t).$$

(ii) Hence, by distributional differentiation, show that
$$\lim_{\tau \to +0} \frac{1}{t + i\tau} = \operatorname{Pv} \frac{1}{t} - i\pi\delta(t).$$

6. (The *primitive* of a distribution) Suppose we seek a functional $g \in \mathscr{D}^*$ whose (distributional) derivative is $f \in \mathscr{D}^*$, i.e.,
$$g(\phi') = f(-\phi).$$

This expression defines g only on that subspace of $\mathscr{D}$ whose elements are the derivatives of testing functions in $\mathscr{D}$, and hence is inadequate as a definition of the primitive or anti-derivative of f. Extend the above relation to the whole of $\mathscr{D}$ in the following way.

(i) First show that a test function ψ can be represented as the derivative of another test function if and only if
$$\int_{-\infty}^{\infty} \psi(t) \, dt = 0.$$

(ii) Then show that if ϕ_0 is a fixed test function such that
$$\int_{-\infty}^{\infty} \phi_0(t) \, dt = 1$$
then any testing function $\phi \in \mathscr{D}$ can be represented in the form
$$\phi = \psi + \gamma \phi_0,$$
where ψ is the derivative of another testing function and
$$\gamma = \int_{-\infty}^{\infty} \phi(t) \, dt.$$

Verify that
$$\int_{-\infty}^{t} \psi(t) \, dt \in \mathscr{D}.$$

(iii) Now show that the functional $D^{-1}f$ defined by

$$D^{-1}f(\phi) = \gamma D^{-1}f(\phi_0) - f\left(\int_{-\infty}^{t}\psi(t)dt\right),$$

where ϕ, ϕ_0 and ψ are related as in (ii), defines an equivalence class of primitives of f on the whole of $\mathscr{D}$ in the sense that any two primitives of f differ by a constant (distribution).

(iv) Show that the primitives of the constant distribution c_1 are the distributions $c_1 t + c_2$, where c_2 is a constant distribution.

7. (The definite integral). How are we to interpret the definite integral

$$\int_I f(t)dt,$$

say, where $f \in \mathscr{D}^*$ and I is a bounded interval in $\mathscr{R}$?

(i) If the support of f is the interval I, verify that we can define the given definite integral of f by

$$\int_I f(t)dt = f(\psi),$$

where $\psi(t) \in \mathscr{D}$ is identically 1 in I and is extended arbitrarily outside I.

(ii) If the support of f is not bounded it is not always possible to assign a meaning to the definite integral.

Suppose the distribution f coincides with the continuous function $f(t)$ except in some subinterval of I (in particular f is continuous in the neighbourhood of the end points of I). Then we define

$$\int_I f(t)dt = f(\psi) - \int_{\mathscr{R}-I} f(t)\psi(t)dt$$

Let $D^{-1}f$ be a primitive of f: show that

$$f(\psi) = -\int_{\mathscr{R}-I} D^{-1}f(t)\psi'(t)dt$$

$$= \int_{\mathscr{R}-I} f(t)\psi(t)dt + [D^{-1}f(t)]_\alpha^\beta,$$

where α, β ($\beta > \alpha$) are the boundary points of I. Hence deduce that

$$\int_\alpha^\beta f(t)dt = [D^{-1}f(t)]_\alpha^\beta,$$

a result which coincides with the fundamental theorem of the calculus.

(iii) Verify that

$$-D\,\mathrm{Pv}\frac{1}{t} = \mathrm{Pf}\frac{1}{t^2} = \mathrm{Pf}\frac{1_+(t)}{t^2} + \mathrm{Pf}\frac{1_+(-t)}{t^2}$$

and hence show that, in the sense of distributions,

$$\int_{-1}^{1} \mathrm{Pf}\frac{1}{|t|^2}\,dt = -2.$$

Sketch a smoothed version of the function $-\ln|t|$, that is to say a function coinciding with $\ln|t|$ except in the vicinity of zero, where it is 'rounded off' at a finite (positive) value.

Now sketch a smoothed version of $-\mathrm{Pv}(1/t)$ using the result

$$\mathrm{Pv}\frac{1}{t} = D\ln|t|,$$

and from this sketch a smoothed version of $\mathrm{Pf}(1/t^2)$; can you now see why the value of the integral might be negative? (These sketches represent regularized versions of singular distributions—see Example 5.2.2(f).)

8. (a) Show that
   (i) $D^k\delta * f = D^k f$;
   (ii) $\delta(t-\alpha)*f(t) = f(t-\alpha)$;
   (iii) $1_+(t)*\mathrm{Pf}(1_+(t)/t) = 1_+(t)\ln t$;
   (iv) $\mathrm{Pf}\dfrac{1_+(t)}{t} * \mathrm{Pf}\dfrac{1_+(t)}{t} = \dfrac{d^2}{dt^2}\int_0^t \ln\tau\ln(t-\tau)\,d\tau.$

   (b) Example 5.2.2(c) indicates that convolution of a distribution f with the unit step 1_+ is 'equivalent to integration': which particular primitive is generated by this process?
   (c) (Translation of a convolution) if f, g are distributions and $h(t) = f(t)*g(t)$ show that

   $$h(t-\alpha) = f(t-\alpha)*g(t) = f(t)*g(t-\alpha).$$

9. (Regularization of $\mathrm{Pf}(1_+(t)/t)$) Let

$$h(t) = \mathrm{Pf}\frac{1_+(t)}{t} * \phi^{(n)}(t),$$

where

$$\phi^{(n)}(t) = \begin{cases} 0, & |t| \geqslant 1/n, \\ \phi(nt) \bigg/ \dfrac{1}{n}\int_{-1}^{1}\phi(\tau)\,d\tau, & |t| < 1/n, \end{cases}$$

with $\phi(t) = \exp(1/(t^2-1))$.

Taking $s = nt$, show that,
(i) for $t \leqslant -1/n$ ($s \leqslant -1$), $h(s) = 0$;
(ii) for $-1/n < t < 1/n$ ($-1 < s < 1$),

$$h(s) = \frac{1}{\Lambda} \int_0^{s+1} \frac{1}{\tau} \left(\exp\frac{1}{(s-\tau)^2 - 1} - \exp\frac{1}{s^2 - 1} \right) d\tau$$

$$+ (\ln(s+1) - \ln n) \exp\frac{1}{s^2 - 1};$$

(iii) for $t \geqslant 1/n$ ($s \geqslant 1$),

$$h(s) = \frac{1}{\Lambda} \int_{s-1}^{s+1} \frac{1}{\tau} \exp\frac{1}{(s-\tau)^2 - 1} d\tau;$$

where

$$\Lambda = \frac{1}{n} \int_{-1}^{1} \exp\frac{1}{\tau^2 - 1} d\tau.$$

10. Let $g(t)$ be a continuous function on the interval $[0, 1]$ and extend $g(t)$ to $(-\infty, \infty)$ by taking $g(t) = 0$ outside $[0, 1]$. Consider the convolutions

$$g^{(n)}(t) = f^{(n)}(t) * g(t) = \int_0^1 f^{(n)}(t - \tau) g(\tau) d\tau$$

where $\{f^{(n)}(t)\}$ is the delta sequence of Exercise 5.2,1 (a)(ii).

Show that the $g^{(n)}(t)$ are polynomials in t of degree $\leqslant 2n$ and hence that we have generated a sequence of polynomials $\{g^{(n)}(t)\}$ which approximate $g(t)$ in $[0, 1]$. (Note: it can be shown [124] that $g^{(n)}(t) \to g(t)$ *uniformly* on $[0, 1]$; this yields a direct proof of Weierstrass' theorem (p. 123).)

11. (a) Show, by repeated distributional differentiation, that $1_+(t)i(t)$ (Example 5.2.2(g)) is the convolution inverse of the stationary, linear differential operator of order n. (Hint: we know that $i(t)$ consists of a sum of exponentials or of products of exponentials and polynomials; hence $i(t)$ is infinitely smooth—cf., Example 5.3.3.)

(b) The result that $1_+(t)i(t)$ is the convolution inverse also holds for nth order, non-stationary, linear differential operators provided the coefficients $a_j(t)$ are infinitely smooth and $a_n(t) \neq 0$. To show what can happen when $a_n(t) = 0$ for some t consider the equation

$$tx' + x = (tx)' = 0, \qquad t \in (-\infty, \infty).$$

Show that this equation has, as solution, the distribution

$$x(t) = \alpha \operatorname{Pv}\frac{1}{t} + \beta \delta(t),$$

where α, β are constants. Thus a first-order equation possesses a two-dimensional solution space!

5.3 Integral transforms

The Fourier transform

We begin by briefly reviewing the properties of the Fourier transform for ordinary functions in $\mathscr{L}_1(-\infty, \infty)$ [47, 133] (Example 3.7.6(c)). For $f(t) \in \mathscr{L}_1(-\infty, \infty)$ we define the Fourier transform $\tilde{f}(\omega)$ by

$$\tilde{f}(\omega) = F(f) = \int_{-\infty}^{\infty} f(t) e^{-2\pi i \omega t} dt.$$

If $\tilde{f} \in \mathscr{L}_1(-\infty, \infty)$ then we have the reciprocal transform

$$f(t) = F^{-1}(\tilde{f}) = \int_{-\infty}^{\infty} \tilde{f}(\omega) e^{2\pi i \omega t} d\omega.$$

Now, in general, $\tilde{f}(\omega)$ is certainly bounded and uniformly continuous for all ω, but is not necessarily an element of $\mathscr{L}_1(-\infty, \infty)$: in that case, we need to interpret the reciprocal transform in a Cauchy principal value sense, that is,

$$f(t) = \lim_{\Omega \to \infty} \int_{-\Omega}^{\Omega} \tilde{f}(\omega) e^{2\pi i \omega t} d\omega.$$

For two functions $f, g \in \mathscr{L}_1(-\infty, \infty)$ we have *Parseval's relation*

$$\int_{-\infty}^{\infty} f(t) \tilde{g}(t) dt = \int_{-\infty}^{\infty} \tilde{f}(t) g(t) dt;$$

both sides converge since $\tilde{f}$ and $\tilde{g}$ are continuous and bounded for all t.

The major result for the Fourier transform and the one which makes it so useful in applications is, of course, the conversion of differentiation with respect to t into multiplication by $2\pi i\omega$. That is, if $D^k f(t) \in \mathscr{L}_1(-\infty, \infty)$ then

$$F(D^k f) = (2\pi i \omega)^k \tilde{f}(\omega).$$

The major drawback with the classical Fourier transform is that it is not applicable to functions which do not tend to zero at infinity fast enough to make them absolutely integrable. We cannot, for example, find the Fourier transform of the unit step function. However, we can apply the theory of distributions to overcome this difficulty.

Because it is the behaviour of functions at infinity that is of particular interest, the space $\mathscr{D}$ of testing functions having bounded support is not suitable for defining a distributional Fourier transform. Instead, we introduce testing functions defined on all of $\mathscr{R}$ which tend, with their derivatives, to zero at infinity faster than any inverse power of t. Let us denote this space of *testing functions of rapid descent* by $\mathscr{S}$: then $\phi \in \mathscr{S}$ if

(i) $\phi(t)$ is infinitely smooth;
(ii) as $|t| \to \infty$, $D^k\phi(t) \to 0$ faster than any power of $1/|t|$.

The sequence $\{\phi^{(n)}\}$ converges in $\mathscr{S}$, $\phi^{(n)} \underset{\mathscr{S}}{\to} \phi$ if $|t|^m |D^k\phi^{(n)}(t)| \leq C_{mk}$ for all m, k and $D^k\phi^{(n)}(t) \to D^k\phi(t)$ uniformly on every finite interval.

Definition 5.3.1 *A continuous linear functional on the space $\mathscr{S}$ is a distribution of slow growth (or tempered distribution).*

The space of distributions of slow growth is $\mathscr{S}^*$, the dual of $\mathscr{S}$.

For a locally integrable function f to assign a scalar $f(\phi)$ to every $\phi \in \mathscr{S}$ through

$$f(\phi) = \int_{-\infty}^{\infty} f(t)\phi(t) \mathrm{d}t$$

it is clearly necessary that f be a *function of slow growth*; that is to say, there exists an m such that

$$\lim_{t \to \infty} |t|^{-m} f(t) = 0.$$

Thus every function of slow growth corresponds to a regular distribution of slow growth.[†]

The operations we defined for distributions in $\mathscr{D}^*$ all apply to distributions of slow growth. (For multiplication we obviously need to use functions of slow growth.)

We can now turn to a consideration of the Fourier transform of distributions of slow growth. However, we should first note that, if $\phi \in \mathscr{S}$, the space of testing functions of rapid descent, then its (ordinary)

[†] Every testing function in $\mathscr{S}$ generates a regular distribution of slow growth. Every distribution in $\mathscr{D}^*$ with bounded support is of slow growth. Every testing function in $\mathscr{D}$ is a testing function of rapid descent. It can be shown that [127]

(i) $\mathscr{D} \subset \mathscr{S} \subset \mathscr{S}^* \subset \mathscr{D}^*$
(ii) $\mathscr{D}$ is dense in $\mathscr{S}$;
(iii) $\mathscr{S}^*$ is dense in $\mathscr{D}^*$.

Fourier transform,

$$\tilde{\phi}(\omega) = F(\phi) = \int_{-\infty}^{\infty} \phi(t)e^{-2\pi i\omega t}dt,$$

is also an element of $\mathscr{S}$. In fact, it can be shown that the mapping $F: \mathscr{S} \to \mathscr{S}$ is continuous and one-to-one on $\mathscr{S}$ (Exercise 5.3,1(a)). If $f(t) \in \mathscr{L}_1(-\infty, \infty)$ and $\phi(t) \in \mathscr{S}$ then $\tilde{f}(\omega)$, $\tilde{\phi}(\omega)$ exist and Parseval's relation reads

$$\int_{-\infty}^{\infty} \tilde{f}(\omega)\phi(\omega)d\omega = \int_{-\infty}^{\infty} f(t)\tilde{\phi}(t)dt.$$

If we now consider f to be a regular distribution in $\mathscr{S}^*$, then this relation can be rewritten as

$$\tilde{f}(\phi) = f(\tilde{\phi}).$$

But the right-hand side is defined for any $f \in \mathscr{S}^*$, i.e., for any distribution of slow growth. This leads to

Definition 5.3.2 *If f is a distribution of slow growth its (distributional) Fourier transform $\tilde{f}$ is the distribution of slow growth[†] given by*

$$\tilde{f}(\phi) = f(\tilde{\phi})$$

for all testing functions of rapid descent.

The same relation obviously also defines the inverse Fourier transform.

The important and familiar rule for the Fourier transform of the (distributional) derivative of f is preserved by the above definition. We have

$$\widetilde{Df}(\phi) = Df(\tilde{\phi})$$
$$= f(-D\tilde{\phi})$$
$$= f(2\pi it\tilde{\phi}).$$

But t is itself a (regular) distribution of slow growth, so by the operation of multiplication we find

$$\widetilde{Df}(\phi) = 2\pi it f(\tilde{\phi})$$
$$= 2\pi it \tilde{f}(\phi).$$

[†] That $\tilde{f}$ is a linear functional on $\mathscr{S}$ is obvious; its continuity is not difficult to prove (Exercise 5.3,1(b)). The Fourier transform is a one-to-one mapping of $\mathscr{S}^*$ onto itself.

Hence, in the sense of distributions

$$\widetilde{Df}(t) = 2\pi i \omega \tilde{f}(\omega),$$

where we have formally assigned the conventional variable names as 'arguments' of the distributions. (Look again at the comment on p. 429.)

One might well ask: 'Can we define the Fourier transforms of arbitrary distributions in $\mathscr{D}^*$?' The answer is, 'Yes, but not within the theory we have just outlined'. To do this it is necessary to introduce another space of testing functions and its dual. The testing functions we need are those whose Fourier transforms are in $\mathscr{D}$: the linear functionals on this space are sometimes called ultradistributions. The interested reader should consult [128, 129, 137].

Example 5.3.1

(a) Let us find the Fourier transform of the unit constant function, $f(t) = 1$. We have

$$\tilde{1}(\phi) = 1(\tilde{\phi}) = \int_{-\infty}^{\infty} \tilde{\phi}(\omega) d\omega = \phi(0),$$

whence

$$F(1) = \delta$$

and reciprocally (Exercise 5.3,2(a))

$$F(\delta) = 1.$$

(b) The function $f(t) = (-2\pi it)^n$ is a function of slow growth. Its Fourier transform is defined by

$$\widetilde{(-2\pi it)^n}(\phi(t)) = (-2\pi i\omega)^n(\tilde{\phi}(\omega))$$

$$= \int_{-\infty}^{\infty} (-2\pi i\omega)^n d\omega \int_{-\infty}^{\infty} \phi(t) e^{-2\pi i\omega t} dt$$

$$= \int_{-\infty}^{\infty} d\omega \int_{-\infty}^{\infty} \phi(t) D^n(e^{-2\pi i\omega t}) dt$$

$$= \int_{-\infty}^{\infty} d\omega \int_{-\infty}^{\infty} D^n\phi(t) e^{-2\pi i\omega t} dt$$

$$= \int_{-\infty}^{\infty} F(D^n\phi) d\omega = D^n\phi(0),$$

whence

$$F((-2\pi it)^n) = D^n\delta(\omega)$$

or
$$F(t^n) = (-2\pi i)^{-n} D^n \delta(\omega).$$

(c) (Poisson's formula) In Example 5.2.1(c) we found that
$$\sum_{n=-\infty}^{\infty} e^{int} = 2\pi \sum_{n=-\infty}^{\infty} \delta(t - 2n\pi).$$

Both sides represent distributions of slow growth, so we may take the Fourier transform to give (*see* Exercise 5.3,2(b))
$$\sum_{n=-\infty}^{\infty} \delta\left(\omega - \frac{n}{2\pi}\right) = F\left(\sum_{n=-\infty}^{\infty} 2\pi\delta(t - 2n\pi)\right),$$

in other words, the Fourier transform of a row of delta distributions is a row of delta distributions. For any $\phi \in \mathscr{S}$ the above relation implies that
$$\sum_{n=-\infty}^{\infty} \phi\left(\frac{n}{2\pi}\right) = 2\pi \sum_{n=-\infty}^{\infty} \tilde{\phi}(2n\pi),$$

which relates the values of $\phi(t)$ at equally spaced points to the values of its transform $\tilde{\phi}(\omega)$ at another set of points.

An important property of the ordinary Fourier transform is that the transform of a convolution is the product of the Fourier transforms of the component functions. We can show that the same result holds for the convolution of two distributions of slow growth whenever the convolution itself exists as a distribution of slow growth.

We need the following result (*see*, for example, [129]). If $f \in \mathscr{S}^*$ is a distribution having a bounded support then its Fourier transform is an infinitely smooth function given by
$$\tilde{f}(\omega) = f(e^{-2\pi i \omega t});$$

that is to say, the pointwise value of $\tilde{f}$ is given from an evaluation of f on the class of functions $e^{-2\pi i \omega t}$.[†]

Let $g \in \mathscr{S}^*$, then
$$F(f * g) = f(g(e^{-2\pi i \omega (t+s)}))$$
$$= f(e^{-2\pi i \omega t} g(e^{-2\pi i \omega s}))$$
$$= g(e^{-2\pi i \omega s}) f(e^{-2\pi i \omega t})$$
$$= \tilde{g}(\omega) \tilde{f}(\omega),$$

[†] Actually on the class of functions $\psi(t) e^{-2\pi i \omega t}$, $\psi \in \mathscr{D}$, where $\psi(t) = 1$ over some neighbourhood of the support of f.

whence

$$F(f*g) = F(f)F(g).$$

If the supports of f and g are bounded the product $\tilde{f}(\omega)\tilde{g}(\omega)$ is the ordinary product of functions. If only the support of f (or g) is bounded then the right-hand side needs to be interpreted as multiplication of a distribution of slow growth by an infinitely smooth function of slow growth. More general conditions of validity exist, but we shall not pursue them here.

Some simple extensions of the distributional Fourier transform to functions of several variables are described in Exercise 5.4,2.

The Laplace transform

The ordinary Laplace transform overcomes the difficulty of convergence at infinity encountered in the Fourier transform by

(i) dealing with functions whose support is bounded on the left, and
(ii) using a 'testing function' which incorporates an exponential decay at $+\infty$.

Let $f(t)$ have support bounded on the left and be locally integrable $[0, \infty)$; then its Laplace transform $\tilde{f}(s)$, given by

$$L(f) = \tilde{f}(s) = \int_{-\infty}^{\infty} f(t)e^{-st}dt,$$

is an analytic function of the complex variable s in the half-plane $\sigma > \sigma_c$, where σ_c is the abscissa of convergence. If we write $s = 2\pi(\sigma + i\omega)$, then we see that

$$L(f(t)) = F(f(t)e^{-2\pi\sigma t}).$$

The formula for inversion of the Laplace transform is

$$f(t) = L^{-1}(\tilde{f}) = \frac{1}{2\pi i}\int_{\mu-i\infty}^{\mu+i\infty} \tilde{f}(s)e^{st}ds,$$

where $\mu > \sigma_c$. As the reader will know, a common way of evaluating this complex integral is to analytically continue $\tilde{f}(s)$ into the complex plane for Re $s < \sigma_c$, convert the line integral into a contour integral and then apply the residue theorem (*see*, for example, [134]).

With this very brief introduction let us now consider the Laplace transform of distributions in $\mathcal{D}_R^*$, the space of right-sided distributions on $\mathscr{D}$. We proceed, in effect, by using the connection cited above between the Laplace and Fourier transforms, treating $f(t)e^{-\sigma t}$ as a distribution of slow growth. That is to say, if $f \in \mathscr{D}_R^*$ we *define*

$$L(f) = \tilde{f}(s) = f(e^{-st}) = f(t)e^{-\mu t}(\psi(t)e^{-(s-\mu)t}),$$

where $f(t)e^{-\mu t}$ is a distribution in $\mathscr{S}^*$ and $\psi(t)$ is an infinitely smooth function with support bounded on the left which equals unity over a neighbourhood of the support of f. For Re $s > \mu$, $\psi(t)e^{-(s-\mu)t}$ is a testing function in $\mathscr{S}$ in which s plays the role of a parameter.

The infimum of all real constants μ such that $f(t)e^{-\mu t} \in \mathscr{S}^*$ is called the *abscissa of convergence*. One can show, as for the ordinary Laplace transform, that $\tilde{f}(s)$ is an analytic function for Re $s > \inf \mu$.

As a consequence it can then be shown that

$$D\tilde{f}(s) = f(t)e^{-\mu t}(t\psi(t)e^{-(s-\mu)t})$$
$$= tf(t)e^{-\mu t}(\psi(t)e^{-(s-\mu)t}),$$

since t is a function of slow growth. This is a particular case of the familiar rule

$$L(t^k f(t)) = (-1)^k D^k \tilde{f}(s)$$

extended to distributions. Conversely, since $D(f(t)e^{-\mu t}) \in \mathscr{S}^*$ whenever $f(t)e^{-\mu t} \in \mathscr{S}^*$, we have

$$L(Df(t)) = D(f(t)e^{-\mu t})(\psi(t)e^{-(s-\mu)t}) + \mu f(t)e^{-\mu t}(\psi(t)e^{-(s-\mu)t})$$
$$= f(t)e^{-\mu t}(\psi(t)(s-\mu)e^{-(s-\mu)t}) + \mu f(t)e^{-\mu t}(\psi(t)e^{-(s-\mu)t})$$
$$= f(t)e^{-\mu t}(\psi(t)se^{-(s-\mu)t})$$
$$= sf(t)e^{-\mu t}(\psi(t)e^{-(s-\mu)t})$$
$$= s\tilde{f}(s),$$

which means to say

$$L(Df(t)) = s\tilde{f}(s).$$

This perhaps doesn't look to the reader like the familiar result for a differentiable function $f_+(t)$ which vanishes for $t < 0$, namely

$$L(Df_+(t)) = s\tilde{f}_+(s) - f_+(0),$$

but the two are readily reconciled if we define a regular distribution by the product $1_+(t)f_+(t)$, for then

$$D(1_+(t)f_+(t)) = 1_+(t)Df_+(t) + \delta(t)f_+(0),$$

and upon taking the transform of both sides,

$$s\tilde{f}_+(s) = L(Df_+(t)) + f_+(0).$$

More generally we have
$$L(D^k f_+(t)) = s^k \tilde{f}(s)$$
corresponding to the classical formula
$$L(D^k f_+(t)) = s^k \tilde{f}(s) - \sum_{j=0}^{k-1} s^{k-1-j} D^j f_+(0).$$

Example 5.3.2

(a) From the basic definition it is easy to see that
$$L(\delta) = 1$$
and hence, by application of the above formula, that
$$L(D^k \delta) = s^k.$$

(b) We can generate the Laplace transforms of certain distributions by treating them as the distributional derivatives of regular distributions. For example, we have
$$D(1_+(t)\ln t) = \mathrm{Pf}\frac{1_+(t)}{t};$$
hence if we can find the Laplace transform of the regular distribution $1_+(t)\ln t$ we can find the transforms for the family of pseudofunctions $\mathrm{Pf}(1_+(t)/t^k)$.

Now
$$L(1_+(t)\ln t) = \int_0^\infty \ln t\, e^{-st}\, dt$$
$$= \frac{1}{s}\int_0^\infty e^{-\eta}(\ln \eta - \ln s)\, d\eta$$
$$= -\frac{1}{s}(\gamma + \ln s),$$
where
$$\gamma = -\int_0^\infty e^{-\eta}\ln \eta\, d\eta = 0.5772\ldots$$
is Euler's constant. Hence
$$L\left(\mathrm{Pf}\frac{1_+(t)}{t}\right) = -(\gamma + \ln s)$$
and (Example 5.2.1(b)(ii))
$$L\left(\mathrm{Pf}\frac{1_+(t)}{t^2}\right) = L\left(-D\,\mathrm{Pf}\frac{1_+(t)}{t} - D\delta(t)\right)$$
$$= s(\ln s + \gamma - 1).$$

(c) The existence of the ordinary inverse $f(t)$ of the transform $\tilde{f}(s)$ is closely bound up with the behaviour of $\tilde{f}(s)$ as $|s| \to \infty$. If $|\tilde{f}(s)| = O(|s|^\lambda)$ as $|s| \to \infty$, where $\lambda \leq -1$, then $\tilde{f}(s)$ is the Laplace transform of a function $f(t)$ which vanishes for $t < 0$. The conditions for the existence of $f(t)$ also allow the evaluation of the inversion integral by contour integration [134]. If $\lambda > -1$ then, as is shown below, $f(t)$ is a distribution with support bounded on the left at $t = 0$. By means of distributional differentiation we can effectively use the familiar contour integral to find distributional inverses in the following way.

Suppose

$$|\tilde{f}(s)| \leq C|s|^\lambda \quad \text{as} \quad |s| \to \infty,$$

where $\lambda > -1$. Now let

$$\tilde{f}(s) = s^{m+1}\tilde{g}(s),$$

where m is an integer greater than or equal to λ; the inverse $g(t)$ of $\tilde{g}(s)$ then exists as a regular distribution and

$$f(t) = D^{m+1}g(t).$$

We give an example of the application of this result in Example 5.5.3.

Using a line of reasoning similar to that which we used for the Fourier transform it is not too difficult to show that if $f, g \in \mathcal{D}_R^*$ then $f*g$, their convolution, is an element of $\mathcal{D}_R^*$ and, furthermore,

$$L(f*g) = \tilde{f}(s)\tilde{g}(s).$$

Example 5.3.3 We have already pointed out in Example 5.2.2(g) that the elements of $\mathcal{D}_R^*$ can be associated with a convolution algebra. Applying the Laplace transform to the convolution equation

$$f*x = g$$

we obtain

$$\tilde{f}\tilde{x} = \tilde{g},$$

giving

$$\tilde{x}(s) = \frac{\tilde{g}(s)}{\tilde{f}(s)}$$

or

$$x = f^{*-1}*g$$

where

$$L(f^{*-1}) = \frac{1}{\tilde{f}(s)},$$

provided $1/\tilde{f}(s)$ is the Laplace transform of a right-sided distribution. For the classical case of the stationary differential operator

$$f \equiv \sum_{k=0}^{n} a_k D^k \delta,$$

we have

$$L(f^{*-1}) = \frac{1}{\sum_{k=0}^{n} a_k s^k},$$

which can be expressed in terms of (real or complex) partial fractions involving the roots of the characteristic polynomial

$$\sum_{k=0}^{n} a_k s^k = 0.$$

The convolution inverse $f^{*-1} = 1_+(t)i(t)$ is then a sum of exponential terms or exponential terms multiplied by polynomials in t when repeated roots occur.

Exercises 5.3

1. (a) Let the function $f(t)$ and its derivatives $D^k f(t)$ be elements of $\mathscr{L}_1(-\infty, \infty)$. Use the classical result (p. 448)

    $$F(D^k f) = (2\pi i \omega)^k \tilde{f}(\omega)$$

    to show that $\tilde{f}(\omega) \to 0$ as $\omega \to \infty$ faster than $1/\omega^k$. Hence show that the Fourier transform of a function of rapid descent is a function of rapid descent. Deduce that the inverse Fourier transform of a function of rapid descent is also a function of rapid descent and hence conclude that $F: \mathscr{S} \to \mathscr{S}$ is one-to-one. (For a proof of the continuity of F, see, for example, [127].)

   (b) Using the fact that $F: \mathscr{S} \to \mathscr{S}$ is continuous prove that $\tilde{f}$ is a continuous linear functional on $\mathscr{S}$ (i.e., a distribution of slow growth). In Definition 5.3.2 put $F(f) = g$ to give

    $$F^{-1} g(\phi) = g(F^{-1} \phi).$$

    Now show that, in the sense of distributions, $F^{-1}(Ff) = f$, and hence deduce that $F: \mathscr{S}^* \to \mathscr{S}^*$ is one-to-one.

2. (a) Show directly that $F(\delta) = 1$.
   (b) Show that, for $f \in \mathscr{S}^*$,
      (i) $F(Ff(t)) = f(-t)$ (hint: $\tilde{\tilde{\phi}}(t) = \phi(-t)$);
      (ii) $F(f(t-\tau)) = \tilde{f}(\omega) e^{-2\pi i \omega \tau}$;

(iii) $F(f(t)e^{-2\pi i \tau t}) = \tilde{f}(\omega + \tau);$
(iv) $F(f(\alpha t)) = (1/|\alpha|)\tilde{f}(\omega/\alpha), \alpha \neq 0.$

(c) (i) Show that

$$\widetilde{\operatorname{sgn} t}(\phi) = 2 \int_0^\infty \tilde{\phi}_0(\omega) d\omega,$$

where $\tilde{\phi}_0(\omega)$ is the odd component of $\tilde{\phi}$.

(ii) Now show that

$$\int_0^\infty \tilde{\phi}_0(\omega) d\omega = \lim_{T \to \infty} \int_{-\infty}^\infty \phi(\omega) \frac{1 - \cos 2\pi i \omega T}{2\pi i \omega} d\omega$$

$$= \int_{-\infty}^\infty \frac{\phi(\omega)}{2\pi i \omega} d\omega.$$

(Hint: look at result concerning the sequence $\{\sin nt\}$ on p. 429.)

(iii) Thence deduce from Exercise 5.2,3(c)(iv)

$$\widetilde{\operatorname{sgn} t} = \operatorname{Pv} \frac{1}{\pi i \omega} \quad \text{and} \quad \widetilde{1_+(t)} = \frac{1}{2}\delta(\omega) + \operatorname{Pv} \frac{1}{2\pi i \omega}.$$

(d) Show that

$$\operatorname{Pf} \widetilde{\frac{1}{t^n}} = -\pi i \frac{(-2\pi i \omega)^{n-1}}{(n-1)!} \operatorname{sgn} \omega$$

(Hint: use Exercise 5.3,2(b)(i) and the above result for sgn t.)

(e) Show that, if $f \in \mathcal{S}^*$,

$$\widetilde{t^n f(t)} = (-2\pi i)^{-n} D^n \tilde{f}(\omega)$$

and use this result to find all possible solutions (in $\mathcal{S}^*$) of the equation

$$t^2 f(t) = 1.$$

(f) An ordinary function is periodic if $f(t + T) = f(t)$. We can define a *periodic distribution* $f \in \mathcal{S}^*$ by the property

$$f(\phi(t)) = f(\phi(t - T)).$$

Show that this relation implies that

$$\tilde{f}(\omega) = \tilde{f}(\omega) e^{2\pi i \omega T}$$

and consequently that $\tilde{f}(\omega)$ is a solution of the equation $(1 - e^{2\pi i \omega T})\tilde{f}(\omega) = 0$. Thence deduce that

$$\tilde{f}(\omega) = \sum_{n=-\infty}^\infty \alpha_n \delta\left(\omega - \frac{n}{T}\right);$$

that is to say, the Fourier transform of a periodic distribution is a row of equally spaced delta distributions.

3. (i) Show that
$$L(\delta(\alpha t + \beta)) = \frac{1}{|\alpha|} e^{s\beta/\alpha}.$$

(ii) Show that
$$L\left(\text{Pf}\frac{1_+(t)\sin \alpha t}{t^2}\right) = \frac{i}{2}\{(s+i\alpha)\ln(s+i\alpha) - (s-i\alpha)\ln(s-i\alpha)\}$$
$$+ \alpha(1-\gamma),$$

where γ is Euler's constant (Example 5.3.2(b)).

(iii) Show that
$$L\left(\text{Pf}\frac{1_+(t)\cosh \alpha t}{t}\right) = -\frac{1}{2}\ln(s^2 - \alpha^2) - \gamma.$$

(iv) Show that if $f(t+T) = f(t), t > 0$, then
$$\tilde{f}(s) = \frac{1}{1-e^{-Ts}} \int_0^T e^{-st} f(t)\,dt.$$

Hence find the Laplace transform of the 'square wave' function
$$f(t) = \begin{cases} 1, & t \in (2n\alpha, (2n+1)\alpha) \\ -1, & t \in ((2n+1)\alpha, (2n+2)\alpha) \end{cases} \quad n = 0, 1, 2 \ldots.$$

thence deduce that
$$L\left(\delta(t) + 2\sum_{k=1}^{\infty} (-1)^k \delta(t-\alpha k)\right) = \tanh\frac{\alpha s}{2}.$$

(For a table of distributional Laplace transforms, see [128].)

5.4 Sobolev spaces

We have often had occasion to mention vector spaces in which the norm includes derivatives of the member functions in addition to the functions themselves. The space of m-times continuously differentiable functions on $[0, 1]$ with the uniform norm (Example 3.3.1(d)),

$$\|x\|_{\mathscr{C}(m)} = \max_{t \in [0,1]} \sum_{k=0}^{m} \left|\frac{d^k x}{dt^k}\right|$$

is complete, but if we norm this space with, say, the mean square norm

$$\|x\|_2^{(m)} = \left\{ \int_0^1 \sum_{k=0}^m \left|\frac{d^k x}{dt^k}\right|^2 dt \right\}^{1/2}$$

then we find that we do not have a complete normed space, since the limit of a sequence of differentiable functions need not converge, in this norm, to a function which is differentiable or even continuous. The situation is exactly analogous to that described in Example 3.2.4(b), where we showed that the space of continuous functions in $[0, 1]$ with norm $\|\cdot\|_1$, is not complete.

The situation is not improved by starting with functions in $\mathscr{L}_2[0, 1]$ because differentiation is not a continuous operator in this space. For example, the functions $x^{(n)}(t) = t^n$ converge to the zero function in $\mathscr{L}_2[0, 1]$ since

$$\lim_n \|x^{(n)}\| = \lim_n \left(\frac{1}{2n+1}\right)^{1/2} = 0,$$

whereas

$$\lim_n \left\|\frac{dx^{(n)}}{dt}\right\| = \lim_n \left(\frac{n^2}{2n-1}\right)^{1/2}$$

does not exist.

To overcome this difficulty we proceed (as we have so often done before) to add 'ideal elements' to the space to make it complete. In this way we arrive at the so-called *Sobolev spaces* which prove to be of great usefulness and significance in any discussion concerning the existence and nature of solutions of partial differential operator equations. Now in contrast to the ordinary derivative operator the distributional derivative *is* a continuous operator on the space of distributions, so it appears possible that one might be able to obtain complete spaces of the type we are discussing if we were prepared to admit distributions as ideal elements. Let us now fill in some detail in this sketchy preamble. We shall carry the discussion through entirely in terms of the $\mathscr{L}_2$ norm, since this leads, naturally enough, to a Hilbert space structure. We shall also, for convenience, restrict ourselves to real spaces.

We shall need to extend our concept of distributions to an n-dimensional space. By the space $\mathscr{D}$ of testing functions we shall now understand the collection of functions $\phi(t)$ of the vector variable $t = (t_1, t_2, \ldots, t_n)$ which vanish outside bounded intervals of $\mathscr{R}_n$ and for which partial derivatives of all orders exist for all t. The concept of convergence in $\mathscr{D}$ is analogous to that for the one-dimensional case, namely $\phi^{(n)} \underset{\mathscr{D}}{\to} \phi$ if the supports of the $\phi^{(n)}$ are contained in some fixed finite interval and if $\phi^{(n)}$ and each of its partial derivatives converges uniformly to ϕ and the derivatives of ϕ, respectively. We denote the space of continuous linear functionals on $\mathscr{D}$ by $\mathscr{D}^*$, as before, and refer to these functionals as distributions on $\mathscr{R}_n$. However, in most applications we are interested not in the whole of $\mathscr{R}_n$ but in some open set $\Gamma \subset \mathscr{R}_n$ with the testing functions ϕ having compact support in Γ (not all have support in the *same* compact set): we then denote the space of testing functions by $\mathscr{D}(\Gamma)$ and refer to the elements of the dual $\mathscr{D}^*(\Gamma)$ as distributions on Γ.[†]

If $f \in \mathscr{D}^*(\Gamma)$ its (partial) derivative $D_j f$ is defined by (cf., Definition 5.2.2)

$$D_j f(\phi) = -f(D_j \phi).$$

Differentiation is a continuous linear operator in $\mathscr{D}^*(\Gamma)$. Furthermore, the order of successive differentiations is always immaterial, in contrast to ordinary partial differentiation. Because of this we can introduce the following shorthand notation for (distributional) partial derivatives on $\mathscr{R}_n$ (cf., Example 3.6.1(f)). Let $k = (k_1, k_2, \ldots, k_n)$ be a vector with integral components and let $|k| = \sum_{i=1}^n k_i$: then we define the symbol D^k to mean (without ambiguity)

$$D^k \equiv \frac{\partial^{|k|}}{\partial t_1^{k_1} \partial t_2^{k_2} \ldots \partial t_n^{k_n}}$$

and distributional differentiation in general is formally defined exactly as in Definition 5.2.2 (Exercise 5.4,1).

Let $\mathscr{H}^{(m)}(\Gamma)$ be the space of distributions $x \in \mathscr{D}^*(\Gamma)$ such that $D^k x \in \mathscr{L}_2(\Gamma)$ for all $|k| \leq m$. We provide $\mathscr{H}^{(m)}(\Gamma)$ with the norm

$$\|x\|^{(m)} = \left(\sum_{|k| \leq m} \|D^k x\|_2^2 \right)^{1/2} = \left(\sum_{|k| \leq m} \int_\Gamma (D^k x)^2 \, d\Gamma \right)^{1/2}$$

and with the (compatible) inner product

$$\langle x, y \rangle^{(m)} = \sum_{|k| \leq m} \langle D^k x, D^k y \rangle_{\mathscr{L}_2(\Gamma)} = \sum_{|k| \leq m} \int_\Gamma D^k x D^k y \, d\Gamma,$$

where the summations are over all possible combinations of indices such that $0 \leq |k| \leq m$.

[†] Unfortunately this accepted notation makes $\mathscr{D}(\Gamma)$ look like the 'domain of an operator Γ'; the context should always make the meaning clear.

Some words of explanation are in order here since the situation is not altogether straightforward. We know already that if we do not allow distributional differentiation then we cannot hope to show that $\mathscr{H}^{(m)}(\Gamma)$ is complete, so in these expressions D^k must be interpreted in this way. But we also know that we cannot, in general, multiply distributions, so what are we to make of the statement $D^k x \in \mathscr{L}_2(\Gamma)$?

Roughly speaking, it means that we are dealing here with those *regular* distributions whose associated functions are square integrable. As usual, we define our operations on functions and by constructing Cauchy sequences of them pass to equivalence classes of ideals which complete the space. We did exactly this in constructing the energy space $\mathscr{H}^T$ for the Rayleigh–Ritz method (*see* p. 330).

But how can we actually tell whether a given distribution $g \in \mathscr{D}^*(\Gamma)$ is in $\mathscr{L}_2(\Gamma)$ or not? Let us recall that g is a continuous, and therefore bounded, functional on $\mathscr{D}(\Gamma)$. If we impose the norm $\|\cdot\|_2$ on $\mathscr{D}(\Gamma)$, then by the definition of $\mathscr{D}^*(\Gamma)$ (Definition 3.5.1)

$$\|g\| = \sup_{\phi \in \mathscr{D}(\Gamma)} \frac{|g(\phi)|}{\|\phi\|_2}.$$

Suppose we take $\Gamma = (-1, 1)$, then is, for example, the delta functional, $\delta \in \mathscr{L}_2(\Gamma) = \mathscr{H}^{(0)}(\Gamma)$? We have,

$$\|\delta\| = \sup_{\phi \in \mathscr{D}(\Gamma)} \frac{|\phi(0)|}{\|\phi\|_2}$$

and a simple application of the integral mean value theorem shows that one can always find a $\phi \in \mathscr{D}(\Gamma)$ such that $|\phi(0)| > M\|\phi\|_2$ for any given M, hence $\delta \notin \mathscr{H}^{(0)}(\Gamma)$. The unit step functional 1_+ is in $\mathscr{H}^{(0)}(\Gamma)$, but not in $\mathscr{H}^{(1)}(\Gamma)$ since $D 1_+ = \delta \notin \mathscr{H}^{(0)}(\Gamma)$.

With this understanding let us show that $\mathscr{H}^{(m)}(\Gamma)$ is complete and is therefore a Hilbert space. If $x^{(r)}$ is a Cauchy sequence in the norm $\|\cdot\|^{(m)}$ then $D^k x^{(r)}$ is a Cauchy sequence in $\mathscr{L}_2(\Gamma)$ for every k with $|k| \leq m$ and since $\mathscr{L}_2(\Gamma)$ is complete $D^k x^{(r)} \to y_k$, say.

Now we may deduce that $\mathscr{D}(\Gamma) \subset \mathscr{L}_2(\Gamma) \subset \mathscr{D}^*(\Gamma)$ so that if $x^{(r)} \to y_0$ in $\mathscr{L}_2(\Gamma)$ then also $x^{(r)} \to y_0$ in $\mathscr{D}^*(\Gamma)$ and, because the derivative operator is continuous on $\mathscr{D}^*(\Gamma)$, $D^k x^{(r)} \to D^k y_0$ in $\mathscr{D}^*(\Gamma)$.

Therefore $y_k = D^k y_0$ for every k with $|k| \leq m$ and the distributional derivative of the limit y_0 is the limit distribution y_k.

It is clear from the definition of $\mathscr{H}^{(m)}(\Gamma)$ that if $m_2 > m_1 > 0$ then

$$\mathscr{H}^{(m_2)}(\Gamma) \subset \mathscr{H}^{(m_1)}(\Gamma) \subset \mathscr{H}^{(0)}(\Gamma) = \mathscr{L}_2(\Gamma).$$

What of the singular distributions which are not in $\mathscr{H}^{(m)}(\Gamma)$?

Since $\mathscr{H}^{(m)}(\Gamma) \subset \mathscr{L}_2(\Gamma) \subset \mathscr{D}^*(\Gamma)$ we should be able to accommodate these as linear functionals on $\mathscr{H}^{(m)}(\Gamma)$ and they will then be endowed with the induced dual norm. The obvious way to do this is to deal with linear functionals on Cauchy sequences in $\mathscr{D}(\Gamma)$ and so define, by the usual limit process, the linear functionals on $\mathscr{H}^{(m)}(\Gamma)$.

Unfortunately there is one difficulty, the space $\mathscr{D}(\Gamma)$ is not dense in $\mathscr{H}^{(m)}(\Gamma)$.[†] However, we can effectively do what we wish by simply defining a space $\mathscr{H}_0^{(m)}(\Gamma)$ which is the completion of $\mathscr{D}(\Gamma)$ in the norm $\|\cdot\|^{(m)}$. The dual of $\mathscr{H}_0^{(m)}(\Gamma)$, namely $\mathscr{H}_0^{(m)*}(\Gamma)$, the space of distributions on $\mathscr{H}_0^{(m)}(\Gamma)$, is dense in $\mathscr{D}^*(\Gamma)$, so we have the inclusions

$$\mathscr{D}(\Gamma) \subset \mathscr{H}_0^{(m)}(\Gamma) \subset \mathscr{H}^{(0)}(\Gamma) = \mathscr{L}_2(\Gamma) \subset \mathscr{H}_0^{(m)*}(\Gamma) \subset \mathscr{D}^*(\Gamma).$$

Now for $m > 0$ we have $\mathscr{H}_0^{(m_2)}(\Gamma) \subset \mathscr{H}_0^{(m_1)}(\Gamma)$ if $m_1 < m_2$ and we could, formally at least, extend this inclusion to 'negative integer values m' if we were to write

$$\mathscr{H}_0^{(m)*}(\Gamma) = \mathscr{H}^{(-m)}(\Gamma), \qquad m > 0.$$

But there is a much more fundamental reason for adopting this notation for the dual of $\mathscr{H}_0^{(m)}$. Let us recall the remark made at the end of Example 5.2.1(b), namely that singular distributions can be generated by differentiating regular distributions a finite number of times. There is a *structure theorem* for $\mathscr{H}_0^{(m)*}(\Gamma) = \mathscr{H}^{(-m)}(\Gamma)$ which embodies this result and tells us exactly what the elements of this space 'look like'. The result is [119]: for integer $m > 0$, every $x \in \mathscr{H}^{(-m)}(\Gamma)$ may be represented (but not uniquely) by

$$x = \sum_{|k| \leq m} D^k x_k, \qquad x_k \in \mathscr{L}_2(\Gamma).$$

Hence the elements of $\mathscr{H}^{(-m)}(\Gamma)$ are those distributions whose mth-order primitives (Exercise 5.2,6) are in $\mathscr{L}_2(\Gamma)$.

For example, with $\Gamma = (-1, 1)$, we would expect δ to be an element of $\mathscr{H}^{(-1)}(-1, 1)$ since it may be generated from the regular distribution $1_+ \in \mathscr{H}^{(0)}(\Gamma)$ by a single differentiation. We can check this directly by computing the norm of δ in $\mathscr{H}^{(-1)}(\Gamma) = \mathscr{H}_0^{(1)*}(\Gamma)$; thus

$$\|\delta\|^{(-1)} = \sup_{x \in \mathscr{H}_0^{(1)}(\Gamma)} \frac{|x(0)|}{\|x\|^{(1)}}$$

and we can show that the right-hand side is bounded (Exercise 5.4,3(a)).

[†] It is not at all easy to show this or the results immediately following and requires much more background than is available to us in this book—*see*, for example, [119, 138, 144].

There is an easier way of characterizing $\mathcal{H}_0^{(m)}(\Gamma)$ when Γ is bounded and the boundary $\partial\Gamma$ is sufficiently smooth; namely, $x \in \mathcal{H}_0^{(m)}(\Gamma)$ is equivalent to

$$x \in \mathcal{H}^{(m)}(\Gamma) \quad \text{with} \quad D_\nu^j x = 0, \qquad 0 \leqslant j \leqslant m-1,$$

where D_ν is the normal derivative on $\partial\Gamma$ [119]. This characterization of $\mathcal{H}_0^{(m)}(\Gamma)$ is clearly of particular importance in the application of Sobolev spaces to partial differential equations of elliptic type; it also makes sense of the suffix 0 on $\mathcal{H}_0^{(m)}(\Gamma)$. The boundary conditions are understood to be satisfied in a distributional sense on $\partial\Gamma$.

It is also possible to show that the semi-norm

$$|x|^{(m)} = \left(\sup_{|k|=m} \|D^k x\|_2^2 \right)^{1/2}$$

for $\mathcal{H}^{(m)}$ is a norm for $\mathcal{H}_0^{(m)}$ which is equivalent to $\|x\|^{(m)}$ (Exercise 4.4,2). In order to discuss these two results properly it would be necessary to consider the behaviour of the elements of $\mathcal{H}^{(m)}(\Gamma)$ on the boundary $\partial\Gamma$, and this would require the setting up of Sobolev spaces $\mathcal{H}^{(p)}(\partial\Gamma)$ together with the determination of the relationship between $\mathcal{H}^{(m)}(\Gamma)$ and $\mathcal{H}^{(p)}(\partial\Gamma)$.[†]

Another type of result is available for Sobolev spaces, the so-called embedding theorem. We have already remarked in Section 4.2, p. 284 that, for a single variable, the existence of the derivative implies continuity, whereas for several variables this is not the case. Thus (Exercise 5.4,3(b)), with $\Gamma = (-1, 1)$, if $x \in \mathcal{H}_0^{(1)}(\Gamma)$ then we can assert that $x \in \mathcal{C}(\Gamma)$, but if $x \in \mathcal{H}_0^{(1)}(\Gamma \times \Gamma)$ we cannot conclude continuity of x. And yet one suspects that if, for a given underlying space $\mathcal{R}_n$, derivatives of sufficiently high order are square integrable, then this might be strong enough to imply continuity of the function and perhaps continuity of some of its derivatives. This is the import of *Sobolev's embedding theorem*: if $x \in \mathcal{H}^{(m)}(\Gamma)$ where Γ is an open set in $\mathcal{R}_n$ then $x \in \mathcal{C}^{(k)}(\Gamma)$ provided $m > n/2 + k$ and (*Sobolev's inequality*)

$$\|x\|_{\mathcal{C}^{(k)}} \leqslant M \|x\|^{(m)}.$$

The embedding operator, that is the operator which associates $x \in \mathcal{H}^{(m)}$ with $x \in \mathcal{C}^{(k)}$, is compact (Definition 3.4.3).

[†] Results of this type are known as *trace theorems*: roughly speaking if $x \in \mathcal{H}^{(m)}(\Gamma)$ then, on $\partial\Gamma, x$ possesses a smoothness consistent with being an element in $\mathcal{H}^{(m-1/2)}(\partial\Gamma)$. Since we have not given any account of Sobolev spaces of fractional order we are unable to appreciate this result [119, 136, 144]. Note that the question of boundary values cannot be pursued in $\mathcal{H}^{(m)}(\Gamma)$ since the boundary $\partial\Gamma$ is a set of measure zero in $\mathcal{R}_n$.

Sobolev's inequality allows us, for example, to decide if the delta functional δ on $\Gamma \subset \mathcal{R}_n$ is in $\mathcal{H}^{(-m)}(\Gamma)$. We have (cf., p. 463)

$$\|\delta\|^{(-m)} = \sup_{x \in \mathcal{H}_0^{(m)}(\Gamma)} \frac{|x(0)|}{\|x\|^{(m)}},$$

which is bounded provided $m > n/2$. Hence in two *and* three dimensions $\delta \notin \mathcal{H}^{(-1)}$.

Let us very briefly illustrate the role that Sobolev spaces and, in particular, the embedding theorem play in the abstract theory of elliptic partial differential equations. Consider the (Dirichlet) boundary-value problem

$$Lx = y \quad \text{in} \quad \Gamma, \quad x = 0 \quad \text{on} \quad \partial \Gamma,$$

where Γ is a bounded open set in $\mathcal{R}_n$ with a sufficiently smooth boundary $\partial \Gamma$ and where

$$Lx = \sum_{i=1}^{n} D_i(D_i x)$$

is the Laplacian in n dimensions. The central problem of the abstract theory then is: 'What restrictions must be placed on y in order that the problem is uniquely soluble and what is the character of the solution?'. An associated question is: 'When is the equation solved in the classical sense and when in a distributional (or weak) sense?'. We have the following result: if $y \in \mathcal{H}^{(m-2)}(\Gamma)$ then there exists a unique solution $x \in \mathcal{H}^{(m)}(\Gamma)$ with $x = 0$ on $\partial \Gamma$. If y is smooth enough in Γ, then the solution will possess the required number of partial derivatives and the solution will be classical: for example, from the embedding theorem we have $x \in \mathscr{C}^{(2)}(\Gamma)$ provided $m - 2 > n/2$.[†] We also have the following inequality relating the norms of the 'data' y and the solution x,

$$\|x\|^{(m)} \leqslant M \|y\|^{(m-2)}.$$

More generally, for the (strongly) elliptic operator of order $2r$ defined on $\Gamma \subset \mathcal{R}_n$,

$$Lx = \sum_{|i|,|j| \leqslant r} (-1)^{|i|} D^i(\alpha_{ij}(t) D^j x)$$

we have $x \in \mathcal{H}^{(m)}$ if $y \in \mathcal{H}^{(m-2r)}$ and $\|x\|^{(m)} \leqslant M \|y\|^{(m-2r)}$.

It is interesting to note that the proof of these results depends on showing that there exists a bilinear coercive form (Exercise 5.4,3(d))

[†] This result can be extended to include the smoothness of boundary data [119].

corresponding to the elliptic operator L in the same way as we showed in Chapter 4 that every positive definite operator defines an energy norm: it can then be shown that the energy norm and the appropriate Sobolev norms are equivalent.

Sobolev spaces and their associated norms also play an important role in the abstract theory of evolutionary equations of the type

$$\frac{\partial x}{\partial t} = Lx + y,$$

where t is a time-like variable defined on an interval of the real line (*see*, for example, [119, 136]). The natural way to approach the solution of this type of problem is to approximate spatially using finite elements and temporally by using finite difference schemes. Error estimates for this type of computation are to be found, for example, in [117, 144].

Exercises 5.4

1. (a) Let $f \in \mathscr{C}^{(m)}(R_n)$ and let $|k| \leq m$ (p. 461): show that the formal definition of distributional differentiation, viz.,

$$D^k f(\phi) = (-1)^{|k|} f(D^k \phi),$$

coincides with ordinary partial differentiation in this case.

(b) Let the unit step distribution in $\mathscr{R}_n$ be defined by

$$1_+(t) = \begin{cases} 1, & \text{all } t_i > 0, \\ 0, & \text{otherwise.} \end{cases}$$

If

$$r(t) = \begin{cases} t_1 t_2 t_3 \ldots t_n, & \text{all } t_i > 0, \\ 0, & \text{otherwise,} \end{cases}$$

show that

$$D_j r(\phi) = \int_0^\infty \ldots \int_0^\infty \prod_{i \neq j} t_i \phi(t) dt_1 \ldots dt_n$$

and hence deduce that, if $\hat{k} = (1, 1, 1, \ldots, 1)$,

$$D^{\hat{k}} r(t) = 1_+(t).$$

(c) Show that

$$D_j 1_+(\phi) = \int_0^\infty \ldots \int_0^\infty \phi(t_1, t_2, \ldots, 0, \ldots, t_n) dt_1 \ldots dt_n,$$

where the zero is in the jth position, and hence deduce that
$$D^k 1_+(t) = \delta(t),$$
where $\delta(t)$ is the n-dimensional delta functional defined by
$$\delta(\phi) = \phi(0, 0, 0, \ldots, 0).$$

(d) The result of Exercise 5.2,5(b) can be extended to $\mathscr{R}_n$: we consider for simplicity the case for $\mathscr{R}_3$. Let f have continuous derivatives in the region $\Gamma \subset \mathscr{R}_3$ with smooth continuous boundary $\partial \Gamma$. Assume that f vanishes outside $\bar{\Gamma}$ and is discontinuous on $\partial \Gamma$.

Show that the (distributional) partial derivative $D_1 f$ is given by
$$D_1 f(\phi) = \int_\Gamma \frac{\partial f}{\partial t_1} \phi \, d\Gamma - \int_{\partial \Gamma} f \phi \cos(v, t_1) \, dS$$
where v is the outward normal to $\partial \Gamma$ with a similar result for $D_2 f$ and $D_3 f$.

Let $\delta(\partial \Gamma)$ be a delta functional with support the two-dimensional manifold (or surface) $\partial \Gamma$ (*see*, for example, [129]) defined by
$$\delta(\partial \Gamma)(\phi) = \int_{\partial \Gamma} \phi \, dS.$$

Then deduce that we may write
$$D_1 f = \frac{\partial f}{\partial t_1} - f \cos(v, t_1) \delta(\partial \Gamma).$$

Taking $\phi(t) = 1$ on a finite region containing $\bar{\Gamma}$, show that
$$\int_\Gamma \frac{\partial f}{\partial t_1} d\Gamma = \int_{\partial \Gamma} f \cos(v, t_1) \, ds$$
and hence deduce the divergence theorem for the vector-valued function $f_i(t)$, viz.,
$$\sum_{i=1}^3 \int_\Gamma \frac{\partial f_i}{\partial t_i} d\Gamma = \sum_{i=1}^3 \int_{\partial \Gamma} f_i \cos(v, t_i) \, dS.$$

2. The Fourier transform of the function $f(t) \in \mathscr{L}_1(\mathscr{E}_n)$ is given by (cf., p. 448)
$$\tilde{f}(s) = F(f) = \int_{-\infty}^{\infty} f(t) \exp(-2\pi i \langle s, t \rangle) \, dt, \qquad s \in \mathscr{E}_n,$$
which is to be interpreted as
$$\tilde{f}(s_1, s_2, \ldots, s_n)$$
$$= \int_{-\infty}^{\infty} \ldots \int_{-\infty}^{\infty} f(t_1, t_2, \ldots, t_n) e^{-2\pi i (s_1 t_1 + s_2 t_2 + \ldots + s_n t_n)} \, dt_1 \ldots dt_n.$$

Here we replace the conventional scalar variable ω by the vector variable s. In $\mathscr{E}_n$ the Fourier transform relates a position vector t to a wavenumber vector s, whereas in $\mathscr{R}$ the common usage relates time to a circular frequency ω (see Example 5.5.2).

(a) Show that (cf., Exercise 5.3,2(b)), for $\tau \in \mathscr{E}_n$,
 (i) $F(f(t-\tau)) = \tilde{f}(s) e^{-2\pi i \langle s, \tau \rangle}$;
 (ii) $F(f(t) e^{-2\pi i \langle \tau, t \rangle}) = \tilde{f}(s+\tau)$;
 (iii) $F(f(\alpha_1 t_1, \alpha_2 t_2, \ldots, \alpha_n t_n)) = \dfrac{1}{|\alpha_1 \alpha_2 \cdots \alpha_n|} \tilde{f}\!\left(\dfrac{s_1}{\alpha_1}, \dfrac{s_2}{\alpha_2}, \ldots, \dfrac{s_n}{\alpha_n}\right)$.

(b) The space of testing functions $\mathscr{S}(\mathscr{E}_n)$ consists of functions which are infinitely smooth in each variable and are such that $|t|^m D^k \phi(t) \to 0$ for all vectors $m, k \in \mathscr{E}_n$ having integral components, where $|t|^m = t_1^{m_1} t_2^{m_2} \cdots t_n^{m_n}$.
 (i) Repeat (a) when $f, \tilde{f}$ are distributions.
 (ii) Show that (cf., Exercise 5.3,2(e))
$$\widetilde{|t|^n f(t)} = (-2\pi i)^{-|n|} D^n \tilde{f}(s), \qquad |n| = \sum_{i=1}^{n} n_i.$$

 (iii) Show that, if we make the change of coordinates
$$t = (t_1, t_2, t_3) \to \hat{t} = (t_1 - \alpha_2 t_2 - \alpha_3 t_3, t_2, t_3),$$
then
$$\widetilde{f(\hat{t})} = \tilde{f}(\hat{s}),$$
where $\hat{s} = (s_1, s_2 + \alpha_2 s_1, s_3 + \alpha_3 s_1)$.

(c) (Radial functions) A distribution $f(t) \in \mathscr{S}^*(\mathscr{E}_n)$ is said to be *radial* if its value depends only on the Euclidean norm of t, $\|t\| = \langle t, t \rangle_{\mathscr{E}_n}^{1/2}$. In that case the Fourier integral is invariant to a rotation of orthogonal axes and it follows that the Fourier transform of f is also radial. It is also possible to give a formula for the Fourier transform of a radial function which does not involve an n-fold integration [47]. We shall quote the result for $\mathscr{R}_3$, which is particularly simple.

Let $f(\|t\|)$ be a radial distribution in $\mathscr{E}_3$ and $\tilde{f}(s)$ its Fourier transform; then
$$\tilde{f}(s) = \dfrac{-\tilde{f}_e'(\|s\|)}{2\pi \|s\|},$$
where $\tilde{f}_e(\|s\|)$ is the (one-dimensional) Fourier transform of the (one-dimensional) function $f(\|t\|)$ extended as an even function into $(-\infty, 0)$ and the prime denotes differentiation with respect to $\|s\|$.

(i) Show that an even solution of the equation

$$(4\pi^2 \|s\|^2 - \alpha^2)\tilde{g}(\|s\|) = -1$$

is (*see* Exercise 5.2,4(d))

$$\tilde{g}(\|s\|) = \frac{-1}{(4\pi^2 \|s\|^2 - \alpha^2)} + \frac{i}{4\pi}\delta\left(\|s\|^2 - \left(\frac{\alpha}{2\pi}\right)^2\right).$$

(ii) Deduce that $F^{-1}(\tilde{g}(\|s\|)) = f_e(\|t\|)$, where f_e is an even function defined, for positive values of its argument, by

$$f_e(\|t\|) = \frac{i}{2\alpha}e^{-i\alpha\|t\|}.$$

(Hint: use Exercise 5.3, 2(c)(iii).)

(iii) Thence deduce that

$$g(t) = F^{-1}(\tilde{g}(s)) = -\frac{e^{-i\alpha\|t\|}}{4\pi\|t\|}$$

satisfies the differential equation

$$(\nabla^2 + \alpha^2)g(t) = \delta(t).$$

3. (a) Verify that $\delta(t) \in \mathcal{H}_0^{(-1)}(\Gamma), \Gamma = (-1, 1)$ by showing that $\|\delta\|^{(-1)}$ is bounded (p. 463). (Hint: use an inequality like that proved in Example 4.4.1, recalling that functions in $\mathcal{H}_0^{(1)}$ vanish outside Γ.)

(b) Infer from the proof of (a) that if $x \in \mathcal{H}^{(1)}(\Gamma), \Gamma = (-1, 1)$, then $x \in \mathcal{C}(\Gamma)$

(c) Let $\Gamma \subset \mathcal{R}_2$: show that the orthogonal complement of $\mathcal{H}_0^{(1)}(\Gamma) \subset \mathcal{H}^{(1)}$ is given by

$$(\mathcal{H}_0^{(1)}(\Gamma))^\perp = \{w \in \mathcal{H}^{(1)} | \nabla^2 w = w\}.$$

This result is formally identical for $\mathcal{R}_n$ when ∇^2 is replaced by the n-dimensional Laplacian (p. 465).

Which vectors, if any, lie in $(\mathcal{H}_0^{(1)}(\Gamma))^\perp$ when $\Gamma = (0, 1) \times (0, 1)$?

(d) Show that the bilinear functional

$$f(x, z) = \int_\Gamma D_1 x D_1 z + D_2 x D_2 z \, d\Gamma$$

is coercive (Exercise 3.7, 3) on $\mathcal{H}_0^{(1)}$. Hence apply the Lax–Milgram theorem (Exercise 3.7, 3) to show that, given $g \in \mathcal{H}_0^{(1)*} = \mathcal{H}^{(-1)}$, there exists $\hat{x} \in \mathcal{H}_0^{(1)}$ such that

$$f(\hat{x}, z) = g(z) \qquad \forall z \in \mathcal{H}_0^{(1)}.$$

This result defines a *weak solution* of $\nabla^2 x = g$ in the space $\mathcal{H}_0^{(1)}$.

We have already seen that the delta functional is not in $\mathcal{H}^{(-1)}(\mathcal{R}_2)$; hence we should expect that the fundamental solution

$\ln\sqrt{t_1^2 + t_2^2}$ of Laplace's equation is not in $\mathscr{H}^{(1)}$: verify that this is the case.

(e) The Friedrichs (or sometimes Poincaré) inequality

$$\int_\Gamma x^2 \, d\Gamma \leq K \int_\Gamma \sum_{j=1}^{2} (D_j x)^2 \, d\Gamma, \qquad x \in \mathscr{H}_0^{(1)}(\Gamma), \quad \Gamma \subset \mathscr{R}_2,$$

is a weak form of one of the Sobolev inequalities [120, 142, 144]. However, a straightforward demonstration of the result can be adduced using an elementary argument along the same lines used to establish the one-dimensional form of the inequality in Example 4.4.1 (p. 328):

(i) Assume that Γ is contained in a rectangle $\Phi = ([\alpha, \beta] \times [\gamma, \delta]) \subset \mathscr{R}_2$.

(ii) Let $\phi \in \mathscr{D}(\Gamma)$ be extended to Φ by taking $\phi = 0$ in $\Phi - \Gamma$.

(iii) Show that

$$\phi^2 \leq (\delta - \gamma) \int_\gamma^\delta (D_2 \phi)^2 \, dt_2$$

$$\leq (\delta - \gamma) \int_\gamma^\delta (D_1 \phi)^2 + (D_2 \phi)^2 \, dt_2.$$

(iv) Now integrate this inequality over Φ. The result is then extended to $\mathscr{H}_0^{(1)}(\Gamma)$ by completion.

5.5 Application examples

Example 5.5.1 In our description of the various forms of projection method in Section 4.5 (pp. 367–369) we pointed out that a collocation solution of the operator equation $Tx = y$ could be interpreted as a form of projection if we could define a functional l_j with the property $l_j(u) = u(t_j)$. Having now dealt with singular distributions we see that the functional we seek is none other than the translated delta functional $\delta(t - t_j)$.

From the point of view of error analysis it is not necessary to adopt this viewpoint (see, for example, [40, 80]) and an alternative approach based on interpolation functions coupled with a numerical quadrature formula is usually employed to obtain error bounds (cf., Exercise 4.5,4). The situation is then somewhat analogous to the error analysis for the finite element method in that the problem reduces to an exercise in approximation theory.

Let us now see how the use of an interpolation basis in conjunction with its dual (or reciprocal) basis is linked to collocation in the projectional sense. As in Section 4.5, p. 365, let $T: \mathscr{V} \to \mathscr{U}$ have domain $\mathscr{D}(T) \subset \mathscr{V}$ and let $\mathscr{U}_n$ be a finite-dimensional subspace of $\mathscr{U}$ with interpolation basis $\{u^i\}$, $i = 1, \ldots, n$, based on a set of points $t_j, j = 1, \ldots, n$. That is to say,

$$u^i(t_j) = \delta_j^i$$

and, for any $u^{(n)} \in \mathcal{U}_n$,

$$u^{(n)} = \sum_{i=1}^{n} u^{(n)}(t_i) u^i.$$

Let $\mathcal{U}_n^* \subset \mathcal{U}^*$ be the dual of $\mathcal{U}_n$ with reciprocal basis $\{u_i\}$, $i = 1, \ldots, n$; by definition (p. 51)

$$u_i(u^j) = \delta_i^j.$$

For any $u^* \in \mathcal{U}^*$ the projection of u^* onto $\mathcal{U}_n^*$ is given by (cf., Example 2.6.5)

$$Q_n u^* = \sum_{i=1}^{n} u^*(u^i) u_i.$$

Now let the closure of $\mathcal{U}^*$ be the space of distributions on $\mathcal{U}$; then we have

$$Q_n \delta(t - \hat{t}) = \sum_{i=1}^{n} u^i(\hat{t}) u_i$$

and in particular, for $\hat{t} = t_j$,

$$Q_n \delta(t - t_j) = u_j$$

and

$$Q_n \delta(t_i - t_j) = u_j(t_i).$$

This result can be interpreted in the following way: the projection of the delta functional with support at $t = t_j$ is the dual basis functional u_j and is thus, in a sense, a smoothed (or regularized) version of $\delta(t - t_j)$. Furthermore, the basis functions u_j have the 'delta-like' property

$$u_j(u) = u(t_j)$$

for any $u \in \mathcal{U}$.

Returning to the projection solution of $Tx = y$ we require (p. 366) the projection, $P_n(Tx^{(n)} - y)$, of $Tx^{(n)} - y$ onto $\mathcal{U}_n$ to vanish, viz.,

$$u_j(Tx^{(n)} - y) = 0, \quad j = 1, \ldots, n,$$

where $x^{(n)} \in \mathcal{V}_n \subset \mathcal{V}$. But from the property of the basis functionals u_j this implies that

$$(Tx^{(n)} - y)_{t=t_j} = 0, \quad j = 1, \ldots, n$$

which is a collocation solution based on the interpolation points t_j.

In practice, the choice of interpolation functions u^j and hence the location of the collocation points t_j has a very significant effect on the accuracy of the approximation, at least when n is small.

The best way to proceed is to derive the interpolation functions from a set of weighted reciprocal polynomials: the collocation points are then the zeros of the chosen set of polynomials. This approach is often referred to as *orthogonal collocation* and was first suggested by Lanczos [139].

We shall illustrate the procedure by seeking a projection solution of the aerofoil integral equation discussed in Example 3.10.3. With reference to our discussion above we shall associate the spaces, $\mathscr{V}$, $\mathscr{U}$ with the spaces $\mathscr{L}_p$ and $\mathscr{L}_q$, respectively (*see* Fig. 3.19). The behaviour of the loading at leading and trailing edges of the aerofoil suggests the use of loading basis functions for $\mathscr{V}$ of the form

$$\sqrt{\frac{1-t}{1+t}} f(t),$$

where $f(t)$ is a smooth bounded function: it is natural to represent $f(t)$ by a polynomial.

With $t = -\cos\alpha$, $\alpha \in [0, \pi]$, it is well known that the Jacobi polynomials [39]

$$g_j(\alpha) = \frac{\cos(2j+1)\alpha/2}{\cos\alpha/2}, \qquad j = 0, 1, 2, \ldots, (n-1),$$

are orthogonal with respect to the weight function

$$\sqrt{\frac{1-t}{1+t}}$$

over the interval $[-1, 1]$ (cf., Example 3.6.5(e)). This suggests the use of the basis functions

$$\sqrt{\frac{1-t}{1+t}} g_j(t), \qquad j = 0, 1, \ldots, (n-1),$$

for $\mathscr{V}_n \subset \mathscr{V}$ with reciprocal basis $g_j(t)$, $j = 0, 1, \ldots, (n-1)$ for $\mathscr{V}_n^* \subset \mathscr{V}^*$.

The interpretation of the dual operator T as referring to a reverse flow suggests the basis $g_j(-t)$, $j = 0, 1, \ldots, (n-1)$ for $\mathscr{U}$ with reciprocal basis

$$\sqrt{\frac{1+t}{1-t}} g_j(-t)$$

for $\mathscr{U}^*$.

We now transform these basis sets into interpolation basis sets. Let $p_j(t)$, $j = 0, \ldots, (n-1)$ be a set of interpolation polynomials whose n nodal points are the zeros of $g_n(t)$, that is, the set of points

$$t_j = -\cos\left(\frac{2j+1}{2n+1}\right)\pi.$$

The $p_j(t)$ are linear combinations of the Jacobi polynomials $g_j(t)$, $j = 0, 1, \ldots, (n-1)$, and we choose these as a basis for $\mathscr{V}_n^*$ with reciprocal basis

$$\frac{1}{\lambda_j} \sqrt{\frac{1-t}{1+t}} p_j(t)$$

for $\mathscr{V}_n$, where

$$\lambda_j = \frac{2\pi}{2n+1}(1-t_j).^\dagger$$

Similarly for $\mathscr{U}_n$, $\mathscr{U}_n^*$ we adopt the basis sets $p_j(-t)$ and

$$\frac{1}{\lambda_j}\sqrt{\frac{1+t}{1-t}}p_j(-t),$$

respectively, where $p_j(-t)$ are interpolation polynomials based on the nodal points

$$\tau_j = -t_j = \cos\left(\frac{2j+1}{2n+1}\right)\pi.$$

The projection form of the aerofoil equation is then

$$\sum_{i=0}^{n-1}\xi_i^{(n)}\left[T\left(\frac{1}{\lambda_i}\sqrt{\frac{1-t}{1+t}}p_i(t)\right)\right]_{t=\tau_j} = w(\tau_j), \quad j=0,\ldots,(n-1).$$

However, the equation can be considerably simplified if we note that

$$T(g_j(t)) = g_j(-t)$$

and the projection equation then reduces to

$$-\frac{1}{\pi}\sum_{i=0}^{n-1}\frac{\xi_i^{(n)}}{t_i-\tau_j} = w(\tau_j), \quad j=0,\ldots,(n-1).$$

There is a direct physical interpretation of this projection. If we replace the continuous vortex sheet representing the aerofoil by a series of line vortices of strength $2\xi_i^{(n)}$ at the points t_i and compute the upwash at the points τ_j using the simple formula for the velocity field due to a line vortex‡ then we arrive directly at the projection equation. This model is a consequence of using interpolation polynomials not only as a basis for $\mathscr{U}_n$ (which would simply lead to collocation on the points τ_j) but also as a basis for $\mathscr{V}_n$.

Clearly if the given downwash $w(t)$ is a polynomial of degree $(n-1)$ or less then the projection solution is exact in the sense that all generalized

---

† It is interesting to note that the λ_j are the weighting numbers in the Gaussian quadrature formula [39]:

$$\int_{-1}^{1}\sqrt{\frac{1-t}{1+t}}f(t)\,dt = \sum_{j=0}^{n-1}\lambda_j f(t_j)$$

‡ The streamlines associated with an isolated line vortex of strength Γ are circles centred at the vortex; the fluid velocity at radius r is given by

$$\frac{\Gamma}{2\pi r}.$$

loadings of the form

$$\int_{-1}^{1} p(t) t^k \, dt$$

with $k \leq n - 1$ are given exactly by treating the loading as being due to a series of line vortices.

Example 5.5.2 (Green's function) Let $T: \mathcal{V} \to \mathcal{U}^*$ be a linear differential operator defined on some region Γ of $\mathcal{R}_n$ with smooth boundary $\partial \Gamma$. Recall from Definition 2.7.1 that the dual (or conjugate) of T is the (differential) operator $T': \mathcal{U} \to \mathcal{V}^*$ defined by

$$T'u(v) = u(Tv) \qquad \forall \begin{cases} v \in \mathcal{D}(T) \subset \mathcal{V}; \\ u \in \mathcal{D}(T') \subset \mathcal{U}: \end{cases}$$

recall also that the boundary values of T specify the (conjugate) boundary values of T'.

Now suppose $\mathcal{V}^*$ is a space of distributions on $\mathcal{V}$ and that we can find $g_\tau \in \mathcal{U}$ such that, in the sense of distributions,

$$T' g_\tau = \delta_\tau,$$

where τ is some interior point of Γ. We can use this function to solve the operator equation $Tx = y$, $x \in \mathcal{D}(T)$, since we have immediately

$$T' g_\tau(x) = \delta_\tau(x) = g_\tau(Tx) = g_\tau(y)$$

or

$$x(\tau) = g_\tau(y).$$

When g_τ is a regular distribution this expression takes the concrete form

$$x(\tau) = \int_\Gamma g(t, \tau) y(t) \, dt.$$

The function $g(t, \tau)$ is called the *Green's function* for the operator T.

Reversing the roles of T and T' we can similarly define the Green's function $h_t \in \mathcal{V}$ for the operator T' such that

$$T h_t = \delta_t,$$

where t is some interior point of Γ.

From the definition of the dual we have

$$\begin{aligned} \delta_\tau(h_t) &= T' g_\tau(h_t) \\ &= g_\tau(T h_t) \\ &= T h_t(g_\tau) = \delta_t(g_\tau) \end{aligned}$$

or

$$h(\tau, t) = g(t, \tau).$$

We have already, in Example 2.5.6 and Exercise 3.4,5 given specific illustrations of Green's functions and pointed out that, formally, the inverse of T is the operator

$$T^{-1}(\cdot) = \int_\Gamma h(\tau, t)\cdot(t)\,dt = \int_\Gamma g(t, \tau)\cdot(t)\,dt.$$

In many applications $\mathscr{V} = \mathscr{U}^* = \mathscr{H}_0^{(m)}$, say, with $\mathscr{V}^* = \mathscr{U} = \mathscr{H}^{(-m)}$ (recall that $\mathscr{H}_0^{(m)} \subset \mathscr{H}^{(-m)}$) and often T is symmetric in the sense that $T = T'$ with $\mathscr{D}(T) \subset \mathscr{D}(T')$. In that case it follows immediately from the foregoing that the Green's function for T (or T') is symmetric in its variables, that is to say,

$$h(t, \tau) = h(\tau, t).$$

If the Green's function is known for a partial differential operator T', then the solution of the homogeneous boundary-value problem $Tx = y$ is available simply through integration. Inhomogeneous boundary values can be accommodated by constructing a suitable coset of the domain space in the usual way or by treating boundary values as point or line distributions on the boundary [46].

Of course, the determination of the Green's function is not at all trivial except for some special regions Γ, and it is often more useful to find instead a so-called *fundamental solution* for the operator T. This is a function f_τ satisfying the (differential) equation

$$Tf_\tau = \delta_\tau$$

but not the boundary conditions. The function f_τ has the correct singularity at the point τ. The Green's function is then of the form

$$g_\tau = f_\tau + f_\tau^0$$

where $f_\tau^0 \in \mathscr{D}(T)$ is a solution of the homogeneous equation $Tf_\tau^0 = \theta$ such that $f_\tau + f_\tau^0$ satisfies the boundary conditions.

There exists a very extensive literature devoted to the exposition and exploitation of Green's functions, much of it based on classical methods which do not employ the notion of distributions [46, 48, 49, 82, 132].

Fundamental solutions can most conveniently be found by using the distributional Fourier transform and we shall now give an example taken from acoustic theory. In this example the pressure, $p(t_1, t_2, t_3, \tau)$, at the space point $t = (t_1, t_2, t_3)$ at time τ is assumed to satisfy the wave equation referred to a coordinate system which moves with steady velocity V in the negative t_1 direction: the wave operator we refer to is [140],

$$Tp \equiv \nabla^2 p \cdot - \frac{1}{a^2}\left(V\frac{\partial}{\partial t_1} + \frac{\partial}{\partial \tau}\right)^2 p,$$

where a is the acoustic speed in the undisturbed fluid and we seek the fundamental solution p satisfying

$$\nabla^2 p - \frac{1}{a^2}\left(V\frac{\partial}{\partial t_1} + \frac{\partial}{\partial \tau}\right)^2 p = \delta(t_1, t_2, t_3, \tau).$$

We take the space–time Fourier transform of this equation, giving

$$4\pi^2\left[-(s_1^2 + s_2^2 + s_3^2) + \frac{1}{a^2}(Vs_1 - \omega)^2\right]\tilde{p} = 1,$$

or

$$4\pi^2 \tilde{p} = \frac{-1}{s^2 - (Vs_1 - \omega)^2/a^2}, \qquad s^2 = s_1^2 + s_2^2 + s_3^2,$$

where (Exercise 5.4,2)

$$\tilde{p} = \int_{-\infty}^{\infty}\int_{-\infty}^{\infty}\int_{-\infty}^{\infty}\int_{-\infty}^{\infty} p(t_1, t_2, t_3, \tau) e^{-2\pi i(t_1 s_1 + t_2 s_2 + t_3 s_3 - \tau\omega)} dt_1\, dt_2\, dt_3\, d\tau.$$

By taking the 'time' Fourier transform of the result in Exercise 5.4,2(c), we may deduce that the Fourier transform of

$$\frac{1}{4\pi}\frac{\delta(\tau - R_1/a)}{R_1}, \qquad R_1^2 = t_1^2 + t_2^2 + t_3^2,$$

is

$$\frac{1}{4\pi^2}\frac{1}{(s^2 - (\omega/a)^2)}$$

and, using the result of Exercise 5.4,2(b)(iii), we find that

$$p(t_1, t_2, t_3, \tau) = -\frac{1}{4\pi}\frac{\delta(\tau - R_{\tau,t}/a)}{R_{\tau,t}},$$

where

$$R_{\tau,t}^2 = (t_1 - V\tau)^2 + t_2^2 + t_3^2.$$

Physically speaking, we have found the pressure field due to a sound source moving with velocity V which emits a pulse at time $\tau = 0$. The pressure is non-zero only on the surface in convected space–time defined by

$$\tau - R_{\tau,t}/a = 0.$$

The roots of this equation are

$$\tau_1 = \frac{-Mt_1 + R}{a\beta^2}, \qquad \tau_2 = \frac{-Mt_1 - R}{a\beta^2}$$

where $M = V/a$ is the Mach number of the source, $\beta^2 = 1 - M^2$ and

$$R^2 = t_1^2 + \beta^2(t_2^2 + t_3^2).$$

Let us examine what the pressure field looks like. We need to distinguish two cases:

(i) $M < 1$ (subsonic) Since $\tau_2 < 0$,

$$p(t_1, t_2, t_3, \tau) = \frac{-1}{4\pi} \frac{\delta(\tau - \tau_1)}{R_{\tau,t}},$$

which means to say that the pressure field consists of a spherical pressure pulse whose centre moves with velocity V and which expands at velocity a; the pressure amplitude is inversely proportional to the instantaneous radius. An observer situated at the point (t_1, t_2, t_3) will observe the pulse at time

$$\tau_1 = \frac{-Mt_1 + R}{a\beta^2}.$$

(ii) $M > 1$ (supersonic) In this case both τ_1 and τ_2 are positive provided $R > 0$, that is, if $t_1^2 > (-\beta^2)(t_2^2 + t_3^2)$. Hence we may write

$$p(t_1, t_2, t_3, \tau) = -\frac{1}{4\pi} \frac{(\delta(\tau - \tau_1) + \delta(\tau - \tau_2))}{R\tau,t} 1_+(t_1 - B\sqrt{t_2^2 + t_3^2}),$$

where

$$B^2 = -\beta^2 > 0.$$

Hence the pressure disturbance is not felt at all by a travelling observer situated outside the cone $t_1 = B\sqrt{t_2^2 + t_3^2}$. An observer within the cone experiences two pressure pulses emanating essentially from the forward- and rearward-going parts of the convecting spherical pulse.

Example 5.5.3 Experiments on the damping in metals suggest that, in an harmonic motion of small amplitude, the damping loss per cycle is, to a large extent, independent of frequency. This is in marked contrast to the case of the familiar viscous damping wherein damping loss is proportional to frequency. There is a simple mathematical representation for the viscous damper namely

$$f_d = d\dot{x},$$

where f_d is the damping force, $x(t)$ the relative linear displacement of the damper and d is a constant. The damping loss per cycle in an harmonic motion $x(t) = \gamma \sin 2\pi\omega t$ is immediately seen to be $2\pi^2\gamma^2\omega d$.

Several models have been suggested, within the context of linear stationary systems, for the so-called *hysteretic damper* in which damping loss per cycle is independent of frequency. For example, either

$$f_d = idx \quad \text{or} \quad f_d = \frac{d}{\omega}x$$

will, for harmonic motion, give the required form of loss. However, these models (and others) are meaningless when applied to a motion which is not simple harmonic.

If we postulate a damper element which is linear and stationary (i.e., invariant to time translation), then the force output $f(t)$ due to the displacement $x(t)$ is most generally given by the convolution,

$$f = g * x,$$

where $g(t)$ is the *force impulse response* of the damper. This convolution can only be replaced by a linear combination of derivatives of $x(t)$ when the Laplace transform of $g(t)$ is a rational function. Whereas a differential expression does not exist to describe the transient motion of a hysteretic damper, we shall show that we are able to suggest several realizable models within the framework of convolution rather than differential equations. In passing, we should note that for the commonly used mechanical elements of mass, viscous damper and spring the corresponding force impulse responses are the singular distributions

$$g_m = mD^2\delta, \qquad g_d = dD\delta, \qquad g_s = k\delta,$$

respectively, where m and k are mass and stiffness parameters.

The convolution inverse $h = g^{*-1}$ of g is the *displacement impulse response* and

$$x = g^{*-1} * f = h * f.$$

Let us seek a 'damping' element whose force impulse response is consistent with the fact that, under harmonic excitation, the quadrature component of the harmonic response is (largely) independent of frequency. Let the force impulse response of the element be $1_+(t)g(t)$: the force transfer function is the Laplace transform of $g(t)$, namely

$$\tilde{g}(s) = \int_0^\infty g(t) e^{-st} dt, \qquad s = 2\pi(\sigma + i\omega),$$

and assuming the element to be stable the frequency response function is

$$\tilde{g}(\omega) = \int_0^\infty g(t) e^{-2\pi i \omega} dt = \tilde{g}_e(\omega) + i\tilde{g}_o(\omega).$$

The real functions $\tilde{g}_e(\omega), \tilde{g}_o(\omega)$ are, respectively, the in-phase and quadrature components of the frequency response function: the suffices indicate that $\tilde{g}_e$ is an even, and $\tilde{g}_o$ an odd function of ω.

Knowing only $\tilde{g}_o(\omega)$ the impulse response can be recovered from

$$g(t) = -1_+(t) \int_{-\infty}^\infty 2\tilde{g}_o(\omega) \sin 2\pi\omega t \, d\omega,$$

but in order that $g(t)$ be causal, $\tilde{g}_o(\omega)$ must be the imaginary part of a function which is analytic when extended into the right half of the complex plane. This means that a specification of $\tilde{g}_o(\omega)$ carries with it a corresponding $\tilde{g}_e(\omega)$.[†] Using the above relation, $g(t)$ is not determined uniquely: a given $\tilde{g}_o(\omega)$ yields an equivalence class of functions whose elements differ by linear combinations of $D^{2m}\delta$ and $D^{-2m}\delta$, the even derivatives and primitives of the delta functional (Exercise 5.2, 6).

The displacement impulse response can be found by applying the Laplace inversion formula to the function $1/\tilde{g}(s)$, but the usual method of contour integration cannot be applied directly since the functions $1/\tilde{g}(s)$ in which we are interested do not usually satisfy the conditions for Jordan's lemma (p. 456). Instead we compute $h(t)$ as the distributional derivative of the *indicial response* (the displacement response to a step force input), viz.,

$$h(t) = \frac{d}{dt}\left(\frac{1}{2\pi i}\int_{\mu-i\infty}^{\mu+i\infty}\frac{e^{st}}{s\tilde{g}(s)}ds\right).$$

In this sense it is convenient to refer to the displacement indicial response as $h^{(-1)}(t)$.

Now suppose we insist that

$$\tilde{g}_o(\omega) = \text{sgn } 2\pi\omega, \quad -\infty < \omega < \infty,$$

that is to say, the quadrature component of our unknown damper element is strictly independent of frequency. The result of Exercise 5.3, 2(d) then implies that

$$g(t) = \frac{-2}{\pi}\text{Pf}\frac{1_+(t)}{t}$$

and the Laplace transform of this (Example 5.3.2(b)) gives, apart from a constant, the transfer function

$$\tilde{g}(s) = \frac{2}{\pi}\ln s$$

so that $\tilde{g}_e(\omega) = (2/\pi)\ln|2\pi\omega|$. Hence, insistence on strict frequency independence of the quadrature component leads to a physically unrealizable damper element because of the singularity of $\tilde{g}_e(\omega)$ at $\omega = 0$.[‡]

By application of the Laplace inversion formula we can show that the

---

[†] The functions $\tilde{g}_e(\omega), \tilde{g}_o(\omega)$ for a causal impulse response function are connected by the Hilbert transform [133]:

$$\tilde{g}_o(\omega) = -\frac{1}{\pi}\int_{-\infty}^{\infty}\frac{\tilde{g}_e(v)dv}{\omega-v}, \quad \tilde{g}_e(\omega) = \frac{1}{\pi}\int_{-\infty}^{\infty}\frac{\tilde{g}_o(v)dv}{\omega-v}.$$

[‡] The unrealizability of this damper is also reflected in the fact that the damping loss per cycle is non-zero at $\omega = 0$.

displacement indicial response is

$$\frac{2}{\pi}h^{(-1)}(t) = 1_+(t)\left(e^t - \int_0^\infty \frac{e^{-\sigma t}d\sigma}{\sigma((\ln\sigma)^2 + \pi^2)}\right).$$

Notice that $h^{(-1)}(t)$ is exponentially unbounded as $t \to \infty$.

The singularity at $\omega = 0$ can be removed by taking

$$\tilde{g}(s) = \frac{2}{\pi}\ln\left(1 + \frac{s}{2\pi\alpha}\right)$$

with α a small positive real number. We now have

$$\tilde{g}_o(\omega) = \frac{2}{\pi}\tan^{-1}\frac{\omega}{\alpha}, \quad \tilde{g}_e(\omega) = \frac{2}{\pi}\ln\sqrt{\frac{\omega^2 + \alpha^2}{\alpha^2}},$$

and $\tilde{g}_o(\omega)$ will be sensibly independent of frequency for all ω greater than about 10α.

The in-phase or stiffness component exhibits a logarithmic singularity as $\omega \to \infty$: the force impulse response is

$$g(t) = -\frac{2}{\pi}\text{Pf}\frac{1_+(t)e^{-2\pi\alpha t}}{t}.$$

The displacement indicial response is given by

$$\frac{2}{\pi}h^{(-1)}(t) = 1_+(t)\left(-1 + 2\pi\alpha t - e^{-2\pi\alpha t}\int_0^\infty \frac{e^{-2\pi\alpha\sigma t}d\sigma}{(1 + \sigma)((\ln\sigma)^2 + \pi^2)}\right),$$

and $h^{(-1)}(t) \sim \pi^2\alpha t$ as $t \to \infty$. This is a reflection of the fact that, for small ω, $\tilde{g}_o(\omega) \sim 2\omega/\pi\alpha$, which is the frequency response of a viscous damper of strength $1/\pi^2\alpha$. This model thus behaves as a viscous damper for $\omega < 10\alpha$ and as a hysteretic damper for all other values.

Further models of increasing sophistication can be developed along the above lines: an extended treatment which also deals with the incorporation of the damping models into mechanical systems can be found in [141].

References

1. M. Kline *et al.*, *Mathematics in the Modern World*, Readings from *Scientific American*, Freeman, San Francisco, 1968.
2. R.G. Bartle, *The Elements of Real Analysis*, Wiley, New York, 1964.
3. J.H. Manheim, *The Genesis of Point Set Topology*, Pergamon Press, Oxford, 1964.
4. B. Noble *et al.*, Papers presented at the symposium on 'Functional analysis in teaching and research', *Bulletin of the Institute of Mathematics and its Applications* **10**, 98–127, 1974.
5. W.W. Sawyer, *A First Look at Numerical Functional Analysis*, Clarendon Press, Oxford, 1978.
6. E. Kreyszig, *Introductory Functional Analysis with Applications*, Wiley, New York, 1978.
7. A.L. Brown and A. Page, *Elements of Functional Analysis*, Van Nostrand-Reinhold, New York, 1970.
8. J.G. Kemeny, H. Mirkil, J.L. Snell, and G.L. Thompson, *Finite Mathematical Structures*, Prentice-Hall, New York, 1959.
9. T.M. Apostol, *Mathematical Analysis*, Addison-Wesley, Reading, Mass., 1963.
10. J. Dieudonné, *Foundations of Modern Analysis*, Academic Press, New York, 1960.
11. D.T. Finkbeiner, *Matrices and Linear Transformations*, Freeman, San Francisco, 1966.
12. G. Birkhoff and S. MacLane, *A Survey of Modern Algebra*, Macmillan, New York, 1965.
13. G.C. Shephard, *Vector Spaces of Finite Dimension*, Oliver and Boyd, Edinburgh, 1966.
14. K. Hoffman and R. Kunze, *Linear Algebra*, Prentice-Hall, New York, 1961.
15. G. Hadley, *Linear Algebra*, Addison-Wesley, Reading, Mass., 1973.
16. A. Mary Tropper, *Linear Algebra*, Nelson, London, 1969.
17. R.A. Frazer, W.J. Duncan and A.R. Collar, *Elementary Matrices*, Cambridge University Press, Cambridge, 1952.
18. E.K. Blum, *Numerical Analysis and Computation: Theory and Practice*, Addison-Wesley, Reading, Mass., 1972.
19. J.H. Wilkinson, *The Algebraic Eigenvalue Problem*, Clarendon Press, Oxford, 1965.
20. F.S. Acton, *Numerical Methods that Work*, Harper International, New York, 1970.

21. C.E. Froberg, *Introduction to Numerical Analysis*, Addison-Wesley, Reading, Mass., 1970.
22. B. Noble, *Numerical Methods*, Vol. I, Oliver and Boyd, Edinburgh, 1964.
23. L.A. Liusternik and V.J. Sobolev, *Elements of Functional Analysis*, Ungar Publishing Co., New York, 1961.
24. C. Goffman and G. Pedrick, *First Course in Functional Analysis*, Prentice-Hall, New York, 1965.
25. G.F. Simmons, *Introduction to Topology and Modern Analysis*, McGraw-Hill, New York, 1963.
26. A.N. Kolmogorov and S.V. Fomin, *Introductory Real Analysis*, Dover, New York, 1970.
27. A.E. Taylor, *Introduction to Functional Analysis*, Wiley, New York, 1958.
28. B.Z. Vulikh, *Functional Analysis for Scientists and Technologists*, Pergamon, Oxford, 1963.
29. A.W. Naylor and G.R. Sell, *Linear Operator Theory in Engineering and Science*, Holt, Rinehart & Winston, New York, 1971.
30. G. Bachman and L. Narici, *Functional Analysis*, Academic Press, New York, 1966.
31. J. Korevaar, *Mathematical Methods*, Vol. I, Academic Press, New York, 1968.
32. L. Collatz, *Functional Analysis and Numerical Mathematics*, Academic Press, New York, 1966.
33. R.F. Curtain and A.J. Pritchard, *Functional Analysis in Modern Applied Mathematics*, Academic Press, New York, 1977.
34. S.G. Mikhlin, *Variational Methods in Mathematical Physics*, Pergamon, Oxford, 1964.
35. L.V. Kantorovich and G.P. Akilov, *Functional Analysis in Normed Spaces*, Pergamon, Oxford, 1964.
36. D.G. Luenberger, *Optimization by Vector Space Methods*, Wiley, New York, 1969.
37. H.L. Royden, *Real Analysis*, Macmillan, New York, 1963.
38. N. Dunford and J.T. Schwartz, *Linear Operators*, Vol. I, Interscience, New York, 1958.
39. V.I. Krylov, *Approximate Calculation of Integrals*, Macmillan, New York, 1962.
40. P.M. Prenter, *Splines and Variational Methods*, Wiley-Interscience, New York, 1975.
41. P. Roman, *Some Modern Mathematics for Physicists and other Outsiders*, Vols 1 and 2, Pergamon, Oxford, 1975.
42. W. Hahn, *Stability of Motion*, Springer-Verlag, Berlin, Heidelberg, New York, 1967.
43. H.F. Harmuth, *Transmission of Information by Orthogonal Functions*,

Springer-Verlag, Berlin, Heidelberg, New York, 1969.

44. R.E. Edwards, *Fourier Series—A Modern Introduction*, Holt, Rinehart & Winston, New York, 1967.

45. J.R. Higgins, *Completeness and Basis Properties of Sets of Special Functions*, Cambridge University Press, Cambridge, 1977.

46. C.D. Green, *Integral Equation Methods*, Nelson, London, 1969.

47. J. Arsac, *Fourier Transforms and the Theory of Distributions*, Prentice-Hall, New York, 1966.

48. C. Lanczos, *Linear Differential Operators*, Van Nostrand, Princeton, N.J., 1961.

49. J.W. Dettman, *Mathematical Methods in Physics and Engineering*, McGraw-Hill, New York, 1969.

50. E.R. Lorch, *Spectral Theory*, Oxford University Press, Oxford, 1962.

51. F. Riesz and B. Sz-Nagy, *Functional Analysis*, Ungar Publishing Co., New York, 1955.

52. W. Flugge, *Tensor Analysis and Continuum Mechanics*, Springer-Verlag, Berlin, Heidelberg, New York, 1972.

53. F.G. Tricomi, *Integral Equations*, Interscience, New York, 1967.

54. R.D. Milne, Application of integral equations to fluid flows in unbounded regions, in *Finite Elements in Fluids*, Vol. II (R.H. Gallagher, J.T. Oden, C. Taylor and O.C. Zienkewicz, eds), Wiley, New York, 1975, pp. 83–100.

55. P. Enflo, A counterexample to the approximation problem in Banach spaces, *Acta Mathematica* **130**, 309–317, 1973.

56. A.H. Flax, Reverse flow and variational theorems for lifting surfaces in non-stationary compressible flow, *Journal of Aeronautical Science* **20**, 120–126, 1953.

57. R. Courant and D. Hilbert, *Methods of Mathematical Physics*, Vol. I, Interscience, New York, 1953.

58. M.M. Day, *Normed Linear Spaces*, Springer-Verlag, Berlin, Heidelberg, New York, 1962.

59. M.J. Mansfield, *Introduction to Topology*, Van Nostrand, Princeton, N.J., 1963.

60. H.R. Pitt, *Intergration, Measure and Probabilty*, Oliver and Boyd, Edinburgh, 1963.

61. G. Helmberg, *Introduction to Spectral Theory in Hilbert Space*, North-Holland, Amsterdam, 1969.

62. M.M. Vainberg, *Variational Method and Method of Monotone Operators*, Wiley, New York, 1973.

63. R.A. Tapia, The differentiation and integration of nonlinear operators, in *Nonlinear Functional Analysis and Applications* (L.B. Rall, ed.), Academic Press, New York, 1971, pp. 45–102.

64. M.Z. Nashed, Differentiability and related properties of nonlinear operators, in *Nonlinear Functional Analysis and Applications* (L.B. Rall, ed.),

Academic Press, New York, 1971, pp 103–310.

65. J.T. Oden, *Finite Elements of Nonlinear Continua*, McGraw-Hill, New York, 1972.

66. C. Lanczos, *The Variational Principles of Mechanics*, University of Toronto Press, Toronto, 1964.

67. G. Leitmann, (ed.), *Optimization Techniques*, Academic Press, New York, 1962, Chaps 1, 4, 5.

68. G.A. Bliss, *Lectures on the Calculus of Variations*, University of Chicago Press, Chicago, 1963.

69. J.C. Clegg, *Calculus of Variations*, Oliver and Boyd, Edinburgh, 1968.

70. P.M. Morse and H. Feshbach, *Methods of Theoretical Physics*, Parts I and II, McGraw-Hill, New York, 1953.

71. L.E. Elsgolc, *Calculus of Variations*, Pergamon, Oxford, 1961.

72. M. Athans and P.L. Falb, *Optimal Control*, McGraw-Hill, New York, 1966.

73. S. Barnett, *Matrices in Control Theory*, Van Nostrand-Reinhold, New York, 1971.

74. S.P. Timoshenko and S. Woinowsky-Krieger, *Theory of Plates and Shells*, McGraw-Hill, New York, 1959.

75. S.G. Mikhlin and K.L. Smolitskiy, *Approximate Methods for Solution of Differential and Integral Equations*, Elsevier, Amsterdam, 1967.

76. V.V. Ivanov, *The Theory of Approximate Methods and their Application to the Numerical Solution of Singular Integral Equations*, Noordhoff, Amsterdam, 1976.

77. M.A. Krasnosel'skii, G.M. Vainikko, P.P. Zabreiko, Ya. B. Rutitskin and V.Ya. Stetsenko, *Approximate Solution of Operator Equations*, Walters-Noordhoff, Amsterdam, 1972.

78. S.G. Mikhlin, *The Numerical Performance of Variational Methods*, Walters-Noordhoff, Amsterdam, 1971.

79. S.G. Mikhlin, *The Problem of the Minimum of a Quadratic Functional*, Holden-Day, San Francisco, 1965.

80. B.A. Finlayson, *The Method of Weighted Residuals and Variational Principles*, Academic Press, New York, 1972.

81. S.H. Gould, *Variational Methods for Eigenvalue Problems*, Oxford University Press, Oxford, 1966.

82. B. Friedman, *Principles and Techniques of Applied Mathematics*, Wiley, New York, 1960.

83. R.D. Milne, An approximation to the influence function for plate-like wings, *Aircraft Engineering* **31**, No. 364, 156–162, 1959.

84. J.H. Argyris and S. Kelsey, *Energy Theorems and Structural Analysis*, Butterworths Scientific Publications, London, 1960.

85. L.V. Kantorovich and V.I. Krylov, *Approximate Methods in Higher Analysis*, Wiley-Interscience, New York, 1958.

86. A.M. Arthurs, *Complementary Variational Principles*, Clarendon Press, Oxford, 1970.

87. P.D. Robinson, Complementary variational principles, in *Nonlinear Functional Analysis and Applications* (L.B. Rall, ed.), Academic Press, New York, 1971, pp. 507–567.

88. B. Noble and M.J. Sewell, On dual extremum principles in applied mathematics, *Mathematics Research Center, University of Wisconsin, Report No. 1119*, April 1971

89. M.J. Sewell, On applications of saddle-shaped and convex generating functionals, in *Physical Structure in Systems Theory* (J.J. Van Dixhoorn and F.J. Evans, eds), Academic Press, New York, 1974, pp. 219–246.

90. D.L. Jones, D.J. Holding and F.J. Evans, The classification of physical variables in network theory and mechanics with applications to variational analysis, in *Physical Structure in Systems Theory* (J.J. Van Dixhoorn and F.J. Evens, eds), Academic Press, New York, 1974, pp. 143–166.

91. R.D. Milne, An oblique matrix pseudoinverse, *SIAM Journal of Applied Mathematics* **16**, 931–944, 1968.

92. M. Becker, *The Principles and Applications of Variational Methods*, MIT Press, Cambridge, Mass., 1964.

93. M.E. Gurtin, Variational principles for linear initial-value problems, *Quarterly Journal of Applied Mathematics* **22**, 252–256, 1964.

94. B. Noble, Variational finite element methods for initial-value problems, in *The Mathematics of Finite Elements and Applications* (J.R. Whiteman, ed.), Academic Press, New York, 1973, pp. 143–152.

95. J.R. Whiteman (ed), *The Mathematics of Finite Elements and Applications*, Academic Press, New York, 1973.

96. V.V. Bolotin, *Nonconservative Problems of the Theory of Elastic Stability*, Pergamon, Oxford, 1963.

97. R.L. Bisplinghoff, H. Ashley and R.L. Halfman *Aeroelasticity*, Addison-Wesley, Reading, Mass., 1955.

98. G. Strang and G.J. Fix, *An Analysis of the Finite Element Method*, Prentice-Hall, Englewood Cliffs, N.J., 1973.

99. A.R. Mitchell and R. Wait, *The Finite Element Method in Partial Differential Equations*, Wiley, New York, 1977.

100. A.K. Aziz (ed.) *The Mathematical Foundations of the Finite Element Method with Applications to Partial Differential Equations*, Academic Press, New York, 1972.

101. O.C. Zienkiewicz, *The Finite Element Method in Engineering Science*, McGraw-Hill, New York, 1971.

102. D.H. Norrie and G. De Vries, *An Introduction to Finite Element Analysis*, Academic Press, New York, 1978.

103 C.S. Desai and J.F. Abel, *Introduction to the Finite Element Method*, Van Nostrand, Princeton, N.J., 1972.

104. H.C. Martin and G.F. Carey, *Introduction to Finite Element Analysis*, McGraw-Hill, New York, 1973.

105. J.J. Connor and C.A. Brebbia, *Finite Element Techniques for*

Fluid Flow, Newnes-Butterworth, London, 1976.

106. E.R. de A. Oliveira, Theoretical foundations of the finite element method, *International Journal of Solids and Structures* **4**, 929–951, 1968.

107. J.R. Whiteman, *A Bibliography for Finite Elements*, Academic Press, New York, 1975.

108. M. Zlamal, Some recent advances in the mathematics of finite elements, in *The Mathematics of Finite Elements and Applications* (J.R. Whiteman, ed.), Academic Press, New York, 1973, pp. 59–81.

109. B.M. Irons and A. Razzaque, Experience with the patch test for convergence of finite elements, in *The Mathematical Foundations of the Finite Element Method with Applications to Partial Differential Equations* (A.K. Aziz, ed.), Academic Press, New York, 1972, pp. 557–588.

110. K.H. Huebner, *The Finite Element Method for Engineers*, Wiley, New York, 1975.

111. L.J. Segerlind, *Applied Finite Element Analysis*, Wiley, New York, 1976.

112. G.J. Fix, Effects of quadrature errors in finite element approximation of steady state, eigenvalue and parabolic problems, in *The Mathematical Foundations of the Finite Element Method with Applications to Partial Differential Equations* (A.K. Aziz, ed.), Academic Press, New York, 1972, pp. 525–556.

113. P.G. Ciarlet and P.-A. Raviart, The combined effect of curved boundaries and numerical integration in isoparametric finite element methods, in *The Mathematical Foundations of the Finite Element Method with Applications to Partial Differential Equations* (A.K. Aziz, ed.), Academic Press, New York, 1972, pp. 409–474.

114. J.H. Bramble and S.R. Hilbert, Bounds for a class of linear functionals with applications to Hermite interpolation, *Numerical Mathematics* **16**, 362–369, 1971.

115. J.H. Bramble and M. Zlamal, Triangular elements in the finite element method, *Mathematics of Computation* **24**, 809–820, 1970.

116. P.G. Ciarlet and P.-A. Raviart, General Lagrange and Hermite interpolation in R^n with applications to finite element methods, *Archives for Rational Mechanics and Analysis* **46**, 177–199, 1972.

117. L.C. Welford Jr and J.T. Oden, Accuracy and convergence of finite-element/Galerkin approximations of time-dependent problems with emphasis on diffusion and convection, *Texas Institute for Computational Mechanics Report 73–8*, 1973.

118. C.A. Brebbia and A.J. Ferrante, *Computational Methods for the Solution of Engineering Problems*, Pentech Press, New York, 1978.

119. J.L. Lions and E. Magenes, *Non-homogeneous Boundary-value Problems and Applications*, Springer-Verlag, Berlin, Heidelberg, New York, 1972.

120. D. Gilbarg and N.S. Trudinger, *Elliptic Partial Differential Equations*

of Second Order, Springer-Verlag, Berlin, Heidelberg, New York, 1977.

121. V.V. Novozhilov, *Foundations of the Nonlinear Theory of Elasticity*, Graylock Press, New York, 1953.

122. A.C. Lock and A.B. Sabir, Algorithm for the large deflection of geometrically nonlinear plane and curved structures, in *The Mathematics of Finite Elements and Applications* (J.R. Whiteman, ed.), Academic Press, New York, 1973, pp. 483-494.

123. L.S.D. Morley, A triangular equilibrium element with linearly varying bending moments for plate bending problems, *Journal of the Royal Aeronautical Society* **71,** 715-719 1967.

124. G.E. Shilov, *Elementary Functional Analysis*, MITPress, Cambridge, Mass., 1974.

125. G. Temple, The theory of generalised functions *Proceedings of the Royal Society A* **228,** 175-190, 1955.

126. M.J. Lighthill, *Introduction to Fourier Analysis and Generalised Functions*, Cambridge University Press, Cambridge, 1959.

127. J.P. Marchand, *Distributions—An Outline*, North Holland, Amsterdam, 1962.

128. A.H. Zemanian, *Distribution Theory and Transform Analysis*, McGraw-Hill, New York, 1965.

129. I.M. Gel'fand and G.E. Shilov, *Generalized Functions*, Vol. 1, Academic Press, New York, 1964.

130. R. Courant and D. Hilbert, *Methods of Mathematical Physics*, Vol. II, Interscience, New York, 1962.

131. G.N. Ward, *Linearized Theory of Steady High-Speed Flow* Cambridge University Press, Cambridge, 1955.

132. M.D. Greenberg, *Applications of Green's Functions in Science and Engineering*, Prentice-Hall, Englewood Cliffs, N.J., 1971.

133. A. Papoulis, *The Fourier Integral and Its Applications*, McGraw-Hill, New York, 1962.

134. N.W. McLachlan, *Complex Variable Theory and Transform Calculus*, Cambridge University Press, Cambridge, 1963.

135. D.K. Cheng, *Analysis of Linear Systems*, Addison-Wesley, Reading, Mass., 1959.

136. R.E. Showalter, *Hilbert Space Methods for Partial Differential Equations*, Pitman, London, 1977.

137. H. Bremerman, *Distributions, Complex Variables and Fourier Transforms*, Addison-Wesley, Reading, Mass., 1965.

138. V.I. Smirnov, *A Course of Higher Mathematics*, Vol. V, Pergamon, Oxford, 1964.

139. C. Lanczos, Trigonometric interpolation of empirical and analytical functions, *Journal of Mathematical Physics* **17,** 123-199, 1938

140. J.W. Miles, *Potential Theory of Unsteady Supersonic Flow*, Cambridge University Press, Cambridge, 1959.

141. R.D. Milne, A constructive theory of linear damping, in *Proceedings of the Symposium on Structural Dynamics* (D.J. Johns, ed.), Loughborough University, Loughborough, 1970, Vol. I, Paper C4.

142. A. Friedman, *Partial Differential Equations*, Holt, Rinehart & Winston, New York, 1969.

143. A. Friedman, *Generalized Functions and Partial Differential Equations*, Prentice-Hall, New York, 1963.

144. J.T. Oden and J.N. Reddy, *An Introduction to the Mathematical Theory of Finite Elements*, Wiley, New York, 1976.

145. J. T. Oden, *Applied Functional Analysis*, Prentice-Hall, New York, 1979.

Answers to selected exercises

Chapter 2

2.2,1   (i) no   (ii) yes   (iii) no   (iv) no

2.3,1   (i) yes   (ii) yes   (iii) no   (iv) yes   (v) yes   (vi) yes

2.3,4 $(0, 1, 2)$ spans $\mathcal{U} \cap \mathcal{W}$; $\mathcal{U} + \mathcal{W} = \mathcal{R}_3$

2.4,5   dimension is $m \times n$: $(m \times n)$ matrices with zero entries save for 1 in (i,j)th place.

2.4,7   (i) 2   (ii) 3

2.5,1   no

2.5,3   (i) $\sum_{i=1}^n \xi_i T x^i$ (ii) $x^1 + x^2 + \ldots + x^n$ (iii) all vectors such that $\sum_{i=1}^n \xi_i = 0$: say, $(1, -1, 0, 0, \ldots), (0, 1, -1, 0, 0 \ldots),$ $(0, 0, 1, -1, 0, 0 \ldots), \ldots$

2.5,11   no

2.6,6 $2(1, 1, 1) + 2i(2, 1, 0)$

2.8,3 $(1, -5, 4)$

2.8,4   (i) $(1, 1), (1, -1)$ (ii) $\begin{bmatrix} 1 & 1 \\ 1 & -1 \end{bmatrix}$ (iv) $\begin{bmatrix} 2 & 1 \\ -2 & -2 \end{bmatrix}$

2.8,6(b)   (i) not in general   (ii) only if either T or S non-singular

2.8,7 $\begin{bmatrix} 3 & 6 & 5 \\ 1 & -5 & -2 \end{bmatrix}$

2.8,8(c) $-1, \alpha, 1$

2.8,9   (i) 2   (ii) 3   (iii) 4

2.8,10   (i) $(1, -1, 2, -2)$ (ii) $(3, -4, -3) + \alpha(-5, 11, 7)$

2.8,12 $\begin{bmatrix} 5 & 4 & 1 & 0 \\ 3 & 2 & 1 & 0 \\ 5 & 4 & 8 & 6 \\ 1 & 0 & 0 & 2 \end{bmatrix}$

2.8,13(a)   (i) 0   (ii) $(\beta - \alpha)(\gamma - \alpha)(\gamma - \beta)$

2.8,16(a)   (i) $1, 3 \pm \sqrt{2}$ (ii) $1, 1 \pm i$

2.8,18   ones on sub-diagonal, otherwise zeros

2.8,19(a)   standard basis

     (b)   (ii) $(0, 1, 1, 0)$ (iii) $(1, 5, 0, -1), (1, 3, 1, 0)$ (iv) $(0, 0, 1, 1)$

Chapter 3

3.2,1(a) $4, \sqrt{10}, 3$

     (d)   (i) $1/6, 1/\sqrt{30}, 1/4$

             (ii) $\dfrac{4}{|\lambda - \mu|}\left(1 - \cos \tfrac{1}{2}|\lambda - \mu|\pi\right), \dfrac{4}{|\lambda - \mu|}(|\lambda - \mu|\pi - \sin |\lambda - \mu|\pi),$

             $\max\limits_{t \in [-\pi, \pi]} 2|\sin \tfrac{1}{2}|\lambda - \mu|t|$

(iii) $\varepsilon, \sqrt{2\varepsilon/3}, 1$.

(e) $\begin{cases} 0, \beta > \frac{1}{2} \\ 1, \beta < \frac{1}{2} \end{cases}, 1 - \beta, \frac{1}{2}$

3.2,7(b) all yes

3.4,2(c) $\|T\| < |\lambda|, (\lambda I - T)^{-1} = \frac{1}{\lambda} \sum_{n=0}^{\infty} \frac{1}{\lambda^n} T^n$

(d) $t^2 - t + \frac{4}{3}, \frac{7}{6}t^2 - \frac{7}{6}t + \frac{251}{180}$

3.4,5 $k(t, \tau) = \begin{cases} \tau(1-t), 0 \leq \tau \leq t \leq 1 \\ t(1-\tau), 0 \leq t \leq \tau \leq 1 \end{cases}$

3.5,1(a) (i) 1 (ii) 1 (iii) $\frac{1}{2}$

3.6,2(j) all even functions

3.6,5(a) equations always consistent, $\tilde{x}$ not necessarily unique but $T\tilde{x}$ is unique

(b) $[t_{ij}]^T [t_{ij}] \{\alpha_j\} = \sum_{k=1}^{M} \xi_i^{(k)} \eta^{(k)}$

3.7,4(a) $T^*x \equiv x'' - (\alpha x)' + \beta x$ with (i) $x(0) = x(1) = 0$ (ii) $x(1) = x'(1) = 0$
(iii) $x(1) = x(0), x'(1) = x'(0)$

(b) $\eta_k = \sum_{j=k}^{\infty} \frac{1}{j} \xi_j$

3.7,5 $k(t, \tau) = \frac{2}{\pi^2} \sum_{n=1}^{\infty} \frac{1}{n^2} \sin n\pi\tau \sin n\pi t, \ T(-1) = -\frac{4}{\pi^2} \sum_{n=0}^{\infty} \frac{1}{(2n+1)^2} \sin(2n+1)\pi t$

3.8,1(b) $[l_{ij}]^T = \begin{bmatrix} c_2 c_3 & s_1 s_2 c_3 - c_1 s_3 & c_1 s_2 c_3 + s_1 s_2 \\ c_2 s_3 & s_1 s_2 s_3 + c_1 c_3 & c_1 s_2 s_3 - s_1 c_3 \\ -s_2 & s_1 c_2 & c_1 c_2 \end{bmatrix}, \ c_i = \cos\theta_i, s_i = \sin\theta_i$

(e) $\frac{1}{42} \begin{bmatrix} 268 & 32 & 50 \\ 448 & 140 & 224 \\ 150 & 48 & 96 \end{bmatrix}$

3.8,4(a) (i) 1, 2, 21 (ii) 1, 1, 7

3.8,5(a) (i) 2.7375 (ii) 1.650

Chapter 4

4.2,1(c) $-\frac{3\xi_1^2}{\xi_2} \sigma_1 - \frac{\xi_1^3}{\xi_2^2} \sigma_2$

4.2,5 (i) $\xi_1(\xi_2 + \xi_1) + \lfloor \xi_2 + 2\xi_1, \xi_1 \rfloor \begin{bmatrix} \sigma_1 \\ \sigma_2 \end{bmatrix} + \frac{1}{2} [\sigma_1, \sigma_2] \begin{bmatrix} 2 & 1 \\ 1 & 0 \end{bmatrix} \begin{bmatrix} \sigma_2 \\ \sigma_2 \end{bmatrix}$

(ii) $\int_0^1 x^2(t)dt + 2\int_0^1 x(t)s(t)dt + \int_0^1 s^2(t)dt$

4.3,3(a) (i) any (smooth) function (ii) circular arc, centre $\left[\frac{\alpha + \beta}{2}, 0\right]$

(iii) first case—arc of rectangular hyperbola, second case—no solution.

(b) $D_1^2 \phi + D_2^2 \phi = f$

ANSWERS TO SELECTED EXERCISES

4.3,5(c)   circular arc centre $(9/4, 0)$, $t_r = 3.6$, $x_r = 9/5$

4.4,5(b) $x(t) = \sin t - \dfrac{2t}{\pi}$, $\min f(x) = \dfrac{2}{\pi} - \dfrac{\pi}{6}$

    (c) $0.2716\, a^4$
    (e) $0.141\, pl^4/EI_0$
    (f) 0.588 for $k = 1$

4.4,8(b)   (ii) 5   (iii) $\dfrac{4.33}{l^2}\sqrt{\dfrac{EI_0}{m}}$ (iv) $\dfrac{3.65}{l^2}\sqrt{\dfrac{K}{m}}$

4.4,11(a) $a(x,w) = \tfrac{1}{2}\langle w,w \rangle_{\mathcal{H}_2} - \tfrac{1}{2}\langle Rx,x \rangle_{\mathcal{H}_1} + \langle y,x \rangle$ is saddle
    (b) $-0.179 < f(x) < -0.164$
    (c) $a(x,w) = \sum_i \int_0^{\phi_i} c_i(\psi_i)\,d\psi_i + b_i\xi_i$

4.5,3(b)   (ii) 16.66
    (c) $z_1 = e^{-\sqrt{10}\,t}$
    (d)   for EI const, $V_c = 3.42\sqrt{EI/k}$
    (f) $(\omega_1/\Omega) = 1, (\omega_2/\Omega) = 2.91$

4.6,2(f)   (i) $[\langle z^i, z^j \rangle_T^{(k)}] = 1/4A\, [\tilde{g}_{ij}]^T[\tilde{g}_{ij}]$ where the (2×3) matrix $[\tilde{g}_{ij}]$ consists of the second and third rows of the matrix $[g_{ij}]$ of p. 388.

    (ii) $[\langle z^{i,} z^j \rangle_T^{(k)}] = \dfrac{1}{4A^2} \int_{\Delta^{(k)}} [D_j z^k]^T [\tilde{g}_{ij}]^T [\tilde{g}_{ij}] [D_j z^k]\, dt_1\, dt_2$

where
$$[D_j z^k] = \begin{bmatrix} 4\sigma_1 - 1 & 0 & 0 & 4\sigma_2 & 0 & 4\sigma_3 \\ 0 & 4\sigma_2 - 1 & 0 & 4\sigma_1 & 4\sigma_3 & 0 \\ 0 & 0 & 4\sigma_3 - 1 & 0 & 4\sigma_2 & 4\sigma_1 \end{bmatrix}.$$

4.6,4(c)   (iii)

$$[d_{ij}] = \dfrac{E}{1-\mu^2} \begin{bmatrix} 1 & \mu & 0 \\ \mu & 1 & 0 \\ 0 & 0 & \tfrac{1}{2}(1-\mu) \end{bmatrix}$$

$$[b_{ij}] = \dfrac{1}{2A} \begin{bmatrix} g_{21} & 0 & g_{22} & 0 & g_{23} & 0 \\ 0 & g_{31} & 0 & g_{32} & 0 & g_{33} \\ g_{31} & g_{21} & g_{33} & g_{22} & g_{33} & g_{23} \end{bmatrix}$$

Index

Abscissa of convergence, 454
Addition,
    of linear transformations, 36
    of vectors, 18
Adjoint,
    eigenvalues of, 229
    formally, 220
    of a linear transformation, 213
    of a mapping, 212
Adjugate of a matrix, 96
Admissible,
    basis functions, 337
    functions, 303
Aerofoil equation, 259
Algebra, 41
    convolution, 440
Almost,
    everywhere, 137, 278
    periodic functions, 210
Annihilator, 54, 174
Approximation,
    best, 140
    Chebyshev, 178
    uniform linear, 285
Area coordinates, 400
Arzela–Ascoli theorem, 129
Asymptotic stability, 408
Augmented matrix, 67

Ball,
    closed, 117
    open, 117
    unit, 117, 133
Banach space(s), 136
    table of, 158, 159
Basis,
    algebraic (Hamel), 30, 138
    change of, 69
    change of, in inner product space, 232
    dual (conjugate, reciprocal), 51
    interpolation, 59, 384, 470
    local, functions, 383
    orthonormal, 193, 197, 200
    Schauder, 138
    standard, 32
Beam (bending),
    differential equation for, 103, 414
    and energy principles, 414
    Euler–Lagrange equation for, 324
    stability of, under end load, 380
    vibration frequencies of rotating, 380
Bending moment,
    in beam, 104, 414
    in thin plate, 325
Bessel's inequality, 198

Best approximation, 140, 166
Bifurcation point, 407
Bilinear functional, 179
    coercive, 327
Boundary conditions,
    essential (principal), 329, 333
    natural, 323, 330, 333
Bounded,
    function, 114
    functions of, variation, 163
    inverse, 150, 155
    linear transformation, 144
    sequence, 113, 115
    subset, 117
    totally, 127
    uniformly, linear transformations, 148
Bubnov–Galerkin method, 367

Canonical Euler equations, 308
Canonical form,
    Jordan, 85
    for orthogonal matrices, 246
    for rectangular matrices, 76
    for square matrices, 84
    with respect to unitary similarity, 238
Cardinal number, 15
Cartesian product,
    of sets, 5
    of vector spaces, 21
Cauchy,
    equivalent, sequences, 124
    sequence, 119
    principal value, 259, 443
Cayley–Hamilton theorem, 98
Characteristic,
    function (eigenfunction), 279
    polynomial, 45, 82
Chebyshev,
    approximation, 178
    polynomials, 205, 266
Closed,
    graph theorem, 216
    interval, 4
    operator, 215
    set, 274
    span, 139
Closure of a set, 123, 275
Codimension, 31
Coercive,
    bilinear functional, 327, 469
    sesquilinear functional, 229
Cofactor of element of a matrix, 79
Collocation method, 368, 470
    and Galerkin method, 381
    orthogonal, 471

INDEX

Compact,
  integral operator, 153
  metric space, 128
  operator, 151
  resolvent, 254
  self-adjoint transformation, 252
  sequentially, 128
  set, image of, 130
  support, 159
Complement,
  direct, 24
  orthogonal, 185
Complementary,
  variational principles, 348, 417
  virtual work principle, 413
Complete,
  metric space, 120
  normed space, 136
  orthonormal set, see maximal
  polynomial, 392
  set, 139
Completely continuous, see compact
Completion of metric space, 125
Components, see coordinates
Composition,
  Fréchet derivative of, 293, 300
  of functions, 12
  of linear transformations, 41
Concave, see also convex
  functional, 351
Condition number, 154
Cone, 27
  convex, 27, 177
Conforming finite elements, 387
Congruent, see conjunctive
Conjugate, see also dual
  directions, 410
  gradient method, 410
  transpose, 218
Conjuctive matrices, 242
Continuous,
  absolutely, 202
  completely, 151
  functions, space of, 113, 114, 118
  linear functional, 155
  operator, 129
  partial derivatives, 285
  spectrum, 248
  uniformly, operator, 130
Contraction (mapping), 267
Contravariant coordinates, 52
Convergence
  componentwise, 115
  of distributions, 429
  in energy, 331
  of finite element method, 392
  of Galerkin method, 372
  of least squares method, 370
  linear, 269
  in the mean of order p, 116
  with respect to metric, 114
  with respect to norm, 136
  pointwise, 117
  quadratic, 404
  in space of testing functions, 425, 449
  strong, 136, 169

weak, 169, 192
weak*, 171
uniform, 115
Convex,
  combination (of vectors), 27
  cone, 27, 177
  functional, 351
  hull, 26
  set 26, 351
  strictly, 363
Convolution,
  algebra, 440
  differentiation of, 438
  of distributions, 437
  Fourier transform of, 452
  inverse, 440
  Laplace transform of, 456
Coordinates,
  area (natural), 400
  contravariant, 52
  covariant, 52
  of a vector, 19, 32
Coset, 25
Countable set, 15
Covariant coordinates, 52
Cramer's rule, 97
Cyclic subspace, 47

Degenerate linear transformation, 230
Degrees of freedom, 309, 385
Delta,
  distribution (functional), 427, 462
  sequence, 422, 441
Dense subset, 124
Denumerable, see countable set
Derivative(s),
  of composition, 293, 300
  covariant, contravariant, 286
  of distribution, 431
  Fréchet, 295
  Fréchet, at a point, 289
  Gateaux, 289
  of norm, 299
  partial, 283
  partial distributional, 461
  partial Fréchet, 349
  second Fréchet, 295
Determinant,
  function, 79
  properties of, 79
  of similar matrices, 80
Diagonizable linear transformation, 84
Differentiability, 285
Differential,
  function, 284
  Gateaux, 287
Dimension of vector space, 28, 30
Direct,
  complement, 24
  integral, 294
  product of distributions, 435
  sum, 24
Dirichlet's principle, 189
Displacement,
  impulse response, 478
  method (see principle of virtual work)

Distance, *see* metric
Distribution(s), 426, 460
  convergence of, 429
  convolution of, 437
  definite integral of, 445
  delta, 427
  derivative of, 431
  direct product of, 435
  Fourier transform of, 450
  Heaviside (step), 427
  periodic, 458
  primitive of, 444
  radial, 468
  rank of, 433
  regular, 426
  regularisation of, 439
  scaling of, 429
  singular, 427
  of slow growth, 449
  support of, 427, 437
  translation of, 429
Divergence theorem, 220, 467
Domain of a function, 11
Dual,
  algebraic, 50
  basis, 51
  extremum principles, *see* complementary variational principles
  linear programming problem, 105
  of linear transformation, 60, 173
  normed, 157
  of product (of transformations), 63
  of projection, 63
  second, 168
  self-, 56
  table of, spaces, 158, 159
Duality, 167

Echelon form,
  of a matrix, 75
  row-reduced, 95
Eigensolution, 44
  of Sturm–Liouville operator, 265
Eigenspace(s), 44
  generalised, 85
  orthogonality of, 236
Eigenvalue(s), 44
  of adjoint transformation, 229
  algebraic multiplicity of, 82
  of dual transformation, 88
  generalised, problem, 261, 361
  geometric multiplicity of, 82
  of Hermitean matrix, 235
  power method for, 99, 244
  and projection methods, 375
  and Rayleigh–Ritz method, 341
  of self-adjoint transformation, 236
  of square matrix, 81
Eigenvector(s), 44
  of dual transformation, 88
  generalised, 85
  of Hermitean matrix, 235
  of square matrix, 81
Element (finite),
  isomparametric, 389
  rectangular, 389
  triangular, 387
Elementary matrix, 75
Elliptic equation, 363, 377, 465
Embedding theorem, 464
Energy,
  convergence in, 331
  norm, 225, 330
  product, 225
Equicontinuous, 129
Equivalence,
  class, 8
  relation, 8, 94
Equivalent,
  Cauchy sequences, 124
  matrices, 74
  norms, 142, 356
Error bounds,
  for finite element method, 394
  for method of orthogonal projections, 347
  for Rayleigh–Ritz method, 336
Essential boundary conditions, 329, 333
Euclidean,
  matrix norm, 241
  norm, 183
  space, 181
Euler–Lagrange equation(s), 304
  canonical, 308
  first integral of, 305
Exchange theorem, 30
Extension,
  as adjoint of linear transformation, 217
  as dual of linear transformation, 175
  of linear functional, 59, 161
  of a linear transformation, 229
Extremal (arc), 303
Extremum, 302

Finite element,
  isomparametric, 389
  rectangular, 389
  triangular, linear, 387
  triangular, quadratic, 388
Finite part of a divergent integral, 428
Fixed point,
  of mapping, 267
  principle, 269
Force,
  impulse response, 478
  method, *see* principle of complementary virtual work
Fourier,
  coefficients, 198
  expansion (series), 198
  series, distribution, 434
  transform, 228, 448
  transform distributional, 450, 468
Fréchet derivative, 295
  of a composition, 293, 300
  partial, 349
  at a point, 289
  second, 295
Fredholm alternative, 254, 373
Frequencies (natural),
  of rotating beam, 380
  of vibrating system, 262
Friedrich's inequality, 356, 470

496 INDEX

Function(s), 11
  bounded, 114
  of bounded variation, 163
  continuous, 113, 118
  convex, 351
  with finite energy, 330
  identity, 12
  integrable, 114
  into, onto, 12
  invertible, 12
  Lipschitz, 129
  with local support, 384
  monotonic, 162
  one-to-one, 12
  testing, with compact support, 425
  testing, of rapid descent, 449
Functional(s),
  bilinear, 179
  coercive, 229, 327
  convex, 351
  Hermitean, 179
  linear, 49
  linear, continuous, 155
  maximum/minimum of, 301
  saddle, 352
  sesquilinear, 179
Fundamental,
  lemma of the variational calculus, 304, 320
  matrix, 72
  sequence, see Cauchy
  soluction, 475
  theorem of calculus, 294, 301, 445

Galerkin,
  method, 367, 371
  method and collocation, 381
Gateaux,
  derivative, 289
  differential, 287
Gauss,
  –Jordan method, 95
  –Seidel method, 271
Gaussian,
  elimination, 95
  quadrature, 172, 473
Generalised,
  eigenvalue problem, 261, 361
  solution, 332, 373
Global nodes, 384
Gradient,
  of functional, 339
  mapping, 339
  in $\mathscr{R}_n$, 291
  strong, 300
  weak, 299
Gram determinant, 188, 335
Gram–Schmidt,
  orthogonalisation process, 194
  reorthogonalisation, 208
Green's function,
  for integral operator, 40, 230, 254
  for linear differential operator, 440, 474
  for Sturm–Liouville operator, 264

Haar system, 204

Hadamard's finite part, 428
Hahn–Banach theorem, 161
Hamel basis, 30, 138
Hamiltonian, 308
Hamilton's principle, 307
Heaviside distribution, 427, 466
Heine–Borel theorem, 128
Hermite,
  delta sequence, 422
  interpolation, 389
  polynomials, 205
Hermitean matrix, 222, 235
Hessian matrix, 297
Hilbert space, 184
Hölder inequality, 273
Hyperplane, 31
  and linear functional, 54

Ideal element, 118
Idempotent transformation, 43
Identity operator, 12, 41
  and compactness, 155
Implicit function theorem, 312
  generalised form of, 315
Implicit rank, 49
Impulse function, 440
  displacement, 478
  force, 478
Incomplete metric space, 120
Incremental load method, 406
Index of a linear transformation, 49
Indicial response, 479
Induced norm of a matrix, 148, 239
Infimum, 9
Inner product, 181
  space, 181
Integrable functions, 114
Integral,
  direct, 294
  of a distribution, 445
  finite part of, 428
  Lebesgue, 114, 137, 278
  Riemann, 114, 277
  Riemann–Stieljes, 164
Internal nodes, 385
Interpolation,
  basis, 59, 384, 470
  Hermite, 389
  Lagrange, 389
Intersection of subspaces, 23
Invariant subspace, 83
Inverse,
  bounded, 150, 155
  convolution, 440
  of linear transformation, 40
  of neighbouring operator, 151
  operator, 149
Inversion by bordering, 93
Invertible linear transformation, 40
Isometric
  mapping, see isometry
  metric spaces, 125
Isometry, 226
Isomorphic spaces, 21, 34, 48
Isoparametric finite element, 389

Isoperimetric problem, 314
Iteration,
  and contraction, 267
  first-order, 269
  Gauss–Seidel, 271
  Jacobi, 271

Jacobi,
  iteration, 271
  method for eigenvalues, 245
  polynomials, 472
Jacobian matrix, 291
Jordan canonical form, 85

Kantorovich method, 359, 374
Kernel, *see* null space,
  see Green's function
Kirchoff delta sequence, 422
Kronecker delta, 14

Lagrange's equations, 307
Lagrange,
  interpolation, 389
  multiplier rule, 312
  multiplers, 312
Lagrangian, 312
Laguerre polynomials, 196, 205
Laplace,
  expansion of determinant, 79
  operator, 223, 465
  transform of distributions, 454
Lax–Milgram theorem, 229, 469
Least squares,
  estimation of data, 209
  method of, 367, 370
  and Ritz method, 369
  solution, 208
  weighted, 378
Lebesgue
  integrable function, 279
  integration 137, 278
Legendre
  polynomials, 195, 205
  transformation, 322
Limit of sequence, 114, 275
Line segment, 26, 290
Linear
  algebraic simultaneous equations, 66
  approximation, 285
  convergence, 269
  dependence of vectors, 28
  differential equations, 38, 39, 106
Linear functional, 49
  continuous, 155
  extension of, 161
  and hyperplane, 54
Linear programming problem, 105
Linear stability, 408
Linear transformation(s), 35
  addition of, 36
  adjoint of, 213
  algebraic dual of, 60
  bounded, 144
  closed, 215
  with compact resolvent, 254
  composition of, 41
  continuous, 144
  degenerate, 230
  diagonalizable, 84
  dual of, 173
  invertible, 40
  norm of, 147, 174
  normal, 238, 243
  nullity of, 39
  null space of, 39
  positive, 225
  positive-bounded-below, 326
  positive definite, 225
  range of, 38
  rank of, 39
  resolvent of, 248
  resolvent set of, 248
  scalar multiplication of, 36
  self-adjoint, 222
  singular, 40
  spectrum of, 248
  symmetric, 222
  uniformly bounded, 148
  unitary, 226
Lipschitz
  condition, 267
  functions, 129
Load vector, 386
Local support, functions with, 384
Lower bound, 9
  greatest, *see* infimum

Matrix,
  adjugate of, 96
  augmented, 67
  canonical form of, 76, 84
  cofactor of, 79
  column echelon form, 76
  congruent, *see* conjunctive
  conjunctive, 242
  determinant of square, 79
  diagonally dominant, 271
  differential equation, 106
  of dual, 69
  eigenvalues of, 81
  eigenvectors of, 81
  elementary, 75
  equivalent, 74
  fundamental, 72
  Hermitean, 222
  Hessian, 297
  induced norm of, 148, 239
  Jacobian, 291
  Markov, 92
  minor of, 79
  multiplication, 68
  norm of, *see* induced norm
  orthogonal, 227, 246
  partitioned, 69
  polar decomposition of, 243
  rank of, 66, 74
  representation of linear transformation, 37
  restriction of, 92
  rotation, 233, 242, 245
  row echelon form, 75
  row-reduced echelon form, 95

similar, 78
stiffness, 386
symmetric, 222
transpose, 69
unitary, 227
Maximal orthonormal set, 200
Maximum/minimum, 10
 of functional, 301
Mean value theorem,
 linear, 293
 quadratic, 296
Measurable,
 function, 278
 set, 279
Measure, 137, 279
 zero, 280
Metric, 112
 convergence with respect to, 114
 discrete, 114
 on product space, 132
 uniform, 116
Metric space 112
 compact, 128
 complete, 120
Minimax approximation, 178
Minimising sequence, 334
Minkowski inequality, 113, 274
Minor of element of a matrix, 79
Moments, method of, 367
Monotonic function, 162
Multiplication,
 of linear transformations, 41
 of matrices, 68
 scalar, 18
Multiplicity of an eigenvalue,
 algebraic, 82
 geometric, 82

Natural,
 boundary conditions, 323, 330, 333
 coordinates, *see* area
 embedding, 168
 frequencies, 262
Nesting sequence of spaces, 335
Net, ε−, 127
Neumann series, 154
Newton's method, 403
Nilpotent transformation, 47
Nodes,
 global, 384
 internal, 385
 multiple, 384
Non-singular, *see* invertible linear transformation
Norm, 135
 of adjoint transformation, 214
 derivative of, 299
 of dual transformation, 174
 energy, 225, 330
 equivalent, 142
 Euclidean, 239
 Euclidean matrix, 241
 induced matrix, 148
 of linear transformation, 147
 of projection, 231
 semi, 141
 spectral, 241

Normal,
 equations, 188, 209
 linear transformation, 238, 243
 modes, 262
Normed,
 dual space, 157
 vector space, 135
Null set, 278
Null space, 39
 of adjoint, 214
 of dual, 174
Nullity of linear transformation, 39

Open,
 interval, 4
 set, 274
Operator,
 adjoint of, 212
 closed, 215
 compact, 151
 completely continuous, *see* compact
 continuous, 129
 inverse, 149
 potential, 339
 Taylor polynomial for, 298
Optimal control problem, 315
Order symbols, 420
Ordering,
 partial, 9, 28
 total, 10
Orthogonal, 185
 collocation, 471
 complement, 185
 matrix, 227, 246
 projection, 188, 336
 projection operators, 231
 projections, method of, 344
 set of vectors, 193
Orthonormal, 193
 basis, 193, 197, 200

Parallelogram law, 183
Parseval's,
 formula, 200
 relation, 448
Partial,
 derivative, 283
 ordering, 9
Partition, 8
 of matrix, 69
Patch test, 387
Periodic distribution, 458
Poisson's formula, 452
Polar decomposition of a matrix, 243
Polynomial(s),
 characteristic, 45
 Chebyshev, 205
 complete, 392
 Hermite, 205
 Jacobi, 472
 Laguerre, 196, 205
 Legendre, 195, 205
 of linear transformation, 41
 space of, 20
 Taylor, 298
 trigonometric, 203

Pontryagin's maximum principle, 318
Positive,
 bounded-below linear transformation, 326
 bounded-below matrix, 377
 definite transformation, 225
 orthant, 28
 transformation, 225
Potential operator, 339
Power method (for eigenvalues), 99
 and Rayleigh quotient, 244
 and sub-dominant eigenvalues, 244
Primitive of a distribution, 444
Principal,
 axes of strain, 258
 boundary conditions, see essential
Principle,
 of complementary virtual work, 413
 of virtual work, 412
Product, see also composition,
 energy, 225
 inner, 181
 space, 21, 132
Projection(s),
 bounded, 176
 dual of, 63
 as linear transformation, 42
 method, 365
 method of orthogonal, 344
 operators, orthogonal, 231
 orthogonal, 188, 253, 336
 theorem, 186
Pseudo,
 function, 428
 -inverse, 377
 metric, 131
 norm, 141
Pythagorean theorem, 185

Quadratic form, 179
Quadrature, 171
Quotient space, 25

Radial distribution, 468
Range of function, 11
Range space, 38
 of adjoint transformation, 214
 of dual transformation, 174
Rank,
 of distribution, 433
 of dual transformation, 63
 of linear transformation, 39, 74
 of a matrix, 66, 74
Rayleigh,
 principle, 341
 quotient, 244, 341, 360
 –Ritz method, see Ritz
Real line, 4
Reciprocal basis, see dual basis
Rectangular finite element, 389
Reduction of a matrix/row, column, 75
Reflexive space,
 algebraically, 53
 norm, 168
Regular,
 distribution, 426
 point, 311, 315

Regularisation of a distribution, 439
Relation, 7
 equivalence, 8
Resolvent set, 248
Restriction,
 of a linear transformation, 48
 of a matrix, 92
Reverse flow theorem, 260, 472
Ricatti equation, 317
Riemann,
 integral, 277
 –Lebesgue lemma, 210
 –Stieltjes integral, 164, 255
Riesz representation theorem, 165, 192
Ritz,
 equations, 335, 342, 347, 361
 method, 334, 341, 346
 method and least squares, 368
 method for potential operators, 339
Rotation, 227
 and orthogonal matrix, 233, 242, 245

Saddle functional, 352
Scalar, 17
 multiplication, 18, 36
Schauder basis, 138
Schwarz inequality, 182, 273
Self-adjoint linear transformation, 222
Seminorm, 141
Separable space, 125
Sequence, 14
 bounded, 113
 Cauchy, 119
 convergent, 113
 delta, 422, 441
 fundamental, see Cauchy
 infinite, 113
 minimising, 334
Sesquilinear functional, 179
 coercive, 229
 Hermitean, 179
 and inner product, 206
 positive, 179
Set(s), 3
 bounded, 9
 Cartesian product of, 5
 closed, 274
 closure of, 123
 complement of, 5
 complete, 139
 convex, 26, 351
 countable, 15
 disjoint, 5
 empty, 4
 finite, 14
 infinite, 15
 intersection of, 5
 maximal, 200
 measurable, 279
 minimal spanning, 28
 null, 278
 open, 274
 orthogonal, 193
 spanning, 23, 139
 uncountable, 15
 union of, 5

Shear force (in beam), 104
Similar,
  matrics, 78
  unitarily, 235
Singular,
  distribution, 427
  linear transformation, 40
Sobolev,
  embedding theorem, 464
  inequality, 464
  space, 460
  structure theorem, 463
  trace theorem, 464
Solution,
  collocation, 471
  fundamental, 475
  generalised, 332
  projection, 365
  of set of simultaneous equations, 66
  weak, 469
Space,
  of almost periodic functions, 210
  Banach, 136
  Euclidean, 181
  finite-dimensional, 30
  Hilbert, 184
  infinite-dimensional, 30, 115
  inner product, 181
  metric, 112
  normed, 135
  null, 39
  quotient, 25
  range, 38
  separable, 125
  Sobolev 460
  solution, 34
  of testing functions, 425
  topological, 275
  unitary, 181
  vector, 18
  zero, 19
Span, closed, 139
Spanning set, 23, 139
  minimal, 28
Spectral,
  norm, 241
  radius, 241
  resolution of transformation, 238
  solution, 108
  theorem, 238
Spectrum of linear transformation, 248
    continuous, 248
    parasitic, 252
    point, 248
    residual, 248
Stability,
    asymptotic, 408
    linear, 408
Stationary point, 302
Steepest descent (method of), 410
Stiffness matrix, 386
Strain,
  displacement relation, 257
  principal axes of, 258
  tensor, 257
Stress,

  equilibrium equation, 257
  tensor, 256
Strict convexity, 363
Strong,
  convergence, 136
  derivative, see Fréchet
  gradient, 300
Structure theorem (in Sobolev spaces), 463
Sturm–Liouville differential operator, 254, 262
    353, 356
Subsequence, 14
Subset, 4
  bounded, 117
  dense, 124
  indexed, 14
Subspace(s),
  cyclic, 47
  direct sum of, 24
  disjoint, 24
  improper, 22
  intersection of, 23
  invariant, 83
  of metric space, 112
  nesting sequence of, 335
  proper, 22
  spanned by a set, 23
  sum of, 23
  of vector space, 21
Sum of subspaces, 23
  direct, 24
Support,
  bounded on left/right, 437
  compact, 159
  of distribution, 427
  local, 384
Supporting plane, 166
Supremum, 9
  principle, 10
Sylvester's law of nullity, 48
Symmetric,
  functional, 179
  linear transformation, 222
  matrix, 222

Tangent,
  manifold, 311
  subspace, 311
Taylor polynomial, 298
Tempered distributions, see distributions of slow
    growth
Testing functions,
  with compact support, 425
  convergence of, 425, 449
  of rapid descent, 449
Torsion problem, 358
Totally bounded, 127
Trace,
  of a matrix, 59
  theorem, 464
Transform,
  Fourier, 228, 448
  Hilbert, 479
  Laplace, 453
Transformation, see function, linear transformation,
    operator
Transpose of a matrix, 69

Transversality condition, 310
Triangle inequality, 112, 135
Triangular finite element,
 linear, 387
 quadratic, 388

Ultradristributions, 451
Uncountable set, 15
Uniform,
 boundedness principle, 179
 countinuity, 130
 convergence, 115
 linear approximation, 285
 metric, 116
Unit,
 ball, 117, 133
 step, see Heaviside distribution
Unitary,
 linear transformation, 226
 matrix, 227
 similarity, 235
 space, 181
Upper bound, 9
 least, see supremum

Variation,
 equations of first, 408
 first, 304
 Gateaux, 287
 total, 163
Vector space, 18

of bounded linear transformations, 147
of contiunous functions, 20
 dimension of, 28, 30
of linear transformations, 36
 normed, 135
of n-tuples, 19
of polynomials, 20
of sequences, 20
Virtual work,
 identity, 101, 256
 principle of, 412
Volterra integral operator, 219

Weak,
 convergence, 169, 192
 *convergence, 171
 derivative, see Gateaux
 differential, see Gateaux
 gradient, 299
 solution, 469
Weighted residuals, method of, 367
 and least squares, 378
Weierstrass',
 M-test, 143
 theorem, 123, 447

Zero,
 space, 19
 transformation, 36
 vector, 18